Wesley H. Manier, Court Illinois. Supreme

Law of Eminent Domain And of Railroads and Warehouses

Wesley H. Manier, Court Illinois. Supreme

Law of Eminent Domain And of Railroads and Warehouses

ISBN/EAN: 9783744690089

Printed in Europe, USA, Canada, Australia, Japan

Cover: Foto ©berggeist007 / pixelio.de

More available books at **www.hansebooks.com**

LAW OF EMINENT DOMAIN,

AND OF

RAILROADS AND WAREHOUSES,

COMPRISING THE

CONSTITUTIONAL AND STATUTORY PROVISIONS,

IN THE

STATE OF ILLINOIS,

TOGETHER WITH THE DECISIONS RELATING THERETO, OF THE SUPREME
COURT AND APPELLATE COURT OF THE STATE OF ILLINOIS,
AND THE COURTS OF LAST RESORT IN OTHER
STATES, AND OF THE SUPREME COURT
OF THE UNITED STATES.

––––––––––

COMPILED AND ANNOTATED BY W. H. MANIER.

––––––––––

D. W. LUSK, PUBLISHER.
SPRINGFIELD, ILLINOIS.
1888,

ILLINOIS STATE REGISTER CO.
PRINTERS AND BINDERS,
SPRINGFIELD, ILL.

PUBLISHER'S ANNOUNCEMENT.

In presenting this volume to the legal profession, the publisher confidently believes that it will be accepted as a valuable contribution to the library of Law, as it forms a convenient and reliable compilation of the statutes of Illinois bearing on the questions of Eminent Domain, Railroads and Warehouses. The well known ability of Mr. MANIER, the compiler, as a lawyer, warrants the assertion that the work will be found both reliable and authentic. The statutes have been taken from publications authorized by the State, and the decisions of the respective courts from official reports, and not from any of the many digests. The revised proof sheets have been read by Mr. MANIER, and carefully compared with his manuscript, which insures accuracy in the printing. Scrupulous care has been observed with regard to the index. In it is cited every section of the statutes and subject contained therein, and every annotation, and it is therefore confidently believed that the publication will meet with the universal approval of the legal fraternity.

D. W. LUSK.

SPRINGFIELD, ILL., April 11, 1888.

PREFACE.

Illinois has been one of the first states which has attempted to regulate and control railway and other corporations by general laws, and many questions have arisen both as to the constitutional power of the legislature to regulate the same, and as to the proper construction of laws enacted for that purpose. These questions have been ably discussed by eminent counsel on both sides, and it would seem that almost every question has been presented and passed upon by our courts. This has resulted in many decisions, which lie scattered through nearly all our State reports, thus making them difficult of access. This fact induced the author to present this compilation of the statutes and decisions in a single volume, hoping thereby to serve the profession and the courts. He has presented all the statute laws on the subject of Eminent Domain, Railways and Public Warehouses, together with the decisions of the Appellate Court and Supreme Court of the State and those of the Federal Courts, and some of the more important decisions of the courts of last resorts in other states.

That part of the work relating to Eminent Domain will be found useful in many if not all the states. The decisions collected include all made since the admission of Illinois into the Union. They embrace much relating to the right to condemn land for public use, and by what bodies or agencies, and the mode of the exercise of the right, both under the Constitution of 1848 and that of 1870. This is especially true as to the proper measure of the compensation and damages to be paid by the bodies seeking to condemn. The cases given relate to condemnation by cities and villages of land for parks and other public uses, and for railway purposes. The cases defining the right of one railway company to condemn a right of way over or along another railway, or to condemn property already devoted to a public use, will be found useful, not only to the profession in this State, but throughout the whole country.

The statutes and decisions relating to the formation of railway corporations, their various powers, rights, duties and liabilities, among which are the right to enter cities and villages and construct their road and tracks in public streets and highways, and the limitations on that right, and the powers of municipal corporations over the location and grade of their tracks, and to compel such companies to make street crossings and approaches thereto, to fence their tracks and keep flagmen at street crossings, are presented in detail.

The statutes and decisions are given relating to the capital stock of corporations, its increase, the forfeiture of stock, its mode of transfer, its sale on execution, and the individual liability of stockholders generally; also those relating to the liability of railway companies for injury to domestic animals from negligence at common law, and from a neglect to fence their tracks, and for injury by the escape of fire from passing locomotive engines, as well as for injury from a neglect to put up warning boards at highway crossings, or to give warning by bell or whistle at such crossings.

The statutes and cases are given which relate to the expulsion of passengers from cars, for what causes and where, and liability for carrying them beyond their stations, the right of the State to regulate and control railways by proper police regulations, to regulate and fix the rates of charges by railways and public warehouses, and to prevent and punish extortion and unjust discrimination, and the right of carriers to limit their liability.

The work is not designed to present all the cases relating to the common law liability of railways for negligence, but rather those of liability for neglect of statutory duties; but as the doctrine of contributory and comparative negligence applies to cases of injury from neglect under the statute, the cases on that subject are presented generally.

The statutes and cases relating to the inspection of grain, and to public warehouses and warehouse receipts are also given, besides many other matters of general importance.

While the work is devoted to the laws of this State alone, yet from the high standing and great learning of our courts of last resort, and the variety and importance of the questions settled by them, it is thought their various rulings will be of service to the profession and courts of other States, especially those having similar statutes.

W. H. MANIER

CARTHAGE, ILL., April 11, 1888.

RAILWAY AND EMINENT DOMAIN LAWS.

CONSTITUTIONAL PROVISIONS.

1. CONSTITUTION 1848—*governs in all cases arising under it.* Although the constitution of 1848 has been suspended by the present one, still all rights acquired under it, or under laws passed while it was in force, must be tried by and enforced as though it was in full vigor. *People* v. *Trustees of Schools*, 78 Ill. 136. See post 44, 45.

2. CONSTITUTION 1870—*acts prospectively.* The constitution of 1870, acts only prospectively, leaving all past transactions unaffected by its provisions. It expressly preserves and continues all prior rights, &c., as they were before its adoption. *Chicago* v. *Rumsey*, 87 Ill. 348; *Garrick* v. *Chamberlain*, 97 Ill. 620. See post 44, 45.

CONSTITUTION OF 1870.

3. SPECIAL LEGISLATION—*in respect to what prohibited.* ART. 4, § 22. The general assembly shall not pass local or special laws in any of the following enumerated cases, that is to say: for— * * *

(*a.*) Granting to any corporation, association or individual the right to lay down railroad tracks, or amending existing charters for such purpose.

(*b.*) Granting to any corporation, association or individual any special or exclusive privilege, immunity or franchise whatever.

(*c.*) In all other cases where a general law can be made applicable, no special law shall be enacted. R. S. 1887, p. 58; S. & C., p. 119, 120; Cothran, p. 9.

DECISIONS.

4. GENERAL LEGISLATION—*what is.* General laws are such as relate to or bind all within the jurisdiction of the law-making power, limited as that power may be in its territorial operation, or by constitutional restraint. *People* v. *Cooper*, 83 Ill. 585.

5. A general law operates alike upon all persons or things of the same class. Its generality is not affected by the number of those within the scope of its operation. *People* v. *Wright*, 70 Ill. 388.

6. Whether a law is general does not depend upon the number of those within the scope of its operation. It is not necessary that it shall operate upon every person in the state; but it is sufficient if every person who is brought within the relations and circumstances provided for, is affected thereby. Nor is it necessary that it shall be

—2

made equally applicable to all parts of the state. It will be sufficient if it extends to all persons doing, or omitting to do an act within the, territorial limits described in the statute. *People* v. *Hoffman*, 116 Ill. 587.

7. A law is general and uniform, and not subject to the objection of being local or special, where it is general and uniform in its operation upon all in the like situation. *People* v. *Hazlewood*, 116 Ill. 319; *Hawthorn* v. *People*, 109 Ill. 302, 312.

8. Law is general, not because it operates on many or few persons in the state, but because every one who is brought within its provisions is affected by it. *Potwin* v. *Johnson*, 108 Ill. 70.

9. LOCAL OR SPECIAL LAWS—*command to enact, general.* This provision against local and special laws is equivalent to a command that general laws alone be enacted. *People* v. *Cooper*, 83 Ill. 585.

10. SAME—*depending on local option.* A law is "local or special," which by reason of local option, is repealed, or has its vitality as a law suspended in one locality, where exists a proper subject matter on which to operate, but remains in full force and vigor in another locality of precisely the same kind, or in the same locality, is law or not law, as shall suit the changing fancies of the local authority. *Ib.*

11. SAME—*not affected by being temporary.* A local or special statute is limited in the object to which it applies. A temporary statute is limited merely in its duration. A local or special law may be perpetual, or a general law may be temporary. The "mayor's bill" is neither local nor special, but is a temporary general law. *People* v. *Wright*, 70 Ill. 388.

12. LAWS HELD LOCAL OR SPECIAL—*act of 1865—sheriff fees.* The act of 1865, as amended, relating to sheriff fees in certain counties, being a special law, is of doubtful constitutionality. *Alexander County* v. *Myers*, 64 Ill. 37.

13. JURY SERVICE. Not competent for the legislature to make an exception in favor of Chicago as to service on juries. *In Re Scranton*, 74 Ill. 161.

14. AMENDING CHARTER. An amendment of a prior special charter can not be made by a local or special law. *Andrews* v. *People*, 75 Ill. 605; *People* v. *Cooper*, 83 Ill. 585.

15. FERRY FRANCHISE. An act to establish a ferry held to be special legislation and void. *Frye* v. *Partridge*, 82 Ill. 267.

16. FOR ONE COUNTY ONLY. A law classifying counties according to population, when only one county can be affected, is special legislation. *Devine* v. *Cook County*, 84 Ill. 590.

17. Act creating each county in the state a justice of the peace district, except Cook county, and making two in it, is in violation of this clause of the constitution. *People* v. *Meech*, 101 Ill. 200.

18. ACTS NOT SPECIAL LEGISLATION—*incorporation of cities.* Art. 9, § 54, of the "act to provide for the incorporation of cities and villages," approved April 10, 1872, held not special legislation. *Guild* v. *Chicago*, 82 Ill. 472.

19. ROAD LAW. The road and bridge law for counties under township organization, is not a local or special law. *Reynolds* v. *Foster*, 89 Ill. 257.

20. WAREHOUSE ACT. Acts in Chap. 114, R. S., classifying warehouses and providing rules for each class, not within the prohibition. *Munn* v. *People*, 69 Ill. 80; *People* v. *Harper*, 91 Ill. 357.

21. PENALTY ON TAXES. The one per cent. per month penalty provided for in the revenue law, Chap. 120, § 177, as amended in 1879, is not a special law regulating the interest on money. *People* v. *Peacock*, 98 Ill. 172.

22. INTEREST ON SPECIAL ASSESSMENT. Fixing the rate of interest installments of special assessments shall bear, is not special legislation regulating interest on money. *McChesney* v. *People*, 99 Ill. 216.

23. LOAN ASSOCIATIONS. The act of April 4, 1872, entitled "An act to enable associations of persons to become a body corporate to raise funds to be loaned only among its members," is not special legislation. *Holmes* v. *Smythe*, 100 Ill. 413; *Freeman* v. *Ottawa Building H. & S. Association*, 114 Ill. 182.

24. DISSOLUTION OF INSOLVENT INSURANCE COMPANIES. The statute for the dissolution of insurance companies for insolvency, is not a special law. *Ch. Life Ins. Co.* v. *Auditor*, 101 Ill. 82.

25. LIMITATION AS TO MUNICIPAL SUBSCRIPTION. Acts limiting the time for the enforcement of corporate liability on municipal subscription in aid of improvement, held not special legislation. *People* v. *Granville*, 104 Ill. 285.

26. AFFIDAVIT OF MERITS IN ATTACHMENT. The requirement in the act for the attachment of boats or water craft, that the defendant shall file an affidavit of merits, is not special legislation. *Johnson* v. *Elevator Co.*, 105 Ill. 462.

27. JUDGMENT AGAINST SURETY. The statutory provision that judgment shall pass against surety without service or appearance, is not special legislation, as it applies to all bonds of that kind. *Johnson* v. *Elevator Co.*, 105 Ill. 462.

28. CITY TAXES. A statute for the assessment and collection of taxes which applies to all incorporated cities and towns in the state, is a general, and not a special law. *People* v. *Wallace*, 70 Ill. 680.

29. CONSTITUTION OF 1818. Under the constitution of 1818, the legislature had the power to pass laws for particular cases. *Edwards* v. *Pope*, 3 Scam. 465; *Lane* v. *Dorman*, 3 Scam. 238.

30. SPECIAL ASSESSMENTS. The provision in the general municipal incorporation act for making and enforcing special assessments by corporations formed under the act, is not special legislation. *Potwin* v. *Johnson*, 108 Ill. 70; *Kilian* v. *Clark*, 9 Bradw. 426.

31. ELECTIONS IN CITIES. The act of 1885, relating to elections in cities, &c., is not a local or special law. The fact that such law has no operation in a city until adopted by the voters thereof, does not render it local or special. A general law may depend on some contingency as to when it takes effect in a particular locality. *People* v. *Hoffman*, 116 Ill. 587.

32. SPECIAL LEGISLATION NOT PROHIBITED—*inspection of grain*. The statute for the inspection of grain in Chicago, is in a certain sense a local and special law, but is not within the constitutional inhibition. The inspection of grain is not enumerated in the clause. *People* v. *Harper*, 91 Ill. 357.

33. SCHOOLS—*providing for system*. This clause prohibits special laws for the management of schools, but not special laws providing for funds for the support of schools; nor does it limit the legislature in the means of providing for a system of schools. *Fuller* v. *Heath*, 89 Ill. 296.

34. Nor does it limit the power of forming districts and providing who shall levy and collect taxes. *Speight* v. *People*, 87 Ill. 595.

35. SALE OR MORTGAGE. The provision forbidding special legislation regulating the sale or mortgage of lands of minors or others under disability, does not apply to a sale or mortgage of land of associations of any kind. *Haps* v. *Hewitt*, 97 Ill. 498.

36. TOWNSHIP ORGANIZATION. The legislature may provide somewhat different means for the government and management of towns lying wholly in the country, and those in an incorporated city, without making the law local or special. *People* v. *Hazelwood*, 116 Ill. 319.

37. SAME—*constitution of 1848 construed.* § 6, Art. 7 of the constitution of 1848, that "the general assembly shall provide by a general law for township organization," &c., relates to the management of the affairs of the several towns of the counties adopting the system, and not to the management of the fiscal affairs of the counties. *Leach* v. *People*, — Ill. —; filed June, 1887.

38. Acts held not local or special legislation or otherwise unconstitutional. *Covington* v. *East St. Louis*, 78 Ill. 548; *Guild* v. *Chicago*, 82 Ill. 472; *People* v. *Cooper*, 83 Ill. 585; *People* v. *Harper*, 91 Ill. 357; *Haps* v. *Hewitt*, 97 Ill. 498; *Ch. Life Ins. Co.* v. *Auditor*, 101 Ill. 82; *Klokke* v. *Dodge*, 103 Ill. 125; *People* v. *Meech*, 101 Ill. 200; *Knickerbocker* v. *People*, 102 Ill. 218; *Hinckley* v. *Dean*, 104 Ill. 630; *People* v. *Granville*, 104 Ill. 285; *Johnson* v. *Ch. & Pac. Elevator Co.*, 105 Ill. 462; *Hawthorn* v. *People*, 109 Ill. 302; *Williams* v. *People*, 121 Ill. 84.

39. SPECIAL OR EXCLUSIVE PRIVILEGE, &C. The prohibition against granting any special or exclusive privilege, &c., extends only to the passing of local or special laws for that purpose. *Munn* v. *People*, 69 Ill. 80.

40. SAME—*applies only to legislature.* The prohibition of the grant of any special or exclusive privilege, &c., is a limitation upon the power of the legislature, and not upon the powers of a city to give leave to build a railroad upon its streets. *Ch. City R. R.* v. *People*, 73 Ill. 541.

41. SAME—*dram shop act.* The dram shop act of 1872, is not unconstitutional as granting special or exclusive privileges. *Streetor* v. *People*, 69 Ill. 595.

42. APPLICABILITY OF GENERAL LAW—*who may decide.* The constitution of 1848 provided that private corporations should not be created by special acts, except where the objects of the corporation could not be attained under general laws. (Art. 10, § 1.) Under this, when a corporation was created by special act, the court held that it would presume, without any recital or preamble, that the general assembly considered the object sought could not be attained by a general law. *Johnson* v. *J. & C. R. R.*, 23 Ill. 202.

43. This clause prohibiting special legislation "where a general law can be made applicable," addresses itself to the legislature alone. When that body has concluded that a special law is necessary, except in the cases prohibited, its conclusion is not the subject of judicial review. *Owners of Land* v. *People*, 113 Ill. 296, 315.

44. NO APPLICATION TO PAST LEGISLATION. This clause of the constitution has no reference to past legislation, but simply prescribes the limits of future legislation in the respects named. *Covington* v. *East St. Louis*, 78 Ill. 548; *Guild* v. *Chicago*, 82 Ill. 475; *People* v. *Cooper*, 83 Ill. 585.

45. It does not invalidate special city charters previously granted. *Covington* v. *East St. Louis*, 78 Ill. 548. See ante 1, 2.

CONSTITUTION OF 1870.

46. SPECIAL LEGISLATION—*prohibited.* ART. 11, § 1. No corporation shall be created by special laws, or its charter extended, changed or amended, except those for charitable, educational, penal or reformatory purposes, which are to be and remain under the patronage and control of the state, but the general assembly shall provide, by general laws, for the organization of all corporations hereafter to be created. R. S. 1887, p. 71; S. & C., p. 160; Cothran, p. 28.

47. CURATIVE LEGISLATION. The legislature has the same power to validate irregularly organized corporations as it has to create a new one. *Mitchell* v. *Deeds*, 49 Ill. 416.

48. NO REPEAL OF GENERAL LAW—*corporations under.* This clause of the constitution does not repeal the general law on the subject of private corporations in force prior to its adoption, and all corporations formed under such laws, after the adoption of the constitution, are valid. *Meeker* v. *Cast Steel Co.*, 84 Ill. 276.

49. Under the constitution of 1848 (Art. 10, § 2), the word "corporators" is used in the sense of shareholders and not that of commissioners or promoters. *Gulliver* v. *Roelle*, 100 Ill. 141.

50. PRIVATE CORPORATIONS—*subject to police power.* Private corporations are subject to the police power of the state, and the legislature may direct and control them in the use of their franchises the same as natural persons. *G. & C. U. R. R.* v. *Loomis*, 13 Ill. 548; *Bank* v. *Hamilton Co.*, 21 Ill. 53, 59; *Reapers Bank* v. *Willard*, 24 Ill. 433; *N. W. Fertilizing Co.* v. *Hyde Park*, 70 Ill. 634; *Ruggles* v. *People*, 91 Ill. 256; *G. & Ch. Union R. R.* v. *Dill*, 22 Ill, 269; *Ward* v. *Farwell*, 97 Ill. 593; *C. & A. R. R.*, v. *People*, 105 Ill. 657; *O. & M. R. R.* v. *McClellan*, 25 Ill. 140.

51. Corporation formed to do "rendering," may under the police power be prohibited from carrying on such business. *N. W. Fertilizing Co.* v. *Hyde Park*, 70 Ill. 634.

52. POLICE POWER—*not unlimited.* The police power is subject to constitutional limitations. Police regulations must have reference to the comfort, safety and welfare of society; and when applied to corporations, they must not be in conflict with any of the rights secured by their charters. *Lake View* v. *Rose Hill Cem. Co.*, 70 Ill. 191.

53. STOPPAGE OF R. R. TRAINS. The statute requiring all regular passenger trains to stop at county seats, is a proper police regulation. *C. & A. R. R.* v. *People*, 105 Ill. 657.

54. Reservation in charter that the legislature may alter or repeal the same, gives the power to change it. *Butler* v. *Walker*, 80 Ill. 345.

55. CHARTER, A CONTRACT—*inviolability of.* The charter of a private corporation is a contract with which the legislature may not interfere, *Bruffett* v. *Great Western R. R.*, 25 Ill. 353; *Ruggles* v. *People*, 91 Ill. 256.

56. REPEAL OF CHARTER. An act which attempts to repeal a railroad charter and confer the powers and property of the corporation upon another body, with a view to declare a forfeiture, or create a dissolution, is unconstitutional. *Bruffett* v. *Great Western R. R.*, 25 Ill. 353.

57. MUNICIPAL CORPORATIONS. Their powers, rights, funds and revenues subject to legislative control. *Pike Co.* v. *State*, 11 Ill. 208;

Richland Co. v. *Lawrence Co.*, 12 Ill. 1, 8; *Trustees of School* v. *Tatman*, 13 Ill. 27, 30; *Dennis* v. *Maynard*, 15, Ill. 477, 480; *People* v. *Power*, 25 Ill., 187, 191; *Greenleaf* v. *Trustees*, 22 Ill. 236; *Mt. Carmel* v. *Wabash Co.*, 50 Ill. 69, 72; *Logan Co.* v. *City of Lincoln*, 81 Ill. 156; *Owners of Land* v. *People*, 113 Ill. 296; *Marion Co.* v. *Lear*, 108 Ill. 343.

58. CORPORATIONS — *limitation as to organizing.* ART. 11, § 2. All existing charters or grants of special or exclusive privileges, under which organization shall not have taken place, or which shall not have been in operation within ten days from the time this constitution takes effect, shall thereafter have no validity or effect whatever. R. S. 1887, p. 71; S. & C., p. 161; Cothran, p. 28. See *People* v. *Lowenthal*, 93 Ill. 191; *Anthony* v. *International Bank*, 93 Ill. 225; *Peoria & Pekin Union Ry.* v. *Peoria & Farmington Ry.*, 105 Ill. 110, 116; *McCartney* v. *C. & E. Ry.*, 112 Ill. 611.

59. CORPORATIONS—*election of directors—minority representation.* ART. 11, § 3. The general assembly shall provide by law, that in all elections for directors or managers of incorporated companies, every stockholder shall have the right to vote in person, or by proxy, for the number of shares of stock owned by him, for as many persons as there are directors or managers to be elected, or to cumulate said shares, and give one candidate as many votes as the number of directors multiplied by the number of his shares of stock shall equal, or to distribute them on the same principle among as many candidates as he shall think fit; and such directors or managers shall not be elected in any other manner. R. S. 1887, p. 71; S. & C. p. 161; Cothran, p. 28. See post 1459.

60. STREET RAILROADS—*consent of public authorities.* ART. 11, § 4. No law shall be passed by the general assembly granting the right to construct and operate a street railroad within any city, town or incorporated village, without requiring the consent of the local authorities having the control of the street or highway proposed to be occupied by such street railroad. R. S. 1887, p. 71; S. & C. p. 161; Cothran, p. 28.` See post 117–171.

61. RAILROADS—*place of office- books—reports.* ART. 11, § 9. Every railroad corporation organized or doing business in this state, under the laws or authority thereof, shall have and maintain a public office or place in this state for the transaction of its business, where transfers of stock shall be made, and in which shall be kept, for public inspection, books, in which shall be recorded the amount of capital stock subscribed, and by whom; the names of the owners of its stock, and the amounts owned by them respectively; the amount of stock paid in, and by whom; the transfer of said stock; the amount of its assets and liabilities, and the names and place of residence of its officers. The directors of every railroad corporation shall, annually, make a report, under oath, to the auditor of public accounts, or some officer to be designated by law, of all their acts and doings, which report shall include such matters relating to railroads as may be prescribed by law. And the general assembly shall pass laws enforcing by suitable penalties the provisions of this section. R. S. 1887, p. 71; S. & C., p. 162; Cothran, p. 29. See post 1174, 1471; see Eminent Domain, Ch. 47, § 1; *Infra* 179–1071.

62. RAILROADS—*rolling stock, &c., personal property.* ART. 11, § 10. The rolling stock, and all other movable property belonging to any railroad company or corporation in this state, shall be considered personal property, and shall be liable to execution and sale in the same manner as the personal property of individuals, and the general assembly shall pass no law exempting any such property from execution and sale. R. S. 1887, p. 72; S. & C., p. 162; Cothran, p. 29. See post 1369.

63. ROLLING STOCK—*changed from realty to personal property.* Prior to the adoption of the constitution, the rolling stock of railway companies was real estate. *Palmer* v. *Forbes*, 23 Ill. 301, 312; *Hunt* v. *Bullock*, 23 Ill. 320; *Titus* v. *Mabee*, 25 Ill. 257; *Titus* v. *Ginheimer*, 27. Ill. 462; *Mich. Cent. R. R.* v. *Chi. &c., R. R.* 1 Bradw. 399. See *C. & A. R. R.* v. *Goodwin*, 111 Ill. 273; *Johnson* v. *Roberts*, 102 Ill. 655; *C. & A. R. R.* v. *People*, 98 Ill. 350; *Maus* v. *L. P. & B. R. R.*, 27 Ill. 77. See post 1369–1375.

64. The doctrine that realty, franchises, &c., of a railway, mortgaged as an entirety, may be sold as an entirety under a decree in equity, without any right of redemption, is not in conflict with this constitutional provision. *Hammock* v. *Loan & Trust Co.*, 105 U.S. 77.

65. Nor does such provision change the rule, that a mortgage made by a railway company, covering after-acquired property, holds such property as against creditors obtaining judgments and executions after the company has received possession of such property. *Scott* v. *Clinton, &c., R. R.*, 6 Biss. 529.

66. The rolling stock of a railroad is a part of the realty so as to pass by a mortgage or conveyance of the road. *M. C. R. R.* v. *C. & M. L. S. R. R.*, 1 Bradw. 399.

67. RAILROAD COMPANIES—*limitation as to consolidation—directors residence.* ART. 11, § 11. No railroad corporation shall consolidate its stock, property or franchises with any other railroad corporation owning a parallel or competing line; and in no case shall any consolidation take place, except upon public notice given, of at least sixty days, to all stockholders, in such manner as may be provided by law. A majority of the directors of any railroad corporation, now incorporated or hereafter to be incorporated by the laws of this state, shall be citizens and residents of this state. R. S. 1887, p. 72; S. & C., p. 163; Cothran, p. 29.

This section cited in *Chicago & Western Indiana R. R.* v. *Dunbar*, 95 Ill. 578. See post 1187, 1386–1421, 1425.

68. RAILWAYS—*declared public highways—fixing maximum rates of charges.* ART. 11, § 12. Railways heretofore constructed, or that may hereafter be constructed in this state, are hereby declared public highways, and shall be free to all persons for the transportation of their persons and property thereon, under such regulations as may be prescribed by law. And the general assembly shall, from time to time, pass laws establishing reasonable maximum rates of charges for the transportation of passengers and freight on the different railroads in this state. R. S. 1887, p. 72; S. & C., p. 163; Cothran, p. 29. See post 1428–1458.

69. HIGHWAYS—*in what sense.* Railroads are highways, not in the sense of public wagon roads, upon which every one may transact his own business with his own means of conveyance, but only in the sense of being compelled to accept of each and all, and take and carry to the full extent of their ability. *T. P. & W. Ry.* v. *Pence*, 68 Ill. 524; *Central Military Tract R. R.* v. *Rockafellow*, 17 Ill. 541, 557.

70. This clause of the constitution does not affect the liability of a railway company for a neglect to fence its road. *T. P. & W. Ry.* v. *Pence*, 68 Ill. 524.

71. A private switch from a railroad to coal lands, which is not owned by the railway company, but by individuals for their own private use, is not a public highway within the meaning of this provision of the constitution. That section applies only to public railroads. *Koelle* v. *Knecht*, 99 Ill. 396.

72. When a railroad track is laid down in a street by authority of the city council, to connect a private manufacturing establishment with other railroad tracks, it becomes a public highway, and the city council have a right to devote a portion of the street to that use. *Parlin* v. *Mills*, 11 Bradw. 396; *Truesdale et al* v. *Grape Sugar Co.* 101 Ill. 567.

73. *Held* applicable to track laid to connect factory with railway. *Parlin* v. *Mills*, 11 Bradw. 396.

74. Railways are public highways only so far as owners and operators are subject to duties of common carriers. *T. P. & W. Ry.* v. *Pence*, 68 Ill. 524.

75. RIGHT OF STATE TO FIX OR LIMIT RATES CHARGED. In an action under the act April 13, 1871, to recover of a railway company for an overcharge of passenger fare made before the railroad commissioners had assigned the defendant's road to any class as required by that act, there was no proof that the charge made was unreasonable, or to what class the road belonged. *Held* that plaintiff could not recover. *Moore* v. *Ill. Central R. R.*, 68 Ill. 385.

76. To hold a railroad company liable to the penalties provided in the act of May 2, 1873, on the ground of extortion, it must be shown that it charged more than the maximum rates fixed by the board of railroad and warehouse commissioners; and until these rates are fixed, no liability can be incurred under the statute, for unreasonable or extortionate charges, and when made, the taking of the rates named, or less rates will not incur the penalty, even though the proof shows them to be more than fair and reasonable rates. *C. B. & Q. R. R.* v. *People*, 77 Ill. 443.

77. An express grant of power to a railway company to fix the rates of tolls to be charged, and to alter and change the same, does not confer unlimited power, but only the right to charge reasonable rates, and what is a reasonable maximum rate may be fixed by statute. *Ruggles* v. *People*, 91 Ill. 256.

78. The legislature has the power to fix a maximum rate of charges by individuals as common carriers, warehonsemen, or others exercising a calling or business public in its character, or in which the public have an interest to be protected against extortion or oppression, and it has the same rightful power in respect to corporations exercising the same business, and such regulation does not impair the obligation of the contract in their charters. *Ib.*

79. The act of April 17, 1871, entitled "An act to establish a reasonable maximum rate of charges for the transportation of passengers on railroads in this state," is not unconstitutional, but is a valid law. *Ib.*

80. The act of April 25, 1871, entitled "An act to regulate public warehouses and the warehousing and inspection of grain, and to give effect to article 13 of the constitution," and which provides a maximum rate of charges, is not in violation of that clause of the bill of rights which declares that no person shall be deprived of life, liberty or property without due process of law, nor of that clause which provides that "private property shall not be taken or damaged for public use without just compensation." *Munn* v. *People*, 69 Ill. 80.

81. The act of May 2, 1873, to prevent extortion and unjust discrimination in railroads, is a constitutional enactment, and is not in violation of the contract between the state and the railroad companies, growing out of the granting and accepting their charters, containing power to establish such rates of toll for the conveyance of persons and property as they shall from time to time, direct and determine in the by-laws. *I. C. R. R.* v. *People*, 95 Ill. 313.

82. The right of a state to reasonably limit the amount of charges by a railroad company for the transportation of persons and property within its jurisdiction, cannot be granted away by its legislature, unless by words of positive grant, or words equivalent in law. *Railroad commission cases*, 116 U. S. 307.

83. A statute which grants to a railroad company the right "from time to time to fix, regulate and receive the tolls and charges by them to be received for transportation," does not deprive the state of its power, within the limits of its authority as controlled by the constitution of the United States, to act upon the reasonableness of the tolls and charges so fixed and regulated. *Ib.; Stone* v. *Ill. Central R. R.*, 116 U. S. 347: *Stone* v. *N. O. & N. E. R. R.*, 116 U. S. 352.

84. It is the settled doctrine in the Supreme Court of the United States that a state has the power to limit the amount of charges by railroad companies for the transportation of persons and property within its own jurisdiction, unless restrained by some contract in the charter, or unless what is done amounts to a regulation of foreign, or inter-state commerce. *Railroad commission cases*, 116 U. S. 307: *R. R.* v. *Maryland*, 21 Wall. 456; *C. B. & Q. R. R.* v. *Iowa*, 94 U. S. 155; *Peik* v. *Ch. & N. W. Ry.*, 94 U. S. 164; *Winona & St. Paul R. R.* v. *Blake*, 94 U. S. 180; *Ruggles* v. *Illinois*, 108 U. S. 526.

85. The act entitled "An act to regulate public warehouses and the warehousing and inspection of grain, and to give effect to article 13 of the constitution of this state," approved April 25, 1871, is not repugnant to the constitution of the United States. *Munn* v. *Illinois*, 94 U. S. 113.

86. For other cases asserting the power of the states to regulate the rates of railroad charges on business not inter-state in its nature, see *Ch., M. & St. P. R. R.* v. *Ackley,* 4 Otto, 179: *Winona & St. Peter R. R.* v. *Blake,* 4 Otto, 180; *Peik* v. *Ch. & N. W. Ry.,* 4 Otto, 164; *Stone* v. *Wisconsin,* 94 U. S. 181; *Union Pacific R. R.* v. *U. S.,* 99 U. S. 700; *Hinckley* v. *Ch., M. & St. P. Ry.,* 38 Wis. 194; *State* v. *Winona & St. Peter R. R.,* 19 Minn. 434; *C. H. & D. R. R.* v. *Cole,* 29 Ohio St. 126; *Iron R. R.* v. *Lawrence Furnace Co.; Id.* 208; *Mobile & M. Ry.* v. *Steiner,* 61 Ala. 559; *Parker* v. *Metropolitan R. R.,* 109 Mass. 506; *Shields* v. *Ohio,* 95 U. S. 319; *American Coal Co.* v. *Consolidation Coal Co.,* 46 Md. 15; *Attorney General* v. *Railroad Companies,* 35 Wis. 435; *L. S. & M. S. Ry.* v. *C. S. & C. Ry.,* 30 Ohio St. 604.

87. RAILWAY COMPANY—*limitation on issue of bonds or stock, or increase of capital stock.* ART. 11, § 13. No railroad corporation shall issue any stock or bonds, except for money, labor or property actually received, and applied to the purposes for which such corporation was created; and all stock dividends, and other fictitious increase of the capital stock or indebtedness of any such corporation, shall be void. The capital stock of no railroad corporation shall be increased for any purpose, except upon giving sixty days' public notice, in such manner as may be provided by law. R. S. 1887, p. 72; S. & C., p. 163; Cothran, p. 30. Post 1376–1385.

88. This clause is intended to prevent reckless and unscrupulous speculators from fraudulently issuing and putting upon the market bonds or stocks that do not, and are not intended to represent money or property of any kind, either in possession or expectancy, the stock or bonds in such case being entirely fictitious. *Peoria & Springfield R. R.* v. *Thompson,* 103 Ill. 187.

89. It was not intended by that provision to interfere with the usual and customary methods of raising funds by railroad companies, by the issue of its stocks, or bonds for the purpose of building their roads or accomplishing other legitimate corporate purposes. *Ib.*

90. Under this provision railroad companies have no right to lend, give away, or sell on credit their bonds or stock, nor have they the right to dispose of either, except for a present consideration and for a corporate purpose. *Ib.*

91. CORPORATIONS—*franchises and property of, subject to right of eminent domain—jury trial.* ART. 11, § 14. The exercise of the power, and the right of eminent domain, shall never be so construed or abridged as to prevent the taking, by the general assembly, of the property and franchises of incorporated companies already organized, and subjecting them to the public necessity the same as of individuals. The right of trial by jury shall be held inviolate in all trials of claims for compensation, when, in the exercise of the said right of eminent domain, any incorporated company shall be interested either for or against the exercise of said right. R. S. 1887, p. 72; S. & C., p. 163; Cothran, p. 30. Post 242–277.

92. This, together with Art. 2, § 13, took effect immediately upon the adoption of the constitution, without the aid of any legislation, and operated as a repeal of so much of the act of 1852, as related to giving of bond on appeal before entry. *Mitchell* v. *I. & St. L. R. R.,* 68 Ill. 286.

93. As to the right of one corporation to condemn the property or franchise of another already devoted to a public use, see *Mills et al* v. *St. Clair County*, 2 Gilm. 147; *Ill. & Mich. Canal* v. *Ch. & Rock Island R. R.*, 14 Ill. 314; *P. P. & J. R. R.* v. *P. & S. R. R.*, 66 Ill. 174; *C. R. I. & P. R. R.* v. *Town of Lake*, 71 Ill. 333; *Metropolitan City Ry.* v. *Ch. West Division Ry.*, 87 Ill. 317; *Central City Horse Ry.* v. *Ft. Clark Horse Ry.*, 81 Ill. 523; *L. S. & M. S. Ry.* v. *Ch. & W. Ind. R. R.*, 97 Ill. 506; *St. L. J. & C. R. R.* v. *S. & N. W. R. R.*, 96 Ill. 274; *E. St. L. Connecting Ry.* v. *E. St. L. Union Ry.*, 108 Ill. 265; *Ch. & N. W. Ry.* v. *Ch. & Evanston R. R.*, 112 Ill. 589; *Ch. & W. Ind. R. R.* v. *Ill. Central R. R.*, 113 Ill. 156; *Ch. & N. W. Ry.* v. *Village of Jefferson*, 14 Bradw. 615; *Ch. & W. Ind. R. R.* v. *Ch. St. L. & Pittsburg R. R.*, 15 Bradw. 587. See Eminent Domain. Post 179–1070.

94. RAILROADS—*duty of passing laws to prevent unjust discriminations and extortions by.* ART. 11, § 15. The general assembly shall pass laws to correct abuses and prevent unjust discrimination and extortion in the rates of freight and passenger tariffs on the different railroads in this state, and enforce such laws by adequate penalties, to the extent, if necessary for that purpose, of forfeiture of their property and franchises. R. S. 1887, p. 72; S. & C., p. 164; Cothran, p. 30. Post 2645–2725.

95. This provision restricts the power of the legislature to the prohibition of such discriminations only as are unjust. *C. & A. R. R.* v. *People*, 67 Ill. 11.

96. An act prohibiting any discrimination under any circumstances, whether just or unjust, and making a difference in charges for the same distance, conclusive evidence of unjust discrimination, and inflicting a forfeiture of franchises, &c., on conviction, is unconstitutional. *Ib.*

97. LIMITATION OF ACTION FOR. The liability imposed by the statute upon railroad corporations for extortion and unjust discrimination, giving triple damages, is a statutory penalty, and actions therefor must be brought within two years after the cause of action accrued. *St. Louis, Alton & Terre Haute R. R.* v. *Hill*, 11 Bradw. 248.

98. DISCRIMINATION MUST BE UNJUST. In an action under the statute prohibiting extortion and unjust discrimination by railroad companies, it must appear not only that the corporation made a discrimination in its rates of toll, but also that such discrimination is unjust, and these facts must be alleged in the declaration. *Ib.*

For laws and decisions as to unjust discriminations and extortion, see post 2645–2725.

99. WAREHOUSES—*what are public warehouses.* ART. 13, § 1. All elevators or storehouses where grain or other property is stored for a compensation, whether the property stored be kept separate or not, are declared to be public warehouses. R. S. 1887, p. 73; S. & C., p. 164; Cothran, p. 31.

100 See Ch. 114, § 126–144, passed in pursuance of this article of the constitution. *Dutcher* v. *People*, 11 Bradw. 312.

101. This article cited in *Munn* v. *Illinois*, 94 U. S. 133, which holds the act passed in pursuance thereof is not in violation of the constitution of the United States.

102. The act of April 25, 1871, providing a maximum rate of charges for warehouses, does not contravene Art. 2, § 2, nor Art. 4, § 22, of the state constitution, but is a constitutional regulation of trade and a valid law. *Munn* v. *People*, 69 Ill. 80; *Munn* v. *Illinois*, 94 U. S. 113.

103. The legislature may commit to a board of warehouse commissioners power to control the inspection of grain. *People* v. *Harper*, 91 Ill. 357.

As to laws on this subject and decisions of the courts, see post 2732–2801.

104. POSTING REPORTS—*mixing grain of different grades.* ART. 13, § 2. The owner, lessee or manager of each and every public warehouse situated in any town or city of not less than 100,000 inhabitants, shall make weekly statements under oath, before some officer to be designated by law, and keep the same posted in some conspicuous place in the office of such warehouse, and shall also file a copy for public examination in such place as shall be designated by law, which statement shall correctly set forth the amount and grade of each and every kind of grain in such warehouse, together with such other property as may be stored therein, and what warehouse receipts have been issued, and are, at the time of making such statement, outstanding therefor; and shall, on the copy posted in the warehouse, note daily such changes as may be made in the quantity and grade of grain in such warehouse; and the different grades of grain shipped in separate lots shall not be mixed with inferior or superior grades without the consent of the owner or consignee thereof. R. S. 1887, p. 73; S. & C.. p. 165; Cothran, p. 31.

105. RIGHT TO INSPECT PROPERTY AND BOOKS. ART. 13, § 3. The owners of property stored in any warehouse, or holder of a receipt for the same, shall always be at liberty to examine such property stored, and all the books and records of the warehouse in regard to such property. R. S. 1887, p. 73; S. & C., p. 165; Cothran, p. 31.

106. WEIGHING GRAIN—*receipt and liability for delivery of grain.* ART. 13, § 4. All railroad companies and other common carriers on railroads shall weigh or measure grain at points where it is shipped, and receipt for the full amount, and shall be responsible for the delivery of such amount to the owner or consignee thereof, at the place of destination. R. S. 1887, p. 73; S. & C., p. 165; Cothran, p. 31. Post 2728—2731; 2802–2811.

107. DELIVERY OF GRAIN AT PLACE DIRECTED—*connections with other roads.* ART. 13, § 5. All railroad companies receiving and transporting grain in bulk or otherwise, shall deliver the same to any consignee thereof, or any elevator or public warehouse to which it may be consigned, provided such consignee or the elevator or public warehouse can be reached by any track owned, leased or used, or which can be used, by such railroad companies; and all railroad companies shall permit connections to be made with their track, so that any such consignee, and any public warehouse, coal bank or coal yard, may be reached by the cars on said railroad. R. S. 1887, p. 73; S. & C., p. 165; Cothran, p. 31.

108. The words *"can be reached"* do not mean reached by physical possibility, but by a track which the company has a right to use. If the place of consignment can be reached by any track of which the railway company is owner or lessee, or which can be lawfully used by it, the company is bound to deliver at that place. *C. B. & Q. R. R.* v. *Hoyt*, 1 Bradw. 374, 386.

109. This section does not require a railway company to do any act it has no right to do: *e. g.*, to use another company's track without license. *Hoyt* v. *C. B. & Q. R. R.*, 93 Ill. 601.

110. It seems that the last clause of the above section, requiring all railroad companies to permit connections to be made with their track, so that any public warehouse may be reached by the cars on such railroad, changes the rule announced in *People ex rel.* v. *C & A. R. R.*, 55 Ill. 95; *Vincent* v. *C. & A. R. R.*, 49 Ill. 33; *People ex rel.* v. *C. & N. W. Ry.*, 57 Ill. 436; *C. B. & Q. R. R.* v. *Hoyt*, 1 Bradw. 387; *Hoyt* v. *C. B. & Q. R. R.*, 93 Ill. 611.

111. FRAUDULENT WAREHOUSE RECEIPTS—*passage of laws to enforce provisions of Art. 13—rule of construction.* ART. 13, § 6. It shall be the duty of the general assembly to pass all necessary laws to prevent the issue of false and fraudulent warehouse receipts, and to give full effect to this article of the constitution, which shall be liberally construed so as to protect producers and shippers. And the enumeration of the remedies herein named shall not be construed to deny to the general assembly the power to prescribe by law such other and further remedies as may be found expedient, or to deprive any person of existing common law remedies. R. S. 1887, p. 73; S. & C., p. 166; Cothran, p. 31.

112. INSPECTION OF GRAIN—*laws to regulate.* ART. 13, § 7. The general assembly shall pass laws for the inspection of grain, for the protection of producers, shippers and receivers of grain and produce. R. S. 1887, p. 73; S. & C., p. 166; Cothran, p. 31.

113. It was competent to delegate to the railroad and warehouse commission the power to control the subject of the inspection of grain, and the law of this state on that subject is a valid law. *People* v. *Harper*, 91 Ill. 357. See post 2756–2763, 2770–2774, 2796–2801.

CHAPTER 22.

CHANCERY.

114. PRIVATE CORPORATIONS—*discovery by.* § 22. When a corporation, other than a municipal corporation, is defendant to a bill or petition praying discovery of any paper or matter alleged to be in the custody or within the knowledge of any officer or agent of the defendant, it shall not be necessary, for the purpose of procuring such discovery, to make such officer or agent a defendant, but the answer touching the paper or matter concerning which discovery is sought, shall be under the oath of such officer or agent the same as if he had been made defendant; provided, no corporation shall be required to procure such answer under the oath of any person not under its control at the time when the bill is filed. R. S. 1887, p. 216, § 22; S. & C., p. 404, § 22; Cothran, p. 188, § 22. In force July 1, 1872. Laws 1871-2, p. 333. This is a new section, not in the prior laws.

115. ANSWER OF CORPORATION—*before this statute.* The answer of a corporation aggregate should be under seal, but not under oath. Before this enactment, if a sworn answer was desired, some managing officer who could answer under oath was required to be made a party. *Fulton Co.* v. *M. & W. R. R.,* 21 Ill. 338, 364.

116. BILL—*charge on information and belief.* Where the matter essential to relief is charged to rest in the knowledge of the defendant, or must of necessity be within the knowledge of the defendant, and is a part of the discovery sought, it may be stated upon the information and belief of the complainant. *Campbell* v. *P. & D. R. R.,* 71 Ill. 611.

CHAPTER 24.

CITIES VILLAGES AND TOWNS.

An act to provide for the incorporation of cities and villages, approved April 10, 1872; in force July 1, 1872; laws 1871-2, p. 218.

117. POWERS OF CITY COUNCIL—*location, grade and crossing of railroads.* ART. 5. § 1. The city council in cities, and president and the board of trustees in villages, shall have the following powers: *   *   *   *   *

Twenty-fifth—To provide for and change the location, grade and crossings of any railroad. R. S. 1887, p. 247; S. & C., p. 465; Cothran, p. 227. See post 1235–1303, 2089–2097.

118. RAILWAY TRACK IN CITY—*consent of council necessary.* The power conferred upon a railway company to select its own route and fix its terminal points, is subject to a proviso affecting its right to construct its road upon or across any street in any incorporated city without the assent of such city. This proviso is a limitation of power, and is an exclusion of such railroad from incorporated cities, except upon compliance with its conditions. Before such railroad can construct its track in or through an incorporated city it must first obtain the consent of the common council acting in a legal manner. *Hickey v. Ch. & W. Ind. R. R.*, 6 Bradw. 172.

119. SAME—*consent, how obtained.* Inasmuch as the railroad act contains no provision as to how such consent may be obtained, the action of the city council must be governed by the provisions of the general statute relating to the incorporation of cities and villages. *Ib.*

120. POWER OF CITY—*to regulate railroads.* Cities have full power to regulate the location and use of railroad tracks within their corporate limits. This is a public power or trust and can be exercised by the corporation when and in such manner as it shall judge best, but such power cannot be delegated to others. *Ib.*

121. ORDINANCE GIVING RIGHT—*certainty in fixing location, &c.* In giving consent to a railway company to locate its track upon or over the streets, the council must prescribe the location of such road with reasonably definite lines; and if it fails to do so, but delegates to the railway company itself a discretion in that respect, the ordinance will be void. *Ib.*

122. An ordinance granting permission to "construct, &c., one or more tracks * * * commencing at the southern boundary line of the city of Chicago, at some point within 100 feet of the west line of Stewart Avenue, and thence northwardly * * * parallel to said avenue to its intersection with Grove street, thence * * * to such terminus as it may establish between the east bank of the south branch of the Chicago river, and the west side of State street, and between Sixteenth street and the south line of Van Buren street," is void for indefiniteness. *Ib.*

123. DELEGATION OF AUTHORITY. The ordinance further provided that the company might permit other railroad companies to use the said railroad track "upon such terms as may be agreed upon by said companies." This delegation of power rendered the ordinance void, because its exercise would result in a deprivation of the city of the control and regulation of a portion of its streets. *Ib.*

124. INJUNCTION. A railroad company having no power to construct its track in a city except by consent of the city council, if such consent is void, a court of equity will have jurisdiction to restrain, by injunction, the company from exercising such power. *Ib.*

125 POWER OF CITY—*to grant right for railroad in a street.* A city has the power to authorize the laying of railroad tracks in its streets; and where a city under a resolution adopted, conveys a street absolutely to a railway company, the resolution and deed will give the company the right to construct, maintain and operate its tracks upon the street, and when such right is exercised, the city cannot resume the grant to the exclusion of the company. *Quincy v. C. B. & Q. R. R.*, 92 Ill. 21.

126. The recognition by a city for over twenty years of a resolution granting a right to lay railroad tracks in certain streets as being in force, and its acquiescence thereunder, affords presumptive evidence of its due publication. *Ib.*

127. MODE OF GRANTING RIGHT. Although a city charter may provide that the city council shall have power to make all ordinances necessary and proper for carrying into execution the powers specified in the charter, the action of the city council, though in the form of a resolution, in connection with its deed granting the use of streets for railroad tracks, will be a sufficient grant of permission to so use the streets. *Ib.*

128. RIGHT TO FIX ROUTE—*consent as to streets.* Under the general law a railway company has authority to select its own route.

to lay out its road and to construct the same; and this power, by necessary implication, carries with it the power of fixing the terminal points of the road, subject only to the limitation that the construction of its road upon or across any street in any city, must be with the assent of the city council. *Ch. & W. Ind. R. R.* v. *Dunbar*, 100 Ill. 110.

129. The lines selected may, without the assent of the city, cross streets, and the company may, without such assent, acquire the right of way, and construct its road on every part of such line, except the parts upon or across the streets. *Ib.*

130. Under the present legislation, it is not necessary as a condition precedent to the location of a railroad within a city, or to its construction within the city, or such parts of its lines as are not within any street, or to the power to condemn private property within the city, that any ordinance should be passed by the city council, either giving assent for the construction of the road upon or across streets, or providing for the location of the road. *Ib.*

131. ORDINANCE—*sufficiency of—certainty.* An ordinance granting permission to construct and operate a railroad within the city limits, is not void because it fails to designate the precise point at which the road may be constructed upon and across the streets to be intersected by it. *Ib.*

132. DELEGATION OF POWER. Permission granted by a city council to a railway company to construct its road across streets at any point to be selected by the company within a given district, is not a delegation to the company of powers which can only be exercised by the council, as the power to locate the line of the road is given by statute to the company alone, and not to the city authorities. The city of Chicago has power to make provision for the location of a railroad within its limits, but no power to locate. That power is in the railway company, subject to such provisions for the location as the city council may make. *Ib.*

133. The mere existence of a power in a city council "to provide for the location, grade and crossings" of railroads within the city, and "to change the location, grade and crossings" of railroads, until exercised, is no limitation upon the power of the railroad company to select its route and locate its road within the city. *Ib.*

134. USE BY OTHER COMPANIES. A provision in an ordinance that the permission to construct a railroad within the city, is upon the condition that the railway company shall permit any other railroad companies, not exceeding two in number, which have not then the right of entrance into the city, to use the main track of the road, therein authorized to be laid, *jointly* with such road so authorized, does not render the ordinance invalid, as it confers upon the railroad company no power not given it by law, nor does it deprive the city of any power whatever. *Ib.*

135. An ordinance giving a railroad company license to construct its track along or across the streets and alleys of a city, upon the condition that it shall permit any other companies, not exceeding two in number, to use its main track upon such fair and equitable terms as may be agreed upon, will not be construed as prohibiting the company from leasing the use of its track within the city to more than two other companies. Such provision is a limitation, not upon the right of the company to admit other companies to a joint use of its track, but upon the exclusive enjoyment of the estate granted by the city. *Chicago* v. *Ch. & W. Ind. R. R.*, 105 Ill. 73.

136. Under the 9th and 25th clauses of § 1, Art. 5, of the general incorporation law, the common council of cities incorporated under that law, is vested with the exclusive control and regulation of the

streets of their cities, and with the power to direct and control the location of railroad tracks within the limits of their cities; and being inconsistent with the 9th clause of § 1, Art. 5, of the amended charter of the city of Chicago, adopted in 1867, must prevail over the latter. *Chicago Dock & Canal Co.* v. *Garrity*, 115 Ill. 155.

137. A city council may grant to private individuals or to a private corporation, the right to lay railroad tracks in the streets connecting with public railway tracks previously laid, and extending to the manufacturing establishments or warehouses of those laying the tracks. They then become part of the railway with which they connect, and are subject to public use and control, as other railway tracks. *Ib.*

138. The only authority that can call in question the right of a railway company to construct its track along or across a street or highway in an incorporated city or village, is such city or village. The county authorities cannot even question the validity of an ordinance of a city or village for the construction of a railroad within such city or village. *Cook Co.* v. *Great Western R. R.*, 119, Ill. 218.

139. GRANT OF USE OF STREET CONSTRUED. An ordinance or resolution of a city appropriated certain streets to a railway company, " so far as said company may require to appropriate them in crossing them in the construction of their railroad tracks, switches, turnouts, &c., and other machinery and fixtures to be used or employed by them in operating their said road, subject, however, to this proviso: that the same shall be occupied with as little detriment and inconvenience as possible," and requiring the crossings to be so graded as to make the embankments no obstruction: *Held* that this was but a provision for a joint use with the public having occasion to use the streets by other modes of travel. *St. L. A. & T. H. R. R.* v. *Belleville*, — Ill. —. Filed June, 1887.

140. VACATION—*of the vote required, &c.* A public street or alley can be vacated or closed only by the city council, and by it only upon a three-fourths majority vote of all the aldermen authorized by law to be elected, to be taken by ayes and noes and entered upon the record of the proceedings of the council or board. *Ib.*

141. PUBLIC BOUND BY LAWFUL GRANT. A city has the power to allow the construction of a railroad upon or over its streets, and the public will be bound by whatever may be lawfully done in regard to the streets by the city. *Ch. & N. W. R. R.* v. *People*, 91 Ill. 251.

142. NUISANCE. A railroad track laid upon a street of a city by authority of law, properly constructed and operated in a careful and skillful manner, is not in law a nuisance. *Ch. & E. Ill. R. R.*, v. *Loeb*, 118 Ill. 203.

143. CONDITIONS—*binding as a contract.* When leave is given to lay a railroad track in a street on conditions which are accepted, this will constitute a contract binding upon the city which it may not disregard, by imposing further conditions and burdens. *People* v. *W Div. Ry.*, 118 Ill. 113.

144. POWER OF CITY—*to compel fencing of railroad track. Twenty-sixth.* To require railroad companies to fence their respective railroads, or any portion of the same, and to construct cattle guards, crossings of streets and public roads, and keep the same in repair, within the limits of the corporation. In case any railroad company shall fail to comply with any such ordinance, it shall be liable for all damages the owner of any cattle or horses or other domestic animal,

—3

may sustain, by reason of injuries thereto while on the track
of such railroad, in like manner and extent as under the general laws of this state, relative to the fencing of railroads;
and actions to recover such damages may be instituted before
any justice of the peace or other court of competent jurisdiction. [R. S. 1887, p. 247; S. & C. 465; Cothran, 227. See
post 1518-1799.]

145. POWERS OF CITY—*railway flagmen—grade of track—
ditches. Twenty-seventh.* To require railroad companies to
keep flagmen at railroad crossings of streets, and provide
protection against injury to persons and property in the use
of such railroads. To compel such railroads to raise or lower
their railroad tracks to conform to any grade which may at
any time be established by such city, and where such tracks
run lengthwise of any such street, alley or highway, to keep
their railroad tracks on a level with the street surface, and so
that such tracks may be crossed at any place on such street,
alley or highway. To compel and require railroad companies to make and keep open and to keep in repair, ditches,
drains, sewers and culverts along and under their railroad
tracks, so that filthy or stagnant pools of water cannot
stand on their grounds or right of way, and so that the natural drainage of adjacent property shall not be impeded. [R.
S. 1887, p. 247; S. & C., p. 465; Cothran, p. 228. See post
2450-2455.]

146. DUTY TO KEEP FLAGMAN—*liability for neglect to do so.*
See *St. L. V. & T. H. R. R.*, v. *Dunn*, 78 Ill. 197; *I. C. R. R.* v. *Ebert*,
74 Ill. 399; *P. & P. U. Ry.* v. *Clayberg*, 107 Ill. 644; *L. S. & M. S. R. R.*
v. *Sunderland*, 2 Bradw. 307; *L. S. & M. S. R, R.* v *Kaste*, 11 Bradw.
536.

147. REGULATION OF USE OF STREETS. The act of 1872, relating
to cities and villages, confers upon them full authority to regulate the
use of streets, to provide for and change the location, grade and crossings of railroads, to require railway companies to fence their roads, to
construct cattle-guards and crossings of streets, to keep the same in
repair, to maintain flagmen at such crossings, to compel the roads to
raise or lower their tracks, &c. This invests incorporated cities and
villages with exclusive authority over the matter of railroad crossings
of streets and highways within their limits, and excludes the jurisdiction of the county or town authorities. *Cook Co.* v. *Great Western
R. R.*, 119 Ill. 218.

148. ALLOWING ICE ALONG TRACKS. A railway company not
being required by law to keep the excavations along the sides of its
track free from water and ice, it will not be liable for stock killed
in consequence of ice therein, so as to prevent escape from the track,
over the same. *P. & R. Q. Ry.* v. *McClenahan*, 74 Ill. 435.

149. LIABILITY OF RAILWAY—*to make safe crossing for new
street.* Long after the construction of a railroad, a street was extended
so as to cross the same, and the city passed an ordinance requiring the
company to make a safe and proper crossing by grading the approaches
of the street at the crossing, there being nothing in the charter of the
company imposing such duty, or any such duty imposed by any gen-

eral law in force at the time the company was created: *Held*, that the ordinance was void, and that the legislature itself could not impose this new burden without making compensation. *I. C. R. R.* v. *Bloomington*, 76 Ill. 447.

150. POWERS OF CITY—*extending streets across railroad. Eighty-ninth.* The city council shall have power by condemnation or otherwise, to extend any street, alley or highway over or across, or to construct any sewer under or through any railroad track, right of way, or land of any railroad company (within the corporate limits); but where no compensation is made to such railroad company, the city shall restore such railroad track, right of way or land to its former state, or in a sufficient manner not to have impaired its usefulness. [R. S. 1887, p. 250; S. & C., p. 472; Cothran, p. 232.]

151. USE OF STREET BY RAILWAY—*petition of lot owners necessary. Ninetieth.* The city council or board of trustees shall have no power to grant the use of, or the right to lay down, any railroad track, in any street of the city, to any steam or horse railroad company, except upon a petition of the owners of the land representing more than one-half of the frontage of the street, or so much thereof as is sought to be used for railroad purposes. [R. S. 1887, p. 250; S. & C., p. 472; Cothran, p. 232. See Horse & Dummy Railroads, Chap. 66, § 3, and Railroads & Warehouses, Chap. 114. See post 1235–1303.]

152. A compliance with this condition is an essential prerequisite to a valid execution of the power. *Hickey* v. *Ch. & W. Ind. R. R.*, 6 Bradw. 172.

153. PETITION. A petition to the common council for the right to construct a railroad track along a public street, is sufficient, if presented by owners representing more than one-half of the frontage of so much of the street as is sought to be used for railroad purposes. *Schuchert* v. *W. C. & W. R. R.*, 10 Bradw. 397.

154. COMPANY TAKES—*subject to damages.* When an incorporated city, by proper ordinance, authorizes a railroad company to construct and operate a railroad in a street, the company acquires the right to build and operate such road without interference by the public or individuals, subject however to the liability to respond to the owners of land abutting on the street, for such injuries sustained by them in consequence thereof as are to be deemed legal elements of damages. *Ch. & W. Ind. R. R.* v. *Berg*, 10 Bradw. 607.

155. HORSE AND DUMMY RAILROAD. The provisions of subdivision 90, § 1, Art. 5, of the act relating to cities and villages, so far as they apply to horse and dummy railroads incorporated under the general law, are repealed by the act of 1874 in relation to horse and dummy railroads, and under that act, no petition of the adjoining property owners is necessary. *Hunt* v. *Ch. & Dummy Ry.*, 20 Bradw. 282.

156. The provision requiring a petition of property holders, has reference only to cases where the city may propose to grant the privilege to a railroad company to run along a street for a given distance, and not to a case where the road merely crosses the street. *Ch. & W. Ind. R. R.* v. *Dunbar*, 100 Ill. 110.

157. USE OF TRACK FOR FREIGHT CARS. When a railway company lays its track in a street of a city, having the right to construct a track for passenger cars only, the city, under § 62, clause 90, of article 5 of the general law, has no power afterwards to grant the use of the track for the operation of freight cars upon it, except upon a petition of property owners upon the street, as required by the statute, and a grant of the use of such track for freight purposes without any petition, being void, such use is unlawful and a public nuisance, which the state may cause to be abated. *McCartney* v. *C. & E. R. R.*, 112 Ill. 611.

158. Clause 90 of § 1, Art. 5, of the general incorporation law, is to be construed as including both corporations and individuals. The word "company," in the clause must be held to embrace natural persons as well as corporations. *Ch. Dock & Canal Co.* v. *Garrity*, 115 Ill. 155.

159. CONDITIONS TO PETITION—*binding on city.* The property owners in their petition for license to a railway company to construct its road in a street, may insert such conditions in their assent as they may see fit, and in such case the city council may not grant the right except upon those conditions. *People* v. *West Div. Ry.*, 118 Ill. 113.

160. WHEN ASSENT OF PROPERTY HOLDERS NECESSARY. Under the general railroad law, it is only necessary to procure the assent of the municipal authorities, to authorize the laying of a railroad track over or along a street. The act as revised in 1874, does not require the assent of the abutting lot owners, and in the absence of any special statutory provisions requiring such assent, it will not be necessary. *Wiggins Ferry Co.* v. *E. St. L. U. Ry.*, 107 Ill. 450.

161. In cities and villages organized under the general incorporation act, or under special charters, requiring the assent of lot owners, this rule does not apply, and the assent of the requisite number of abutting lot owners, will be required, as well as that of the municipality. Not so, however, in a city under a special charter containing no such provision. *Ib.*

CHAPTER 27.

An act to fix the liability of common carriers receiving property for transportation, approved March 27, 1874; in force July 1, 1874.

162. COMMON CARRIERS—*limitation of common law liability.* § 1. *Be it enacted by the People of the State of Illinois, represented in the General Assembly:* That whenever any property is received by a common carrier, to be transported from one place to another, within or without this state, it shall not be lawful for such carrier to limit his common law liability safely to deliver such property at the place to which the same is to be transported, by any stipulation or limitation expressed in the receipt given for such property. [R. S. 1887, p. 316; S. & C., p. 562; Cothran, p. 301. This act is substantially re-enacted as to railroad corporations in the railroad and warehouse act. See § 33 of the act entitled "An act in relation to the fencing and operating railroads," approved March 31, 1874; in force July 1, 1874, and notes thereto. See post 2339–2442.]

CRIMINAL CODE, CHAPTER 38, DIVISION 1.

163. CANADA THISTLES—*bringing into state—allowing to seed.* § 40. Whoever shall bring into this state, whether in the packing of goods, or in grain or grass seed, or otherwise, any seed of the Canada thistle, and permit the same to be disseminated so as to vegetate on any land in this state, and whoever shall permit any Canada thistle to mature its seed on any land owned or occupied by him, so that the same is or may be disseminated, shall be fined not less than $10 nor more than $100; the fine to be paid to the commissioners of Canada thistles, if any is appointed in the town, precinct, city or village, or otherwise as directed by law. [Laws of 1867, p. 79, §§ 1, 2, re-written with penalties altered. R. S. 1887, p. 435; S. & C. p. 765; Cothran, p. 448.]

164. *Quere* whether this is not an attempt to regulate inter-state commerce. See Animals, R. S. 1887, Chap. 8, p. 140; S. & C., p. 279; Cothran, p. 104.

165. CANADA THISTLES—*railroads to destroy.* § 41. If any company, association or person owning, controlling or operating a railroad shall refuse or neglect to dig up and distroy, or take other certain means of exterminating Canada thistles and other noxious weeds that may at any time be growing upon the right of way or other lands of such roads, or appertaining thereto, they shall be fined for each offense not less than $50 nor more than $200; the fine to be paid as in the preceding section. [In lieu of L. 1869, p. 326, §§ 1, 2; R. S. 1887, p. 436; S. & C., p. 765; Cothran, p. 448.]

166. COMMON CARRIERS—*liability for gross negligence.* § 49. Whoever, having personal management or control of or over any steamboat, or other public conveyance used for the common carriage of persons, is guilty of gross carelessness or neglect in, or in relation to, the conduct, management or control of such steamboat, or other public conveyance, while being so used, for the common carriage of persons, whereby the safety of any person shall be endangered, shall be imprisoned in the penitentiary not exceeding three years, or fined not exceeding $5,000. [R. S. 1887, p. 440; S. & C., p. 768; Cothran, p. 451.]

167. *As to criminal liability for negligence,* see *C. B. & Q. R. R.* v. *Triplett,* 38 Ill. 487.

168. CRUELTY—*by railroads to animals.* § 51. No railroad company or other common carrier in the carrying or transportation of any cattle, sheep, swine or other animals, shall allow the same to be confined in any car more than twenty-eight consecutive hours (including the time they shall have been upon any other road), without unloading for rest, water and feeding, for at least five consecutive hours, unless delayed by storm or accident, when they shall be so fed and watered as soon after the expiration of such time as may reasonably be done. When so unloaded they shall be properly fed, watered and sheltered during such rest by the owner, consignee or person in custody thereof, and in case of their default, then by the railroad company transporting them, at the expense of said owner, consignee or person in custody of the same; and such company shall have a lien upon the animals until the same is paid. A violation of this section shall subject the offender to a fine of not less than $3 nor more than $200. [Laws of 1869, p. 115, 116, §§ 5, 6, 7, re-written; R. S. 1887, p. 440; S. & C., p. 769; Cothran, p. 451.]

169. EMBEZZLEMENT—*by officers and agents of corporations.* § 75. If any officer, agent, clerk, or servant of any incorporated company; or if a clerk, agent, servant or apprentice of any person or copartnership, or society, embezzles or fraudulently converts to his own use, or takes and secretes with intent so to do, without the consent of his company, employer or master, any property of such company, employer, master, or another, which has come to his possession, or is under his care by virtue of such office or employment, he shall be deemed guilty of larceny. [In place of § 70, R. S. 1845, p. 162; R. S. 1887, p. 446; S. & C., p. 776; Cothran, p. 457.]

170. EMBEZZLEMENT—*of railroad ticket.* § 77. Whenever any person in the employ of any railroad company, whether such company is incorporated by this or any other state, shall fraudulently neglect to cancel or return to the proper officer, company or agent, any coupon or other railroad ticket or pass, with the intent to permit the same to be used in fraud or injury of any such company, or if any person shall steal or embezzle any such coupon or other railroad ticket or pass, or shall fraudulently stamp, or print, or sign any such ticket, coupon or pass, or shall fraudulently sell or put in circulation any such ticket, coupon or pass, the person so offending shall be punished by imprisonment in the penitentiary for the term of one year. [Laws of 1859, p. 154, § 2; R. S. 1887, p. 447; S. & C., p. 777; Cothran, p. 457.]

171. MALICIOUS MISCHIEF TO RAILROAD—*murder for causing death by.* § 186. Whoever willfully, and maliciously, displaces or removes, any switch, signal, or rail of any railroad, or displaces, or removes, any signal or signal-light, from any bridge that is built across any navigable stream in this state, or breaks down, rips up, injures or destroys any track, bridge or other portion of any railroad, or places obstructions thereon, or places any false signal upon or along the line of any railroad track, or upon any bridge built across any navigable stream in this state, or does any act to any engine, machine or car of such railroad, with intent that any person or property being or passing on or over such railroad, or over or through, or under such bridge built across any navigable stream of this state, should be injured thereby, shall be imprisoned in the penitentiary not less than one year nor more than five years. Or if in consequence of any such act done with such intent, any person being or passing on or over such railroad, or over, through or under such bridge, built across any navigable stream of this state, suffers any bodily harm, or any property is injured, the person so offending, shall be imprisoned in the penitentiary not less than three nor more than ten years, and if in consequence of any such act, done with such intent, any person is killed, the person so offending, shall be deemed guilty of murder and punished accordingly. [Laws of 1853, p. 217, §§ 1, 2, 3; Laws of 1877, p. 86, § 1, as amended Laws of 1879, p. 118; R. S. 1887, p. 464; S. & C., p. 805; Cothran, p. 482.]

172. CONSPIRACY—*combination to injure railroad.* § 187. If any two or more persons shall conspire or combine to break down, take up, injure or destroy any railroad track, or railroad bridge, or to burn or destroy any engine, engine house, car house, machine shop, or any other building or machinery necessary to the free use of any railroad, every such person shall be punished by imprisonment in the penitentiary not less than two nor more than five years. [2d Laws of 1861, p. 8, § 1, re-written; R. S. 1887, p. 465; S. & C., p. 806; Cothran, p. 483.]

173. MALICIOUS MISCHIEF—*attempt to commit as to railroad.* § 189. Whoever shall maliciously make any attempt, although the same may not succeed, to place obstructions on any railroad track, to burn, blow up or destroy any railroad bridge, or in any other way prevent the free and safe passage of trains on any railroad, shall be imprisoned in the penitentiary not less than one, nor more than ten years. [2d Laws 1861, p. 8, § 3, re-written; R. S. 1887, p. 465; S. & C., p. 806, § 241; Cothran, p. 483, § 189.]

174. MALICIOUS MISCHIEF—*influencing others to injure railroad.* § 190. Whoever shall maliciously hire, persuade or induce, attempt to hire, induce or persuade any person to burn, or in any way injure or destroy any railroad bridge, to take up, injure or destroy any railroad track, or any machine shop, engine house, car house, engine or car, or other machinery or property necessary for the operation of any railroad, shall be imprisoned in the penitentiary not less than one nor more than ten years. [2d Laws of 1861, p. 9, § 4, re-written,

and punishment increased; R. S. 1887, p. 465, § 190; S. & C., p. 806, § 242; Cothran, p. 483, § 190.]

175. RAILROAD ENGINEERS, &C.—*willful injury to stock.* § 191. Any engineer or person having charge of and running any railroad engine or locomotive, who shall willfully or unnecssarily kill, wound or disfigure any horse, cow, mule, hog, or other useful animal, shall, upon conviction, be fined in a sum not less than the value of the property so killed, wounded or disfigured, and confined in the county jail for a period of not less than ten days; and any such engineer who shall wantonly or unnecessarily blow the engine whistle so as to frighten any team shall be liable to a fine of not less than $10 nor more than $50. [See act of 1874 in relation to fencing and operating railroads. R. S. 1887, p. 1014, § 6½; also § 203 of Criminal Code; Laws 1845, p. 179, § 156; and 2d Laws of 1861, p. 9, § 4; R. S. 1887, p. 465, § 191; S. & C., p. 807, § 243; Cothran, p. 484, § 191. Post 2084–2086.]

176. MALICIOUS MISCHIEF—*injury to baggage.* § 193. If any baggage master, express agent, stage driver, hackman or any other person, whose duty it is to handle, remove or take care of trunks, valises, boxes, packages or parcels, while loading, transporting, unloading, delivering or storing the same, whether or not in the employ of a railroad, steamboat or stage company, shall wantonly or recklessly injure or destroy the same, he shall be fined not exceeding $200. [R. S. 1887, p. 465, § 193; S. &. C., p. 807, § 245; Cothran, p. 484, § 193. See act in relation to fencing and operating railroads, R. S. 1887, p. 1007, § 68; S. & C., p. 1944, § 93; Cothran, p. 1115, § 79.]

177. RAILWAY PROPERTY—*taking without consent.* § 242. If any person shall purchase or receive for sale from any other person any link, pin, bearing, journal, or other article of iron, brass or other metal which has been manufactured and is used exclusively for railroad purposes, and which shall have stamped thereon the name of some railroad company, or the initial letter thereof, without the consent in writing of the president, general manager or general superintendent of such railroad company, such person shall be fined in a sum not less than $100 nor more than $500, and be imprisoned not less than ten days nor more than ninety. [R. S. 1887, p. 474, § 242; S. & C., p. 820, § 298; Cothran, p. 499, § 242.]

178. JURISDICTION—*offense on railroad car or water-craft.* § 11. When any offense is committed in or upon any railroad car passing over any railroad in this state, or any water-craft navigating any of the waters within this state, and it cannot readily be determined in what county the offense was committed, the offense may be charged to have been committed and the offender tried in any of the counties through or along or into which such railroad car or watercraft may pass or come, or can reasonably be determined to have been on or near the day when the offense was committed. [R. S. 1887, p. 494, sec. 402; S. & C., p. 856, § 462; Cothran, p. 525, § 402.]

EMINENT DOMAIN.

179. CONSTITUTION OF 1848. ART. 13, § 11. No person shall, for the same offense, be twice put in jeopardy of his life or limb; nor shall any man's property be taken or applied to public use without the consent of his representatives in the general assembly, nor without just compensation being made to him. [R. S. 1887, p. 50, § 11; S. & C., p. 90, § 11.]

DECISIONS UNDER.

180. LIMITATION—*on legislature.* The constitution is a limitation upon the powers of the legislative department of the government. *Field* v. *People,* 2 Scam. 79; *Sawyer* v. *Alton,* 3 Scam. 127; *Prettyman* v. *Supervisors,* etc.. 19 Ill. 406; *Mason* v. *Wait,* 4 Scam. 127; *Edwards* v. *Pope,* 3 Scam. 465; *People* v. *Marshall,* 1 Gilm. 672; *People* v. *Reynolds,* 5 Gilm. 1; *People* v. *Wilson,* 15 Ill. 388; *Fireman's Benevolent Assoc.* v. *Lounsbury,* 21 Ill. 511.

181. RAILWAY COMPANY—*right to condemn.* Under the general law of 1849, a railway company had no right to condemn land for right of way without a law approving of the route and *termini* of its road. *Gillinwater* v. *M. & A. R. R.,* 13 Ill. 1.

182. COMPENSATION—*for property taken by contractors.* A corporation having the right to take materials for the construction of a public work by making compensation, will be liable to the owners for property taken by its contractors, although they were to furnish all materials. *Lesher* v. *Wabash Nav. Co.,* 14 Ill. 85; *Hinde* v. *Wabash Nav. Co.,* 15 Ill. 72.

183. NATURE OF POWER—*limitation, public use and compensation.* The right of eminent domain is an inherent sovereign power of the state. The exercise of the power is unlimited, except that it must be invoked for a public use, and only when required by public necessity, and that just compensation be made. *Johnson* v. *J. & C. R. R.,* 23 Ill. 202.

184. COMPENSATION—*by jury not necessary.* Not necessary that the compensation be assessed by a jury. The clause in the constitution of 1848, securing the right of trial by jury, has no application to a proceeding to condemn. *Ib.*

185 SAME—*when to be paid.* When the statute does not otherwise direct, if the condemnation price is paid when demanded by suit or otherwise, the parties entering upon the right, of way will not be trespassers *ab initio. Ib.*

186. NOTICE. Unless the act authorizing the condemnation so directs, a notice of the intention to condemn need not be given. *Ib.*

187 EMINENT DOMAIN—*not applicable to municipal subscriptions.* This clause of the constitution of 1848 was designed to regulate the exercise of the right of eminent domain, and in no wise relates to or affects the taxing power of the state. It does not prevent the legislature from authorizing counties and cities to take stock in railway corporations. *Johnson* v. *Stark Co.*, 24 Ill. 75.

188. APPORTIONMENT OF TAXES—*between county and city.* This constitutional provision is not violated by an amendment to a city charter requiring an apportionment of county taxes between the county and a city. *People* v. *Power*, 25 Ill. 187.

189. COMPENSATION—*when to be paid.* The constitution of 1848 does not require that compensation shall be made before the land is taken and used. It is sufficient if provision is made for its payment. *Shute* v. *Ch. & M. R. R.*, 26 Ill. 436.

190. INJUNCTION—*till compensation is paid.* If the compensation awarded is not paid, the company condemning may be restrained by injunction from using the right of way until it is paid. But non-payment will not make the condemnation invalid. *Ib.*

191. POWER—*necessity of compensation.* The power of eminent domain can only be exercised by making just compensation; and the compensation required is a matter of substance and not of form. *Chicago* v. *Larned*, 34 Ill. 203.

192. DIVESTITURE OF TITLE. When a condemnation is effected, and the damages are assessed and accepted by the owners, who declare their assent to the proceedings, the title thereby becomes divested. *Rees* v. *Chicago*, 38 Ill. 322.

193. SPECIAL ASSESSMENTS. The power to levy and collect special assessments is derived under the right of eminent domain, and not under the taxing power. *Chicago* v. *Larned*, 34 Ill. 203.

194. SAME—*compensation in benefits.* The just compensation for special assessments may be either in money or in benefits. *Ib.* See also *Chicago* v. *Baer*, 41 Ill. 306.

195. PUBLIC USE—*private road.* The legislature cannot provide for the laying of a private road over the land of another without his consent. His right is supreme, except when such land is needed for the *public* use, and then he must be compensated. *Nesbit* v. *Trumbo*, 39 Ill. 110; *Crear* v. *Crossly*, 40 Ill. 175.

196. PARTIAL TAKING—*property damaged.* This clause of the constitution of 1848 applies as well to secure the payment for property partially taken for the use of a street as when wholly taken and converted into a street. *Nevins* v. *Peoria*, 41 Ill. 502, 511.

197. INJUNCTION—*use before payment.* An attempt to open a road over improved land before the owner's damages are adjusted and paid may be restrained by a court of equity. *Coms. Highways* v. *Durham*, 43 Ill. 86.

198. PUBLIC USE. To authorize the taking of private property under the constitution (1848), the use must be such as is public in its character, and not public merely because called such. *E. St. Louis* v. *St. John*, 47 Ill. 463.

199. LIMITATIONS. The constitution of 1848 recognized the power of the state to take and apply private property to public use upon two indispensable conditions: *First*, that it must be by the consent of the general assembly, manifested by a law regularly adopted, and *secondly*, that just compensation shall be paid for the property taken. *Ib.*

200. DELEGATION OF POWER. This power is lodged alone in the general assembly, and its exercise is dependent upon the action of that body exercised in a proper case, or in such a case delegated to a body capable of its exercise. Without legislative authority it cannot be exercised. *Ib.*

201. PAYMENT—*must precede occupation—park.* Until the damages assessed for land condemned for a public park are paid, it cannot be occupied for the purposes intended. *People* v. *Williams*, 51 Ill. 63.

202. SAME—*provision for.* The act of February 24, 1869, in reference to South Park in Chicago, provides the means of making compensation for land condemned by adopting the mode provided in the act of 1852, in which ample provision is made for payment of the condemnation money. *Ib.*

203. EMINENT DOMAIN—*not applicable to taxing power.* The doctrine of eminent domain is strictly applicable only to the condemnation of property, and not to the levy and collection of a tax. *Harward* v. *St. Clair & Monroe Levee & Drainage Co.*, 51 Ill. 130; *Hessler* v. *Drainage Commissioners*, 53 Ill. 105.

204. SAME—*no application to special assessments.* An attempt to give a private corporation power to levy and collect a tax upon lands for supposed benefits by a drainage system, cannot be sustained under the doctrine of eminent domain, because the just compensation required under that right, must be determined by some impartial agency. *Harward* v. *St. C. & M. L. & D. Co.*, 51 Ill. 130; *Hessler* v. *Drainage Commissioners*, 53 Ill. 105.

205. JUDGMENT—*divesting of title.* The final judgment of the circuit court approving of the report of the commissioners appointed under petition under the general law of 1859, relating to plank, gravel and McAdamized roads, passes the title to the lands condemned, to the corporation. *Skinner* v. *Lake View Avenue Co.*, 57 Ill. 151.

206. COMPENSATION—*fixing, a judicial act.* The determination of what is a "just compensation" for private property taken for public use, is a judicial act, which can properly be performed only by the judicial department, and former decisions holding the award of persons not of the judicial department conclusive, is overruled. *Rich* v. *Chicago*, 59 Ill. 286; *Cook* v. *S. Park Commissioners*, 61 Ill. 115.

207. TRIAL BY JURY. An act giving a city council and board of public works power to assess the damages on the condemnation of land for the widening of a street, is not unconstitutional under the constitution of 1848. *Rich* v. *Chicago*, 59 Ill. 286.

208. COMPENSATION—*payment necessary to complete condemnation.* Park commissioners can not take and occupy land condemned for a public park until the damages assessed are paid the owner. *Cook* v. *South Park Commissioners*, 61 Ill. 115; *Ch. & Milwaukee R. R.* v. *Bull*, 20 Ill. 218; *Johnson* v. *Joliet & Ch. R. R.*, 23 Ill. 202; *Shute* v. *Ch. & Milwaukee R. R.*, 26 Ill. 436.

209. JUDICIAL PROCEEDING NECESSARY. The right of the state to take private property for public use cannot be asserted by mere enactment. The constitution providing that the citizen shall not be deprived of property except by due process of law, or in conformity to the law of the land, requires a trial, or judicial proceeding and a judgment. *Cook* v. *South Park Commissioners*, 61 Ill. 115.

210. COMPENSATION—*pecuniary.* The compensation required must be pecuniary in its character. *Weckler* v. *Chicago*, 61 Ill. 142.

211. JURY TRIAL—*on appeal.* On an appeal to the circuit court from an assessment of damages for a right of way for a railroad, the statute act of 1852, gives a trial by jury. *T. P. & W. R. R.* v. *Darst*, 61 Ill. 231.

212. CONDEMNATION—*before assessment and payment of damages.* Under constitution of 1848, charter power authorizing the taking of lands by condemnation before the ascertainment or payment of damages, was not unconstitutional. *Townsend* v. *C. & A. R. R.*, 91 Ill. 545.

213. STATE ALONE CAN CONFER THE POWER. The right of a

corporation to condemn is derived solely from the state law, not from the consent of city authorities. *Metro. City Ry.* v. *Ch. W. Div. Ry.,* 87 Ill. 317.

CONSTITUTION OF 1870.

214. EMINENT DOMAIN—*limitations on the right.* ART. 2, § 13. Private property shall not be taken or damaged for public use without just compensation. Such compensation, when not made by the state, shall be ascertained by a jury, as shall be prescribed by law. The fee of land taken for railroad tracks, without consent of the owners thereof, shall remain in such owners, subject to the use for which it is taken. [R. S. 1887, p. 55; S. & C., p. 105, 1037; Cothran, p. 3.]

215. EMINENT DOMAIN—*property and franchises of corporations—jury trial.* ART. 11, § 14. The exercise of the power, and the right of eminent domain, shall never be so construed or abridged as to prevent the taking, by the general assembly, of the property and franchises of incorporated companies already organized, and subjecting them to the public necessity the same as of individuals. The right of trial by jury shall be held inviolate in all trials of claims for compensation, when, in the exercise of the said right of eminent domain, any incorporated company shall be interested either for or against the exercise of said right. [R. S. 1887, p. 72; S. & C., p. 163, 1037; Cothran, p. 30.]

DECISIONS.

216. EMINENT DOMAIN—*when clause took effect.* These provisions were not merely prospective in their effect, but operated *in presenti* without legislative action. *People* v. *McRoberts,* 62 Ill. 38; *Mitchell* v. *Ill. St. L. R. R. & Coal Co.,* 68 Ill. 286.

217. SAME—*repeal of former laws.* The provision abrogated all existing laws for the assessment of damages by commissioners, appraisers or supervisors, so that the assessment of damages in such old ways, after it took effect, was void. *Kine* v. *Defenbaugh,* 64 Ill. 291; *People* v. *McRoberts,* 62 Ill. 38. It repealed so much of the act of 1852 as authorized the land to be entered upon before an assessment of damages by a jury. *Mitchell* v. *Ill. & St. L. R. R. & Coal Co.,* 68 Ill. 286.

218. SAME—*effect on prior rights and unfinished proceedings.* A constitution operates only prospectively unless clearly expressed otherwise, leaving all past transactions unaffected. Hence, when work was commenced on a street and it was mostly completed when the constitution took effect, it was held that a lot owner whose lot was not taken, could not recover compensation for damages to it. *Chicago* v. *Rumsey,* 87 Ill. 348.

219. SAME—*not conferred by the constitution.* The right of eminent domain is not conferred by the constitution, but only recognized and limited. But the power to declare under what circumstances it may be exercised, and to provide for the mode of its exercise, is conferred upon the general assembly by that clause vesting in it the legislative power. *L. S. & M. S. R. R.* v. *Ch. & W. Ind. R. R.,* 97 Ill. 506.

220. SAME—*limitation.* This constitutional provision is a limitation upon the exercise of the power, which, but for such limitations, is plenary, and might be exercised *ad libitum. Chicago* v. *Larned,* 34 Ill. 203; *Johnson* v. *Joliet & Ch. R. R.,* 23 Ill. 202; *E. St. L.* v. *St. John,* 47 Ill. 463.

221. SAME—*laws repealed by.* The first six sections of the act of 1852, which provide for the filing of a petition, due notice to the persons interested, the appointment of commissioners, their inspection of the premises, and a report of the compensation assessed by them to be filed with the clerk of the circuit court, are in no sense in conflict with the constitution of 1870, and are not abrogated by it. *People* v. *McRoberts,* 62 Ill. 38.

222. But the seventh section making the decision of the commissioners conclusive upon the parties before they can have a trial by jury, is inconsistent with the constitution of 1870. Their decision does not conclude the owner or confer any right upon the corporation, unless he assents by an acceptance of the compensation, or in some other manner. *Ib.*

223. EMINENT DOMAIN—*effect on taxing power.* The limitation in the constitution of 1870 (Art. 2, § 13), relates entirely to the subject of eminent domain, and has no reference to the taxing power. *White* v. *People,* 94 Ill. 604; *Johnson* v. *J. & Ch. R. R.,* 23 Ill. 202; *Johnson* v. *Stark Co.,* 24 Ill. 75; *Harward* v. *St. Clair Drain. Co.,* 51 Ill. 130; *Hessler* v. *Drainage Commissioners,* 53 Ill. 105.

224. SAME—*special assessments not affected by.* The levy of special assessments for building sidewalks, &c., is not a taking of private property under the right of eminent domain, but is the exercise of the right of taxation. *White* v. *People,* 94 Ill. 604.

225. CONSTITUTION—*application to—completed proceedings.* The constitutional provision that the fee to lands taken for right of way, shall not pass, but remain in the land-owner, has no application to proceedings completed before the adoption of the constitution. *T. P. & W. Ry.* v. *Pence,* 68 Ill. 524.

226. SAME—*not applicable to proceeding commenced under old one.* A proceeding to condemn land by a railway company for a right of way was commenced prior to the adoption of the constitution of 1870, under a charter which gave the land taken in fee simple to the company, but the assessment of damages was had after its adoption: *Held,* that the proceeding was governed by the charter under which it was commenced. *P. & R. I. R. R.* v. *Birkett,* 62 Ill. 332.

227. NATURE OF POWER. The right of eminent domain being an inherent attribute of sovereignty, exists independently of written constitutions or statutory laws, though its exercise is usually regulated by appropriate legislation. *Sholl* v. *German Coal Co.,* — Ill. —. Filed Jan. 25, 1887.

228. The right of eminent domain is founded upon public utility and necessity, and its exercise is a strictly legislative function, but subject to the right of the courts to determine whether the use for which property is sought to be taken, is a public one, and whether the proceedings have been conducted according to the law made on the subject. But the legislature is the exclusive judge of the necessity or emergency justifying the exercise of the power. *Ib.*

229. GRANT IN RESTRAINT OF THE RIGHT. The right of eminent domain is an element of sovereignty, and a legislative grant or contract in restraint of a free exercise of this right, is not binding on the state, and does not fall within the inhibition of the federal constitution relating to laws impairing the obligation of contracts. *Hyde Park* v. *Oakwoods Cem. Assoc.,* 119 Ill. 141.

WHAT CONSTITUTES A TAKING OR DAMAGING.

230. PARTIAL. The constitutional provision (1848) applies as well to secure the payment for property partially taken for the use of a street, as when wholly taken and converted into a street. The degree to which property is taken makes no difference in the application of the principle. *Nevins* v. *Peoria*, 41 Ill. 502. See *O. & M. Ry.* v. *Wächter*, — Ill. —. Filed Jan. 20, 1888.

231. BY CROSSING RAILROAD TRACK. The construction of a railroad track across a street upon which another railroad has its track, though built on the same grade, is a taking of the latter's property within the constitution. *Ch. & W. Ind. R. R.* v. *Ch. St. L. & P. R. R.*, 15 Bradw. 587.

232. NEW BURDEN—*highway for telegraph.* The use of a highway for a telegraph is a new and additional burden on the fee not contemplated on the assessment of damages, for which the owner of the fee is entitled to compensation. *Board of Trade Tel. Co.* v. *Barnett*, 107 Ill. 507.

233. WHAT IS A TAKING. The right of the state to take private property for public use cannot be asserted by mere enactment. This is not a taking. *Cook* v. *South Park Comrs.*, 61 Ill. 115.

234. BY DAMAGING. Until the adoption of the constitution of 1870, it was the settled doctrine of this court that any actual physical injury to private property by reason of the erection, construction or operation of a public improvement in or along a public street or highway, whereby the appropriate use or enjoyment of property on the street, was materially interrupted or its value substantially impaired, was regarded as a taking of private property to the extent of the damages thereby sustained. But the remedy was restricted to cases of direct physical injury. *Rigney* v. *Chicago*, 102 Ill. 64.

235. NEW REMEDY—*damage to property not touched.* The constitution of 1870, providing that private property shall not be "damaged for public use," gives redress in cases not provided for in the constitution of 1848, and embraces every case where there is a direct physical obstruction or injury to the right of *user*, or enjoyment of private property, by which the owner sustains some special damage in excess of that sustained by the public generally. *Ib.*

236. PROPERTY DEFINED. Property, in its appropriate sense, means that dominion or indefinite right of *user* and disposition which one may lawfully exercise over particular things or objects, and generally to the exclusion of all others, and doubtless this is substantially the sense in which the word is used in the constitution as to the taking or damaging of private property for public use. But the word is often used to indicate the subject of the property or the thing owned. *Rigney* v. *Chicago*, 102 Ill. 64.

237. VACATION OF STREET. The vacating of a public street not adjoining or contiguous to a particular lot, which does not deprive the owner of access to or egress from such lot, can in no sense be construed as either taking or damaging private property for public use. *E. St. Louis* v. *O'Flynn*, 119 Ill. 200.

238. The law will not regard the land as taken or acquired until the last act in the proceeding—that is, payment is performed. *Cook* v. *S. Park Comrs.*, 61 Ill. 115.

239. TAKING PROHIBITED—*requiring new duty.* A municipal corporation cannot by ordinance require a railway company to make proper crossings of its road over a new street laid out and opened long after the completion of the railroad, where no such duty is imposed by its charter or the general law in force when the company was created. Even the legislature cannot impose such burden without making compensation. *I. C. R. R.* v. *Bloomington,* 76 Ill. 447.

240. Under proceedings to condemn for public use the filing of the petition is not a taking of the property, and it would be a trespass to take possession before the damages are ascertained. *South Park Comrs* v. *Dunlevy,* 91 Ill. 49.

241. NEW AND ADDITIONAL BURDEN—*protection against.* The use of a street or highway for a telegraph is a new and additional burden upon the fee not contemplated on the assessment of damages in case the easement was obtained by condemnation, or had in view by the land owner in case of a dedication for ordinary highway purposes, and for such additional burden the owner of the fee is entitled to compensation, and if entry be made without an agreement with the owner or a condemnation, the owner may have his action. *Board of Trade Tel. Co.* v. *Barnett,* 107 Ill. 507.

WHAT MAY BE TAKEN.

242. FERRY PRIVILEGE. A grant of a franchise or privilege by the state to a person or corporation, such as a ferry, is subject to an implied reservation in favor of the sovereign power that, when the public good requires it, all the rights and privileges conferred may be resumed upon adequate compensation being made therefor in the manner required by law. *Mills* v. *St. Clair Co..* 2 Gilm. 197, 227.

243. STATE GRANTS. All grants made by the state, whether to the canal trustees or others, although irrevocable, are subject to the right of eminent domain, unless that right is expressly relinquished. *Ill. & Mich. Canal* v. *Ch. & R. I. R. R.,* 14 Ill. 314.

244. A railway charter giving the power to condemn the right of way over lands granted by the state to the canal trustees for a railroad, and the effect of the contemplated road in diminishing the revenues and business of the canal, is not in violation of the contract of the state with the trustees. *Ib.*

245. RAILROAD PROPERTY—*subject to.* The lands of a railway corporation, not absolutely necessary for the enjoyment of its franchise, are subject to the right of eminent domain, under legislative authority, the same as those of individuals, though they may be taken from the actual and profitable use of the corporation. *P. P. & J. R. R.* v. *P. & S. R. R.,* 66 Ill. 174.

246. Property of a railway or other corporation, though acquired by condemnation, is subject to be taken for the public use the same as that of private persons. *C. R. I. & P. R. R.* v. *Town of Lake,* 71 Ill. 333; to same effect, *Richmond R. R.* v. *Louisa. R. R.,* 13 How. 74; *West River Bridge Co.* v: *Dix,* 6 How. 529; *Boston Water Power Co.* v. *Boston & Worcester R. R.,* 23 Pick. 360.

247. PROPERTY IN PUBLIC STREET—*injunction.* A horse railway company has no right to condemn and take for its joint use a part of a previously constructed railway of another company in successful operation, and thus render the fragments not so taken unproductive, and make the franchise of such other company of but little value: and if such an attempt is made, a court of equity will enjoin the same. *Central City Horse Ry.* v. *Ft. Clark Horse Ry.,* 81 Ill. 523.

248. The statute authorizing the condemnation by horse and dummy roads (R. S. 1887, ch. 66,) contemplates private property alone, and not property occupied and used by the public. *Ib.*

249. By a very liberal construction of the statute, and of the eminent domain act, it may be that a newly organized horse railway company may condemn the entire road of a similar company previously incorporated, and appropriate it to its own use. *Ib.*

250. The right of a horse railway company, under a contract with a city providing against having a similar railway on certain streets running parallel with its road, is property within the sense of the eminent domain act, and may be condemned for the use of a new company when the public necessity so requires. *Metropolitan City Ry.* v. *Ch. W. Div. Ry..* 87 Ill. 317.

251. Under the constitution the property and franchises of incorporated companies may be appropriated to the public use as well as the property of individuals, and the exercise of the right of eminent domain can never be so construed or abridged as to prevent the general assembly from appropriating such property when the public exigency demands it. Whatever exists in any form, tangible or intangible, is subject to the exercise of this power. *Ib.*

252. PROPERTY IN PUBLIC USE—*taking for same use.* When property has already been appropriated to public use, and is in fact in such use in the hands of one railway corporation, it cannot rightfully be taken from such corporation, even by authority of a statute, for the purpose of subjecting it to the same public use in the hands of another corporation. *L. S. & M. S. R. R.* v. *Ch. & W. Ind. R. R.,* 97 Ill. 506.

253. SAME—*of the new use.* To warrant the taking of property of one party already appropriated to public use. and placing it wholly or in part in the hands of another party, it is essential that the new use shall be for the benefit of the public. Whether the new use be different from the present one is a judicial question for the court to decide, but whether a public benefit, and the change will be for the benefit of the public, are political questions to be decided by the law-making power. *Ib.*

254. In a proceeding to condemn a part of the property of one railway company for the use of another leading from other and different points and regions of country, the use is not the same as that of the prior road, but is rather a joint or co-operative use, to be exercised and enjoyed by both companies, so as to furnish the public an additional line of travel and transportation, and may be properly granted by the legislative action. *Ib.*

255. CORPORATE PROPERTY—*how far private.* The property of corporations as to the ownership thereof and the profit and gain to be made from its use, is to all intents private property, although applied to a use in which the public have an interest, and § 14, Art. 11, of the constitution, simply places such property, like that of natural persons, within the power of eminent domain, as it was before any such declaration, and protects it the same as any other private property. *Ib.*

256. The power of eminent domain is not conferred by the constitution, but is an inherent attribute of sovereignty. § 13, Art. 2, recognizes the power, and its purpose is to limit and regulate its exercise. § 14, Art. 11, recognizes the same power, but does not profess to grant or add to it. It is only an authoritative explanation of the nature and extent of this power, and it is but declaratory of the power the state would have had without it. *Ib.*

256a. POWER TO CONDEMN ONE RAILWAY FOR USE OF ANOTHER—*taking part of another road.* The legislature, subject to the constitutional limitations, has the power by a general law, to authorize one railway company to condemn a part of the right of way of another longitudinally, several miles, when necessary for the construction and use of a new road; but without such legislative authority this cannot be done. *Ill. C. R. R.* v. *C., B. & N. R. R.,* — Ill. —. Filed Sept. 26, 1887.

256b. The general grant of power given in § 17, chap. 114 relating to railroads, to take and condemn real estate for railroad purposes, is not intended to extend to property already applied to a public use. *Ib.*

256c. While the legislature has provided by law for the crossing and intersection of one railroad over and across the track and right of way of another, and required the company whose road is crossed or intersected, to unite with the new railway company in forming such intersections and connections, and grant the proper facilities therefor, it has not given a new railway corporation the right to condemn the right of way of a prior company longitudinally for a number of miles in length, or any part thereof, to the exclusion of such prior company. *Ib.*

256d. PROPERTY APPROPRIATED TO PUBLIC USE—*limited to crossings and connections.* The power to take the right of way, or any part of the right of way of another railway company, is expressly limited by the statute to the purposes of crossing, intersecting and uniting, or more shortly stated, to the connections of the two roads. *Ib.*

256e. The petitioning company has no power under the statute to take any part of the right of way of another company, except for the purpose of some connection resulting from a crossing or intersection, or the uniting and joining of the two roads at some point on the line of the new road, selected by the petitioning company. *Ib.*

257. CROSSING ANOTHER ROAD. One railroad company is entitled to have condemnation under the statute for its right of way across the right of way of a previously constructed railroad, but the company whose right of way is condemned is entitled to be fully compensated for all damages it may sustain in consequence thereof. *St. L., J. & Ch. R. R.* v. *S. & N. W. R. R.,* 96 Ill. 274.

258. PRESUMPTION AGAINST MONOPOLY. The public welfare requires that the business of carrying shall be open to competition as far as possible, and no monopoly in that regard, however limited, can be presumed to have been intended by the legislature in the enactment of the general law for the formation of railroads. *E. St. L. Connecting Ry.* v. *E. St. L. Union Ry.,* 108 Ill. 265.

259. The mere grant of the right to build a railroad between given points, creates no implied obligation on the state not thereafter to grant the right to build other railroads, parallel with it between the same *termini;* nor does it imply an obligation on the part of the state that other railroads with their tracks and switches shall not thereafter be granted the right to cross the state in a different direction, and thus pass over its tracks and switches. *Ib.*

260. RIVAL RAILROAD—*injunction.* Under the laws of this state, a railway corporation already organized and operating its road, cannot enjoin another such corporation organized under the same general law, from building a rival road between the same *termini,* and parallel with the track of the former for the transaction of the same business, although the main and lateral tracks and switches may be intersected and crossed by the proposed new road, no continuous portion of its track being sought to be taken. *Ib.*

261. The fact that the construction of the new road may damage the business of the old one, and cause delay in operating its trains, affords no ground for enjoining proceedings to condemn for a right of way by the new corporation. Legal damages assessed, as is provided by law, will afford the old company an adequate remedy for all the injury it may sustain. *Ib.*

—4

262. PROPERTY DEVOTED TO PUBLIC USE. In the absence of a clearly expressed intention to the contrary, the courts will not so construe a railway charter as to authorize one company to take the property of another already devoted to a particular public use. When there is no change in the use, it becomes a matter of mere private concern, without at all affecting the public interests. This is when the taking merely changes the ownership and not the use. *Ch. & N. W. Ry.* v. *Ch. & E. R. R.*, 112 Ill. 589.

263. WHEN USE IS DIFFERENT. The condemnation of a piece of ground for a right of way, and the construction of an abutment thereon for a bridge essential for its use as a right of way, which piece of ground had before been used by another railway company for a wharf or dock for the receiving and discharge of freights, is not a condemnation for the same public use as that to which the property was already applied. *Ib.*

264. RIGHT TO CONDEMN A CROSSING. The sixth clause of § 19 of the railroad law of 1872 confers power upon any railroad corporation formed under that act, to cross, intersect, &c., any other railroad before constructed at any point in its route, and upon the grounds of such other company; and provides that if the two companies cannot agree upon the compensation to be made, or the points and manner of such crossings, the same shall be ascertained and determined in the manner provided by law, which means by a proceeding under the eminent domain act. *Ch. & W. Ind. R. R.* v. *Ill. C. R. R.*, 113 Ill. 156.

265. RIGHT TO ADDITIONAL CROSSINGS. The fact that a railway company has acquired a strip of land thirty feet wide across another railway company's right of way for a crossing upon which to lay two tracks at the expense of the former, by mutual agreement and for a consideration paid, will not preclude the former from obtaining by condemnation an additional right of way of twenty feet across the latter company's road, when rendered necessary by the increased business of the former road, where there is no restriction of such right in the agreement; and it matters not that such increased business is brought about by its contracts of connection with other roads. *Ib.*

266. SAME—*effect of prior contract.* Where a right of one railway company to cross another's road by two tracks has been acquired by purchase, and limited to a right of way thirty feet wide, and it does not appear that the relinquishment for the future of any right of further application for additional facilities of crossing in any way entered into the amount of the compensation which was arranged, and the right to lay additional tracks across the same road is sought by condemnation, the company seeking to condemn will not be required to surrender its rights acquired by the purchase in order that it may have the condemnation sought, and have compensation assessed for the four tracks in that proceeding. *Ib.*

267. SAME—*ground of necessity immaterial.* It matters not that the necessity for an increase of the right of way for additional tracks is caused by the use of its road by other companies acting under its lease or by contract; nor does it matter by what corporation, or corporations, its road is actually operated. It is still a public use, and in such case the need of the lessees is that of the lessor company, and the lessees may proceed to condemn in the name of the lessor when the public necessity so requires. *Ib.*

268. EASEMENT IN HIGHWAY. Where a railway company acquires in perpetuity an easement in so much of a public street as it occupies for its road, this easement is property, and it is as much protected from unlawful invasion as any other property, and cannot be taken or damaged for public use without just compensation. *Ch. & N. W. Ry* v. *Village of Jefferson*, 14 Bradw. 615.

269. SAME—*new burden on.* Where a railway company acquires an easement or right of way in a public street or highway, subject to the public right to use the same for the ordinary purposes of a highway, the occupation of a considerable portion of the street for the construction of a ditch for the purpose of draining adjacent land, is a new use of the street, for which compensation must be made in case the property of the railway company therein is damaged thereby. If the city or village makes no provision to pay such damages, the company may enjoin the construction of the ditch. *Ib.*

270. If a railway company, under permission from village authorities, constructs its road in a public street, it will thereby acquire a perpetual easement in the street, which consists in the right to maintain, use and enjoy its railroad free from hindrance or molestation, except such as is incident to the proper and ordinary use of the street; and this right will be as much protected from unlawful invasion as any other property. *Ch. & W. Ind. R. R.* v. *Ch. St. L. & Pittsburgh R. R.*, 15 Bradw. 587.

271. The construction of a railway track across a public street upon which another railway company has its tracks, although built on the same grade, is a taking of the latter's property within the meaning of the constitution, and the latter may have the construction enjoined until compensation is made. *Ib.*

272. A leasehold interest for 999 years in a perpetual easement of another railway company in a public street, is private property, and cannot be taken or damaged by another railway corporation without the payment of compensation. If attempted to be taken or damaged by condemnation proceedings against the lessor company alone, the proceeding may be enjoined. *Ch. & E. R. R.* v. *Englewood connecting Ry.*, 17 Bradw. 141.

273. Where a railway company has, by agreement, acquired the right to lay two railroad tracks over a railroad previously constructed, and it seeks to condemn an additional strip on which to construct two others of its tracks across the same road, the fact that it will produce an obstruction and inconvenience to the company whose road is sought to be crossed, is no reason for enjoining the proceeding to condemn, as all the damages caused thereby will have to be paid, and it will be presumed that they will be fully awarded. *Ch. & W. Ind. R. R.* v. *I. C. R. R.*, 113 Ill. 156.

274. RAILROAD RIGHT OF WAY—*damages for.* A right of way for a railroad is not a corporate franchise, but is property acquired in the exercise of such franchise; and if it is sought to be condemned, the party in whom the same is vested in trust, will have the right to present his claim for compensation and be heard in support thereof. *Johnson* v. *F. & M. River R. R.*, 116 Ill. 521.

275. LEASEHOLD ESTATE. The estate of a tenant for years may be taken for the public use upon precisely the same terms as any other estate in lands may be, on payment of compensation. *Chicago* v. *Garrity*, 7 Bradw. 474.

276. EASEMENT. Where a party conveys land reserving in his deed the privilege of a water power, and the right to enter upon so much of the land as may be needful for an abutment on the bank, he has such an interest in the land as may be affected by the construction of a railroad, and the company cannot appropriate the land to its own use without ascertaining, in the mode pointed out in the statute, what damage he will sustain. *Galena & S. Wis. R. R.* v. *Haslam*, 73 Ill. 494.

277. RAILROAD CROSSING—*change of use—right to select place of crossing.* Under the present legislation, a railroad company is expressly authorized, in locating and constructing its road, to cross and intersect any intervening railroads at any point in its route, and this by necessary implication is a legislative declaration that the subordination of premises already occupied by a railroad company to the uses of another for a crossway, is a change in the use which the public good demands; but the corporation seeking the right of way, when the parties cannot agree, must select the place and manner of the proposed crossing, and the character and condition of the use sought, and this should be stated in the petition to afford the proper basis for ascertaining the compensation to be paid. *L. S. & M. S. R. R.* v. *Ch. & W. Ind. R. R.*, 97 Ill. 506. See ante 91–93.

RIGHT TO JURY TRIAL.

278. UNDER CONSTITUTION OF 1848. The provisions in the constitution of 1848 relating to jury trials—§ 8, Art. 13—that no one shall be deprived of his life, liberty or property, except by the judgment of his peers, or the law of the land, have no application to condemnation proceedings under the right of eminent domain. *Johnson* v. *Joliet & Ch. R. R.*, 23 Ill. 202.

279. It is not necessary that a jury shall be called to aid in condemnations for right of way. *Ib.*

280. ON APPEAL—*act of 1852.* On an appeal to the circuit court, in a case for the assessment of damages on the condemnation of the right of way of a railroad, the statute gives a trial by jury. This by act of 1852. *T. P. & W. Ry.* v. *Darst*, 61 Ill. 231.

281. CONSTITUTION OF 1870. The just compensation for property taken or damaged for public use must be ascertained by a jury. This requirement of the constitution is affirmative in its character, and implies an exclusion of any other mode of fixing the compensation. *People* v. *McRoberts*, 62 Ill. 38.

282. The compensation for property damaged, as well as taken, must be ascertained by a jury, and there can be no entry upon or possession of land for public use until the compensation for property damaged, as well as taken, has been paid. *Ib.*

283. CONDEMNATION—*under act of 1852.* There is enough in the act of 1852, not abrogated by the new constitution, to enable private property to be acquired for public use. After notice of the filing of the report of the commissioners the land owners may by appeal bring the proceedings before the circuit court, and if satisfied with the compensation fixed, it may be accepted and an adjustment made. If not satisfied, a trial can be had in the circuit court by a jury. *Ib.*

284. LAYING OUT HIGHWAY. The statute authorizing the assessment of damages by commissioners of highways and supervisors on appeal for land taken for highways, was repealed and rendered inoperative by § 13, Art. 2 of the constitution of 1870. *Kine* v. *Defenbaugh*, 64 Ill. 291.

285. The assessment of damages to a party by reason of the laying out and construction of a highway over his land by the commissioners of highways, or by the supervisors on appeal in a proceeding commenced after the adoption of the new constitution, will be void, although no mode for an assessment by a jury had then been provided. *Ib.*

286. Where certain persons obligated themselves by their bond to procure certain grounds for the state, or to pay the compensation required to be paid to the owners upon condemnation thereof by the state: *Held*, that as the compensation was not to be paid by the state, the parties in interest had a right to have the compensation determined by a jury. *People* v. *Stuart*, 97 Ill. 123.

287. LESS THAN TWELVE. Under § 5, Art. 2, of the constitution the legislature is authorized to provide for a jury of less than twelve men in the trial of civil causes before justices of the peace. Under this, a general law for the assessment of damages for land condemned by commissioners of highways for roads, may constitutionally provide for a jury·of six men, and their assessment will be valid and binding on the land owners. *McManus* v. *McDonough*, 107 Ill. 95.

288. RIGHT TO—*waiver*. Either party has the right to have the compensation assessed by a jury, and it is error to deny the right; but the parties may waive or dispense with a jury, and the finding of the court will be valid. *C. M. & St. P. Ry* v. *Hock*, 118 Ill. 587.

WHAT IS A PUBLIC USE.

289. It is a settled doctrine that the appropriation of property to the construction or use of a railroad, is an appropriation of such property to the public use. *C. R. I. & P. R. R.* v. *Joliet*, 79 Ill. 25.

290. PUBLIC GROUNDS—*taking by state*. Where directors of a railway company, under legislative authority, locate and construct their road along and across the public grounds and streets of an unincorporated town, in so doing they act as public agents, and the location is the act of the state, unless such use is inconsistent with the use to which such public grounds had been previously applied. *Ib.*

291. SAME—*estoppel*. Case and facts stated from which the city authorities were held estopped from disputing the right of a railway company to use streets and public grounds for right of way. *Ib.*

292. The right to take private property for railroad purposes by the exercise of the right of eminent domain, rests wholly upon the doctrine that the railroad use is a public use. The corporation itself is private and it has private rights: still its uses are public. *Ib.*

293. To authorize the taking of private property under the constitution (of 1848) the use must be such as is public in its character, and not public merely because so called. *E. St. Louis* v. *St. John*, 47 Ill. 463.

294. FOR STREETS. The taking and appropriating of property for a public street or highway by a municipality is a public use in its nature. *Dunham* v. *Hyde Park*, 75 Ill. 371, 375; *C. R. I. & P. R. R.* v. *Town of Lake*, 71 Ill. 333.

295. Property taken for a railroad or damaged by the construction and operation of a railroad, is taken or damaged for a public use. *Ch. & W. Ind. R. R.* v. *Ayres*, 106 Ill. 511.

296. RAILROAD PURPOSES. Although a railway company may be a private corporation, yet the road is to be regarded as a public improvement, made to subserve the public interests. Such roads are of such public use as to justify the exercise of the right of eminent domain. *Ch. Dan. & Vin. R. R.* v. *Smith*, 62 Ill. 268, 275.

297. Mere convenience is not sufficient to justify the exercise of the right. The public use must be necessary and pressing. *Ib.*

298. STREETS. The taking of land for a public street or highway by a municipality is a public use in its nature, and cannot be questioned or denied. *C. R. I. & P. R. R.* v. *Town of Lake*, 71 Ill. 333, 336.

299. POWER OF COURT. On application to condemn land the court has the right to determine whether the proposed use is public in its nature or not. *Ib.*

300. The question whether the use to which the property is to be taken is a public use or purpose, and whether such use or purpose will justify the exercise of such compulsory taking, and, where the power is attempted to be exercised by a corporation, whether the power is delegated to it by the legislature, and whether the uses and purposes for which such power is sought to be exercised falls within the legislative grant of powers, are proper subjects of judicial determination. *Ch. & E. Ill. R. R. v. Wiltse*, 116 Ill. 449.

301. PUBLIC USE—*determined from nature of business done.* The business proposed to be done, and the manner of doing it, must be looked at in determining whether the use to which property is to be devoted will be a public or private one. If, from the nature of the business and the way in which it is to be conducted, it is clear no obligation will be assumed to the public, or liability incurred other than such as pertains to all strictly private enterprises, then the use is private and not public. *Sholl v. German Coal Co.,* — Ill. —. Filed Jan. 25, 1887.

302. The use of a strip of land by a coal company upon which to construct a trainway leading from the coal works to a railway track is a private use, and such strip cannot be condemned under the act for such use. *Ib.*

CHAPTER 47.

EMINENT DOMAIN.

An act to provide for the exercise of the right of eminent domain. Approved April 10, 1872. In force July 1, 1872. L. 1871-2, p. 402.

303. COMPENSATION—*jury.* § 1. *Be it enacted by the people of the State of Illinois, represented in the general assembly,* That private property shall not be taken or damaged for public use without just compensation; and that in all cases in which compensation is not made by the state, in its corporate capacity, such compensation shall be ascertained by a jury, as hereinafter prescribed. [R. S. 1887, p. 646; S. & C., p. 1041; Cothran, p. 646. See post 1213-1219, 1220-1225a, 1512.]

304. LEGISLATURE—*its power in respect to eminent domain.* The power to declare under what circumstances this right may be exercised, and to provide the mode of its exercise, is conferred upon the general assembly by that clause of the constitution which vests in that body the "legislative power" of the state. *L. S. & M. S. Ry. v. Ch. & W. Ind. R. R.,* 97 Ill. 506.

305. STATUTE REMEDIAL—*liberal construction.* The statute is a remedial one, and should be liberally and beneficially construed. *Ch. & Eastern R. R. v. Englewood Con. Ry.,* 17 Bradw. 141.

306. STATUTE MANDATORY—*strict compliance.* A statute providing how the property of an individual shall be condemned for public use is not merely directory, but is mandatory. A strict compliance with its provisions is necessary. *Mitchell v. Ill. & St. Louis R. R. & Coal Co.,* 68 Ill. 286.

307. DEPRIVATION OF PROPERTY—*only under power of eminent domain.* A person cannot be deprived of his property except by the exercise of the right of eminent domain, in which case just compensation must be made. *Lake View* v. *Rose Hill Cemetery Co.*, 70 Ill. 191, 199.

308. ACT OF 1852—*how far a repeal of act of 1845.* The act of 1852, concerning the right of way, did not repeal any of the provisions of the act of 1845 not inconsistent therewith. Authority to acquire right of way under the act of 1852, or any other act, carried with it the power conferred by the act of 1845. *Taylor* v. *Pettijohn*, 24 Ill. 312.

309. WHAT LAW GOVERNS—*change pending proceeding.* After the institution of a proceeding in 1852 to condemn a right of way for a railroad, and during its pendency, the law under which it was commenced was amended, excluding the consideration of benefits in common with other lands: *Held*, that the case was governed by the law as it stood when the petition was filed. *A. & S. R. R.* v. *Carpenter*, 14 Ill. 190.

310. ACT OF 1845—*not repealed by act of 1852.* The act of 1852 being intended only as an amendment of the law concerning right of way, did not repeal any more of the act of 1845 than was repugnant thereto. They both might stand. *Taylor* v. *Pettijohn*, 24 Ill. 312.

311. UNDER DIFFERENT ACTS. A corporation was authorized to acquire a right of way under the act of 1852, or as authorized by any other act: *Held*, that its authority embraced the act of 1845, or such part thereof as was not repealed by the act of 1852 amendatory thereof. *Ib.*

312. UNDER WHAT LAW—*election.* A railway company instituted a proceeding under the act of 1852 to condemn a right of way, and, on appeal from the award of the commissioners to the circuit court, asked to amend so as to make the proceeding under the act of 1845: *Held*, not admissible. The petitioner is bound by his election. *P. P. & J. R. R.* v. *Black*, 58 Ill. 33.

313. REPEAL. The act of 1845, entitled "right of way," was not repealed by the act of 1852 on the same subject, except in so far as it was repugnant to the latter act. The general law of 1849, relating to railroads, did not affect the act of 1845. *P. & R. I. Ry.* v. *Warner*, 61 Ill. 52.

314. ACT OF 1852—*how far repealed by new constitution.* There is enough in the act of 1852 not abrogated by the new constitution to enable private property to be acquired for public use. A jury may be had on appeal. *People* v. *McRoberts*, 62 Ill. 38.

315. WHAT LAW GOVERNS. A railway company having commenced proceedings for condemnation under the statute of 1852 must adhere to it throughout, and cannot resort to other statutes. The rights of the parties must be controlled by the act under which the proceedings are begun. *P. P. & J. R. R.* v. *Laurie*, 63 Ill. 264.

316. CHANGE IN LAW—*effect on proceeding.* The state has the right to say on what terms it will allow the right of eminent domain to be exercised, so long as there remains anything to be done by the corporation to complete the condemnation of the land. *S. & I. S. Ry* v. *Hall*, 67 Ill. 99.

317 A proceeding to condemn land for a right of way under the act of 1852 was brought, but before a trial was had the act of 1872 had taken effect, and the damages were assessed according to the latter act, which expressly repealed all conflicting laws: *Held*, that the assessment was properly made under the latter act, as the proceedings were *in fieri* when it took effect. *Ib.*

318. Where land was taken for a right of way for a railroad, and a proceeding to fix the compensation commenced before the act of 1872 on that subject went into effect: *Held*, that the compensation should be ascertained under the law in force at the time the proceeding was begun. *Emerson* v. *Western Union R. R.*, 75 Ill. 176.

319. CONDEMNATION—*strict compliance.* A proceeding to condemn land for a right of way being an extraordinary and summary remedy, the party exercising the power must strictly observe all the requirements of the statute under which he acts. *C. & A. R. R.* v. *Smith*, 78 Ill. 96.

320. WHAT LAW GOVERNS. The laws in force at the time a city enters upon a public improvement of a street by changing its grade will fix and determine the rights of a property holder to damages, and it cannot be altered by subsequent legislation. *City of Elgin* v. *Eaton*, 83 Ill. 535.

321. CONDEMNATION—*strict compliance.* To divest a person of his property by proceedings against his will, there must be a strict compliance with all of the provisions of the law which are made for his protection and benefit. Those provisions must be regarded as in the nature of conditions precedent, which must not only be complied with before the right of the property owner is disturbed, but the party claiming under the adverse proceedings must affirmatively show such compliance. *Hyslop* v. *Finch*, 99 Ill. 171.

322. The rule which requires great strictness in a statutory proceeding has application only in summary and *ex parte* cases where the person whose right is to be affected is not a party. It is not enough to require the rigid rules of strictness, merely that the proceeding is a statutory one. The rule does not apply to a proceeding to condemn land for a right of way under the statute. *Bowman* v. *V. & C. Ry*, 102 Ill. 472.

323. The taking of private property under the eminent domain act being in derogation of common right, the grant of power to corporations for its exercise will be strictly construed. *Ch. & E. Ill. R. R.* v. *Wiltse*, 116 Ill. 449.

324. ACT OF 1852—*parts repealed.* § 9 of the act of 1852, which requires the execution of an appeal bond on taking an appeal from the award of the commissioners to the circuit court, and § 12, which permits the land to be entered upon pending the appeal, being inconsistent with the bill of rights, are repealed by the constitution. *People* v. *McRoberts*, 62 Ill. 38.

325. JURISDICTION—PETITION—PARTIES. § 2. That in all cases where the right to take private property for public use, without the owner's consent, or the right to construct or maintain any public road, railroad, plankroad, turnpike road, canal or other public work or improvement, or which may damage property not actually taken, has been heretofore or shall hereafter be conferred by general law or special charter upon any corporate or municipal authority, public body, officer or agent, person, commissioner or corporation, and the compensation to be paid for or in respect of the property sought to be appropriated or damaged for the purposes above mentioned cannot be agreed upon by the parties interested, or in case the owner of the property is incapable of consenting, or his name or residence is unknown, or he is a nonresident of the state, it shall be lawful for the party authorized to take or damage the property so required, or to construct, operate and maintain any public road, railroad, plankroad, turnpike road, canal or other public work or improvement, to apply to the judge of the circuit or county court, either in vacation or term time, where the said property or any part thereof is situate, by filing with the clerk a petition, setting forth, by reference, his or their authority in the premises, the purpose for which said property is sought to be taken or damaged, a description of the property, the names of all persons interested therein as owners or otherwise, as appearing of record, if known, or if not known stating that fact, and praying such judge to cause the compensation to be paid to the owner to be assessed. If the proceedings seek to affect the property of persons under guardianship, the guardians, or conservators of persons having conservators, shall be made parties defendant, and if of married women their husbands shall also be made parties. Persons interested, whose names are unknown, may be made parties defendant by the description of the unknown owners; but in all such cases an affidavit shall be filed by or on behalf of the petitioner, setting forth that the names of such persons are unknown. In cases where the property is sought to be taken or damaged by the state for the purpose of establishing, operating or maintaining any state house or state charitable or other state institutions or improvements, the petition shall be signed by the governor or such other person as he shall direct, or as shall be provided by law. [R. S. 1887, p. 646, § 2; S. & C., p. 1042, § 2; Cothran, p. 646, § 2.]

326. JUDICIAL PROCEEDING. The right of the state to take private property for public use, cannot be asserted by a mere enactment. A condemnation by a judicial proceeding and judgment is necessary. *Cook* v. *South Park Coms.*, 61 Ill. 115.

327. WHO MAY CONDEMN AND FOR WHAT PURPOSES—*incorporated town, for a street.* The town of Mt. Sterling has the power to lay out and open streets and to condemn land therefor. *Curry* v. *Mt. Sterling*, 15 Ill. 320; *Dunlap* v. *Mt. Sterling*, 14 Ill. 251.

328. SAME—*strict construction.* The rule of strict construction is applied only in cases of ambiguity, or where a power is claimed by implication. *Newhall* v. *Galena & Ch. Union R. R.*, 14 Ill. 273.

329. SAME—*length of lateral railroad.* Where the legislature has given a railway company power to build lateral roads without fixing any limits to the length of such roads, the courts will not, as a general rule, fix any limits. *Ib.*

330. POWER TO CONDEMN—*for branch or lateral road.* Where a right is conferred to build lateral roads, the presumption follows that the company has the same authority to obtain the right of way of such roads as is conferred for the main line. *Ib.*

331. SAME—*extension, applies to lateral roads.* An extension of time to a railway company which has the right to build lateral routes, for completing its road, will embrace the lateral branches as well as the main line. *Ib.*

332. RIGHT TO CONDEMN—*forfeiture of power.* The failure of the Illinois Central Railroad company to locate its road within the limits of the city of Chicago by the first day of January, 1852, as required by its charter, did not work a forfeiture of its right to condemn lands to its use where the assent of the city to such location was not given until after that day had expired. *Ill. C. R. R.* v. *Rucker,* 14 Ill. 353.

333. WHO MAY ENFORCE EMINENT DOMAIN. The right of eminent domain may be exercised either directly by the agents of the government or through the medium of corporate bodies. *Beekman* v. *S. & S. R. R.,* 3 Paige 45. It may be exercised by the United States, even within states, within its constitutional powers and purposes. *Kohl* v. *United States,* 91 U. S. 367; *Darlington* v. *U. S.,* 82 Pa. St. 382. But the general government cannot control the states in the exercise of this power. *Boone Co.* v. *Patterson,* 98 U. S. 403.

334. SAME—*under law of 1849.* Railway companies organized under the general railroad law of 1849 cannot condemn lands for right of way until they have obtained an act of the legislature approving of the route and *termini* of their roads. *Gillinwater* v. *Miss. & Atlantic R. R.,* 13 Ill. 1.

335. SAME—*under act of 1852.* § 19 of the general railway act of 1852 was intended to reserve power in the legislature to fix the route and *termini* of all roads organized under its provisions, and not to repeal the law of 1845 relating to right of way. The sole object of that section was to continue the reservation of power in the legislature to fix the route and *termini* of all roads before the corporations should exercise the right of eminent domain. *P. & R. I. R. R.* v. *Warner,* 61 Ill. 52, 55.

336. SAME—*power not exhausted by its exercise.* The power to condemn land for railroad purposes is not exhausted by an apparent completion of the road, if an increase of business shall demand other appendages or more room for tracks. *C. B. & Q. R. R.* v. *Wilson,* 17 Ill. 123, 127.

337. SAME—*for workshops, &c.* A grant of power to a railway company to construct a road with such appendages as may be deemed necessary for the convenient use of the same will authorize it to acquire land by condemnation for workshops, &c., these being necessary appendages. *Ib.*

338. POWER TO CONDEMN—*for paint shops, &c.* The Galena & Chicago Union Railroad Company, under its charter, has the power to condemn lands for depot grounds or on which to erect a paint shop and lumber sheds for its use. *Low* v. *G. & Ch. Union R. R.,* 18 Ill. 324.

339. RIGHT TO CONSTRUCT ROAD IN CITY. Power to a railway company to bring its road to a city and acquire property within it, also carries with it the power to enter the city and acquire right of way therein. *Moses* v. *P. Ft. W. & Ch. R. R.,* 21 Ill. 516.

340. POWER OF CITY—*to condemn for a street.* The city of Peoria, under its special charter of 1844, as amended in 1855, had the right to extend a public street, and have the benefits and damages assessed. *Peoria* v. *Kidder,* 26 Ill. 351.

341. POWER OF RAILWAY COMPANY—*to take public property.* A railway charter authorizing the company to enter upon, take possession of, and use all and singular any lands, streams and materials of any kind for the location of the road, depot, &c., and for the construction of the road, contained this provision: that "all such lands, materials and privileges belonging to the state are hereby granted to said corporation for said purposes:" *Held,* that the grant did not include the ground connected with and used by the state for the education of the blind, although adjoining the road and convenient for its use. *St. L. J. & C. R. R.* v. *Trustees, &c.,* 43 Ill. 303.

342. POWER OF CITY—*for what purposes.* A municipal corporation has no power to condemn private property for purposes not specifically named in the law, and which is not within the proper scope and meaning of the delegated authority. *E. St. Louis* v. *St. John,* 47 Ill. 463.

343. SAME—*for city prison.* Power given a city in its charter to "take private property for opening, altering and laying out any street, lane, avenue, alley, public square, or other public grounds," does not confer the power to condemn property on which to erect a city prison. *Ib.*

344. POWER OF COMPANY—*to use a street.* The grant in a charter of a railroad company to run its road through a town, cannot by any reasonable and fair intendment, operate as a grant of the use of the streets, or either of them, to the company. *St. L. V. & T. H. R. R.* v. *Haller,* 82 Ill. 208.

345. SAME—*by legislative recognition.* A consolidated railway may acquire property within the city of Chicago for right of way, by condemnation, if its authority to construct its road within the city and its existence is recognized by the legislature in an amendatory act, although the companies before consolidation had no such power. *McAuley* v. *C. Ch. & Ind. C. Ry.,* 83 Ill. 348.

346. POWER TO CONDEMN—*derived from the state alone.* The right of a corporation to condemn property for the construction and operation of a horse or dummy railway in the streets of a city, is derived solely from the state law; and the consent of the city to the construction of such road is not a condition precedent to proceedings to condemn. *Metropolitan City Ry.* v. *Ch. W. Div. Ry.,* 87 Ill. 317.

347. SAME—*whether termini are fixed.* A railway company was authorized by its charter to construct its road "from A to some eligible and convenient point in the county of DuPage, there to connect with the G. & C. U. Railroad." By an amendatory act, the company was authorized to construct a branch road from its main line from A to and in the city of C by way of N: *Held,* that under either of these acts, the *termini* of the road were so far fixed as to authorize the company to condemn land for its use under the act of 1852, which was confined in its operation to railroads, the *termini* of which were fixed by the legislature. *C. B. & Q. R. R.* v. *Chamberlain,* 84 Ill. 333.

348. FOR BOULEVARD. Under § 12 of the act of 1869, in relation to parks, &c., the West Park Commissioners have the power to condemn land for a boulevard within their district, to connect the park and boulevard under their control with those under the control of the South Park Commissioners. *Park Coms.* v. *Western Union Tel. Co.,* 103 Ill. 33.

349. SECOND TIME—*exhaustion of power.* The law does not require a railway company to acquire by condemnation all the lands necessary for the construction and operation of its road, at the same time. It may increase its facilities as the business of the country may require. *Fisher* v. *Ch. & S. R. R.,* 104 Ill. 323.

350. So, when a railway company had a side track for many years before, connecting its main track with a public warehouse and elevator in a town, over the land of another, but without having the right of way therefor, except by the mere consent or license of the owner, it was held that the company had the right to institute proceedings to condemn the land over which such branch ran, for right of way. *Ib.*

351. FOR TELEGRAPH. Authority is given by statute to all telegraph companies to erect poles on which to place their wires on all highways or public roads, by first obtaining the consent in writing of the county board of the county in which the highway is situated. But this permission by the county board is subject to the constitutional inhibition that private property shall not be taken or demanded for public use without just compensation. *Board of Trade Tel. Co.* v. *Barnett*, 107 Ill. 507.

352. POWER OF RAILWAY TO CONDEMN—*fixing route and termini.* The general railroad act requires the persons incorporating a company to name the places from and to which it is intended to construct the proposed railway, but no limitation is laid down as to the places where switches, turnouts, or side tracks shall be constructed. *South Ch. R. R.* v. *Dix*, 109 Ill. 237.

353. SAME—*for branch road.* Under the power of an incorporated railway company to condemn land for side tracks, turnouts or switches, it has no right to take land for the construction of an independent branch road to subserve only new private interests. *Ib.*

354. But it is no valid objection that the proposed track may serve private use, if in addition thereto, it is one also necessary for the successful and convenient operation of the main line of the road. *Ib.*

355. SAME—*additional tracks in city.* Where a railway corporation is limited by village or town authorities to thirty feet in the centre of a public street on which to locate its main track, and it becomes necessary to construct a switch or side track, it is no objection to the condemnation of land for that purpose, that it runs parallel to the main track, there not being room enough in the right of way along the street for the side track in addition to its two main tracks. *South Ch. R. R.* v. *Dix*, 109 Ill. 237.

356. SAME—*for switches, etc.* A railway corporation organized under the general act of 1872, and the amendment thereto of 1877, is expressly empowered to condemn land for the purpose of switches, turnouts and side tracks when necessary for the successful operation of the road. *Ib.*

357. RIGHT TO BUILD ROAD IN CITY. A grant of power to a railway company "to locate, construct and maintain and operate with horse or locomotive, cars from the city of Chicago to any point in the town of Evanston, a railroad," &c., without any express or implied restrictions, will authorize the grantee, so far as the state is concerned, to locate its tracks and fix its Chicago *terminus* at any point in the city. *Ch. & N. W. Ry.* v. *Ch. & E. R. R.*, 112 Ill. 589.

358. STREETS. The power of an incorporated town to open streets, extends to all lands within the corporation. *Curry* v. *Mt. Sterling*, 15 Ill. 320.

359. Under the present legislation it is not necessary, as a condition precedent for the location of a railroad in a city, when not over a street, or to the power to condemn private property within the city for right of way, that an ordinance shall be passed by the city giving its assent. *Ch. & W. Ind. R. R.* v. *Dunbar*, 100 Ill. 110.

360. CITY OR VILLAGE—*special assessments.* A city or village may make and collect special assessments for a public improvement before acquiring private property necessary therefor, by condemnation. *Hyde Park* v. *Borden*, 94 Ill. 26.

361. POWER OF CITY—*to condemn for a sewer—ordinance.* The statute does not require that an ordinance for the construction of a sewer by a city or village shall make any provision for acquiring the right of way, but after the passage of an ordinance for an improvement which requires the taking or damaging of private property the statute requires the city or village to file a petition to ascertain the compensation to be paid if it cannot be agreed upon. *Hyde Park* v. *Borden*, 94 Ill. 26.

362. POWER OF RAILWAY—*to condemn—de facto corporation.* The fact that a railway company has been organized under a valid charter, and is shown to have done corporate acts under it, is sufficient to establish a *prima facie* right to take private property under the act; and this *prima facie* right cannot properly be questioned in a collateral proceeding. That must be done by *quo warranto.* *Ch. & N. W. Ry.* v. *Ch. & E. R. R.*, 112 Ill. 589.

363. SAME—*for lateral branches.* The fact that the building of lateral branch roads may add to the earnings of the main line of a railway company, or increase its business, will not authorize such corporation to build the same under its charter, which fails to so provide. *Ch. & E. Ill. R. R.* v. *Wiltse*, 116 Ill. 449.

364. POWER OF VILLAGE—*laying street—ordinance.* The statute does not require that an ordinance to establish a street by a village shall be published. *Village of Byron* v. *Blount*, 97 Ill. 62.

365. VALIDITY OF ORDINANCE. Where the first section of an ordinance established a street upon the defendant's land, and the second section provided that contiguous property should be taxed one-fourth of the cost incurred in establishing and opening the same: *Held*, on application to condemn the land for the street and fix the compensation, that the validity of the first section was not involved. *Village of Byron* v. *Blount*, 97 Ill. 62.

EXTENT OF LAND TAKEN.

366. WIDTH OF RIGHT OF WAY. The right of way for a railroad track is not limited to any given width. It may vary in different localities; but obviously a railway company may appropriate and use for its right of way such width of ground as may be reasonably necessary for the economical and convenient transaction of its business. *C., R. I. & P. R. R.* v. *People*, 4 Bradw. 468.

367. A judgment condemning a strip of land 120 feet wide for a right of way for a railroad, will not be reversed because the land condemned exceeds 100 feet in width, when it does not appear from the pleadings or the record that the additional 20 feet was not necessary and no such objection was raised before the court below either by demurrer or reasons assigned in arrest of judgment. *Booker* v. *V. & C. Ry.*, 101 Ill. 333.

368. Under our statute the amount of land which a railway company is allowed to take for right of way is measured by the necessities of the case only, and is not limited to a strip 100 feet wide. When the petition states the amount of the land necessary for the road, and such allegation is not controverted, no question can arise as to whether more land is sought to be taken than is necessary. *Bowman* v. *V. & C. Ry.*, 102 Ill. 459.

369. The statute does not designate the width of the strip of land that may be condemned for telegraph purposes, but only authorizes such companies to acquire such an amount of land as may be necessary; and when only one line of poles is specified in the petition, and the evidence does not show that a half of a rod in width is an unreasonable amount of land, the judgment condemning that much will be sustained, and will be construed to authorize the erection of but one set of poles. *Lockie* v. *Mutual Union Tel. Co.*, 103 Ill. 401.

370. DEPOTS AND SIDE TRACKS. A charter giving a railroad company a right to acquire a strip of land not exceeding 100 feet in width has reference to the right of way for a single or double track, and does not prohibit it from acquiring more land for depot grounds and side tracks at stations. *Carmody* v. *C. & A. R. R.*, 111 Ill. 69.

371. On the trial, evidence that it is not necessary to take a strip of land 150 feet wide through the defendant's land is not admissible. That is not a question for the jury to pass upon. *DeBuol* v. *F. & M. Ry.*, 111 Ill. 499.

372. AMOUNT OF LAND. A railway company cannot, even by making compensation, take more land than is necessary for the purpose of its road, and the same principle applies to damaging land. *O. & M. Ry.* v. *Wachter*, — Ill. —. Filed Jan. 20, 1888.

OF THE EXPEDIENCY, NECESSITY AND PROPRIETY.

373. NOT A JUDICIAL INQUIRY. Whether private property, and if so, how much shall be taken for the public use, is a matter which, of necessity, rests alone in the discretion of the legislature, and no appeal from, or review of its decision can be had. It is not for the court to say whether more land than is necessary for a ferry landing and a public road is sought to be condemned or not. *Mills* v. *St. Clair Co.*, 2 Gilm. 197, 238.

374. HIGHWAY. Whether the public interests require, and whether the fiscal condition of the county will justify the payment of the damages awarded, is for the county court alone to decide. On appeal from an order laying out a public road, the propriety or expediency of the road is not involved. *Sangamon Co.* v. *Brown*, 13 Ill. 207.

375. STREET. On appeal from an order laying out a street the circuit court cannot inquire into the expediency of opening the street. That is left solely to the judgment of the board of trustees. *Curry* v. *Mt. Sterling*, 15 Ill. 320; *Dunlap* v. *Mt. Sterling*, 14 Ill. 251.

376. WHO MUST DETERMINE. The necessity and expediency for the exercise of the right to condemn private property in making public improvements, either for the benefit of all the people of the state, or of a particular municipality, must be determined by the legislature. Mere convenience is not sufficient to justify the exercise of the right of eminent domain. *Ch., Dan. & Vin. R. R.* v. *Smith*, 62 Ill. 275.

377. If the use for which private property is sought to be taken is a public use, the courts cannot inquire into the necessity or propriety of the exercise of the right of eminent domain. That right is political, and belongs to the legislative branch of the government. *C. R. I. & P. R. R.* v. *Town of Lake*, 71 Ill. 333.

378 WHO MAY JUDGE OF EXPEDIENCY. The legislature is the proper body to determine the necessity or expediency of the exercise of the power of eminent domain, and the extent of its exercise. Mills on Em. Domain, § 11; Pierce on Railroads, 146; 1 Rohrer, on Railroads, 286, 291; *St. Louis Co.* v. *Griswold*, 58 Mo. 175; *Brooklyn Park* v. *Armstrong*, 45 N. Y. 234; *Secombe* v. *Milwaukee R. R.*, 23 Wall. 108; *Weir* v. *St. Paul R. R.*, 18 Minn. 155; *Tyler* v. *Beacher*, 44 Vt. 648; *Bankhead* v. *Brown*, 25 Iowa 540; *Brayton* v. *City of Fall River*, 124 Mass. 95; *Hold* v. *Somerville*, 127 Mass. 408.

379. POWER OF COURT OVER QUESTION. Where the power of condemnation is delegated to corporate bodies, and their jurisdiction depends upon the existence of a necessity, their decision is not conclusive on the courts. *Milwaukee & St. Paul Ry.* v. *City of Faribault*, 23 Minn. 167. In some of the states the constitution requires a jury to determine the necessity. *Paul* v. *Detroit*, 32 Mich. 108; *M. & St. P. Ry.* v. *Faribault*, 23 Minn. 167; *R. & S. R. R.* v. *Davis*, 43 N. Y. 137; *In Re N. Y. Central R. R.*, 66 N. Y. 407.

380. Of the necessity or expediency of appropriating private property to public uses, the opinion of the legislature, or of the corporate body or tribunal upon which it has conferred the power to determine the question, is conclusive upon the courts. *Ib.*

381. If the use for which private property is proposed to be taken is public, or if it be so doubtful that the court cannot pronounce it not to be such as to justify the compulsory taking, the decision of the legislature, embodied in the enactment, giving the power, that a necessity exists for taking the property, is final and conclusive. *Ib.*

382. DELEGATION OF POWER TO DECIDE. Where the case is such that it is proper to delegate to individuals, or to a corporation the power to appropriate private property, it is also competent to delegate the authority to decide upon the necessity of the taking. *Ib.*

383. The power to determine in any case whether it is needful to exercise the power of eminent domain, must rest with the state itself; and the question is always one of strict political character, not requiring any hearing upon the facts or any judicial determination. *Ib.*

384. The law authorizing the condemnation of private property for railroad purposes is limited to such property as is necessary for the purpose in question, and no condemnation proceedings, can lawfully be had of property not necessary for the construction or use of the road. But this necessity need not be made certain before it is lawful to proceed with the condemnation. *Ch. & W. Ind. R. R.* v. *Dunbar,* 100 Ill. 110.

385. POWER OF COURT TO DECIDE. Where the description of the land and the purpose for which it is sought to be taken are stated in the petition, as they must be in every case, whether the land is reasonably necessary for the purpose stated, depends mainly upon the facts thus stated in the petition. But the court, in passing upon this question, as it must, before submitting the question of damages or compensation to the jury, should take into consideration the section of the country and the particular locality in which the improvement is to be constructed:—whether in an obscure country village, or in a great commercial center; and acting upon its own knowledge of the commerce and business necessities of the country, must upon the facts stated in the petition, determine this question for itself. The jury impanneled can find no fact except what is just compensation to the owner. *Smith* v. *Ch. & W. Ind. R. R.,* 105 Ill. 511.

386. NO EVIDENCE HEARD ON QUESTION. The law does not contemplate that when the petitioner has brought itself within the provisions of the statute, the right of condemnation can be defeated by simply showing, in the opinion of witnesses who have no interest in or connection with the objects of the proceeding, that the land sought to be condemned is not necessary for the purpose stated. *Ib.*

387. HOW FAR CORPORATION MAY DETERMINE. Every company seeking to condemn land for a public improvement must, in a modified degree, be permitted to judge for itself as to the amount that is necessary for such purpose. This right is subject to the constitutional and statutory restrictions, and to the further limitation that the courts are clothed with ample power to prevent any abuse of the same. *Ib.*

388. WHEN COURT MAY INTERFERE. If the court can see from the facts, and its general knowledge of the locality and the public wants, that the land sought to be taken is manifestly in excess of what is reasonably necessary for the purpose stated in the petition, it will be fully warranted in denying the application, otherwise not. *Ib.*

389. To deny a petition of a railway company for the condemnation of land for a side track, it should appear that the object sought is clearly an abuse of power, or a taking of private property for an object not required for the convenient operation of the road. *South Ch. R. R.* v. *Dix,* 109 Ill. 237.

890. The exercise of the right of eminent domain is subordinate to all constitutional and statutory restrictions on the subject, and to the further limitation that the courts which are authorized to enter tain applications for its exercise, are clothed with ample power to prevent any abuse of the right. *Ch. & E. Ill. Ry* v. *Wiltse*, 116 Ill. 449.

891. The question of the necessity of the exercise of the right of eminent domain, and in what cases it may be exercised within constitutional restrictions, is legislative and not judicial; and when this power has been delegated to a corporation, its exercise within the scope and for the uses and purposes named in the legislative grant, will not be a proper subject for judicial interference or control, unless to prevent a clear abuse of power. *Ch. & Ill. E. R. R.* v. *Wiltse*, 116 Ill. 449.

892. INJUNCTION -- *condemnation by city.* A court of equity will not enjoin the exercise of the right of eminent domain by a city for a street, it being a political question of expediency, and not a judicial one. *Chicago* v. *Wright*, 69 Ill. 318.

893. EXERCISE OF RIGHT BY CITIES—*expediency.* The authorities of cities and villages are the exclusive judges of the propriety and necessity of the widening or laying out of streets, and unless there is manifest injustice, oppression or gross abuse of power in their action, a court of equity will not interfere with the exercise of the discretion vested in them. *Dunham* v. *Hyde Park*, 75 Ill. 371.

894. The necessity or propriety of exercising the right of eminent domain is a political question which belongs exclusively to the legislature to determine. Hence, the legislature may properly withhold from municipal authorities the power to condemn land for public purposes as against some other body or corporation on its lands already devoted to some other public use. It may confer power to use the right upon one corporation in preference to another. *Hyde Park* v. *Oakwood Cemetery Assoc.*, 119 Ill. 141.

JURISDICTION.

895. SUBJECT MATTER—*description of land necessary in petition.* To give the court jurisdiction to condemn land for a right of way, the land sought to be taken must be described in the petition, and in the several orders, where it should properly occur, and in the final order. *G. & Ch. Union R. R.* v. *Pound*, 22 Ill. 399.

896. SAME—*how conferred.* It is sufficient, to give the judge jurisdiction, that the facts requiring him to act appear in the petition, or in the order of the court, or, indeed, in any part of the record. *Ib.*

897. HOW SHOWN. .Where application for the appointment of commissioners to assess damages for a right of way is to be made to the county judge, on the absence of the circuit judge, it is not indispensable to the jurisdiction of the county judge that the petition should allege the absence of the circuit judge; but this fact must exist, and must appear in some part of the proceeding. *Shute* v. *Ch. & Mil. R. R.*, 26 Ill. 436.

898. HOW CONFERRED—*petition—appearance.* The presentation of a petition seeking a condemnation of land for a plank or gravel road, properly describing the land and praying for the appointment of commissioners to assess damages, confers on the court jurisdiction of the subject matter, and the appearance of the land owners that of the parties defendant. *Skinner* v. *Lake View Avenue Co.*, 57 Ill. 151.

899. SUPERIOR COURT OF COOK COUNTY. The superior court of Cook county, being in law a circuit court, may lawfully act in a proceeding authorized to be brought in the circuit court. Such court has jurisdiction of a proceeding to condemn land. *Ch. & N. W. Ry.* v. *Ch. & E. R. R.*, 112 Ill. 589.

400. COUNTY COURT—*no equity powers*. County courts have no other jurisdiction in proceedings to condemn than that conferred by the eminent domain act. They have no general chancery jurisdiction. *McCormick* v. *W. Ch. Park Comrs.*, 118 Ill. 655.

401. OF THE PETITION—*its sufficiency*. The statute having determined specifically what facts must appear on the face of the petition, the court or judge is powerless to take any action in the premises until a petition is filed containing the statutory requirements, for it is by the petition that jurisdiction is obtained of the subject matter. *Smith* v. *Ch. & W. Ind. R. R.*, 105 Ill. 511.

OF THE PETITION AND ITS PRESENTATION.

402. IN WHAT NAME FILED. Where the statute declares that "the directors may present a petition," it is fully complied with when the petition is signed by the corporation by its attorney. In suits by corporations the corporate name is used, and not the name of the directors. *Skinner* v. *Lake View Avenue Co.*, 57 Ill. 151.

403. MUST DESCRIBE THE LAND. In order to condemn the right of way of another *de facto* railway company for even a qualified or conjoint use, it must be described in the petition, which must show an inability to agree upon the compensation. The describing of the tracts of land over which such prior way is located, making no reference to the other road or company, will not authorize the condemnation of such prior right of way. *Cin., Laf. & Ch. R R.* v. *Dan. & Vin. Ry*, 75 Ill. 113.

404. AS SHOWING INABILITY TO AGREE. In a proceeding by a railway company under the act of 1852, the petition alleged that the company "has not been able to acquire the title to said several tracts, &c., from the persons interested therein by voluntary grant or otherwise:" *Held*, a sufficient averment that the title to the land sought to be condemned could not be acquired by purchase. *C. B. & Q. R. R.* v. *Chamberlain*, 84 Ill. 333.

405. TIME FOR PRESENTING. When notice is given under the act of 1852 of an application on a certain day in term time to the court to appoint commissioners, the petitioner is not restricted to such day, but may apply on a subsequent day of the term. *Ib.*

406. NEED NOT DESCRIBE PROPERTY NOT SOUGHT. The petition need not describe property not sought to be taken or damaged. *Hyde Park* v. *Dunham*, 85 Ill. 569.

407. A statute authorizing the appointment of commissioners to ascertain the damages which the owners of lands taken for right of way *have* sustained, means also such as the owners *will* sustain thereafter. Therefore a petition to have assessed the damages the owners *will* sustain is not invalid in failing to use the words "*have sustained.*" *Townsend* v. *C. & A. R. R.*, 91 Ill. 545.

408. TO CONDEMN FOR A STREET. A petition by a village, showing the passage of an ordinance establishing a street within the corporate limits, and alleging an inability to agree with the owner of the land sought as to his compensation and damages, and praying that on a final hearing the just compensation be paid be ascertained according to law, and that when the same was paid to him or deposited according to law, an order be made for possession of the land condemned, and for other relief, is in strict conformity to the law. *Village of Byron* v. *Blount*, 97 Ill. 62.

409. SUFFICIENCY OF DESCRIPTION OF LAND. There is no rule of law that requires any greater degree of certainty in the description of land sought to be condemned for a street than will enable a surveyor to find and locate it from the description given. *Ib.*

410. INABILITY TO AGREE. An allegation in a petition by a railroad company to condemn land for a right of way, that the petitioner "has not been able to acquire the title nor the right of way over the land by purchase or by voluntary grant from" the defendants, although not formal, is substantially sufficient under the statute, as showing an inability to agree as to the compensation to be paid. *Booker* v. *V. & C. Ry*, 101 Ill. 333; *Bowman* v. *V. & C. Ry*, 102 Ill. 459.

411. REQUISITES OF. Every petition properly framed must contain all of the statutory requirements, and will therefore, of necessity, show in every case where such petition is sufficient to confer jurisdiction, the authority of the company seeking the condemnation to take the specific land sought to be taken, and the object or purpose for which it is required; and from this statement of facts it must clearly appear that the use for which the land is sought to be condemned is a public one. *Smith* v. *Ch. & W. Ind. R. R.*, 105 Ill. 511.

412. STATEMENT OF THE USES. It is not necessary that the petition should state the petitioners' purposes fully and completely, giving the number of tracks and a purpose to allow other companies to use the same. It is sufficient for the petition to show generally that the land is needed for railroad purposes. *C. R. I. & P. Ry* v. *Smith*, 111 Ill. 363.

413. AS SHOWING PUBLIC USE. A petition to condemn a strip of land 150 feet wide over defendant's land, alleged that petitioner was a corporation, organized and existing under the general railroad laws of the state, from F to G; that it was authorized to exercise the right of eminent domain, and that in accordance with the purposes of its organization, petitioner had surveyed, staked off and located its railway: *Held*, that from these averments it was apparent the proceeding was to condemn private property for a public purpose, and for no other. *De Buol* v. *Freeport & Miss. River Ry.*, 111 Ill. 499.

414. INABILITY TO AGREE. When the petition states that the petitioner is unable to agree with the owner as to the compensation to be paid for the right of way, if this need be proved it may be shown by the defendant's own evidence. *Ib.*

415. WIDTH OF RIGHT OF WAY—*estoppel*. A railway company seeking condemnation for a right of way, especially if it proceeds under the act incorporating the Mississippi Railroad Company, is not bound to take and pay for all the land described in the petition, if less is needed for its purposes. It is not estopped by the allegations in its petition as to the width of the right of way. *Peoria & R. I. Ry.* v. *Bryant*, 57 Ill. 473.

416. AS SHOWING PURPOSE. A petition, after stating that the company had located its line of road over certain tracts described, averred "that a part of each of said lands is necessary to petitioner for its right of way, side tracks, depot and depot grounds, freight yards, shops and appurtenances for the construction and operation of its said line of road:" *Held*, a sufficient statement of the purposes for which the property was sought. *Seever* v. *Ch. S. F. & Cal. Ry.*, — Ill. —. Filed Nov. 11, 1887.

417. PRACTICE—*defects in petition—how reached*. If the petition is defective in stating the purpose of the taking and the manner in which the land is to be used, the proper course is for the defendant to demur. *Ib.*

418. SHOWING MANNER OF USE. It is not necessary to state in the petition the particular manner in which the land is to be used. To show that, the petitioner may give in evidence its plans and specifications, and the defendant may have them produced on motion. *Ib.*

PLEADINGS SUBSEQUENT TO PETITION.

419. STRIKING OUT PLEAS—*benefit under answer.* The correctness of striking a defendant's pleas to a petition from the files will not be inquired into, where the defendant, by answer subsequently filed, has had the full benefit of all the matters of defense presented by his pleas. *Metropolitan City Ry.* v. *Ch. W. Div. Ry.*, 87 Ill. 317.

420. NO ANSWER OR PLEAS ALLOWED. There is no rule of law or practice authorizing the filing of an answer of any kind to a petition for the condemnation of land, and it is not the proper practice to allow one to be filed. If one is filed it may be stricken out. *Smith* v. *Ch. & W. Ind. R. R.*, 105 Ill. 511.

421. Under the eminent domain act an answer or plea to the petition is not allowable, and if a special plea is filed there is no error in striking it from the files. *Johnson* v. *F. & M. R. Ry.*, 111 Ill. 413.

422. There being no rule of law or practice authorizing the filing of any kind of answer or plea to the petition, there is no error in compelling the land owner to proceed with the trial before disposing of a plea of *nul tiel* corporation. *Henry* v. *Centralia & Chester R. R.*, 121 Ill. 264.

OF THE PARTIES.

423. PLAINTIFF—*for consequential injury.* A permanent injury to adjacent property by the construction and operation of a railroad in a public street, dates from the time when the road was constructed and first put in operation, and any right of action therefor is vested in the then owner of the premises. His grantee cannot maintain such action. *Ch. & E. Ill. R. R.* v. *Loeb*, 8 Bradw. 627.

424. HEIRS. On the death of the land owner his title descends to his heirs, and they must be made parties to a proceeding to condemn, when it is sought to pass the title, in order to bind them. *P. & R. I. R. R.* v. *Rice*, 75 Ill. 329.

425. DEATH OF OWNER BEFORE FINAL JUDGMENT. In a proceeding to condemn under a law which passed the fee upon payment of the damages, the land owner died after reversal of the judgment: *Held*, that his heirs were necessary parties before taking any further steps, and that it was error to dismiss the proceedings on the administrator's motion. *Ib.*

426. The remainde-man, as well as the tenant for life is a necessary party defendant, to bind him. *C. & A. R. R.* v. *Smith*, 78 Ill. 96.

427. Proceedings against a former owner, who has, by a deed duly recorded, conveyed the land, are invalid and do not bind the true owner. *Smith* v. *C. A. & St. L. R. R.*, 67 Ill. 191.

428. TENANTS IN COMMON—*jurisdiction.* Where the land belongs to two or more tenants in common, it is not essential to the jurisdiction of the court that all the owners shall be brought into court, but the court has power to hear and determine the case as to those before it. *Bowman* v. *V. & C. Ry.*, 102 Ill. 459.

429. If the right of any one in any sense should depend upon the disposition of the case as to the others, then each party in interest would have the right to insist on all the parties being before the court before proceeding to a trial. *Ib.*

430. NON-RESIDENT—*right to remove cause to United States* *court.* A non-resident defendant may remove his part of the case to the United States court, but not the whole case. *Chicago* v. *Hutchinson,* 11 Biss. 484.

431. LESSEE. Where one railroad company gives another one a lease of a portion of its track between a certain place and its *terminus,* but reserves its franchise and the right to exercise its corporate powers and the general control and management of the main line, and of the management, use, location and repair of the same, the lessee company will not have such an interest in the line of the road leased, as to make it a necessary party to a proceeding by another company to condemn a right of way across the track of the lessor company. *Englewood Con. Ry.* v. *Ch. & E. Ill. R. R.,* 117 Ill. 611.

432. PETITION IN VACATION—*fixing hearing—summons and publication.* § 3. If such petition be presented to a judge in vacation, the judge shall note thereon the day of presentation, and shall also note thereon the day when he will hear the same, and shall order the issuance of summons to each resident defendant, and the publication of notice as to each non-resident defendant, and the clerk of the court shall at once issue the summons and give the notices accordingly. [R. S., 1887, p. 646, § 3; S. & C., p. 1043, § 3; Cothran, p. 647, § 3.]

433. COURTS ALWAYS OPEN. The condemnation of private property for public use being a judicial proceeding, it can only be instituted and prosecuted to a final determination in either the circuit or county court, and hence, whether commenced in vacation or term time, it is equally a proceeding in court. Under the statute these courts are always open for proceedings to condemn for right of way. *Bowman* v. *V. & C. Ry,* 102 Ill. 459.

434. SAME—*alias summons.* Under the statute the circuit and county courts are always open for proceedings to condemn for right of way, and when the summons is quashed the court may order an *alias* summons returnable in vacation, and when so issued and served ten days before the return day, the court will acquire jurisdiction to assess the compensation to be paid. *Liebengut* v. *L. N A. & St. L. Ry,* 103 Ill. 431.

435. A proceeding to condemn land for public use, whether instituted in term time or in vacation, is a judicial one, and the judge before whom the same is had has the same powers in either case, and may grant new trials to correct errors. The circuit and county courts are always open for such proceedings, and their judicial powers are the same in vacation as in term time. *Centralia & Chester R. R.* v. *Rixman,* — Ill. —. Filed Jan., 1887.

436. SERVICE—PUBLICATION. § 4. Service of such summons and publication of such notice shall be made as in cases of chancery. [R. S., 1887, p. 646, § 4; S. & C., p. 1043, § 4; Cothran, p. 648, § 4.]

OF THE NOTICE UNDER PRIOR STATUTES.

437. OF LAYING OUT STREET. The publication of the ordinance authorizing the opening of a new street was all the notice required. *Curry* v. *Mt. Sterling,* 15 Ill. 320.

438. OF CONDEMNATION—*right of way.* Unless the act authorizing the condemnation of property for right of way so directs, a notice of the proceeding need not be given. *Johnson* v. *Joliet & Ch. R. R.*, 23 Ill. 202.

439. APPEARANCE. If the parties whose lands are sought to be condemned appear at the hearing before the commissioners, a notice to them of the time and place of the hearing is unnecessary. *Skinner* v. *Lake View Avenue Co.*, 57 Ill. 151.

440. RECORD MUST SHOW. Where the commissioners are required to view the land sought to be condemned and hear evidence as to the damage, it is indispensable to their action that they give personal notice of the time and place of meeting to assess damages; and a recital in their report that they have given notice is not sufficient. It should appear in the report or order approving the same. *Ib.*

441. OF TIME OF FILING REPORT. Under § 13 of the act of 1859 the court has the power to modify the report, and for such purpose evidence may be heard. Therefore the land owner should have notice of the time of the filing of the report. *Ib.*

442. JURISDICTIONAL. A party must have notice of the proceeding before he can be deprived of his property by condemnation, whether it is under the act of 1845 or 1852. The notice is indispensable to the validity of the condemnation. *P. & R. I. Ry* v. *Warner*, 61 Ill. 52.

443. MUST BE TO THE OWNER. The Belleville & Illinoistown Railroad company, chartered in 1852, gave notice of proceedings to condemn land for right of way to a former owner of a life estate therein, but who had previously conveyed his title, and whose deed was recorded: *Held*, that the proceedings were invalid for the reason they were not instituted against the owner. *Smith* v. *C., A. & St. L. R. R.*, 67 Ill. 191.

444. ON TENANT FOR LIFE—*not binding on remainder-man.* The charter of a railway company required notice by publication to the owner or occupier or unknown owners of land sought to be condemned, of the application to appoint commissioners, and the company published such notice as to one who had held a life estate only, but who was dead, not naming the remainder-man: *Held*, that the subsequent proceedings of condemnation were not binding upon the latter, and that he might recover the land taken by ejectment. *Ch., A. & St. L. R. R.* v. *Smith*, 78 Ill. 96.

445. WHEN MUST NAME OWNER. A charter authorizing the condemnation of land for right of way, upon giving notice by publication for thirty days, "to the owners or occupiers or unknown owners, as the case may be, of the intention to apply for the appointment of commissioners," &c., requires that the notice shall be given specifically to the owner or occupier, if known—if not, to unknown owners by that designation. A general notice in such case will not be sufficient. *Ib.*

446. HOW TO BE GIVEN. Where a notice is required by statute, and the mode of service is not specified, it must be personal; and usually, where notice is required by publication, it must be directed to the person by name who is required to be notified. *Ib.*

447. WHEN ESSENTIAL. It is a rule of general application that a party cannot be deprived of his rights without having notice and an opportunity of being heard. When the proceeding is summary, and the notice only constructive, the courts will never abridge the right to notice or substitute another for it. *Ib.*

448. PROOF OF PUBLICATION—*certificate of publisher.* Publisher's certificate of the publication of notice required by law to be published, after he has ceased to be the publisher, is not admissible as evidence of the publication. *Smith* v. *C., A. & St. L. R. R.*, 67 Ill. 191.

449. Service of summons on land-owner less than ten days before the day set for hearing the petition will give the court jurisdiction of the person of the defendant, and the court may continue the case to a subsequent day. *Bowman* v. *V. & C. Ry.*, 102 Ill. 472.

450. HIGHWAYS—*notice—certificate.* Where the notice given to the land-owner by the commissioners of highways of the presentation of their certificate that they are about to establish a road, fixed the time on March 11, while the justice's docket entries in the case were dated March 13, it was held that it might be shown by other evidence that the certificate was presented on the first named day and the jury selected on that day. *Hankins* v. *Calloway*, 88 Ill. 155.

451. FIXING TIME OF HEARING—*continuance—jurisdiction.* Where the justice fixes the time for the assessment of damages within ten days from the filing of the certificate with him, he may continue the case to a later day, and beyond the ten days, if for any cause notice has not been served on all the parties in time, and in such case he will not lose jurisdiction, and there is no error in taking the statement in his docket, and in the final order establishing the road, to show such fact. *Ib.*

452. CERTIFICATE—*nature of petition—description.* The statute does not require the certificate of the commissioners of highways filed with the justice to give a minute description by courses and distances of the whole road, or even the portion for which damages are claimed, but a general description of the portion for which damages are claimed will suffice. *McManus* v. *McDonough*, 107 Ill. 95.

453. HEARING. § 5. Causes may be heard by such judges in vacation as well as in term time, but no cause shall be heard earlier than ten days after service upon defendant, or upon due publication against non-residents.

454. PETITION INCLUDING SEVERAL TRACTS—*separate assessment.* Any number of separate parcels of property, situate in the same county, may be included in one petition, and the compensation for each shall be assessed separately, by the same or different juries, as the court or judge may direct.

455. AMENDMENTS. Amendments to the petition, or to any paper or record in the cause, may be permitted whenever necessary to a fair trial and final determination of the questions involved.

456. NEW PARTIES—PRACTICE—*process to execute judgments, &c.* Should it become necessary at any stage of the proceedings to bring a new party before the court or judge, the court or judge shall have the power to make such rule or order in relation thereto as may be deemed reasonable and proper; and shall also have power to make all necessary rules and orders for notice to parties of the pendency of the proceeding, and to issue all process necessary to the execution of orders and judgments as they may be entered. [R. S. 1887, p. 646, § 5; S. & C., p. 1044, § 5; Cothran, p. 648, § 5.]

457. HEARING IN TERM. The petition though filed in vacation may be tried in term time, and a motion to dismiss and a challenge of the array of jurors on the ground the petition was filed in vacation is properly disallowed. *Johnson* v. *F. & Miss. River Ry.*, 111 Ill. 413. See also *Bowman* v. *V. & C. Ry.*, 102 Ill. 468; *Harlam* v. *V. G. & S. Wis. R. R.*, 64 Ill. 353.

458. Where no hearing was asked or fixed for a day in vacation, but the cause was commenced at the March term of the county court, and continued to the April term, and process served for that term: *Held*, that the court had jurisdiction to try the case at the April term. *DeBuol* v. *F. & M. R. R. R.*, 111 Ill. 499.

459. RIGHT TO DISMISS. The petitioner in a proceeding to condemn land for right of way, even after possession wrongfully taken of the land, and after the filing of a cross petition for damage to the part not taken, has the right to dismiss the proceeding and it is error to refuse that right. *Ch., St. L. & Western R. R.* v *Gates*, — Ill —. Filed March 23, 1887.

460. SEVERAL TRACTS—*assessment as to each tract.* Where the petition embraces several tracts of land, and avers that they are owned by several persons named, in the absence of anything to the contrary in the record, it will be presumed that the several persons named hold as tenants in common, so that it will not be necessary for the jury to make a separate assessment upon each tract. But when several tracts belonging to different owners are embraced in one petition, doubtless it will be the duty of the jury to make a separate assessment for each tract. *Grayville & Mattoon R. R.* v. *Christy*, 92 Ill. 337.

461. SEPARATE ASSESSMENT. Where several tracts of land belonging to different persons are included in the same petition, the statute provides that the compensation for each shall be assessed separately by the same or different juries, as the court or judge shall direct, and the same principle may be extended to cases where different persons have several and distinct interests in the same tract. *Bowman* v. *V. & C. Ry.*, 102 Ill. 459.

462. In a proceeding against several land owners, each separate owner may have his damages assessed before a separate jury, and is entitled to a separate appeal from the judgment on the verdict. *Johnson* v. *F. & Miss. Ry.*, 116 Ill. 521.

463. Under the statute the compensation may be assessed by the same or different juries as to two or more separate tracks of land described in the same petition, although belonging to different owners. This is a matter within the discretion of the trial judge, and unless such discretion is shown to have been abused, it will not be interfered with. *Concordia Cem. Assoc.* v. *Minn. & N. W. R. R.*, 121 Ill. 199; — Ill. —. Filed June, 1887.

464. AMENDMENTS—*in vacation making new parties.* Whether the proceeding has been brought in term time or in vacation, there is no impropriety in allowing such amendments to be made as are by law allowable as of course and as a matter of right, whether with or without notice, when the adverse party is not taken by surprise, or otherwise prejudiced. There is no error in allowing an amendment of the petition in vacation, making new parties, as part owners of the land, and ordering summons and publication as to them. *Bowman* v. *V. & C. Ry.*, 102 Ill. 459.

465. It is the duty of courts to allow amendments, when it is necessary to bring all parties before the court, who may have an interest in the premises sought to be taken. *Ch. St. L. & Western R. R.* v. *Gates*.

466. But where the case is called for trial, the court will not delay the hearing for the purpose of bringing in other parties not shown to have any interest in the premises. Amendments are not allowed as a matter of course on the eve of a trial on the motion of a party, except for good cause shown. *Ib.*

467. If the jury fail to find any damages for the diversion of a stream, the court may direct them orally, on the return of their report to find some sum as such damages under § 9 of the act, and this is not a violation of the statute requiring the court to instruct in writing. *Kiernan* v. *Ch., S. F. & Cal. Ry.,* — Ill. —. Filed Nov. 11, 1887.

OF THE JURY.

468. Jury in vacation—*listing*—*venire.* § 6. In cases fixed for hearing of petition in vacation, it shall be the duty of the clerk of the court in whose office the petition is filed, at the time of issuing summons or making publication, to write the names of each of sixty-four disinterested freeholders of the county on sixty-four slips of paper, and, in presence of two disinterested freeholders, cause to be selected from said sixty-four names twelve of said persons to serve as jurors—such selection to be made by lot and without choice or discrimination; and the said clerk shall thereupon issue *venire,* directed to the sheriff of his county, commanding him to summon the twelve persons so selected as jurors to appear at the court house in said county, and at time to be named in the *venire.* [R. S. 1887, p. 646, § 6; S. & C., p. 1044, § 6; Cothran, p. 648, § 6.]

469. Construction of statute. The words, "in cases fixed for hearing in vacation," should be construed with reference to the fact that a hearing is fixed, and a day appointed for the hearing, when such is the case, so that it may read, "in cases fixed in vacation for hearing. *Haslam* v. *Galena & S. Wis. R. R.,* 64 Ill. 353.

470. Array chosen from part of the county greatly interested in the improvement, one being a subscriber and two not freeholders: *Held,* not fairly chosen. *Ib.*

471. Evidence of selection. When the final order laying out a road has the positive statement that the commissioners' certificate was presented to the J. P. on a certain day, and a jury was selected by them and the land owners from the list presented by the J. P., it will afford evidence of the facts. *Hankins* v. *Calloway,* 88 Ill. 155.

472. Mode of selecting second one—*re-writing names of by clerk.* When a second jury is called for, the clerk need not re-write the names of the persons selected who were not drawn, but he may write the names of enough to make up sixty-four. *Kiernan* v. *Ch., Santa Fe & Cal. Ry.* Filed Nov. 11, 1887.

SELECTION OF COMMISSIONERS.

UNDER PRIOR STATUTES.

473. Mandamus—*to compel appointment.* When a county judge improperly refuses to make an order appointing appraisers to assess damages under the act of 1852, concerning right of way, the supreme court will compel him to do so by mandamus. *I. C. R. R.* v. *Rucker,* 14 Ill. 353.

474. On application to a judge for the appointment of commis-

sioners to condemn lands, he is compelled to act if such a case is made as the statute requires. He is rather a ministerial than a judicial officer, having no discretion in the matter. *C., B. & Q. R. R.* v. *Wilson,* 17 Ill. 123.

475. APPOINTED BUT ONCE. Appraisers for the condemnation of land for the use of the Galena & Chicago Union Railroad company receive but one appointment, and when once sworn under it, their proceedings will be valid, although they may be directed to make a re-appraisal. *Low* v. *G. & Ch. Union R. R.,* 18 Ill. 324.

476. WHO MAY APPOINT. An act which provides for the appointment of commissioners to assess damages for a right of way by the senior county commissioner, in the absence of the circuit judge, may be executed by the judges of the county court, they being the successors of the county commissioners' court. In such case, before the county judge can act, it must appear that the circuit judge is absent from the county. *Shute* v. *Ch. & Mil. R. R.,* 26 Ill. 436.

477. MANDAMUS—*to compel appointment.* The act for the location and maintenance of a park being held valid, this court awarded a mandamus requiring the circuit court to appoint three commissioners to assess the damages for the land sought to be condemned for the park. *People* v. *Williams,* 51 Ill. 57.

478. EVIDENCE OF APPLICATION FOR APPOINTMENT—*recital in record.* A recital in the record of the appointment of commissioners that, "it appearing to the court that the said defendants have had due notice of the filing of said petition and of this application," is sufficient to show that the application was made in conformity with the notice, and at the time stated in such notice. *C., B. & Q. R. R.* v. *Chamberlain,* 84 Ill. 333, 341.

479. FIXING TIME OF MEETING. The act of 1852 provides that the court appointing commissioners shall fix the time and place of their first meeting, but it is not explicitly required that this shall be done in the order appointing them. An omission to fix the time is a mere error not available in a collateral proceeding. *Ib.*

480. Where the order appointing commissioners left the day of their meeting blank, but the copy of the order annexed to their report designated the day of the first meeting, and the report showed that the meeting was held on that day and the appearance of the parties: *Held,* as tending to show that the day of the meeting was fixed. *Ib.*

481. APPOINTMENT—*under the act of 1852—mandamus.* After the adoption of the constitution of 1870, and before any legislation thereunder in respect to the condemnation of private property for public use, the supreme court awarded a peremptory *mandamus* requiring a circuit judge to appoint commissioners to estimate the compensation for right of way, as provided by the act of 1852. *People* v. *McRoberts,* 62 Ill. 38.

482. JURY—CHALLENGES—*filling panel.* § 7. The petitioner, and every party interested in the ascertaining of compensation, shall have the same right of challenge of jurors as in other civil cases in the circuit courts. If the panel be not full by reason of non-attendance, or be exhausted by challenges, the judge hearing such petition shall designate by name the necessary number of persons, of proper qualification, and the clerk or justice shall issue another *venire,* returnable *instanter,* and until the jury be full. [R. S., 1887, p. 647, § 7; S. & C., p. 1044, § 7; Cothran, p. 648, § 7.]

483. CHALLENGE—*as to each defendant.* On a proceeding to condemn lots for a sewer, all the objections are properly submitted to one jury, each objector having a separate right of challenge; and, if after having exercised his right, new jurors are introduced into the panel by challenges from others, or by the petitioner, he will have the right to challenge again, if he has not before exhausted his rights. *Fitzpatrick* v. *Joliet*, 87 Ill. 58.

484. COMPETENCY. Commissioners appointed to assess damages are *quasi* jurors, and like them should be free from interest. or legal disability. *R. I. & A. R. R.* v. *Lynch*, 23 Ill. 645.

485. REPORT OF DISQUALIFIED PERSONS—*a nullity.* The report of persons disqualified by statute is a nullity. *Daggy* v. *Green*, 12 Ind. 303. The fact of the commissioners being disinterested freeholders must appear in the record of appointment, this being jurisdictional. *Judson* v. *Bridgeport*, 25 Conn. 426; *State* v. *Jersey City*, 25 N. J. (Law) 309.

486. JURY—*oath of.* § 8. When the jury shall have been so selected, the court shall cause the following oath to be administered to said jury:

You and each of you do solemnly swear that you will well and truly ascertain and report just compensation to the owner (and each owner) of the property which it is sought to take or damage in this case, and to each person therein interested, according to the facts in the case, as the same may be made to appear by the evidence, and that you will truly report such compensation so ascertained: so help you God.

[R. S., 1887, p. 647, § 8; S. & C., p. 1045, § 8; Cothran, p. 648, § 8.]

487. SWEARING OF JURY—*waiver of objection.* An objection that the jury were not sworn in the manner directed by the statute, comes too late after verdict. *R., R. I. & St. L. R. R.* v. *McKinley*, 64 Ill. 338.

488. SAME—*error in, does not affect jurisdiction.* An oath administered to the jury by a justice of the peace on a proceeding to establish a road in the form prescribed by the statute, with the addition of the words, "if any," as to the damages, does not render the proceeding void. The error, if any, does not go to the jurisdiction. *Hankins* v. *Calloway*, 88 Ill. 155.

489. JURY—VIEW OF PREMISES—*verdict—benefits.* § 9. Said jury shall, at the request of either party, go upon the land sought to be taken or damaged, in person, and examine the same, and after hearing the proof offered make their report in writing, and the same shall be subject to amendment by the jury, under the direction of the court or the judge, as the case may be, so as to clearly set forth and show the compensation ascertained to each person thereto entitled, and the said verdict shall thereupon be recorded: *Provided,* that no benefits or advantages which may accrue to lands or property affected shall be set off against or deducted from such compensation, in any case. [R. S. 1887, p. 647, § 9; S. & C., p. 1045, § 9; Cothran, p. 649, § 9.]

VIEW OF PREMISES.

490. CHANGE OF VENUE—*bar of the right.* When a company seeking to condemn land for right of way procured the venue of the cause to be changed to another county: *Held,* that it thereby barred itself of the right under the statute of 1872 to have the jury personally view the premises. *R., R. I. & St. L. R. R.* v. *Coppinger.* 66 Ill. 510.

491. TIME FOR VIEWING. The statute makes it the duty of the jury to examine in person the land sought to be taken or damaged; but at what time in the progress of the trial they shall go, is left to the discretion of the court. *Galena & S. Wis. R. R.* v. *Haslam,* 73 Ill. 494.

492. OBSERVATION, AS EVIDENCE. The jury have the right to view the premises and draw their own conclusions from such observation as well as from the testimony given in the case. *Mitchell* v. *Ill. & St. L. R. R. & Coal Co.,* 85 Ill. 566.

493. AFTER ARGUMENT CLOSED. The statute, giving the right to have the jury go on the land sought to be taken or damaged and examine the same, is imperative, but fixes no time when it shall be allowed; and it is error to refuse a motion to have the jury view the premises, even after the evidence has been closed and the arguments heard, but before the instructions are given. *Kankakee & Seneca R. R.* v. *Straut,* 102 Ill. 666.

494. The right of either party to have the jury go upon and examine the premises, may be exercised at any stage of the case before the court gives its instructions, and it is error to deny this right. *Ib.*

495. When the jury at the request of both parties view the premises, and no other evidence is offered, every presumption will be indulged in favor of the correctness of the verdict. In such case it cannot be known the damages assessed are excessive. *P. & F. Ry.* v. *Barnum,* 107 Ill. 160.

496. Where the jury view the premises, and no other evidence is given, the instructions given can only be considered as abstract propositions of law. But were this not so, there being no evidence preserved it cannot be known whether any of the instructions were calculated to mislead on the facts of the case. *Ib.*

HEARING, PRACTICE AND EVIDENCE.

NECESSITY OF CLAIMING DAMAGES UNDER PRIOR STATUTES.

497. LAYING ROAD—*waiver.* A claim for damages for the location of a public road is not to be presumed, but must be expressly made, and at the proper time, so that if the state or county thinks the benefits will not equal the costs, it may abandon the project or locate the road elsewhere. *Ferris* v. *Ward,* 4 Gilm. 499.

498. The land owner must object to the location of a road over his land in the first instance, or he will be concluded from insisting on damages. He must claim damages at the proper time, so that the county may abandon the project if the damages are considered too great. *Sangamon Co.* v. *Brown,* 13 Ill. 207.

499. STREET—*waiver of claim.* If a party having notice by publication of the ordinance suffers a street to be opened through his land without objection, he cannot afterwards interpose a claim for compensation. *Curry* v. *Mt. Sterling,* 15 Ill. 320.

500. WAIVER OF RIGHT—*delay to claim.* Where a railway company entered upon land and built its road without procuring a right of way or license from the owner, and occupied it for twelve years, and then instituted proceedings to condemn: *Held,* that the owner was not estopped from claiming damages. He can be barred only by the statute of limitations, and not by a mere non-claim for a period short of that fixed by the statute of limitations. *T. P. & W. Ry.* v. *Darst,* 61 Ill. 231.

501. PUBLIC ROAD—*damages to be adjusted.* § 56 of the township organization law of 1861 imperatively required the commissioners of highways to adjust the question of damages to the owners of land before opening a road across it. *Com. of Highways* v. *Durham*, 43 Ill. 86.

502. The question of damages must be satisfactorily adjusted by release or assessment, or in some other recognized mode, before the owner can be dispossessed of his property He is not required to claim as under prior laws. On failure to agree the damages must be assessed. *Ib.*

503. An effort on the part of the commissioners of highways to agree with the land owners as to their compensation is not indispensable to a proceeding to have the same assessed. *Hall* v. *People*, 57 Ill. 307.

504. To excuse a failure to condemn land for a highway, and the assessment of the owner's compensation, on the ground of there being no claim for the same, his release in writing should be filed in the town clerk's office and recorded with the order laying out the road. *Hyslop* v. *Finch*, 99 Ill. 171.

505. ESTOPPEL—*to deny title to land.* Where town authorities, in a proceeding to condemn land for a street, describe the land as A's, they cannot afterwards deny his right to be heard on the question of damages on account of his want of title. *Mt. Sterling* v. *Givens*, 17 Ill. 255.

506. PROOF OF TITLE—*when not necessary.* Where the petition to condemn a tract of land describes it as the property of the defendant, and the report of the commissioners shows it to be his land, and it appears he was in possession when the proceeding was begun, he will not be required to establish his title by proof in order to contest the amount of the compensation. The rule is different when his title is not admitted, and he applies for the assessment against the corporation. *P. & R. I. Ry.* v. *Bryant*, 57 Ill. 473.

507. The petition must state the names of the owners of the land sought to be condemned, and those interested therein, and notice must be given them; and the company will be estopped from proving before the commissioners that the party alleged to be the owner has not title. The commissioners cannot consider the question of title, but only the extent of the damages. *P., P. & J. R. R.* v. *Laurie*, 63 Ill. 264.

508. On the assessment of damages for right of way under the act of 1852, the land owner is not bound to prove title to entitle him to compensation. By instituting the proceeding against the defendant, the petitioner admits his ownership. *St. L. & S. E. Ry.* v. *Teters*, 68 Ill. 144.

509. The filing of a petition by a railway company to condemn "whatever property, rights, interest or privileges" a defendant corporation may have in certain streets by contract with the city, admits the legality of that contract, at least for the purposes of the proceeding, and estops the petitioner from insisting that the defendant has no interest in that which is sought to be condemned. *Metropolitan City Ry.* v. *Ch. W. Div. Ry.* 87 Ill. 317.

510. A proceeding to condemn land as the property of the defendant, and asking to have his compensation assessed, is an admission of his title and right to compensation. *Ch. & Iowa R. R.* v. *Hopkins*, 90 Ill. 316.

511. Where park commissioners proceed to condemn land for park purposes as the property of a person named as owner in the petition, they will be estopped from afterwards disputing his title, in the absence of any adverse claimant of the condemnation money, and more especially so where such alleged owner's title is not put in issue in the suit to recover such money. *South Park Coms.* v. *Todd*, 112 Ill. 379.

512. CORPORATE EXISTENCE—*de facto corporation.* In a proceeding by a railway company to condemn land for the use of its road, it is sufficient that it is *de facto* a corporate body. *McAuley* v. *Columbus, Ch. & Ind. Central Ry.*, 83 Ill. 348.

513. A proceeding by a railway company to condemn land for a right of way is a collateral proceeding, so far as it concerns the question of the corporate existence of the company, and it is sufficient to show it is a *de facto* corporation. *P. & P. U. Ry.* v. *P. & F. Ry.*, 105 Ill. 110.

514. By going to trial on the merits, the defendant waives the necessity of the preliminary proof of the corporate existence of the petitioner. *Ward* v. *M. & N. W. R. R.*, 119 Ill. 287.

515. Proof of the corporate existence of the railway company, if required to be made, is addressed to the court, and not to the jury called to assess the damages. The right to exercise the right of eminent domain is a question exclusively for the court to determine. The defendant may raise the question of the petitioner's right without plea or answer. *Ib.*

516. Requiring plans an admission of *de facto* corporation. *Ib.*

517. TRIAL OF OTHER ISSUES—*evidence working no injury.* The land-owner, not being injured by proof of issues tendered by co-tenants, questioning the right to condemn, he having insisted upon the same thing, cannot have a reversal on account of the admission of such evidence. *McAuley* v. *Columbus, Ch. & Ind. Central Ry.*, 83 Ill. 348.

518. ISSUES—*compensation only.* The jury impanneled can find no fact, except what is just compensation to the owner. *Smith* v. *Ch. & W. Ind. R. R.*, 105 Ill. 511, 520.

519. If the truth of any of the averments of the petition may depend upon the existence or non-existence of facts not appearing upon the face of the petition, and hence their truth or falsity is open to extrinsic proof, such proof may be made on the part of the land-owner, as well without an answer as with it, for the inquiry in such case will be directed to the truth or falsity of the petition. *Ib.*

520. DEFENCES—*Railroad track over another track.* In a proceeding to condemn the right of way for a railroad across the track and right of way of another company, questions as to the sufficiency of a city ordinance in respect to the right of the petitioner, and as to the right to cross the track of the defendant company, and as to injury to the franchise of the latter, and as to the proposed crossing being a continuing nuisance to the defendant from the operation of the new road, are all of a character, if available at all, such as may be interposed at law in the condemnation proceeding. *L. S. & M. S. Ry.* v. *Ch. & W. Ind. R. R.*, 96 Ill. 125.

521. EVIDENCE—*inability to agree.* Where, from the contest and the acts of the parties, it is evident that they could not agree as to the compensation for the land sought for a right of way, the judgment will not be reversed because no direct testimony was offered to show such inability to agree. *Ward* v. *M. & N. W. R. R.*, 119 Ill. 287.

522. RIGHT TO OPEN AND CLOSE. The party against whom judgment would be given as to a particular issue, whether affirmative or negative, in case no proof is offered on either side, has the burden of proof, and the right to open and close the case, whether plaintiff or

defendant. Therefore the party seeking to condemn on the question of the assessment of the damages to be paid, has the right to open and close. *McReynolds* v. *B. & O. River Ry.*, 106 Ill. 152.

523. CROSS PETITION—*demurrer.* The fact that a cross petition avers only the evidence of title and not any actual present title in the party filing it, and is uncertain in the description of his interest in the property, may afford ground of demurrer, but not any for dismissing the same. *Johnson* v. *F. & M. R. Ry.*, 116 Ill. 521.

524. WIDTH OF THE WAY. The question whether it is necessary for the petitioner to take a strip of land one hundred and fifty feet wide through the defendant's land, is not one for the jury to pass upon. The only question for the jury is the amount of the damages to be assessed. *De Buol* v. *F. & M. R. Ry.*, 111 Ill. 499.

525. ORDER OF EVIDENCE—*cross petition.* When real estate is sought to be condemned for widening a street, and the petition only describes the property to be taken, and a cross petition is filed to recover compensation for damages to parts of the property not sought to be taken, it is error to require the plaintiff to enter upon proof as to the question of damages to the property described in the cross petition before the land owner has given any testimony in support of his claim. *Hyde Park* v. *Dunham*, 85 Ill. 569.

526. PROOF OF ORDINANCE—*laying street.* On petition to condemn land for a street established by ordinance, if the land owner in his answer shall deny that the ordinance was duly passed, the petitioner will be required to prove that every step necessary to make it a valid ordinance has been taken. Otherwise, no such proof is required. *Village of Byron* v. *Blount*, 97 Ill. 62.

527. EVIDENCE BY TENANT IN COMMON—*available to all.* In a proceeding to condemn land owned by tenants in common, the appearance and testimony adduced by one of them will be for the benefit of all the others. *C., B. & Q. R. R.* v. *Chamberlain*, 84 Ill. 333.

528. PROOF OF CORPORATE EXISTENCE. In a proceeding to assess the owner's compensation for land taken for right of way and damages to lands not taken, it is not necessary to show that the capital stock of the railway company plaintiff has all been subscribed, or in other words, to show it is a corporation *de jure.* It is sufficient to show it is a corporation *de facto. Henry* v. *Centralia & Chester R. R.*, 121 Ill. 264.

529. Evidence showing the petitioner is a *de facto* corporation should not go to the jury called to estimate the compensation and damages; but if it is sufficient to satisfy the court of the petitioner's right to proceed, the error in letting it go to the jury is so small and harmless as not to require a reversal. *Ib.*

OF THE REPORT OR VERDICT.

530. RECITAL OF APPOINTMENT. The act requiring a copy of the appointment of the appraisers to be recited in the report will be complied with if the appointment is attached to the report and is made a part of it. *Low* v. *G. & Ch. U. R. R.*, 18 Ill. 324.

531. SHOWING BASIS—*allowance for fencing.* Under the act of 1855, in relation to fencing by railway companies, the record of the proceeding to condemn land for right of way should show the amount allowed for fencing as a component part of the damages. This should be done as a protection of the company against any future claim to fence the right of way. *R. I. & A. R. R.* v. *Lynch*, 23 Ill. 645.

532. The finding of the jury should show on what basis the damages are assessed in order that the record may show thereafter the rights of the parties, as to who shall keep up the fences. *St. L., J. & Ch. R. R.* v. *Mitchell*, 47 Ill. 165.

533. The verdict in a proceeding under the act of 1852 should find the compensation for the land taken and the damages separately. *Hayes* v. *O. D. & F. R. V. R. R.*, 54 Ill. 373.

534. The cost of erecting and maintaining fences along the line of the proposed road is a proper element of damages to be considered by the jury, yet if no evidence is offered on the subject the jury will not be required to find in their verdict anything in respect to it. *P. & R. I. R. R.* v. *Birkett*, 62 Ill. 332.

535. If the amount assessed is to cover damages to other parts of the property not taken, all this should be distinctly stated in the order. *Bloomington* v. *Miller*, 84 Ill. 621.

536. CERTAINTY IN. A verdict in a proceeding to condemn land for right of way by a railway company, which finds that the land owner "is entitled as compensation to the sum of \$420, and as damages the sum of \$411.25, a total sum of \$831.25," is sufficiently certain. *Ill. W. Extension R. R.* v. *Mayrand*, 93 Ill. 591.

537. FINDING SEPARATELY AS TO EACH TRACT—*waiver.* Where both parties on the trial treat the several tracts over which the right of way is sought as one farm in the examination of witnesses and in the instructions, and the compensation and damages are fixed as upon one tract, the objection that the finding should have been as to each tract separately, comes too late on appeal, or even on motion for a new trial. *Kankakee & Ill. River R. R.* v. *Chester*, 62 Ill. 235.

538. DESCRIPTION OF LAND—*sufficiency.* The petition described the land over which a strip was sought as lot 1 of n. w. qr., &c. The part sought was described as "a strip of land * * * 200 feet wide for a distance of 1,151 feet across the tract (fifthly) above described, commencing," &c. The verdict and judgment described the land as "the land taken for right of way across" lot 1, as described in the petition: *Held*, that the description of the land taken was sufficiently certain by reference to the petition. *Suever* v. *Ch., S. F. & Cal. Ry.* — Ill. —. Filed Nov. 11, 1887.

539. GROSS SUM FOR COMPENSATION AND DAMAGES—*presumption.* In the absence of a bill of exceptions showing the evidence, a verdict awarding a gross sum for compensation for the land taken, and damages to the part not taken, will not be held erroneous. It will be presumed the evidence justified such a finding. *Ib.*

540. FINDING, WHEN SEPARATELY AND WHEN IN GROSS. Where the petition shows that each tract belongs to separate owners, the verdict should find the compensation and damages as to each defendant separately; but if it avers that a particular tract is owned by several persons, they will be presumed to be tenants in common, in the absence of proof to the contrary, and the verdict may find a gross amount to be paid to the defendants. *Ib.*

NEW TRIAL.

541. ON THE EVIDENCE. The verdict of the jury, unless manifestly against the weight of the evidence, will not be disturbed. *Ill. & Wis. R. R.* v. *Von Horn*, 18 Ill. 257.

542. Where the question of damages is fairly submitted, no benefit being likely to result to the owner of the land, and the company not being absolutely bound to erect and maintain a fence, etc., the supreme court will not disturb the verdict. *T. & P. R. R.* v. *Unsicker*, 22 Ill. 221.

543. The verdict will not be set aside merely because the damages are large, when the land-owner is not to receive any particular benefit by the location of the road. *T. & P. R. R.* v. *Roberts*, 22 Ill. 224.

544. EXCESSIVE DAMAGES. Over ten acres of land in the city of Peoria were sought to be condemned by a railway company for a right of way, and twenty-five witnesses estimated the damages to the owner at various sums ranging from $1800 to $18,000, and the jury assessed the damages at $5,500: *Held*, not excessive. *P. & R. I. R. R.* v. *Birkett*, 62 Ill. 332.

545. PRESUMPTION—*that jury followed instructions.* Where a railway company has, under proceedings to condemn, made embankments and constructed its road before the final hearing, and the court instructs the jury that the land-owner is entitled to the value of the land, with the improvements put thereon, it will be presumed that the jury were governed by the instruction. *Mitchell* v. *Ill. & St. L. R. R. & Coal Co.*, 85 Ill. 566.

546. PERSONAL VIEW. Where the jury go upon the land in person and examine the same, such examination is in the nature of evidence, and in such case, even though the preponderance of the evidence preserved in the record is clearly against so large an assessment as found, a new trial will not be granted, as the facts ascertained by the personal examination may have fully justified the verdict. *Ch. & Iowa R. R.* v. *Hopkins*, 90 Ill. 316.

547. PRESUMPTION IN FAVOR OF VERDICT. Where the evidence is not preserved in the record, every presumption will be indulged in favor of the findings of the court upon all questions of fact. *Fisher* v. *Ch. & Spr. R. R.*, 104 Ill. 323.

548. ON THE EVIDENCE—*conflicting.* Where the evidence is conflicting in a condemnation proceeding as to the damages and compensation to be paid, and consists chiefly in the opinions of witnesses, some of whom sustain the finding and some do not, this court will not feel justified in reversing, unless it is able to say the verdict is clearly against the evidence. *Smith* v. *Ch. & W. Ind. R. R.*, 105 Ill. 511.

549. Where the evidence as to damages from locating a railway across a farm is conflicting and widely variant, and the jury go upon the premises and examine for themselves, their assessment of damages will not be set aside, though not so large as the preponderance of the evidence preserved in the record might justify, when the finding is not manifestly wrong upon the proofs. *McReynolds* v. *B. & O. R. Ry.*, 106 Ill. 152.

550. Where the evidence is conflicting as to the value of the property sought to be condemned for railroad purposes, and the jury have examined the premises in person, this court will not reverse on the ground alone that the damages assessed may be considered high, unless they are clearly excessive. *Ch. & E. R. R.* v. *Jacobs*, 110 Ill. 414.

551. In an eminent domain proceeding where the evidence as to the value of the land taken is conflicting, and the jury take a view of the land, and then find a verdict for an amount larger than that supported by the evidence for the petitioner and smaller than that supported by the evidence for the respondent, this court will not disturb the verdict. *Ch. & E. R. R.* v. *Blake*, 116 Ill. 163.

552. INSTRUCTION—*singling out certain evidence.* An instruction which singles out and calls attention to the testimony of the land-owner is erroneous and unfair, and calculated to mislead the jury by seemingly giving undue importance to such testimony. *J. & S. E. Ry.* v. *Walsh*, 106 Ill. 253.

552a. NEW TRIAL—*amount of damages*. Where there is great disparity in the evidence as to the value of land sought to be condemned for a railroad, and as to the damages to the parts not taken, this court will not reverse, unless it appears the verdict is unreasonable, and the damages are so grossly excessive as to evince that the verdict is the result of passion or undue and improper motive or influence. *Cal. Riv. Ry.* v. *Moore*, — Ill. —. Filed March 26, 1888.

OF THE ELEMENTS AND MEASURE OF DAMAGES.

(a) IN CASE OF ACTUAL TAKING.

553. LAND FOR ROAD AND FERRY—*damages to a prior ferry*. On the condemnation of land for a ferry landing and a public road thereto, neither the value of the ferry of the land owner across the same stream, nor that of the ferry privilege, should be considered. *Mills* v. *St. Clair Co.*, 3 Scam. 53.

WHEN ONE RAILROAD CROSSES ANOTHER.

554. DAMAGES BESIDE LAND TAKEN—*loss and inconvenience*. A railway company whose right of way is condemned by another company is entitled to not only just compensation for the land actually taken, but for all such incidental loss, inconvenience and damage which may be reasonably expected to result from the construction and use of the crossing in a legal and proper manner. This is the true measure contemplated by § 13, Art. 2, of the constitution of 1870, and § 14, Art. 11, places corporations upon the same footing. *C. & A. R. R.* v. *Spr. & N. W. R. R.*, 67 Ill. 142.

555. KEEPING ROAD IN REPAIR—*evidence of cost—cutting through embankment*. When one railway company acquires, by condemnation, the right to run its road through a high embankment of another railroad, twenty feet below the track of the latter, it will be under no legal obligation to erect and maintain a bridge to support the track of such other company; and therefore proof of what it will cost to build such a bridge and keep it in repair is proper on the assessment of damages. The defendant company, in such case, is entitled to have such sum for damages as will enable it to construct and keep in repair all such works as may be necessary to keep its track in a safe and secure condition, and also for all such incidental loss and inconvenience as may be a necessary result. *Ib.*

556. SAME—*expectations of party not bound, not evidence*. The expectations of a contractor for the construction of a railroad across that of another company to keep the proposed work in repair, is not proper evidence on the question of damages to the company whose road is to be intersected, there being nothing to bind him to make the repairs. *Ib.*

557. EVIDENCE—*opinion of witness—matter of law*. It is improper to ask a witness whose duty it would be to keep a railroad crossing and bridge at the intersection of two roads in repair, as calling for an opinion on a matter of law. *Ib.*

558. SAME—*opinion of experts as to damages*. It is competent for experts, such as engineers, to give their opinions as to matters which may form the proper ingredients of a verdict, but not to usurp the province of the jury. The witness must first be shown to be competent to give an opinion. *Ib.*

559. On the assessment of damages in a proceeding by one railway company to condemn a right of way across that of another company, which made it necessary to cut through a heavy embankment

—6

twenty feet below the grade of the defendant company, and thus remove the support of its road for the space of sixty feet, the petitioner, after proving by one of its contractors that he proposed to support defendant's track by timbers, which he described, asked him: "If you put in the cut work you propose to do and have described, what would be the damage to the defendant?" He answered: "There would be no damages." *Held*, that the evidence was improper, on the ground that the question called for an opinion based upon an assumption that petitioner would put in supports which it was not obliged to put in, and because the answer was an opinion covering the very question to be settled by the jury. *Ib.*

560. DAMAGES WHEN NEW ROAD CUTS THROUGH EMBANKMENT —*elements.* When a right of way is sought across or under the track of another company, or through its embankment, the latter company will be entitled to receive such sum as will enable it to place its track over the point at which the ground is condemned in as safe a condition, as near as can be, as it was before making the excavation. The damages should cover additional expense for watchmen when travel over the excavation is hazardous; the expense of building and maintaining permanent abutments or retaining the walls; losses incident to rebuilding or repairing, and contingent losses by fire or otherwise; and if any other kind of bridge over the excavation is more safe than a wooden one, the compensation should be sufficient to enable the company to erect and perpetually maintain a bridge of that degree of safety, and likewise to reimburse it for all inconvenience and expense incident to the erection and maintenance of such a bridge. *St. L., J. & C. R. R. v. S. & N. W. R. R.*, 96 Ill. 274.

561. DAMAGES WHEN PROPERTY ADAPTED TO SPECIAL USE— *evidence.* Where land has no market value from the fact of its being used as a right of way for a railroad, and devoted to a special use of making railroad transfers, estimates of its value with reference to such use, by those competent to speak in that regard, should be received on the question of compensation to be paid for its condemnation for the use of another railroad company for its right of way, and it is error to refuse such evidence. *L. S. & M. S. Ry. v. Ch. & W. Ind. R. R.*, 100 Ill. 21.

562. DAMAGES TO USE, AS AN ENTIRETY. Although a right of way of a railroad company is limited to the use of the land for the construction, maintenance and operation of a railroad upon it, this limited use is property, and any interference with it at any point by condemnation by another railroad, whereby the use is impaired, may be considered in connection with and as affecting its use as an entirety. *Ib.*

563. DAMAGES TO PART NOT TAKEN. On a proceeding to condemn a strip of land across the right of way of a railroad company, a limitation of the damages to those for physical injury to the land sought to be condemned for another railroad will be too restricted. The defendant should be allowed to recover for the obstruction to the use of its remaining property, and for all damage to it resulting from the operation of the second railroad on the strip so taken. *Ib.*

564. SAME—*elements of.* In a proceeding by a railroad company to condemn a right of way across the prior right of way of another company upon certain blocks, the company whose franchise is sought to be taken in part, will not be restricted in its compensation to the damages of its right of way, or railroad property within the blocks. In such case, it will be competent for the defendant company to show and recover for damages it will be subjected to by placing obstructions upon its right of way, in maintaining and operating the proposed new road, whereby access to different parts of its line will be

interfered with, and its capacity for the transaction of business impaired. *Ib.*

565. DIRECT AND REMOTE DAMAGES—*diminution of capacity to do business.* Direct and immediate damages alone are recoverable in this class of cases, and remote and merely incidental damages cannot be considered. It is that injury which depreciates the value of the property, whether by taking a portion of it or rendering the portion left less useful, or, in case of a railroad company or other corporate body, less capable of transacting its business,—such a hinderance and inconvenience as to occasion loss or diminish and limit its capacity to transact its business by decreasing the power to transact as much, or necessarily increasing the expense of what may be done, although not diminished; and this hinderance must produce immediate or future loss. If the new structure, when made, does not abridge the owner's capacity without increased expense to transact an equal volume of business, then though there may be inconvenience and annoyance, unless the property is depreciated in value, these are not elements of damages. *Peoria & Pekin Union Ry.* v. *Peoria & Farmington Ry.,* 105 Ill. 110.

566. ELEMENTS OF DAMAGE—*stoppage of trains at crossings*-The law requiring railroad trains to stop before crossing another railroad, being a mere police regulation and subject to repeal at any time, the damages sustained by a railway company for the delay, inconvenience and trouble in stopping before crossing another road seeking a condemnation for a right of way, are too vague, indefinite and contingent to be an element in the assessment of damages in favor of the road so to be crossed. *Ib.*

567. INCREASED DANGER FROM CROSSING—*too remote and uncertain.* Nor is the increased danger arising from the crossing of the track of one railroad by the trains of another to be considered as an element of damage in such proceeding. To allow damages on such a claim would violate the rule that they cannot be allowed on mere conjecture, speculation, fancy or imagination. *Ib.*

568. This rule is not in conflict with what was said in *L. S. & M. S. R. R.* v. *Ch. & W. Ind. R. R.,* 100 Ill. 21, where it was held that not only such injury and inconvenience as reduce the capacity of the corporation to transact its business, and necessarily result in damage and loss, are elements of damage. *Ib.*

569. SEVERING CONNECTION WITH ELEVATOR—*when no damage.* Where an elevator used for the deposit of grain stands on ground considerably above and some distance from a river upon which grain was carried from the elevator, and it appearing that the grain was transferred from the elevator to boats at the wharf through an inclined chute, or tube, called a conductor, and that a railroad seeking a condemnation was proposed to be located between the elevator and the river, and was to be constructed on trestles and elevated entirely above the chute or conductor, so as not to interfere with the transferring of grain from the elevator to the river, it was *held*, that there was no loss to the owner of the elevator, and therefore could be no damage. *Ib.*

570. EVIDENCE—*plan of proposed road on question of damages.* In a proceeding to condemn a right of way for a railroad over a strip of land between an elevator and a river, the plans by which the company proposes to build the road, as showing the track is to be laid upon trestles elevated so high as not to interfere with the transfer of grain from the elevator to the river in chutes or conductors, are admissible in evidence on the question of damages and compensation. *Ib.*

571. STIPULATION TO MAINTAIN AND KEEP IN REPAIR FROGS AND CROSSING—*evidence as to damages.* In a proceeding to condemn a right of way across that of another railway company, the petitioners offered in evidence a stipulation or covenant, properly executed by it, that it would, and should, at its own expense, put in and thereafter maintain in suitable and proper repair, the frogs and crossing across two main tracks of the defendant, and that this stipulation should be binding on the successors and assigns of the petitioner so long as a grade crossing should be maintained at the crossing of the right of way sought to be condemned: *Held,* that this was a valid obligation, enforcible against the petitioner and its successors and assigns, and was properly admissable in evidence. *C. & A. R. R.* v. *Joliet, Lockport & Aurora Ry.,* 105 Ill. 388.

572. The obligation being a valid one, securing the construction and maintenance of the proposed crossing at the expense of the petitioner, its successors and assigns, the cost thereof could not become an element of damages in favor of the defendant corporation, and would operate to exclude any evidence on behalf of the defendant on that subject. *Ib.*

573. STIPULATION—*right of action on.* The stipulation is sufficiently definite as to the manner in which the work of making the crossing was to be done, and as to what extent it would affect the defendant. A "suitable and proper crossing" is a phrase well understood by civil engineers and practical railroad men. Any marked departure from the stipulation in that regard would afford the defendant a right of action for the recovery of any damages caused thereby. *Ib.*

574. The force and effect of the obligation as an instrument of evidence, and as excluding all question of damage arising from the expense of constructing and maintaining the crossing, is not at all impaired from the fact that it is a mere promise, which may not be performed. The covenant is thought to run with the land, and for any breach thereof a right of action is given, which will afford complete indemnity to the defendant company. It cannot be presumed, in the absence of testimony, that the petitioner will be unable from any cause to perform its obligation. *Ib.*

575. CASES DISTINGUISHED—*crossing on or under grade.* In this case the crossing was upon grade, and it would be the duty of both parties to see that the crossing was properly constructed and maintained in a safe condition, and in this respect is to be distinguished from the cases of *C. & A. R. R.* v. *Springf. & Northwestern R. R.,* 67 Ill., 142, and 96 Ill. 274. In that case the crossing was not upon grade, but was an under crossing made by cutting through a high embankment under the track of defendant's road, thereby removing all the support it had. It did not appear that the petitioner was under any obligation by its duty to the public as a common carrier, or by any stipulation or otherwise, to keep defendant's track above its own in a suitable and safe condition, and so the expense incurred by the defendant in that regard was a very proper element of damage. *Ib.*

576. STOPPAGE OF TRAINS—*no element of damages.* The fact that the defendant corporation is required by statute to bring its trains to a halt upon an ascending grade before crossing the new road, and thereby the hauling capacity of its engines will be impaired, affords no element of damages. The statute requiring such stoppage is simply a police regulation, the existence of which is subject to the legislative will. *Ib.*

577. It is a principle underlying all conduct that neither a natural person nor a corporation can claim damages on account of being compelled to render obedience to a public regulation. Obedience to law is a service all citizens and corporations are bound to render to the state, and no damages can grow out of such act of obedience. *Ib.*

578. DEPRECIATION OF VALUE—*impairing capacity to do business.* A railway company seeking a right of way across the track of a defendant company is liable for all damages directly resulting to the latter from the making or the using of the crossing, whereby the value of its property is diminished, or its facilities are materially impaired for the transaction of its business. If the crossing abridges the defendant company's capacity to transact an equal volume of business, it is an element of damages, even though it does not increase its expenses. *Ch. & W. Ind. R. R.* v. *Englewood Connecting Ry.*, 115 Ill. 375.

579. SAME—*stipulation.* On a proceeding by a railway company to condemn a right of way over the track of another company, the latter will have the right to show that the value of its road and its capacity to do business will be impaired, notwithstanding a stipulation of the former that it will, at its own expense, put down and keep in repair all necessary frogs and crossings for its main tracks, and it is error to exclude such evidence. *Ch. & W. Ind. R. R.* v. *Englewood Connecting Ry.*, 115 Ill. 375.

580. Wherever a condemnation and subsequent use of a right of way across a railroad track will injuriously affect the strength, permanency and durability of the defendant company's structures, and their adaptability and capacity of doing railroad business, the injury thus occasioned will form a proper basis for the assessment of damages in a proceeding to condemn. *Ib.*

LAND TAKEN FOR PUBLIC USE.

581. BENEFITS—FARM LAND—*set off.* The rule for the assessment of damages for land taken is an equitable one. Special benefits to the land may be deducted from the damages, but benefits in common with other lands growing out of the enhanced value by the public improvement should not be. *State* v. *Evans*, 2 Scam. 208.

582. SAME—*from location of railroad.* In assessing damages for right of way for a railroad over a tract of land under the act of 1833, only the benefits resulting to the land from the construction of the road, and not those arising from the location of the road, may be considered by the jury. *Ib. State* v. *Wilson*, 2 Scam. 225.

583. COMPENSATION—*value of land taken, and dividing farm.* The measure of damages is not merely the value of the land taken, but also such other damages as may result, as the breaking up of the convenient arrangement of the farm, the necessity for additional fencing, &c. *State* v. *Evans*, 2 Scam. 208.

584. SAME—*may be in benefits.* The word "compensation" in the constitution of 1848, means that which is given as an equivalent for a loss, but that instrument does not determine how that equivalent shall be made up. *A. & S. R. R.* v. *Carpenter*, 14 Ill. 190.

585. BENEFITS—*under law of 1845.* Under Chap. 92, R. S. 1845, "concerning the right of way," in assessing damages to the owners of land, it is proper to consider all appreciable advantages and disadvantages accruing to them from whatever cause *Ib.*

586. From this statute the payment of damages to the owner of a tract of land for the right to construct a railroad through it, was never intended, where the additional value given to the land is equal to any injury sustained. *Ib.*

587. BENEFITS—*from any reason, set off.* If [additional value is given to the land by the construction of public works, it matters not whether it be by draining the land which was before wet, by affording additional facilities for taking its produce to market, or by the general enhancement in value of the land occasioned by its contiguity to the public works. *Ib.*

588. FARM LAND—*elements of damage.* In estimating damages for constructing a railroad through a farm, the injuries which the proprietor suffers by having his farm divided so as to make it inconvenient to pass from its different parts, and to compel him to erect additional fences, are as proper elements of damages to be considered as the value of the land taken. *A. & S. R. R.* v. *Carpenter*, 14 Ill. 190.

589. BENEFITS—*by laying a street.* Under a statute requiring the jury to take into consideration the benefits as well as the injury caused by the opening of a street, if the benefits are equal to the injury, or the land will sell for as much with the proposed street as without it, it is the duty of the jury to find no damages. *Curry* v. *Mt. Sterling*, 15 Ill. 320.

590. ELEMENTS OF DAMAGES—*extra fencing.* The cost of erecting and maintaining a fence along the right of way of a railroad, is a proper element of damage. *St. L., J. & Ch. R. R.* v. *Mitchell*, 47 Ill. 165; *R. I. & A. R. R.* v. *Lynch*, 23 Ill. 645; *Tonica & Petersburg R. R.* v. *Unsicker*, 22 Ill. 221.

591. BENEFITS—*land for a park.* The compensation to be paid the owner of land for a public park may be raised by special assessments upon the land benefited by the location and construction of the park, including the owners of other lands. The benefits to his remaining land may be the compensation contemplated by the constitution of 1848. *People* v. *Williams*, 51 Ill. 63.

592. BENEFITS—*not set off as to land taken.* The owner of land taken for a railroad, under the act of 1852, must be paid in money alone the full value of the land taken, irrespective of any benefits or advantages to his remaining land by the construction and use of the road. *Hayes* v. *Ottawa, Oswego & Fox River Valley R. R.*, 54 Ill. 373.

593. BENEFITS—*against damages for land not taken.* But in estimating his damages by reason of the construction and use of the road apart from the question of the land taken, such benefits and advantages are to be taken into consideration and estimated. *Ib.*

594. The decision in *Alton & Sangamon R. R.* v. *Carpenter*, 14 Ill. 190, holding that compensation could be made in benefits, was made under the act of 1845, and does not control the construction of the act of 1852 on that subject. *Ib.*

595. ELEMENTS OF DAMAGE—*loss of a spring on farm.* Where the owner of land over which it is sought to condemn a right of way claims that he will thereby lose the beneficial use of a spring on the land, that is a proper subject for the consideration of the jury in adjusting the compensation. *Peoria & Rock Island Ry.* v. *Bryant*, 57 Ill. 473.

596. BENEFITS—*under act of 1852.* Under the act of 1852, in estimating the benefits to the land owner, the jury should not consider such as he receives on his other land in common with owners of other lands, but assess to him only such benefits as he will receive over such common benefit. *P., P. & J. R. R.* v. *Black*, 58 Ill. 33.

597. SAME—*to other lands of same owner.* Where a railway charter provided that in condemning the right of way, the commissioners should view the premises and assess the value of the same and all damages to the owner and the benefits of the road, taking into consideration the advantages and disadvantages by reason of the construction of the road, and report the amount of damages, if any, over and above the benefits: *Held*, that these provisions did not authorize the estimation of the benefits the other lands of the owner over which the road did not run would receive. *St. L., V. & T. H. R. R.* v. *Brown*, 58 Ill. 61.

598. Where the road was located over two forty acre tracts of the same person, and he released the right of way over one of them, it would be error to estimate the benefits that that forty acres would receive by the building of the road and deduct them from the damages to the other tract. *Ib.*

599. ALL APPRECIABLE DAMAGES. All injuries which are appreciable, and which result to the land owner from the construction of a railroad over the land, are legitimate subjects in the estimation of damages. *St. L., V. & T. H. R. R.* v. *Mollett*, 59 Ill. 235.

600. DAMAGES—FRUIT TREES—*separate assessment.* If fruit trees which are upon the land taken are not included in the damages for the land itself, they may properly be the subject of a separate assessment. The mode of assessment is immaterial so that the damages are fairly and truly assessed. *Ib.*

601. ELEMENTS OF—*ditching made necessary.* When ditching the adjacent land becomes necessary by means of embankments thrown up for the road, the expense thereof is a proper element of damages. *Ib.*

602. SAME—*cattle guards.* But cattle guards are not proper subjects for such an assessment, because they could enter into the estimate only on the hypothesis that the proprietor of the land may construct them, which he would have no right to do, except by permission of the company. *Ib.*

603. BENEFITS—*under act of 1852.* Where private property is condemned under the act of 1852 for the use of a railroad, the land taken must be paid for without regard to the benefits accruing to the owner by reason of the construction and operation of the road. *Wilson* v. *R., R. I. & St. L. R. R.*, 59 Ill. 273.

604. COMPENSATION—*market value of land taken.* The measure of compensation in such case, and as guaranteed by the constitution of 1848, is the market value of the land taken. *Ib.*

605. BENEFITS—*set off against damages in act of 1852.* But as to damages to land not taken, resulting from the construction and operation of the road, there may be set off the benefits accruing to him thereby. *Ib.*

606. CUTTING OFF STRIP OF FARM—*elements of damage.* Where the right of way severs a strip of about two acres from a farm, while compensation cannot be demanded for such a strip, it will form an element in assessing the damages to the owner by the operation of the road. Such strip or its value, the inconvenience of the owner, and the danger to which he and his family and his stock are exposed in passing from one part of the farm to the other, are proper elements of damages, against which should be set off the facilities afforded by the road and a convenient depot for getting the products of the farm to market, as also the actual increase in the market value of the farm occasioned by the road. *Wilson* v. *R., R. I. & St. L. R. R.*, 59 Ill. 273.

607. DAMAGE TO OTHER LANDS—*road made without authority.* Where a railway company, without license of the owner or authority of law, enters upon land and constructs its road over the same, on a proceeding to condemn the right of way, the company will be held liable for damages resulting to other lands of the owner from the construction of the road. *T., P. & W. Ry.* v. *Darst*, 61 Ill. 231.

608. BENEFITS—*under acts of 1845 and 1852.* The law of 1845 permitted the general benefits received to be estimated against the damages, though conferred upon other lands and in other ways, while the law of 1852 restricts the offset of benefits against the particular tract benefited. *P., P. & J. R. R.* v. *Laurie*, 63 Ill. 264.

609. FARM LAND—*various elements of damages.* In a proceeding to condemn land for a railroad track, the jury are entitled to know the amount of land taken, how it affects the remainder, how it divides the farm as to water, pasturage, improvements, &c., and also the danger and inconvenience in the perpetual use of the track for moving trains over the same, and what injury, if any, to stock kept on the farm, and many other things connected therewith, better understood and better to be explained by persons of large experience in such matters; and, as a general rule, any evidence that tends to illustrate these various subjects is admissable. *R., R. I. & St. L. R. R.* v. *McKinley*, 64 Ill. 338.

610. LAND TAKEN—*market value and capabilities.* The true test as to the damages to be paid for land taken is its market value; but in estimating the damages, reference may be had not merely to the uses to which the land is actually applied, but its capabilities, so far as they add to its market value, may also be taken into consideration. If the land has a mine under its surface, that fact may be considered, if the mine adds to the market value of the land even though such mine has never been used. So of a water power, even though it has never been utilized. *Haslam* v. *G. & S. W. R. R.*, 64 Ill. 353.

611. BUILDING DESTROYED—*measure.* The law requires that for all property taken by a railway company for its use, or damaged by it, just compensation shall be made to the owner. If a building stands in the way which it is necessary to destroy, its value must be paid by the corporation; and the jury, in estimating its value, will take into consideration, not the value of the material composing the same, but the value of the building as such. Should any of the *debris* remaining on its removal or destruction be appropriated by the owner of the land, to the extent of its value will the claim of the owner be lessened. *L., B. & M. R. R.* v. *Winslow*, 66 Ill. 219.

612. EASEMENT—*considered on assessment of damages.* If asked, the court should instruct the jury to take into consideration the fact that the corporation acquires only an easement in the land condemned, and they should allow to that fact such importance as they may deem proper. *Ib.*

613. COMPENSATION—*in money alone.* The compensation to be ascertained by a jury for the taking of land must be, in terms, money; and the jury have no power to prescribe the performance of other acts, such as fencing the road, making crossings, &c. *Ch., Mil. & St. P. Ry.* v. *Melville*, 66 Ill. 329.

614. FARM LAND—*elements of damages—fire.* Evidence as to the danger of killing stock and the escape of fire by reason of the construction of a railroad through a farm, is proper to be considered by the jury. Such damages are as much proximate as those growing out of the danger and inconvenience of crossing the road from one part of the farm to another. *St. L. & S. E. Ry.* v. *Teters*, 68 Ill. 144.

615. MEASURE IN GENERAL—*all injuries.* The design of the law is to fully compensate a party for all injury he may sustain by reason of the appropriation of his land for railroad purposes and which shall grow out of, or be occasioned by the location and use of the road. *Ib.*

616. BENEFITS—*set off—only against damages.* The owner of land taken for public improvement is entitled to the value of the land actually taken without any regard to supposed benefits arising from the proposed improvement. If he claims damage to the part of his land not taken, and it has received special benefits, they may be considered in arriving at the owner's damages. The same rule applies in a proceeding to condemn land for a street under Art. 9 of the act relating to cities and villages, as it does under the eminent domain act. *Harwood* v. *Bloomington*, — Ill. —. Filed March 28, 1888.

617. It is hardly practicable to state any inflexible rule for estimating damages to the land owner. The amount should be sufficient to cover all the actual damages sustained by reason of the construction of the road, the land taken, all physical injuries to the residue, and for all inconveniences of every character actually produced, but nothing should be allowed for imaginary or speculative damages, or such remote and inappreciable damages as may be imagined, but never occur. *Jones* v. *Ch. & Iowa R. R.*, 68 Ill. 380.

618. FENCING. When the railway company has fenced its track through land it is seeking to condemn for right of way, it is not error to instruct the jury not to consider the failure to maintain the fences as an element of damages. *Ib.*

619. FARM CROSSINGS. The statute not having given the land owner any remedy to compel the erection and maintenance of farm crossings, and they not being any part of the fence, the failure to erect and maintain such crossings may be considered as an element of damages. *Ib.*

620. DANGER OF FIRE. Damage from fire in most cases may be reckoned among imaginary dangers that may or may not occur, and if they do the law affords a speedy and effectual remedy. But if the road is constructed so near the owner's buildings as that the danger from fire is real, it may constitute an element of increased damages. *Ib.*

621. DIVIDING FARM. The fact that a portion of a farm is cut off by a railroad, is in very many, if not in all cases, a permanent injury to the whole farm and materially diminishing its value and is a legitimate source of damages. *G. & S. Wis. R. R.* v. *Birkbeck*, 70 Ill. 208.

622. DAMAGE TO PART CUT OFF—*must be to entirety.* When a railroad is located over a tract of land, and compensation is allowed the owner for the land actually taken, he cannot recover damages to a small part of the tract not taken, if the whole is not damaged when taken together. *Page* v. *Ch., Mil. & St. P. Ry.*, 70 Ill. 324.

623. The jury, under the act of 1872, are not required to assess the damages to a strip of land lying within a few feet of the right of way of a railroad, but the damages, if any, to the entire tract by reason of the construction and operation of the road. The effect must be considered upon the market value of the entire tract, and not a distinct part. *Ib.*

624. SAME—*true measure—difference in value.* The true compensation for land not taken by a railway company for a right of way, is the difference between what the whole property would have sold for unaffected by the railroad, and what it would sell for as affected by it, if it would sell for less. The damages must be for an actual diminution of the market value of the land and not speculative. *Ib.*

625. BENEFITS—*against damage to land not taken.* If the inconvenience of the road to a certain selected part of the tract is outweighed by the additional convenience of the road to the residue of the tract, it will not be damaged. This is not deducting benefits from damages. *Ib.*

626. LOTS NOT TAKEN—*difference in market value.* Where land is not taken by a railway company for its right of way, but damaged only, the question should be, will the property be of less value when the road is constructed than it was when it was located. If so, then the difference is the measure of damages. To ascertain this, the opinion of intelligent witnesses is proper. *Eberhart v. Ch., Mil. & St. P. Ry.,* 70 Ill. 347.

627. SAME—*damages must be actual.* The damages contemplated by the constitution, where the property is not taken, must be actual, real and present damage to the property. *Ib.*

628. DIVIDING FARM—*elements of damages.* The cutting off of a portion of a person's farm by a railroad through it, requiring him to travel a greater distance to reach the part cut off, and the danger to him, his family and stock in crossing the track from one part of the farm to another, are proper elements of damages. *P., A. & D. R. R. v. Sawyer,* 71 Ill. 361.

629. DAMAGE TO PART NOT TAKEN—*must be direct and physical.* The damage to lands not taken, but injured by the taking of other land of the same owner, must be direct and physical, and result from the taking of a portion of his land. *Stetson v. Ch. & E. R. R.,* 75 Ill. 74.

630. RAILWAY STRUCTURE—*put on land by consent.* Where it is alleged in a petition to condemn land for a right of way by a railway company that a railroad had been previously constructed on the same with the owner's consent, which is not denied in the pleadings, the land owner will not be entitled to the value of the road structure as a part of his land. *Emerson v. Western Union R. R.,* 75 Ill. 176.

631. BENEFITS—*as against land taken.* Under § 13, Art. 2, of the constitution of 1870, the full value of land taken for a highway must be paid in money alone, disregarding all benefits and advantages that may result to the portion not taken by reason of the establishing of the road, and it is not in the power of the legislature to provide otherwise. *Carpenter v. Jennings,* 77 Ill. 250.

632. Where the proceedings to lay out a public highway show that the jury in assessing the compensation to be paid to the owner, undertook to pay him in part in benefits to his other land by the construction of the road, and not wholly in money, it was *held,* that the jury transcended their powers, and that their action was void. *Ib.*

633. BENEFITS—*set off under act of 1852.* Under the act of 1852 the owner of land taken by a railroad is entitled to compensation at all events to the extent of the value of the land taken, without any deduction for benefits the land may receive from the location or construction of the road; but such benefits may be set off against any damage the remaining land may sustain by the construction of the road. *Todd v. K. & I. R. R. R.,* 78 Ill. 530.

634. The damages done to one piece of land through which a railroad is run cannot be compensated by benefits accruing to another and separate piece of land through which it does not run, although belonging to the same person. *Ib.*

635. BLOCKS—*when treated as distinct tracts.* Where a town has been laid out into blocks and streets for many years, and the same has always been treated as blocks and streets, the blocks will be treated as distinct tracts for the purposes of assessing damages for right of way, although the plat may not be made according to law. *Todd v. K. & Ill. River R. R.,* 78 Ill. 530.

636. BENEFITS—*set off against damages—elements of, to farm land.* In assessing damages under the act of 1872 to the owner for land taken by a railway company for right of way, the jury may take into consideration not only the value of the land taken, but all the facts which contribute to produce the damages to that not taken, as that the farm is put in a worse shape for cultivation or pasturage; that some portion of it is more dangerous for use; that there is danger of fire from passing engines, and all other actual inconvenience and damage the property may sustain in its use, not only for the present, but for the future; and against such damages it is proper to set off or allow for any benefits or advantages received by the owner of the land in common with others from the construction of the road. *K. & E. R. R.* v. *Henry*, 79 Ill. 290.

637. FARM OF SEVERAL TRACTS—*damages to whole when proper.* Where a farm through which a railroad ran consisted of 240 acres, and the petition for the condemnation of the right of way described the road as running through both the quarter section and the 80 acre piece: *Held*, that in assessing the damages the jury should consider the damage to the whole farm by reason of the construction of the road. *Ib.*

638. DAMAGE TO PART NOT TAKEN—*land to widen street.* It is a question of fact whether the diminution of a lot for the purpose of widening a street, impairs its value; and if the taking of a part of the lot sensibly impairs the relative value of that remaining, the owner is entitled to compensation, not only for the part taken, but also as to the remaining part. *Hyde Park* v. *Dunham*, 85 Ill. 569.

639. SAME—*effect on balance as a whole—benefits.* But in determining whether the act of diminution has impaired the relative value of the remaining part, a partial effect only is not to be considered, but the whole effect, and the effect not upon any selected part of the lot, but the whole. In such case it is error to exclude the consideration of special benefits to the property not taken. *Ib.*

640. DIVIDING FARM—*elements of damage.* As elements of damage, the fact that the railroad separates the wood, water and timber from the balance of the farm, the inconvenience to the owner from the perpetual use of the track for moving trains over it, danger to stock kept on the farm, and many other things may be considered, as well as the actual increase or decrease in the market value of the farm occasioned by the road. *Ch. & Iowa R. R.* v. *Hopkins*, 90 Ill. 316.

641. BENEFITS—*increase of value by improvement.* In estimating the compensation to be paid for land taken for a public park, the jury may consider the location and situation of the land at the time of the taking, without regard to the possible increase of value thereafter by reason of the prospective improvement in the vicinity. *South Park Coms.* v. *Dunlevy*, 91 Ill. 49.

642. PARTIAL TAKING—*market value of part taken.* In every case of a partial taking, the proper inquiry is as to the true value of the part taken, without regard to whether the remaining part is benefited or damaged. If the part taken is of such size and shape as to be available for purposes of business or habitation, and by reason thereof has a market value, that must control. If it is of such a size and shape as not to be available for either of these purposes, then its relative value as a part of the entire lot, and other considerations must be looked to in determining its actual value. *Green* v. *Chicago*, 97 Ill. 370.

643. DAMAGES TO PART LEFT—*relative value as an entirety.* When the owner claims compensation for damages to the part not taken, its value after such taking as compared with the value of the entire lot before the taking, is not only an important, but a necessary factor in determining what, if any, compensation he is to receive. *Ib.*

· **644.** BENEFITS—*not allowed against value of land taken—measure of value.* Where land is taken for a public improvement, the compensation required to be made to the owner by both the statute and the constitution, is the value of the land taken, without regard to any supposed benefits or damages that may result to adjacent property by reason of the proposed improvement, and the compensation in no case should be less than the land will sell for in a fair and open market, when it has a marketable value. *Ib.*

645. BENEFITS—*location of a highway.* The owner of land condemned for a highway is entitled to be paid in money for the full value of the land actually taken, and he cannot be paid therefor in benefits to result from the laying out of the highway. As to damages to the remaining land he may be thus compensated. *Hyslop* v. *Finch,* 99 Ill. 171.

646. SAME—*railway in street—set off.* Where a lot is divided by a street through the same, benefits to one part of the property cannot be set off against damages to the other part on the other side of the street by the laying of railroad tracks in the street so as to prevent access to the same and excluding ordinary travel on the street. *Pittsburg, Ft. Wayne & Ch. R. R.* v. *Reich,* 101 Ill. 157.

647. TO LESSEE—*future profits of land too uncertain.* In a proceeding to condemn land for a right of way, the jury allowed a lessee of the land taken, whose lease had three years to run, the amount of rent he was to pay per acre for the whole term, as to the land condemned, he contending that for gardening purposes it might yield much more. There was no proof that it would be used for such purpose, and no other damages were shown, and it appeared that the lessee had the option of terminating the lease at any time: *Held,* that the verdict would not be set aside as against the evidence, and that future profits of the land taken were too uncertain to be depended upon as a measure of damages. *Booker* v. *V. & C. Ry.,* 101 Ill. 333.

648. DAMAGES FOR TAKING FOR A TELEGRAPH—*excessive.* A telegraph company sought to condemn a strip of land 18 inches wide and 3 feet deep every 150 feet from the point of beginning, of sufficient width to erect telegraph poles to be set along the line of the right of way of a railroad which was fenced, and the proof showed that there would be eleven poles on defendant's land which was worth $60 per acre. Three witnesses of the defendant testified that the damages would be $10 a pole, arising from their interfering with the use of farming implements, while three witnesses for the petitioner testified, one that fifty cents, and the other two, that $1 a pole would be full compensation, and that when the poles pursued the line of the right of way of the railway company, as in this case, in their judgment there could be no other damage than the value of the land taken. It also appeared that a strip of land six feet wide across the whole tract would be two-tenths of an acre of the value of $12 at $60 per acre, and such a strip 18 feet wide would be but six-tenths of an acre, and worth but $36. The jury gave the defendant $38.50. *Held,* that the verdict was manifestly too high. *Mut. Union Tel. Co.* v. *Katkamp,* 103 Ill. 420.

649. FARM LAND—*fencing road first six months.* In a proceeding to condemn land for a railroad across a farm, the court instructed the jury for the land-owner, that under the law, the company was not bound to fence its road until six months after its completion, and that in estimating the damages, the jury might consider the damage the keeping open of the road for that time would be to the farm: *Held,* no error. *St L., J. & S. R. R.* v. *Kirby,* 104 Ill. 345.

650. SAME—*farm thrown open.* The inconvenience of having one's land temporarily thrown open in the progress of constructing a railway over the same, may be a material element of damage and justly require compensation. *Ib.*

651. TRAINING TRACK—*profits from nature of use.* The value of a training track which will be destroyed, is allowable as a part of the compensation to be paid for the right of way through the farm. The value of land consists in its fitness for use, present or future; and before it can be taken for public use the owner must have just compensation. If he has adopted a peculiar mode of using the land by which he derives profit, and he is deprived of that use, justice requires that he be compensated for his loss. It is the value which he has and of which he is deprived, that must be made good to him. *St. L., J. & S. R. R.* v. *Kirby,* 104 Ill. 345.

652. BENEFITS—*set off against damages.* In assessing the damages to another portion of a farm, aside from the value of the land taken for a right of way for a railroad, the jury should consider the road as running only through the farm, and not consider any general benefit which the road may prove in making a better market or convenience for travel; and in some cases they would be justified in estimating the damages to the farm the same as though the road commenced on one side of it and ran across to the other side and no further. *Ib.*

653. FARM LAND—*dividing farm.* The inconvenience of carrying on a farm divided into two parts by a railroad, is a legitimate element of damages to be considered by the jury in assessing damages for right of way, although such damages may be largely conjectural and not susceptible of anything like definite ascertainment. *McReynolds* v. *B. & O. R. Ry.,* 106 Ill. 152.

654. But damages from danger in crossing the road with teams and from danger to children and members of the family of the owner, are so unreliable and uncertain as not to form a proper basis in the assessment of damages. The assessment should be confined to such damages only as are reasonably probable. *Ib.*

655. BENEFITS—*instruction to find only for land taken.* It is not error to instruct on the assessment of damages, that if by the construction of the road the defendant's lands will be specially benefited, the jury should find only the compensation for the land actually taken, when there is evidence on which to base such an instruction. *McReynolds* v. *B. & O. R. Ry.,* 106 Ill. 152.

656. LAND TAKEN—*cash value.* In a proceeding to condemn land for a railroad depot, the cash value of the property is the only proper measure of damages. All evidence tending to show that value is proper, and all evidence tending to enhance the damages above, or reduce them below that sum, is improper. *J. & S. E. Ry.* v. *Walsh,* 106 Ill. 253.

657. SAME—*elements of damages—business and profits.* In such a case, the purpose for which the property was used and designed, its location and advantages as to situation, are proper matters for the consideration of the jury; but the profits of the business part, and conjectural profits for the future, are too speculative and uncertain upon which to ascertain the market or cash value of the property. The evidence should be confined to the market value of the property, and all evidence of the amount of business that was or could be done on it, or the probable profits arising therefrom, should be rejected. *Ib.*

658. COST OF IMPROVEMENTS. The question of the cost of erecting such buildings as were upon the premises is not an element of damages, unless it is shown that they would actually increase the value of the premises to the extent of their cost. Such improvements may or may not enhance the value of the land to the amount of their cost. The true question is, not what the property cost, but for how much would it sell. *Ib.*

659. LEASEHOLD PROPERTY—*damage to property as an entirety.* Where a party is using fourteen lots as an entirety, holding four of them under a lease for two years, and owning the other ten, in a proceeding to take a portion of the leasehold lots for a right of way, if the market value of the whole tract is lessened for the two years the lease has to run, the owner and occupant should be allowed damages to the extent that the market value of the entire property was thereby depreciated. *Ch. & E. R. R.* v. *Dresel,* 110 Ill. 89.

660 SAME—*loss of profits in business.* On application to condemn for right of way a part of four lots held by the defendant under a lease, which lots were occupied by him in connection with adjacent lots, of which he was the owner, and which were all used in carrying on an extensive hot-bed system of flower gardening, the court instructed the jury that there could be no recovery for loss of business or loss of profits: *Held,* correct. *Ch. & Evanston R. R.* v. *Dresel,* 110 Ill. 89.

661. LOSS OF BUSINESS AND PROFITS. On application to condemn a part of four lots held by defendant under a lease, and which, with other lots, are used in carrying on a hot-bed system of flower gardening, no recovery can be had for loss of business or loss of profits. *Ib.*

662. LAND TAKEN—*market value—special use.* The true test is the market value of the property taken for any purpose to which it is adapted or may be applied. If the lots are in use for market gardening purposes, and are more valuable for that than for any other purpose, the owner has the right to show that fact. No error to admit proof of the value of manure or compost on the land per load. *Ch. & E. R. R.* v. *Jacobs,* 110 Ill. 414.

663. The case of *L. S. & M. S. R. R.* v. *Ch. & W. Ind. R. R.,* 100 Ill. 21, fixing the damages for lots condemned for a right of way on a different basis than their market value, is to be applied only to property which, in its use or condition, has no market value. *Ib.*

664. SAME—*market value—instruction misleading.* On the assessment of damages for lots sought to be taken, the court instructed the jury that if they found from the evidence that there was no market value for such property in such condition, they should determine the actual value from the evidence in the case: *Held,* misleading and erroneous. The jury should have been so instructed as to direct their inquiry to the market value of the property. *Ch. & E. R. R.* v. *Jacobs,* 110 Ill. 414.

665. STRUCTURE ON LAND—*put on by company under license.* Where a railway company under license of the life tenant enters upon land and constructs its road over the same with costly embankments, and enjoys the use of the same without objection, on the application of the company after the termination of the life estate, to condemn a strip of land on which such road and structures are built, for a right of way, the law will not require it to pay the owner of the land for the structures so placed upon the same at its own expense. *C. & A. R. R.* v. *Goodwin,* 111 Ill. 273.

666. A railway company seeking a condemnation of land for a right of way already occupied by it, is not required by law to pay the land owner for structures placed upon the land at its own expense with a view of subsequently acquiring the right of way, even though its original entry may have been without license or tortions. *Ib.*

667. So, in a proceeding by a railroad company to condemn a strip of land for right of way then and previously occupied by it, and upon which strip the company had before constructed its road, consisting of costly embankments and structures, the court instructed the jury that in estimating the compensation of the owner, they should consider the whole property, including all the structures upon it, as well as the soil to which they were affixed, and award such sum as compensation as said property was reasonably worth for the purpose for which it was intended, although of no practical value to the land owner in connection with his farm: *Held*, that the instruction was erroneous in requiring more than a just compensation. *Ib.*

668. TRESPASS—*no damages for.* In a proceeding to condemn land for a right of way, the land owner cannot recover damages for a prior trespass by entering upon his premises. *Ib.* and *L. B. & M. R. R.* v. *Winslow,* 66 Ill. 219.

669. LAND FOR DEPOT—*special value beyond market.* On a proceeding to condemn lots for a depot and other railroad uses, the defendant offered to prove that the property had a special value beyond its general market value, and also that certain prices had been offered for the property within a few months of the time of the trial, above the general market value, all of which was excluded: *Held*, that the court erred in excluding the proposed evidence. *Johnson* v. *Freeport & Miss. River Ry.,* 111 Ill. 413.

670. If property has a special value from any cause that value belongs to the owner, and he is entitled to be paid it by the party seeking condemnation. *Ib.*

671. LAND TAKEN—*improvements on.* If the land sought has upon it an improvement which materially adds to its market value, the owner will have the right to show its character and extent and its value, for the purpose of enhancing its market value. The market value is not confined to any one particular use, but the value for any purpose for which the land may be adapted, may be shown. *De Buol* v. *Freeport & Miss. River Ry.,* 111 Ill. 499.

672. SAME—*abandoned improvements of another company.* Where some grading and excavations have been made on defendant's land by a different railway company and abandoned, and such improvement is sought to be taken in a proceeding to condemn, it is error to refuse to allow the defendant to testify how many cubic yards of grading and filling are on the land, and the present value of the grading and filling on the line of the proposed road over his land. *Ib.*

673. SAME—*value to owner.* The value of the land to the railroad company seeking its condemnation, is not a matter to be considered in estimating the damages to be allowed, as the value of the land to the petitioner, whether great or small, cannot affect the true compensation which the owner is entitled to receive. *Ib.*

674. SAME—*market value—uses of land—profits.* In ordinary cases the question to be determined is the market value of the land to be taken, and in order to arrive at that value, it is proper to show that the land is valuable for grazing, for raising corn, wheat, oats, grapes or any other product for which it may be used; but the probable profits arising from a wine cellar or otherwise, are too remote. *Ib.*

675. So, in a proceeding to condemn land used as a farm and a vineyard, it is proper to instruct the jury not to take into consideration the profits of the land-owner in his business, in estimating the damages. *Ib.*

676. Where the proposed right of way took the defendant's wine cellar, the court refused to let him testify what damage he would suffer by the taking of his cellar, but he was allowed to testify to the value of his land for any and all purposes: *Held*, no error in refusing the evidence as to the cellar. *Ib.*

677. FARM LAND—*incidental damages—dividing farm.* Where a strip of land through a farm is sought to be condemned, it is error to instruct the jury that "incidental damages" should not be considered by them, as being calculated to confine the jury to the value of the land actually taken. Damages may be allowed where one part of a farm is cut off from the other, and where it is rendered more inconvenient to reach a highway. Such damages may be regarded as incidental. *Ib.*

678. LAND TAKEN—*market value—instruction where it has none.* Where the property has a market value and is not devoted to any particular use making it more valuable to the owner than to any one else, such value affords the true measure of compensation; but where the proof tends to show the property has no market value by reason of the particular use to which it is applied, it is error to instruct the jury that the compensation should not be more nor less than its fair market value, and to refuse all instructions based on the theory it has no market value. *Ch. & N. W. Ry.* v. *Ch. & E. R. R.,* 112 Ill. 589.

679. SAME—*no market value—value, how found.* Where there is no market value of a piece of property by reason of its being used with and as a part of some extensive business or enterprise, its value must be determined by the uses to which it is applied. In such case the market value of neighboring lands differently circumstanced may be shown as throwing some light on the question, but it falls far short of furnishing a true or adequate test of the value of the property. *Ib.*

680. FARM LAND—*farm crossings.* In condemning land for right of way for a railroad across a farm, the necessities and conveniences of location for farm crossings should be taken into consideration; and after the condemnation they will be presumed to have been considered, and that damages were estimated upon the hypothesis that a farm crossing would not be constructed and maintained at any particular point where it would directly and seriously affect the safe and efficient operation of the road. *Chalcraft* v. *L., E. & St. L. R. R.,* 113 Ill. 86.

681. LAND TAKEN—*market value.* The measure of damages for land taken is its cash value at the time of the filing of the petition, if it has a market value. *Dupuis* v. *Ch. & N. W. Ry.,* 115 Ill. 97.

682. SAME—*profits—uses, as adding to value.* It may be true that the supposed profits arising from the business carried on upon the lands taken are not proper elements of damages, but it is also true that in determining the market value of such lands it is proper for the jury to consider the purposes for which the lands, were used, and in so far as the particular use to which the lands were or had been appropriated, added to their market value. An instruction which confuses these elements, and excludes both, is erroneous. *Dupuis* v. *Ch. & N. Wis. Ry.,* 115 Ill. 97.

683. NO MARKET VALUE—*worth of special use.* The correct measure of damages of land condemned is its market value, if it has one. But if devoted to some particular use which gives it an intrinsic value, the owner is entitled to receive its worth for such use or purpose. *Ib.*

684. LAND TAKEN—*cash value depending on use.* In order to determine the fair cash value of the lands taken, the jury may consider the purpose for which they are used—whether they are adapted to that use, and whether they are valuable or profitable for that use— and in so far as such use adds to their market value, this may be considered. *Ib.*

685. DAMAGE TO PART NOT TAKEN—*difference in value.* When other land of a party not sought to be taken is damaged by the right of way, the measure of damages as to it is the difference between its value before and after the construction of the road. *Ib.*

686. LAND TAKEN—*strip as of the value of the whole.* A strip of land often has a greater value as a part of the large tract of which it is a part than when considered alone; and it is proper for the court to tell the jury that if they find from the evidence that such is the case to allow such larger value in assessing the damages. *C. & E. R. R.* v. *Blake,* 116 Ill. 163.

687. PARTIAL TAKING—*value as part of the whole.* Where the part of the lot sought to be taken is of greater value as a part of the entire lot than as a distinct part, its compensation should be its fair cash value when considered in its relation to and as a part of the whole lot. *Ib.*

688. BENEFITS—*only special, set off.* Benefits or advantages which may accrue to the part not taken, in common with all other lands along the proposed railroad, cannot be set off or deducted from the compensation for the property taken and damaged. *Ch. & Evanston R. R.* v. *Blake,* 116 Ill. 163.

689. The question of damages in a condemnation proceeding is to be determined with reference to the special benefits only to the property not taken. Any mere general and public benefit or increase of value received by the land in common with other lands in the neighborhood, is not to be taken into consideration in assessing the damages. *Hyde Park* v. *Washington Ice Co.,* 117 Ill. 233.

690. DAMAGES—*destruction of pond for use of mill.* In a proceeding to condemn a strip of land for a railroad track which crossed a pond supplying the owner's steam mill with water, on the question of damages to property not taken, the defendant gave evidence on the basis that the pond would be destroyed as a source of supply of water for his mill, and there would be no other means of such supply. The petitioner then offered to show that a certain waterworks company would furnish the mill regularly with all the water it might require at a less cost than that of pumping from the pond, and also that a creek flowing nearer the mill than the pond had a capacity to furnish better water, and an abundance for the use of the mill, which the court refused to admit: *Held,* that the court erred in refusing the evidence. *Ill. & St. L. R. R. & Coal Co.* v. *Switzer,* 117 Ill. 399.

691. FOR PROPERTY PARTIALLY TAKEN—*special use as a pond for ice.* In a proceeding to condemn a strip of land for a street through premises made into a pond for freezing ice thereon, the proof showing that the property could not be devoted to any other use without a cost much in excess of its value, the court instructed the jury to ascertain from the evidence, after their own view, the fair market value of the property sought to be taken, and also the damages to the property from which the strip was to be taken, and that if they believed from the evidence the property of the defendant in its (then) present condition, had a special capacity as an entirety for the purpose of ice freezing, cutting and transporting, and as an entirety was devoted to such purposes, and that the value of such tract would be depreciated and lessened by the taking of the strip, then the owners of the property were entitled to receive a sum equal to such depreciation in value. *Held* correct. *Hyde Park* v. *Washington Ice Co.* 117 Ill. 233.

692. ENTIRE TRACT TAKEN—*injury to business—cost of removal.* Where an entire lot of ground upon which the owner is engaged in business is condemned for the use of a railway company, the cost and inconvenience of a removal of the business to some other place, are proper elements of compensation. *C. M. & St. P. Ry.* v. *Hock,* 118 Ill. 587.

—7

693. SAME—*market value.* Where an entire strip of land is taken for railroad purposes so that the owner has no adjoining property to be damaged, the measure of compensation is the market value of the property. *Ch. E. & L. S. R. R.* v. *Catholic Bishop*, 119 Ill. 525.

694. SAME—*particular use of the property.* When the owner of land elects to use it for one purpose rather than another, or assumes a restriction as to the character of use he will permit, in no wise binding on him by the nature of his tenure, this will not prevent his recovering in a proceeding to condemn the same, its value, from its capacity and adaptability for other uses. *Ib.*

695. So, in determining the market value of land sought to be taken, reference may be had not merely to the uses to which the land is actually applied, but its capacity for other uses, so far as the same may be shown by the evidence. *Ib.*

696. COMPENSATION—*when governed by particular use of the property.* When the owner of land is restricted by statute or by the provisions of the deed under which he holds title, or in any other binding way, to a particular use of it, so that he cannot lawfully apply it to any other use, the measure of his compensation will be its value to him for such special use. *Ib.*

697. SAME—*to a tenant.* A verdict in a proceeding to condemn land, which gives a tenant in possession of a part of the premises the full value of his improvements thereon, and also allows him to remove the same, is so manifestly unjust as to call for a reversal. *Ib.*

698. BENEFITS—*against land taken.* Since the present constitution came into force, the statute allowing the jury to consider or disregard benefits to the owner in the matter of laying out roads, does not apply to the matter of damages for taking the land. *Deitrick* v. *Highway Comrs.,* 6 Bradw. 70.

699. FARM LAND—*dividing same.* When a railroad crosses a farm, the inconvenience in operating the farm thus divided, is proper to be considered in fixing the damages to the part not taken. *L., E. & St. L. Ry.* v. *Chalcraft*, 14 Bradw. 516.

700. TENANT—*good will in business.* Whether a tenant will be entitled to damages for loss of good will in his business, is not decided. If the jury are of the opinion that the evidence establishes any damages of that character they may possibly assess them in a separate item; but evidence of such damages cannot be resorted to in support of a general assessment of damages for other items of property taken. *Chicago* v. *Garrity,* 7 Bradw. 474.

701. OBSTRUCTING STREET—*elements of damages—condemning railroad track.* In condemning a right of way across a previously constructed railroad in a street, the total obstruction of the old road while the tracks of the new one are being laid, and the permanent interference, by means of the crossing, with the business of the old road, are proper elements of damages. *Ch. & W. Ind. R. R.* v. *Ch., St. L. & P. R. R.*, 15 Bradw. 587.

702. ADDITIONAL DAMAGES—*change of plan.* Land owner has a claim for additional damages caused by a material change of the work. *W., St. L. & P. Ry.* v. *McDougall*, 118 Ill. 229; *J. & S. R. R.* v. *Kidder*, 21 Ill. 131; *P. & R. I. R. R.* v. *Birkett*, 62 Ill. 332.

703. DAMAGES TO TENANT. A tenant whose term expires during the proceeding, and whose lease secures him no right of renewal, cannot acquire any new rights in the property adverse to the petitioner. Any rights acquired by him thereafter are subordinate to the rights of the petitioner. *Schreiber* v. *Ch. & Evanston R. R.*, 115 Ill. 340.

704. SAME. If property is taken before the expiration of the term, the tenant is entitled to compensation therefor, but if he enjoys the property for the entire term before compensation paid, he will be entitled to none. *Ib.*

705. DAMAGE BEFORE ASSESSMENT—*lawful acts—remedy.* Where a railway company in exercising the right of eminent domain, commits an injury to the land of another, by entry upon it to make preliminary surveys, or by taking materials, therefrom, or the like, in pursuance of the powers vested in it, and the law under which it acts, prescribes a mode of assessing damages for such injuries, an action of tort will not lie therefor, but the statutory remedy must be pursued. But this is only where the authority conferred has been followed. *Smith* v. *Ch., A. & St. L. R. R.,* 67 Ill. 191.

706. DISTINCTION BETWEEN COMPENSATION AND BENEFITS. The act of 1877, concerning roads and bridges in counties under township organization, as respects the matter of awarding compensation and assessing benefits, makes no discrimination between the value of the land actually taken and damages otherwise resulting to the land owner in consequence of laying a highway. But the eminent domain act of 1874 does make such discrimination and is to be construed *in pari materia* with the former act, supplementing the same. *Hyslop* v. *Finch,* 99 Ill. 171.

707. FARM LAND—*fencing.* In a proceeding by a railway company to condemn a right of way through farm land, it is proper for the court to instruct the jury that the company is not required to fence its road for six months after the same is open for use, and that the damages attending the keeping open of the right of way for that length of time, may properly be considered as an element of damages. *Centralia & Chester R. R.* v. *Rixman,* 121 Ill. 214.

708. BENEFITS. It is competent to consider special benefits to property claimed to be damaged, but not taken, for the purpose of reducing, or rather to the extent of the special benefits, of showing there are no damages. *Concordia Cem. Asso'c.* v. *Minn. N. W. R. R.,* 121 Ill. 199.

709. SAME—*instruction.* Where the court instructed the jury that the defendant was entitled as compensation to the cash market value of his land sought to be taken as of the date of the petition, and damages to the remainder of his land described in his cross petition, and then instructed that the total compensation and damages to which the defendant was entitled, must be equal to, but must not exceed the difference between the fair market value of the whole land described in the petition and cross petition as it was on the date of the petition, and the fair market value of what remained after the taking of part by the petitioner and the appropriation thereof to its use. The jury awarded $2,380 for the land taken and $6,450 for damages to land not taken: *Held,* that it must be presumed that the $6,450 was in excess of any and all special benefits to the lands damaged and not taken, and consequently no benefits were allowed against the value of the land taken. *Ib.*

710. MARKET VALUE. The proper measure of damages in the case of the location of a railroad over a farm, is the actual fair cash value of the land taken and the decrease in the actual fair cash value of that not taken. *Kiernan* v. *Ch., Santa Fe & Cal. Ry.,* — Ill. —. Filed Nov. 11, 1887.

711. BASIS OF ASSESSMENT. In assessing the value of land taken, and the damage to the remaining part, the jury should not assess the same on the basis of what the owner would take for the same or any part thereof, or what the jury would take if they were the owner. *Ib.*

712. DAMAGES TO PART NOT TAKEN—*speculative.* In assessing

the damages to the land not taken, the jury should not take into consideration anything as an element of damages which is remote, imaginary or speculative, even though testified to by witnesses. The only elements they should consider are those which are appreciable and substantial, and which will actually lessen the market value of the land, and the jury may be so instructed. *Ib.*

713. MARKET VALUE. The fair market value of land proposed to be taken, having proper regard to the location and advantages as to situation and the purposes for which it was designed and used, is the proper measure of compensation. *C., B. & Q. R. R.* v. *Bowman*, — Ill. —. Filed Nov. 11, 1887.

714. PART TAKEN HAVING A VALUE AS A WHOLE. Where a part is taken, and that part has a greater value in connection with the whole than as a separate parcel, the measure of damages will be the fair cash value of the part taken, as a part of the whole. *Ib.*

715. DAMAGES TO THE PART NOT TAKEN. *On cross petition.* Where a cross petition is filed for damages to land not sought to be taken, the jury should award to the owner such damages in cash as his lands not taken will sustain, if any, by the construction of the proposed railroad and its continued use and operation through his farm. In such case it is proper for the jury to give damages for all actual and appreciable injuries resulting from the construction and operation of the road. *C., B. & Q. R. R.* v. *Bowman*, — Ill. —. Filed Nov. 11, 1887.

716. SAME—*difference in market value.* If the land not taken will be depreciated in value, the measure of damages will be the difference in their market value before and after the construction of the road. In determining this, the jury may consider the injury to the land arising from inconveniences actually brought about by the construction of the proposed road, or incidentally produced by dividing the land as to water, pastures and improvements, although such injury may not be susceptible of definite ascertainment, and also for such incidental injury as will result from the perpetual use of the track for moving trains, or from danger of killing stock, or escape of fire, and generally for such damages as are reasonably probable to ensue from the construction and operation of the road. *Ib.*

717. EVIDENCES OF DAMAGE. The physical condition of land over which a right of way is sought for a railroad, whether effected by another railroad, a water course, or other natural or artificial object, must be considered, not in respect to the damage or depreciation caused by such other railroad, water course, &c., but for the purpose of determining the damages occasioned to the owner by the proposed improvement. *Ib.*

718. While it is true that only real, tangible and proximate damages are recoverable, yet it is all such damages as are reasonably probable, as distinguished from possible, speculative or remote damages that form the proper basis of recovery. *Ib.*

719. CASH VALUE—*basis of assessment.* In a proceeding to condemn for a public use the compensation and damages to be awarded the owner must be based upon the fair cash value of the land at the time of the comdemnation. *Cal. Riv. Ry.* v. *Moore*, — Ill. —. Filed March 26, 1888.

720. ASSESSMENT—*matters for the finding of the jury.* The questions ordinarily to be found by the jury are: (1) What is the present market value of the land taken; and, (2) to what extent, if at all, will the remainder of the tract be depreciated in its market value by reason of the appropriation of the part taken for the proposed use. *Cal. Riv. Ry.* v. *Moore*, — Ill. —. Filed March 26, 1888.

721. USES AND CAPABILITIES—*an element of value.* The compensation is to be estimated with reference to the uses for which the property is suitable in its then condition, having regard to its location, situation and quality, and to the business wants in that locality, or such as may reasonably be expected in the near future. *Cal. Riv. Ry.* v. *Moore,* — Ill. —. Filed March 26, 1888.

722. PROSPECTIVE VALUE—*possible future demand—too remote.* If lots abutting upon a river are suitable for dock purposes, of which there is no present demand, their value when improved by the building of docks, the profits that may be derived therefrom, or the value of the lots at some future time, as when business or the wants of the community may make profitable the making of docks or slips in the lots, is merely conjectural and remote, forming no proper element in estimating the damages to be paid. *Cal. Riv. Ry.* v. *Moore,* — Ill. —. Filed March 26, 1888.

723. WHEN FUTURE USE MAY FORM AN INGREDIENT OF VALUE. If the fact that lots are located with a frontage on a river, at a place where they can at some future time, when demanded, be made available as dock property, enhances their present market value in their present condition and state as to improvement, that fact will be proper to be shown and considered by the jury on the assessment of the damages. *Ib.*

724. In such case it can make no difference that there may be no present demand for docks upon the lots, if in consequence of their supposed adaptation to such use they have an increased market value above what they otherwise would have. Such value may form a proper basis of a recovery. *Ib.*

OF THE EVIDENCE ON ASSESSMENT.

725. VALUE OF LOTS—*opinions of witnesses.* Lands and city lots have no standard value, and to arrive at their proper valuation it is right to take the opinion of witnesses and to hear the facts upon which such opinions are founded. *I. & W. R. R.* v. *Von Horn,* 18 Ill. 257.

726. WITNESSES—*credibility and weight.* In estimating damages for a right of way across a farm, where there is a conflict of evidence as to the damages, the jury will be justified in giving greater weight to the testimony of farmers than to that of persons engaged in other pursuits. *J., A. & St. L. R. R.* v. *Caldwell,* 21 Ill. 75.

727. PLANS AND ESTIMATES—*of work.* On an assessment of damages resulting from the construction of a railroad over a farm, the plans and estimates of the company for that part of the road should be admitted in evidence. *J. & S. R. R.* v. *Kidder,* 21 Ill. 131.

728. The company will be bound to construct the road substantially according to the plans and estimates thus given in evidence. Should it deviate from them so as to cause additional damages, they may be recovered by the land-owner in an action on the case, or a court of equity may enjoin the work until such damages are assessed and paid. *Ib.*

729. PLANS—*explaining.* The engineers and officers of a railway company on the assessment, may be examined for the purpose of explaining the plans and estimates for the construction of the road. *Ib.*

730. VERBAL PROMISES OF AGENTS. The verbal representations and promises of the engineer of the company and others, which may not be binding on the company, should not go to the jury to influence their finding, unless sworn to and in proper explanation of the plans for constructing the road. *Ib.*

731. CONTRACT TO FENCE ROAD. It is error to refuse evidence on the part of the petitioner that, at the time of the trial, it was in the act of building a fence along its right of way; that the lumber and posts were on the ground and the contract let to build the fence. *St. L., J. & Ch. R. R.* v. *Mitchell,* 47 Ill. 165.

732. STIPULATION TO BUILD DEPOT. A stipulation of a railway company seeking a condemnation, that it will erect a depot near the land, is admissible in evidence in behalf of the company, although the location of the depot had not been fixed before the trial. *Hayes* v. *O. O. & Fox River Valley R. R.,* 54 Ill. 373.

733. OPINIONS OF WITNESSES—*as to benefits.* Upon the question of damages and benefits arising from the construction of a railroad over a tract of land, the opinions of witnesses are admissible as to the benefits that will probably result to the land by the location of a depot within a certain distance of it. *Ib.*

734. CHANGE OF PLANS—*additional damages.* The company must construct its road as indicated by its maps and plans introduced on the trial. A subsequent alteration will give the land-owner the right to recover for damages resulting therefrom. *P. & R. I. R. R.* v. *Birkett,* 62 Ill. 332.

735. CROSS-EXAMINATION—*as to other matters.* A witness having testified to the damages to a particular tract of land touched by the track of a railroad company, cannot on cross-examination be required to testify to the effect upon other tracts owned by the same party. *P., P. & J. R. R.* v. *Laurie,* 63 Ill. 264.

736. OF TRESPASS AND VIOLENCE. On the assessment of damages for the right of way, it is error to admit evidence of the violent entry upon the premises by the agents and servants of the company, showing a willfull trespass, and the error is not cured by instructing the jury to disregard it. *L. B. & M. R. R.* v. *Winslow,* 66 Ill. 219.

737. OPINIONS AS TO VALUE. In a proceeding to condemn land and city lots for railway purposes, it is necessary and proper to take the opinions of witnesses, and to have the facts upon which such opinions are founded, to enable the jury to fix the compensation. *Ib.*

738. Where witnesses are allowed without objection to give their opinions as to the extent of the damages in a proceeding to condemn, as well as to testify to the facts, the jury may rightfully consider such evidence. *R., R. I. & St. L. R. R.* v. *Coppinger,* 66 Ill. 510.

739. DEEDS—*as evidence of value.* Where the land-owner gave in evidence the deeds for his land, it was held no ground for reversal to instruct the jury for the petitioner that they could take into account the consideration recited in the deeds in determining the value of the land taken. If the land had been recently purchased, the price paid might tend to enlighten the jury upon that issue. *Jones* v. *C. & I. R. R.,* 68 Ill. 380.

740. VIEW OF LAND—*treated as evidence—instruction.* In a proceeding to condemn land for a right of way, under a law allowing the jury to view the premises, it is not improper to instruct the jury to fix the compensation from the evidence, as the facts learned by the examination is part of the evidence upon which the jury may act. *P. A. & D. R. R.* v. *Sawyer,* 71 Ill. 361.

741. OPINIONS OF WITNESSES—*as to damages.* Witnesses may give their opinion as to the amount of damages occasioned to the owner of land by the construction of a railroad; and where they possess peculiar knowledge of the facts, such evidence is often valuable. *G. & S. Wis. R. R.* v. *Haslam,* 73 Ill. 494.

742. On an assessment of damages under a proceeding by a railway to condemn the right of way through a farm, it is competent for witnesses who are acquainted with the farm and familiar with the use and production of such property, and its value, to give their opinion as to the extent of the damages which the construction of the road over the same will occasion. *K. & E. R. R.* v. *Henry*, 79 Ill. 290.

743. EVIDENCE AS TO VALUE. If the land has a market value for the purpose of sub-division into lots and blocks, it may be properly proved. The jury may take into consideration each and every element that may enter into the true market value of the property. *South Park Coms.* v. *Dunlevy*, 91 Ill. 49.

744. The amount of compensation for land taken is a question of fact to be found by the jury from an actual survey of the premises, where that is practicable, their own knowledge of values and the opinions of witnesses who are familiar with the subject of inquiry, and whose business in life has afforded them opportunities of acquiring information and judging accurately upon the question. *Green* v. *Chicago*, 97 Ill. 370.

745. JURY NOT CONFINED TO OPINION. While it is proper on the examination of witnesses as to the value of property sought to be condemned for public use, to call out the various theories and processes upon which their conclusions are based, to ascertain their correctness, yet the jury after all must determine the question of value according to their own judgment of what seems to be just and proper from all the evidence before them. *Ib.*

746. OPINIONS—*jury not bound by.* The opinions of witnesses upon the question of damages in a proceeding to condemn, are not to be passively received and blindly followed, but they are to be weighed by the jury and judged in view of all the testimony in the case, and the jury's own general knowledge of affairs, and have only such consideration given to them as the jury may believe them entitled to receive. *McReynolds* v. *Burlington & Ohio River Ry.*, 106 Ill. 152.

747. EXPERTS—*weight of their evidence.* In the assessment of damages the jury will be warranted in giving but slight, if any weight, to the evidence of mere experts, based simply on theory and conjecture as to the damages the construction of a railroad between an elevator and a river, would be to the owner of the elevator *P. & P. U. Ry.* v. *P. & F. Ry.*, 105 Ill. 110.

748. OPINIONS—*weight—competency.* Persons familiar with land sought to be condemned who have opinions of its value, though not shown to be experts, are competent witnesses to express their opinions. But the weight of such evidence presents a different question. On that point where there is equal credibility, superior opportunity and intelligence are entitled to the greater weight. *Johnson* v. *Freeport & Miss. River Ry.*, 111 Ill. 413.

749. Such opinions as to the value of the land are not however to be passively received and blindly followed, but should be weighed by the jury and judged of in view of all the evidence in the case and the jury's own general knowledge of affairs and have only such consideration given to them as the jury may believe them entitled to receive. *Ib.*

750. PLANS AND PROFILES—*production compelled.* Where land is sought to be condemned for a right of way over a river upon which the land abuts, and upon which to build an abutment for a railroad bridge across the river, and the owner (another railway corporation) has other lands adjoining that sought to be taken that may be injured more or less, depending upon the character and nature of the structure to be erected on the land sought to be condemned, it is error to refuse the defendant's motion to require the petitioner, before the trial is begun, to exhibit its plans and profiles of its proposed railroad across the land, and to file such plans as will show to what use the petitioner designs devoting the land it seeks to condemn, and what it proposes to put upon said land, as tracks, bridges, abutments or otherwise. *Ch. & N. W. Ry.* v. *Ch. & Evanston R. R.*, 112 Ill. 589.

751. EVIDENCE—*as showing value for special use.* Evidence showing that the lands are valuable as located, bordering on or near a river, for saw-mill, planing-mill or factory, or for any other purpose, is proper on the question of their market value. *Dupuis* v. *Ch. & N. Wis. Ry.*, 115 Ill. 97.

752. SAME—*of state of the improvements.* Where the value of a mill on property sought to be condemned for railroad purposes, is involved, evidence that the mill is of an old pattern that has gone out of use, and therefore less valuable, is proper on the assessment of the compensation to be awarded. *Ib.*

753. OPINION—*competency of witness.* Preliminary proof of personal knowledge of the witness as to the value of land, based on actual sales, is not indispensable. The lack of such acquaintance or proof thereof, goes to the weight rather than to the admissibility of the evidence. *C. & E. R. R.* v. *Blake*, 116 Ill. 163.

754. EVIDENCE—*plan of proposed building.* The plan of a proposed building rendered impossible by the taking, is inadmissible to prove future probable profits, and so enhance the damages, but it is proper to show the uses to which the property might be put. It should be so limited by the court. *Ib.*

755. EVIDENCE—*plan of proposed improvement by owner.* On the assessment of compensation and damages in a proceeding to condemn a railroad right of way across lots abutting upon a river, the court allowed the lot owners to give in evidence a plat of a proposed improvement on the property, showing water fronts of proposed docks along the river. The court in admitting the plat and in an instruction limited this evidence to the question of what uses the lots might be adapted: *Held*, no error. *Cal. Riv. Ry*, v. *Moore.* — Ill. —. Filed March 26, 1888.

756. OPINION—*competency to give.* Real estate brokers acquainted with the value of real estate in the neighborhood, are competent to give their opinion of the value of property sought to be condemned, although their knowledge is not shown to be based on actual sales. *Ch. & Evanston R. R.* v. *Blake*, 116 Ill. 163. •

757. PLANS OF THE ROAD—*preserving in record.* In a proceeding to condemn land for a right of way, it is competent on the question of damages for the company to show the plan of construction of its road over the premises sought to be taken. But where such plan will materially affect the question of damages, the plan should be presented and preserved in the records of the court; so that if there should be a departure from the plan to the defendant's injury he may have his remedy for any increased damages resulting from such departure. *Ill. & St. L. R. R. Coal Co.* v. *Switzer*, 117 Ill. 399.

758. CHANGE OF PLANS—*liability for.* While a purchaser of land cannot recover for an injury by the construction of a railroad over the same, yet if the company, after his purchase, adopts a new feature in the construction and operation of its road in the future by making an opening in an embankment for the passage of water, and constructing a bridge over the opening, such purchaser will, in a proceeding to condemn, be entitled to compensation for any damages growing out of the change or alteration in the nature of the work. *W., St. L. & P. Ry.* v. *McDougall*, 118 Ill. 229.

759. STIPULATION OF PETITIONER—*evidence on question of damages.* In a proceeding by one railway company to condemn a right of way across the track of another company, a stipulation or covenant of the petitioner, properly executed, that it will, at its own expense, put in and maintain in proper repair the frogs and crossings over two main tracks of the defendant company, expressed to be binding on its successors and assigns, is proper evidence for the petioner on the question of damages. *C. & A. R. R.* v. *Joliet, Lockport & Aurora Ry.*, 105 Ill. 388.

760. PROFILE OF GRADE OF STREET—*as evidence on question of damages.* In a suit by the owner of a house and lot to recover damages growing out of a change in the grade of a street, after the work is commenced and before its completion, the profile of the proposed improvement is proper evidence against the city. *City of Elgin* v. *Eaton*, 83 Ill. 535. As to plans, profiles, specifications, &c., being proper evidence, see also *Hyde Park* v. *Andrews*, 87 Ill. 229; *Peoria & R. I. R. R.* v. *Birkett*, 62 Ill. 332; *St. L., J. & Ch. R. R.* v. *Mitchell*, 47 Ill. 165; *Hayes* v. *O. O. & F. R. V. R. R.*, 54 Ill. 373; *Mix* v. *L. B. & M. Ry.*, 67 Ill. 319; *Wilkin* v. *St. Paul R. R.*, 16 Minn. 271; *Rippe* v. *Ch. R. R.*, 23 Minn. 18.

761. AVERAGING THE EVIDENCE. The jury may take an average of the testimony on the question of compensation or damages, if properly done by a consideration of all the elements and circumstances referred to in the law as proper, to aid in determining the weight of evidence, and they should not be told that they have no right to average the testimony without explanation. *Peoria & Rock Island R. R.* v. *Birkett*, 62 Ill. 332.

762. AVERAGING EVIDENCE. The jury have not the right to take the gross amount as sworn to and divide it by the number of the witnesses to obtain their verdict, unless there is afterwards full and free consultation and the judgment assents to the sum uninfluenced by any previous agreement. *P. & R. I. R. R.* v. *Birkett*, 62 Ill. 332.

763. DAMAGES—*when nominal.* The amount of the damages must be shown, not necessarily with precision, but approximately. If damage is shown but the amount is not approximately made to appear, no more than nominal damages can be allowed. *P., P. & U. Ry.* v. *P. & F. Ry.*, 105 Ill. 110.

764. OF THE USE OF THE LAND.—If property sought to be condemned by a railway company for a right of way is claimed by a cemetery company, it may be shown on the question of the compensation and damages that the land is not used for burial purposes and is not susceptible of being used for cemetery purposes. The owner of the land is entitled to have the highest price for which the same can be sold for any purpose. *Concordia Cem. Assoc.* v. *Minn. & N. W. R. R.*, 121 Ill. 199.

765. OF OTHER SALES. Evidence in regard to sales of prairie land one mile distant from the land sought to be condemned, may be received as tending in some measure to show the value of the land involved, where there is no evidence of any actual present-market value, nor of sales of like property nearer. Where the land sought is not laid out into lots and improved as cemetery property, proof of sales of other cemeteries is not competent evidence on the assessment. *Ib.*

766. OF OTHER SALES OF LAND. On the question of the damage of a railway to a farm, the defendant gave in evidence the opinions of witnesses as to the amount of the depreciation of its market value, and thereupon evidence was admitted in rebuttal to show how the selling values of other farms in the county crossed by railroads were affected: *Held*, that the latter evidence was improper. *Kiernan* v. *Ch., S. F. & Cal. Ry.*—Ill. —. Filed Nov. 11, 1887.

767. DAMAGE BY DIVERSION OF STREAM. If damages are claimed

for the division of a stream, evidence tending to show it was a receptacle of all the sewerage of a city near by, and had become so foul as to be worthless for stock water, is proper as bearing on the question of damages. *Ib.*

768. PERSONAL VIEW—*its weight*, The result of the jury's personal view of the land over which a railroad is sought to be laid, is proper evidence upon which they may act, and give it greater weight than the opinion of witnesses. *Ib.*

DAMAGES TO OTHER LANDS NOT DESCRIBED IN PETITION.

769. CROSS PETITION NECESSARY. On an assessment of damages to certain lots abutting upon a street caused by the location of a side track of a railroad in a public street, the owner will not have the right to prove damages to his entire land, consisting of many lots lying together, and with those named constituting an entire tract, unless he files a cross petition setting up that the other lots will be damaged. *Mix.* v. *L. B. & Miss. Ry.*, 67 Ill. 319.

770. The inquiry as to damages should be confined to the tract of land described in the petition in the absence of a cross bill by the defendant showing that he owns contiguous lands which will be damaged. *Jones* v. *Ch. & Iowa R. R.*, 68 Ill. 380.

771. Where the petition describes only one tract of the defendant's land, a portion of which the right of way cuts off from the entire farm, also consisting of another tract, the correct practice, in order to recover damages as to the whole, is to file a cross petition; but when this is not done, and the damages are assessed without objection to the whole farm, and the court protects the petitioner from further proceedings for the recovery of damages to the balance of the farm, by requiring the owner to execute a release as to it, the judgment will not be reversed for the error. *Galena & S. Wis. R. R.* v. *Birkbeck*, 70 Ill. 208.

772. Where the petition describes only one tract of the defendant's farm which is cut off from the rest, and damages are assessed in respect to that tract, the owner may afterwards cause the damages to be assessed as to the balance of the land. *Ib.*

773. Where a petition is filed to condemn land for right of way and there is no cross petition to include other land within it, it is improper to permit evidence to be introduced in regard to land adjoining that described in the petition and belonging to the same owner. *P. A. & D. R. R.* v. *Sawyer*, 71 Ill. 361.

774. The owner may by cross petition have the damages to his other contiguous land assessed in addition to the compensation for the land taken. *Stetson* v. *Ch. & E. R. R.*, 75 Ill. 74.

775. Where a part of a lot is sought to be condemned by a city for a street, damages as to the part not sought to be taken may be allowed without any cross petition by the owner. *Bloomington* v. *Miller*, 84 Ill. 621.

776. The ascertainment of the just compensation to the owner for taking away a part of his lot of necessity involves the consideration of the value of the whole property intact, and the value of that part not taken after the proposed part shall have been taken. *Ib.*

777. The petition need not describe the property not sought to be taken or damaged, and if other property is brought in by cross petition, it is incumbent on the defendant to show, in the first instance, that it was taken or damaged, and the petitioner is entitled to give evidence in rebuttal. *Hyde Park* v. *Dunham*, 85 Ill. 569.

778. Where the defendant filed a pleading, stating that "he is the owner of the lands mentioned in the petition and other lands contiguous thereto, making a farm of 730 acres in a compact body; that the railroad company takes about 12 acres out of his farm, dividing wood, water and timber from the balance of the farm; that the land thus taken is of the value of $150 per acre, and the damage by reason of the cutting the farm is $10,000; and he respectfully asks that this, his compensation, may be awarded to him as shall be just and proper:" *Held*, sufficient to answer the purpose of a cross petition for damages to contiguous lands, and gave the court jurisdiction as to the claim of such damages. *Ch. & Iowa R. R.* v. *Hopkins*, 90 Ill. 316.

779. Where the petition for right of way shows that the defendant is the owner of an entire tract of land, and that petitioner proposes to take a strip through the same, a cross petition is not necessary to enable the defendant to have damages assessed for land not taken. *Ill. Western Extension R. R.* v. *Mayrand*, 93 Ill. 591.

780. The evidence will be confined to the particular lands described in the petition, unless the defendant files a cross petition, setting up that he is the owner of other land not described in the original petition which will be damaged, and makes claim to have the damages thereto likewise assessed. *Ch. & Iowa R. R.* v. *Hopkins*, 90 Ill. 316.

781. Where the owner by cross petition claims damages to other parts of the same tract, an instruction confining the assessment of the jury to the strip of land actually taken, and excluding consideration of damages to the remainder of the farm, is properly refused. *Ib.*

782. CROSS PETITION—*right to file—defects, how reached.* On a petition to condemn land by a railway company, the defendant has a right to file a cross petition where his interests are not accurately or fully stated in the petition, and thereby recover compensation for damages to the adjacent property not sought to be taken, and it is error to strike such a petition from the files. If it be defective, or the property damaged is insufficiently described, or the cross petition does not show how the property will be damaged, the proper course is to demur to it, so as to afford an opportunity to amend. *Johnson* v. *Freeport & Miss. River Ry.*, 111 Ill. 413.

DAMAGES AS OF WHAT DATE.

783. ACT OF 1852—*facts at date of trial govern.* Under the act of 1852, in assessing the damages above the benefits, the jury is not confined to a consideration of the facts as they existed at the time the land was taken, but may consider the subject in the light of the facts as they exist at the time of the trial. *Hayes* v.*O.,O.& Fox River Valley R. R.*, 54 Ill. 373.

784. LAND FOR PARK—*value at date of condemnation.* In assessing damages for land taken for a public park, its value at the time of the condemnation should be considered, the owner being entitled to the benefit of an advance caused by the prospective establishment of a public park. *Cook* v. *South Park Comrs.*, 61 Ill. 115.

785. SAME—*suit against owner.* Where the public authorities in a proceeding to condemn land for a public park, have not acquired either the title or the possession of the land, it is error to award rent against the owner for the use of the premises from the date of the law, or time it took effect. *Ib.*

786. Where the witnesses on both sides in a proceeding to condemn property testified as to its value at the date of the institution of the proceeding, except three, and from their testimony it did not appear that the property was worth more at the time of the trial, it

was *held*, that a modification of an instruction confining the jury to its value at the first date, was not of sufficient importance to affect the right of the land-owner. *McAuley* v. *Col., Ch. & Ind. Cent. Ry.*, 83 Ill. 348.

787. TAKEN BEFORE CONDEMNATION—*advance in value.* Where land has been taken and occupied for railroad purposes prior to the institution of proceedings to condemn, the value of the land taken, at the time of the condemnation, is the value to be ascertained, the owner being entitled to any advance between that time and the actual taking of the land; and when the land is sold after its occupation for a right of way and before proceedings to condemn, the purchaser will be entitled to the advance in value. *Ch. & Iowa R. R.* v *Hopkins*, 90 Ill. 316.

788. DATE OF FILING PETITION. On petition to condemn land for public use, the compensation to be paid must be fixed by the valuation of the property at the date of the filing of the petition and not at the time of the trial. *South Park Coms.* v. *Dunlevy*, 91 Ill. 49.

789. Where compensation is paid, the rights of the petitioner relate to the time of filing the petition, and the amount of compensation is determined by the valuation at that time. *Schreiber* v. *Ch. & E. R. R.*, 115 Ill. 340.

790. The compensation to be paid is fixed by the value of the property taken at the time of the filing of the petition. *Ib.*

WHO ENTITLED TO DAMAGES.

791. SUBSEQUENT PURCHASER. Where a railway company without any authority locates and operates its road over a tract of land belonging to an estate, on a judicial sale, the whole land with the right of way will pass to the purchaser, and he will be entitled to compensation for the land taken and damages for any injury to the residue. *Ch. & Iowa R. R.* v. *Hopkins*, 90 Ill. 316.

ASSESSMENT COVERS ALL FUTURE DAMAGES.

792. BAR TO FURTHER ACTION. All damages, present and prospective, resulting, or to result to the land owner from the proper construction, maintenance and operation of a railroad over or upon his land, constitute one single, indivisible cause of action, whether enforced under the eminent domain act, or by action. After the recovery of damages for right of way the land-owner and his subsequent grantee are barred as to any subsequent damages that might have been reasonably anticipated. *O. & M. Ry.* v. *Wachter*, — Ill. —. Filed Jan. 20, 1888.

793. Where a right of way is condemned for public use over a tract of land, the owner will be entitled to compensation, not only for the value of the land taken, but also for all damages to the residue of the tract, past, present and future, which the public use may thereafter reasonably produce. *C., R. I. & P. Ry.* v. *Smith*, 111 Ill. 363.

794. GRANT OF RIGHT OF WAY—*increased use.* The grant of a right of way to a railway company "for all uses and purpose, or in any way connected with the construction, preservation, occupation and enjoyment of said railroad," is broad enough to embrace all uses for railroad purposes, however much increased and by other companies authorized by law. *Ib.*

795. RECOVERY—*when a bar to future damages.* In an action for deterioration in the value of real estate from a nuisance of a permanent character, all damages for past and future injury may be recovered, and one recovery is a bar to all future actions for the same

cause. *Ottawa Gas Co.* v. *Graham,* 28 Ill. 73; *I. C. R. R.* v. *Grabill,* 50 Ill. 244; *Cooper* v. *Randall,* 59 Ill. 321; *Decatur Gas Co.* v. *Howell,* 92 Ill. 19; *C. & A. R. R.* v. *Maher,* 91 Ill. 312; *C. & E. Ill. R. R.* v. *McAuley,* 121 Ill. 165; *Troy* v. *Cheshire R. R.,* 3 Fost. N. H. 83; *Stodghill* v. *C., B. & Q. R. R.,* 53 Iowa 343; *Powers* v. *Council Bluffs,* 45 Iowa 652; *C. & E. Ill. R. R.* v. *Loeb,* 118 Ill. 209; *Fowle* v. *N. H. & N. R. R.,* 112 Mass. 334; *Kansas R. R.* v. *Mihlman,* 17 Kan. 224; *Fowle* v. *N. H. & N. R. R.,* 107 Mass. 352; *Warner* v. *Bacon,* 8 Gray 397; *I. C. R. R.* v. *Allen,* 39 Ill. 205; *C., B. & Q. R. R.* v. *Schaffer,* — Ill. —. Filed March 28, 1888.

RELEASE, AS A BAR.

796. CONTRACT FOR. Where a party executes a contract with a railway company, agreeing to release and convey a right of way for its road over any land owned by him, as soon as the road is located, he will not be entitled to any damages by the construction of the road over any of his lands. *Conwell* v. *Spr. & N. W. R. R.,* 81 Ill. 232.

797. CONSTRUCTION OF. Where a deed is given a railway company for a right of way 100 feet wide through the grantor's land, releasing all claim for damages by reason of the location and completion of the road over the same or any part thereof, it will confer the same right on the grantee as it might have acquired by condemnation, and an immunity from all damages that the grantee might have claimed. *St. L., V. & T. H. R. R.* v. *Hurst* 14 Bradw., 419.

798. Unless the acts complained of were a departure from or were not embraced in the purposes for which the deed was given there can be no recovery, and it is error to refuse to admit such deed in evidence. *Ib.*

DAMAGES TO PROPERTY WHERE NONE OF IT IS TAKEN.

799. CONSTITUTION OF 1848. Under the constitution of 1848 and the statutes in force in March, 1870, a party is not entitled to damages by reason of the construction of a highway adjoining and abutting against his land, when no part thereof has been taken. *Hoag* v. *Switzer,* 61 Ill. 294.

800. At that date the commissioners of highways had no power to assess or award consequential or remote damages to a party by reason of the construction of a highway, when no part of his land was taken. The road law of 1861,. §§ 55, 56 and 68 does not conflict with this view. *Ib.*

801. DAMAGES CONTEMPLATED. The word "damaged" in this clause of the constitution is used in its ordinary and popular sense, which is "hurt," "injury" or "loss." The damage contemplated is an actual diminution of present value, or of price, caused by the construction of the road, or a physical injury to the property that renders it less valuable in the market. *Ch. & P. R. R.* v. *Francis,* 70 Ill. 238.

802. SAME—*depreciation of value.* Where the property is not taken, the damages must be real and not speculative. If the property is not worth less in consequence of the construction of the railroad in its vicinity, or upon a street upon which the lots abut than if no road were constructed, the owner will not be entitled to damages. *Ib.*

803. The words in the act of 1872 "which may damage property not actually taken," relate to contiguous lands of the same owner, a part of which only is taken. The damages to land not taken must be direct and physical and result from the taking of a portion of his land. *Stetson* v. *Ch. & E. R. R.,* 75 Ill. 74.

804. The constitution of 1870 was intended to afford redress in cases not provided for before, and embraces every case where there is a direct physical injury to the right of use or enjoyment of private property, by which the owner sustains some special damage in excess of that sustained by the public generally. *Rigney* v. *Chicago*, 102 Ill. 64.

805. While the present constitution was intended to afford redress in a class of cases for which there was no remedy under the old constitution, still it was not intended to reach every possible injury occasioned by a public improvement. The building of a jail, police station, or the like, will generally cause a direct depreciation in the value of neighboring property, but that is a case of *damnum absque injuria*. *Ib.*

806. Any expressions used in *Stetson* v. *Chicago & Evanston R. R.*, 75 Ill. 74; and *C. M. & St. P. R. R.* v. *Hall*, 90 Ill. 42, which may seem to restrict the remedy of owners of private property as given by the present constitution to cases where there has been a direct physical injury, are not to be accepted as embodying the views of the court on that subject. *Ib.*

807. The right to recover damages for injury to private property occasioned by the taking of other property for public use, if not conferred, is secured by § 13 Art. 2 of the constitution of 1870. *Ch. & W. Ind. R. R.* v. *Ayres*, 106 Ill. 511.

808. PROPERTY NOT TAKEN. Prior to the constitution of 1870, no compensation was required to be paid for property not taken for public use, but which was damaged by the construction and maintenance of public improvements. Under that constitution an action by a lot owner for a physical injury to his property by constructing and operating a railway in a public street near his lot, may be regarded as a proceeding to recover just compensation for private property damaged for the public good, and one recovery will bar any subsequent action for the same cause. *Ch. & E. Ill. R. R.* v. *Loeb*, 118 Ill. 203.

DAMAGE TO CONTIGUOUS PROPERTY.

LIABILITY OF MUNICIPAL CORPORATION.

809. CHANGE OF STREET GRADE. Municipal authorities have the undoubted power to alter the grade of streets at their discretion and compel property owners to conform thereto; and if the work is done with reasonable care and diligence, the town or city will not be liable to such owners for damages growing out of obstructing the streets, but if they act wrongfully, or with bad intent, damages may be recovered. *Roberts* v. *Chicago*, 26 Ill. 249.

810. SAME—*creating nuisance*. If a city in fixing the grade to a street turns a stream of water and mud upon the grounds or cellar of a citizen, or creates in his neighborhood a stagnant pond that generates disease, it will become liable to him in damages. *Nevins* v. *Peoria*, 41 Ill. 502; *Aurora* v. *Gillett*, 56 Ill. 132; *Aurora* v. *Reed*, 57 Ill. 29; *Shawneetown* v. *Mason*, 82 Ill. 337.

811. DEFECTIVE SEWER—*surface water*. The liability of a city for an injury to private property resulting from drains and sewers constructed by the city, being defective or having become obstructed, by reason whereof surface water from the streets is thrown upon the premises of another, is correctly stated in *Nevins* v. *Peoria*, 41 Ill. 502; *Aurora* v. *Gillett*, 56 Ill. 132.

812. DRAINING STREETS. If it becomes necessary for the public interest in the process of grading or draining the streets that the lot of an individual shall be rendered unfit for occupancy, either wholly or in part, the public should pay for it to the extent it deprives the owner of its legitimate use. *Nevins* v. *Peoria*, 41 Ill. 502.

813. GRADE OF STREET. A city has full control over the grade of its streets, and may lower or elevate it at pleasure, and for the inconvenience and expense of adjusting their lots with the streets the owners thereof will have no right of action. *Aurora* v. *Reed*, 57 Ill. 29.

814. A city, under the plea of public convenience, cannot be allowed to exercise its powers over the public streets to the injury of private property in such a mode as would render a private owner liable. *Ib.*

815. GRADE—*throwing water on lot.* Where a city fixes the grade of a public street, and has the same so improved that water from rains and melting snows runs to and discharges itself upon a private lot, the city will be liable to the owner in damages, although the street may have been improved before the lot was. It is no defense that the owner might have protected his lot by digging ditches. *Ib.*

816. DEPRIVING OF SIDEWALK. City authorities have no power to appropriate such part of land dedicated for a public street as will deprive the owners on one side of the street of a sidewalk, and if they attempt to do so they may be enjoined. *Carter* v. *Chicago*, 57 Ill. 283.

817. INJURY TO SEWERAGE. If, in abating or removing a nuisance, by a system of sewerage or drainage, a city unavoidably inflicts an injury upon private property, it should, by condemnation or otherwise, make compensation for the injury. *Jacksonville* v. *Lambert*, 62 Ill. 519.

818. CHANGE OF GRADE. Municipal corporations may regulate and establish the grade of their streets, but this must be so done as to do no serious injury to the owners of abutting lots. They have no right to change the natural flow of water and throw it upon the lands of another. *Dixon* v. *Baker*, 65 Ill. 518.

819. Where a city, by elevating the grade of a street, caused the surface water to flow upon the plaintiff's lot and into the basement of his cellar, whereby the building thereon was injured, and the walls were cracked, and it appeared that the injury might have been avoided by proper sewerage: *Held*, that the city became liable. *Ib.*

820. While the corporate authorities are vested with power to grade their streets, yet the manner of its exercise is limited in the same way and to the same extent as the power of a private person in the use of his property, unless such authorities call to their aid the right of eminent domain, in which case compensation must be made. *Pekin* v. *Brereton*, 67 Ill. 477.

821. GUTTER OUT OF REPAIR. If a city suffers a gutter in a street it has constructed, to get out of repair, so that the water which it should have carried off, is thrown upon the lot of an individual near by, and his buildings are damaged thereby, the city will be liable for the injury. *Alton* v. *Hope*, 68 Ill. 167.

822. CHANGE OF GRADE. A city may elevate or lower the grade of its streets, when done in good faith with a view to fit them for use, and cannot be held responsible for errors of judgment in that respect, or made liable for the inconvenience and expense of adjusting the adjacent property to the grade as changed. *Shawneetown* v. *Mason*, 82 Ill. 337.

823. STREET FOR LEVEE. But if the street is appropriated to another use than that contemplated when it was laid out, as for a levee to prevent a river from overflowing the town, and the grade is raised for *such* a purpose *only*, then under the constitution of 1870, the owners of property damaged thereby, are entitled to just compensation. *Ib.*

824. Under the constitution of 1870, if injury to private property is sustained by changing the grade of a street, the municipal corporation causing the same, will be liable to the owner in damages. *Elgin* v. *Eaton*, 83 Ill. 535.

825. Excavation in street. The distinction between an excavation made in a street and one made by an individual upon his own adjoining land, as respects the right of recovery by the owner of abutting premises, is that such owner has the legal right to use the street. If his right of ingress and egress is disturbed, he may have damages therefor, while if the adjoining proprietor excavates upon his own land, no harm is done, unless his neighbor's lot has been disturbed thereby. *Elgin* v. *Eaton*, 2 Bradw. 90.

826. Bridge in street. In an action by an adjacent lot-owner for damages caused by the construction of approaches to a bridge, evidence of damages caused by the bridge employes throwing dust and dirt from the bridge in baskets, is not admissible; nor is evidence of damage arising from the diversion of travel and trade. *E. St. Louis* v. *Wiggins Ferry Co.*, 11 Bradw. 254.

827. Railroad in street. A city is not liable for damages resulting from the proper exercise of authority in permitting railroad tracks to be laid in the streets, or in raising the grade of streets. Unless the authorities exceed their power there is no liability. *Murphy* v. *Chicago*, 29 Ill. 279.

828. Depriving of sidewalk—*injunction.* Where city authorities undertake by ordinance from fraudulent and malicious motives to appropriate so much of one side of a street to the purposes of a roadway, as will deprive the adjacent property owners of any sidewalk, a court of equity has jurisdiction to restrain the execution of the ordinance. *Carter* v. *Chicago*, 57 Ill. 283.

829. Excavations in streets. A municipal corporation while acting within the scope of its authority in making excavations in a street for the purpose of opening and improving it, using proper care and skill, is not liable to the lot-owner for an injury to his buildings caused by removing the lateral support of the soil in the streets. *Quincy* v. *Jones*, 76 Ill. 231.

830. Allowing railroad excavations. If a railway company under a right conferred by a city, constructs its track along a public street, and makes excavations along such street, so that a lot owner is thereby deprived of convenient access to and from the street and to his lot, and the lot and building thereon are subject to injury by the caving and falling in of the lots, the city will be liable to the owner in an action on the case for the injury caused by such excavations. *Pekin* v. *Brereton*, 67 Ill. 477.

831. Obstructing access to lots. Where a city had established no grade of a street upon which the plaintiff had a house and lot, and a railway company by permission of the city, filled up the space between an original embankment and the plaintiff's lot, so as to prevent access to his lot by wagons and carriages from the street, as had been his custom: *Held*, that as this was a special injury to the plaintiff and peculiar to him, he was entitled to damages from the city. *Pekin* v. *Winkel*, 77 Ill. 56.

832. Railroad in streets. A city or village may authorize the laying of railroad tracks in its streets, and such use is not inconsistent with the trust for which they are held, but in so doing the city has no right to so obstruct the streets as to deprive the public and adjacent property holders from their use as a highway. *Stack* v. *East St. Louis*, 85 Ill. 377

833. If the authorities of a town or city authorize a structure upon a public street, or other obstruction that causes injury to adjacent lot holders, it will be liable for the damages. *Ib.*

834. A city has no right to so obstruct its streets, or to authorize the same to be done, as to deprive property holders from free access to and from their lots abutting on the same. If it permits the use of a street for an approach to a bridge, it must see that the approach is so constructed as not to produce injury to adjacent property holders. *Ib.*

835. BRIDGE APPROACHES IN STREET. If a city authorizes a bridge company to construct an approach to a bridge in a public street, whereby the street is obstructed in front of and along a party's lot abutting on the same, rendering the use of the street in front of the lot impassable and useless, and whereby ingress and egress to the lot from the street is prevented, and water is caused to drain and flow upon the lot and fill the cellar thereon, and by reason of the noise, confusion, shaking and the falling of dirt and dust caused by teams and wagons passing over the approach, the plaintiff's tenants occupying the houses on the lot are driven out, the city will be liable to such lot owner for all the damages thus caused to his premises. *Ib.*

836. TUNNEL IN STREET. Where a city, under legislative authority, constructs a tunnel in a street in a proper manner and without unreasonable delay, no action lies against it in favor of an adjoining lot owner whose property has received no physical injury. *Chicago v. Rumsey*, 87 Ill. 348.

837. Where the city owns the fee in its streets, it is not liable under the constitution of 1848 to the owner of a lot abutting on a street for damages claimed on account of constructing a tunnel in the street in front of his property, when the work is properly planned and executed under the sanction of law, and no physical injury is done to his property, and there is enough of the street left for ordinary travel. *Ib.*

838. WATER TANK IN STREET. The erection of a water tank in the center of a street, occupying one-half of the width thereof, and the erection and operation of a steam engine in connection therewith, even for the purpose of supplying the city and its residents with water, is not an use to which the street can appropriately be put, and the owner of an adjoining lot does not take subject to such easement, and may maintain an action against the city for any damage to his property. *City of Morrison v. Hinkson*, 87 Ill. 587.

839. VIADUCT OR BRIDGE IN STREET—*physical injury.* To authorize a recovery by an individual for an injury to his property by the construction of a public improvement under the authority of a statute, it must appear that there has been some direct physical disturbance of a right, either public or private, which the plaintiff enjoys in connection with his property, and which gives to it an additional value, and that by reason of such disturbance he has sustained a special damage with respect to his property in excess of that sustained by the public generally, and which by the common law in the absence of any constitutional or statutory provision, would have given a right of action. *Rigney v. Chicago*, 102 Ill. 64.

840. Where a city constructed a viaduct or bridge on a public street near its intersection with another street, thereby cutting off access to the first named street from the plaintiff's house and lot over or along the street intersected, except by means of a pair of stairs, whereby the plaintiff's premises fronting on the latter street and near the obstruction, were permanently damaged and depreciated in value by reason of being deprived of such access, it was held that the city was liable to the plaintiff in damages. *Ib.*

841. The owners of property bordering upon streets, have as an incident to their ownership, a right of access by way of the streets,

—8

which cannot be taken away, or materially impaired by the city without incurring legal liability to the extent of the damage thereby occasioned; and to this extent it may be said there is a special trust in favor of adjoining property holders. *Chicago* v. *Union Building Assoc.*, 102 Ill. 379.

842. USE OF STREET FOR RAILROAD. The grant of the use of a street to a railway company, whereby access to and egress from a lot is not prevented, will not render a city liable for damages to the owner. *East St. Louis* v. *O'Flynn*, 119 Ill. 200.

843. For an injury to, or an obstruction of a public and common right, no private action will lie for damages of the same kind as those sustained by the general public, although the private property may be injured much greater in degree. *Ib.*

844. So, a lot owner in a city cannot maintain an action against the city for the vacation of a portion of a public street, not bordering on his lot and not necessary to afford him access thereto. *Ib.*

845. Where a railway company is authorized by ordinance to build its road within a part of the street, which is thereby legally vacated, the city cannot be held liable to a lot-owner whose property is not adjacent to the vacated street for any act done by the company not authorized by such ordinance. *Ib.*

LIABILITY FOR INJURY BY USE OF STREETS.

846. PRIOR TO CONSTITUTION OF 1870—*injunction—damages.* Municipal authorities having the exclusive control over the streets may give permission to a railway company to locate its tracks along a street, and the owner of lots along such street cannot enjoin the laying of the track or receive any damages or compensation for such use of the street. *Moses* v. *P., Ft. W. & Ch. R. R.*, 21 Ill. 516.

847. NEW BURDEN—*damages for.* Where the public has acquired an easement over a person's land for an ordinary street or highway, the location of the track of a railroad on the same, is an additional burden and servitude upon the land. which will entitle the owner to additional compensation. *I., B. & W. R. R.* v. *Hartley*, 67 Ill. 439.

848. But where the fee of the street is in the municipality granting the right of way in the same to a railway company, the owners of lots fronting on such streets, cannot enjoin the laying of the track in the street. nor receive compensation for the use of the street so appropriated. *Ib.*

849. FOR WHAT INJURY LIABLE. Where the fee of a street is in the adjacent land-owner, the town or city may grant the right to a railway company to lay its track along or across the same, but the company avails of its privilege at its peril. If in laying its track, it causes a private injury to him who owns the fee in the adjoining premises, it will be liable to him for the damages. *Ib.*

850. The clause of the constitution that "private property shall not be taken or *damaged* for public use without just compensation" must receive a reasonable and practicable interpretation. Where the property is not taken, the damages must be real and not speculative. If the property is not worth less in consequence of the construction of a railroad in its vicinity, or upon a street upon which the lot abuts, than if no road were constructed, the owner will not be entitled to damages, and cannot enjoin the construction of the road. *Ch. & Pac. R. R.* v. *Francis*, 70 Ill. 238.

851. While a town or city may rightfully permit a railway company to occupy and use a public street for right of way, yet under the

organic law of the state the company must be held responsible to property owners upon the street for such direct and physical damage as shall result from the construction of the road, or its operation after completion. *Stone* v. *F., P. & N. W. R. R.*, 68 Ill. 394.

852. THROWING SURFACE WATER ON LAND. If a railway company, in constructing its road along a public street, under license from the corporate authorities, turns waste and surface water and mud upon the adjacent premises of another, it will be liable to the owner in damages for the injury thereby done. *St. L., V. & T. H. R. R.* v. *Capps*, 72 Ill. 188.

853. RIGHT TO HAVE DAMAGES ASSESSED. Where no part of a person's land is taken or sought to be condemned by a railway company, he will not be entitled to have proceedings instituted to ascertain what damages his property will sustain by the construction and operation of a railway upon adjacent lands, but will be left to his action at law. *Stetson* v. *Ch. & E. R. R.*, 75 Ill. 74.

854. INJUNCTION. A court of equity will not enjoin the construction and operation of a railroad upon a public street or other lands not belonging to the complainant until the damages to lots owned by him abutting upon the street are ascertained and paid, but will leave him to his action at law. *Ib.*

855. Where the fee of the street is in the adjacent lot-owner, subject to public easement, the rule is different, for the reason that the railroad is an additional burden on his land. *Ib.*

856. INJUNCTION. In case of a claim of consequential damages to land on account of the operating of a railroad where no part of the land claimed to be affected is taken for the use of the road, a court of equity will not enjoin the use of the railroad until such damages are assessed and paid, nor will it, at the suit of an individual, enjoin a railway company from operating its road laid in a public street without leave of the city, but will leave the redress to the public authorities. *Patterson* v. *Ch., Dan. & Vin. R. R.*, 75 Ill. 588.

857. Where, after the construction of a railroad over a portion of a lot, the owner erected a dwelling house upon the lot in close proximity to the road, and occupied the same as a residence, it was held that the owner having built the house with full knowledge that it would be affected by the road, could not in an action against the railroad company recover for the loss he thus knowingly and voluntarily incurred by building the house near the road, but that so far as the house sustained a direct physical injury by the company, which it was its duty to avoid, as against all adjacent property, the owner was entitled to recover. *I., B. & W. Ry.* v. *McLaughlin*, 77 Ill. 275.

858. INJUNCTION. Where city authorities grant permission to a railway company to lay its track along a street, the owners of property fronting on such street cannot enjoin the laying of such tracks, nor be allowed any damage or compensation for such use of the street. *C., B. & Q. R. R.* v. *McGinnis*, 79 Ill. 269.

859. PHYSICAL INJURY. The liability of a railway company to a lot-owner in consequence of its use of a public street in front of the lot under license from the city, is confined to the direct physical injury done to the property by the operation of the road. *Ib.*

860. RIGHT TO USE OF STREET. The grant to a charter to a railroad company to run its road through a town, cannot by any reasonable or fair intendment operate as a grant of the use of the streets, or either of them, to the company. *St. L., V. & T. H. R. R.* v. *Haller*, 82 Ill. 208.

861. RIGHT TO HAVE CONDEMNATION. Under the eminent domain

act of 1872, an adjoining land-owner, where no part of his land is actually taken, or sought to be condemned for public use, is not entitled to have proceedings instituted to ascertain what damages his property may sustain in consequence of the construction and operation of a railway upon contiguous or adjacent lands in which he has no interest. *P. & R. I. Ry.* v. *Schertz*, 84 Ill. 135.

862. INJUNCTION. A court of equity will not enjoin the use of a railroad track upon a public street until the adjoining land-owner's damages have been assessed and paid, even though the company may be insolvent. *Ib.*

863. The rule is well settled that for any obstruction to streets not resulting in special injury to the individual, the public only can complain. If a special injury results to a person, he may have his action against the wrong-doer. *McDonald* v. *English*, 85 Ill. 232.

864. POWER OF CITY OVER STREETS. A city has the power to allow the construction of a railroad upon or over its streets, and the public will be bound by whatever may be lawfully done in regard to the streets by the city. *C. & N. W. Ry.* v. *People*, 91 Ill. 251.

865. ADDITIONAL TRACKS. A lot owner has a right of action to recover damages to his lot from the unauthorized laying of additional railroad tracks in the street fronting his lot, whereby the use of the street for all ordinary purposes of a highway is destroyed, and access to his lot is cut off, and for the creating a nuisance by allowing stock cars to stand in the street adjoining the lot. *P., Ft. W. & Ch. R. R.* v. *Reich*, 101 Ill. 157.

866. ACTION—*cutting off access to lot.* Where railway tracks are constructed in a public highway on ground thrown up considerably above the common level, under proper license, in front of a person's land, whereby he is cut off from access and egress from the same, he cannot recover of the company for any injury or damage he thereby sustains in common with the public generally, but may recover for any damages he may have sustained individually in respect to his private property separate and distinct from the disturbance of the public easement. *Ch. & W. Ind. R. R.* v. *Ayres*, 106 Ill. 511.

LIABILITY FOR OTHER ACTS THAN USE OF STREETS.

867. OBSTRUCTING FLOW OF WATER—*bridges and culverts.* An individual or corporation constructing a road under legislative authority over water-courses on private land, is bound to make suitable bridges, culverts or other provisions for carrying off the water effectually, and to keep them in suitable repair. *I. C. R. R.* v. *Bethel*, 11 Bradw. 17.

868. If the construction over a water-course is not properly done, and it is washed out by an extraordinary flood leaving the *debris* upon the land of an adjacent owner beyond the line of the company's right of way, the company will not be bound to remove the same. If by reason of its being so lodged, the waters of the stream are diverted in a subsequent freshet, whether extraordinary or only ordinary, it will give no cause of action to the adjacent owner for damages resulting from the last flood. *Ib.*

869. FLOODS CHOKING UP CHANNEL—*damage to adjacent owner.* Where a corporation has exercised ordinary care in the construction or repair of bridges and culverts over water-courses on private land, and is not otherwise guilty of negligence, it cannot be made liable for damages occasioned to an adjacent proprietor by extraordinary floods choking up or washing out the channel of the stream. *Ib.*

870. OBSTRUCTING FLOW OF WATER. If a railway company makes an embankment near the land of another whereby the water is thrown back on such other's land, leaving no opening for the water to escape, it will be liable in an action on the case to the owner of such land for all the injury caused thereby. *Gillham* v. *Madison Co. R. R.*, 49 Ill. 484.

871. Where a horse railway company constructs its road as required by its charter and the license of the city, whereby the water is obstructed and the premises of another overflowed, it will be liable to the owner of the land so overflowed, the same as if the road had been constructed under the directions of its own engineer. *A. & U. A. Horse Ry.* v. *Dietz*, 50 Ill. 210.

872. Where a corporation accepts its charter and constructs a railway as therein authorized, it will be implied that it will not injure others by its construction and maintenance, and if injury results therefrom, it must be held responsible for the damages. *Ib.*

873. FLOODING PRIVATE LAND. A railroad company has no right by an embankment or other artificial means to obstruct the natural flow of the surface water, and thereby force it in an increased quantity upon the lands of another, and if it does so, it will be liable for the damages thereby caused to the owner. *T., W. & W. Ry.* v. *Morrison*, 71 Ill. 616.

874. SAME—*from manner of use.* The fact that a railway company owns a right of way over the plaintiff's land, does not authorize it to make such a change thereon by structures or otherwise as to flow water back upon the land of the plaintiff, or others, and thereby inflict an injury. *C., R. I. & P. R. R.* v. *Carey*, 90 Ill. 514.

875. Where a railway company in constructing a second track on its right of way across land, obstructs a prior drainage, so as to dam up and throw the water back on the plaintiff's land, the depreciation in the value of the land caused solely by the structure may be considered as the measure of damages as to the real estate injured thereby. *Ib.*

876. SAME—*right to remove obstruction.* If the obstruction causing the injury is upon the company's right of way, the owner of the land injured has no right to enter thereon to remove the same, and the law will not require him to commit a trespass to remove the same, even if it would cost but a trifle; nor can the company require such owner to enter its right of way for such purpose. In such case, the party injured by the obstruction has the right to claim it as a permanent injury, and the jury to allow damages as such. But if the obstruction is on plaintiff's land, he may remove the same, and the cost to remove the same will constitute the depreciation to his land. *Ib.*

877. A railway company has no right to stop, by its embankment, the natural and customary flow of the surface water from higher grounds, and by its ditch along its track, convey the same upon the premises of another over whose land the road is constructed, without providing some sufficient outlet for it to pass off; and where such person's land is injured in consequence of the accumulation of such surface water on his land, the company will be liable to him for all the damages occasioned thereby. *J., N. W. & S. E. R. R.* v. *Cox*, 91 Ill. 500.

878. A land-owner, by giving a deed for a right of way over his land to a railway company, will not be estopped from recovering damages occasioned by the wrongful construction of its road. Such a deed gives no right to flood his remaining land with water brought from other land, the natural flow of which would have carried it another way, where the consideration is only for the land conveyed. *Ib.*

879. Commissioners of highways are individually liable in an action on the case for making a drain or a ditch or an embankment. so near the land of a party, and in so unskillful and careless a manner as to cause the rain and surface water running from such drain, to flow upon the plaintiff's premises to his injury. *Tearney* v. *Smith*, 86 Ill. 391.

880. NEGLIGENCE IN CONSTRUCTING DRAINS. The acquisition of land for a highway, gives the public the right to construct a highway over it in the mode and manner deemed most expedient, and the owner cannot afterwards recover for injuries then shown that he must unquestionably suffer. But such condemnation is no bar to a suit by the land-owner for a subsequent injury growing out of negligence and unskillfulness in the public authorities in constructing drains in the highway, resulting in serious injury to the land-owner. *Ib.*

881. The maxim that no one has the right to use his own so as to injure another, applies as well to townships as to incorporated cities and natural persons. They must exercise their right in such a manner as to inflict no avoidable injury upon an individual. *Ib.*

882. NEW BURDEN—*injunction*. Where a railway company acquiring its easement in a highway, takes it subject to such rights as the public have therein, that is, subject to the right of the public to subject the street to the ordinary and proper uses of a highway, the occupation of a considerable portion of the street for the construction of a ditch, not for the improvement of the street, but for the purpose of draining adjacent lands, is a new use of the street, for which compensation must be made in case property is damaged thereby. If the city has made no provision to pay damages, the railway company may have the construction of the ditch enjoined until provision for its payment is made. *Ch. & N. W. Ry.* v. *Village of Jefferson*, 14 Bradw. 615.

, MEASURE OF DAMAGES.

TO CONTIGUOUS PROPERTY NONE OF WHICH IS TAKEN.

883. WHEN TOO REMOTE. The injury resulting to lots not taken for the purpose of widening a street, by making lots on the enlarged street more attractive and desirable, either for residence or business purposes, and thus diminishing the value of the former, is too remote to form the basis of a recovery. *Hyde Park* v. *Dunham*, 85 Ill. 569.

884. Municipal authorities of cities and villages are vested with complete control over streets, and they may contract or widen them when, in their opinion, the public good so requires; and any damage sustained in consequence of the exercise of such power where property is neither taken nor directly damaged thereby, is too remote and contingent to be allowed. *Ib.*

885. NUISANCE. Although it is true that a municipal corporation cannot authorize that which is deemed a legal injury to the property of another without making compensation, yet the individual cannot recover for every technical nuisance to the streets of a city without regard to whether he has sustained special injury. *McDonald* v. *English*, 85 Ill. 232.

886. CHANGE OF GRADE—*pecuniary loss*. If private property is damaged by a change in the grade of a street, the recovery must be measured by the extent of the pecuniary loss. If it is benefited as much as it is damaged, there can be no recovery, and it is error to refuse testimony to show that fact. *City of Elgin* v. *Eaton*, 83 Ill. 535.

887. EVIDENCE—*profile of grade*. In a suit by the owner of a house and lot to recover damages growing out of a change in the

grade of a street, after the work is commenced and before its completion, the profile of the proposed improvement is proper. *Ib.*

888. EVIDENCE—*depreciation of value.* In an action against a railway company for damages arising from a direct physical injury to the plaintiff's dwelling, by reason of running its trains along a public street in front of his premises, evidence of the general depreciation of the value of his property is not admissible, where the witness is unable to distinguish between damages, such as were the result of the injury complained of, and such as arose from other general causes. *Ch. & E. Ill. R. R.* v. *Hall,* 8 Bradw. 621.

889. EVIDENCE—*what required.* In an action against a railroad company for injuries to adjacent property caused by the running of trains, where the declaration alleges an injury to the possession of the plaintiff, he must prove possession of the premises, the injurious act alleged to have been done, and the damages resulting therefrom. *Ch. & E. Ill. R. R.* v. *Loeb,* 8 Bradw. 627.

890. ONLY DAMAGES PECULIAR TO PROPERTY. Where a railroad is built and operated through a street, the owner of land abutting on such street is not entitled to recover of the railway company all the damages sustained by him by the location and operation of the road, including the loss by depreciation in the market value of his property, and which are common to other owners or the public, but his right to recover must be limited to such damages as are peculiar to his property, and which are of a physical nature, such as the cutting off of access to his premises, jarring of his buildings, casting cinders and smoke upon his dwelling, &c. *Ch. & W. Ind. R. R.* v. *Berg,* 10 Bradw. 607.

891. ELEMENTS—*obstructing street.* On an assessment of damages to an adjoining lot owner by the location and building of a side track of a railroad in a public street, it is error, by an instruction, to exclude from the estimate of damages, the obstruction of the street by the running of trains. *Mix* v. *L. B. & M. Ry.,* 67 Ill. 319.

892. An ordinance prohibiting the obstruction of streets by railway trains for more than fifteen minutes will not legalize such obstruction for that length of time so as to exclude it from the estimate of damages to contiguous property that may be injuriously affected thereby. *Ib.*

893. EVIDENCE—*as to uses of property.* On the assessment of damages to lots abutting upon a street sought to be taken for a side track of a railroad, the owner gave evidence that the proposed location would render his lots useless for business purposes: *Held,* competent for the railway company in rebuttal, to show that the property could be beneficially used for warehouse purposes, or for any other purpose. *Mix* v. *L. B. & M. Ry.,* 67 Ill. 319.

894. INSTRUCTION—*as to measure.* In such a case, the court instructed the jury that the damages to be allowed the lot-owner could only be such, in kind, as lots not lying or abutting on the same street, but in the vicinity, did not sustain in any degree: *Held,* erroneous, as virtually cutting off all claim for damages. *Ib.*

895. In the same case, the court instructed that the law did not give indemnity for all losses or damages occasioned by the building of a railroad, such as inconvenience arising from the crossing of railroad tracks by the public or by individuals, or from noise and confusion of passing trains, smoke from the same, or frightening horses, &c.: *Held,* as applicable to the case where the track was along a street within ten to eighteen feet of the front line of the lots abutting on the street, that the instruction was improper, and calculated to mislead the jury. *Ib.*

896. EVIDENCE—*ordinance*. On the assessment of damages to lots by the location of a side track in an adjoining street, where the petition states that such track is to be constructed and maintained according to an ordinance, the ordinance is proper evidence on the question of damages, as tending to show the nature of the work and the probable use of the street. *Ib.*

897. DECLARATION—*statement of the injury*. In a suit against a railway company for damages caused to plaintiff's lots and property, the declaration averred in substance, that the plaintiff owned and occupied as a residence certain property fronting on a certain public street; that the defendant constructed along, upon and over such street its railroad, and run daily its locomotives and trains thereon, and that smoke and cinders were cast and thrown from the engines and locomotives in and upon the property of the plaintiff, thereby greatly damaging the same: *Held*, that the declaration showed a good cause of action. *Stone* v. *F. P. & N. W. R. R.*, 68 Ill. 394.

898. DAMAGE BY BRIDGE. *To property on river*. The state can not take or damage a party's land fronting upon or in the bed of a river without first making compensation therefor, nor can it authorize a railway company to do the same. If such a company, under its charter, erects a bridge across a river, and the property of another bounded by the stream is taken or damaged thereby, a right of action exists in his favor; but he can only recover for damages which are special to his property, and not for such as are incidental to and are shared by the public at large. *Ch. & Pac. R. R.* v. *Stein*, 75 Ill. 41.

899. MEASURE—*injury to market value*. Where the erection of a railroad bridge across a river causes a permanent injury or depreciation in the value of a lot in the immediate vicinity, which is used for dock purposes, such injury is a proper element of damages in a suit by the owner against the company, and it is proper to allow the lot-owner to show such damage by proving the value of his property before the erection of the bridge and its value after; or, in other words, to prove how much less the property would sell for in consequence of the building of the bridge. *Ib.*

900. DAMAGE BY APPROACH TO BRIDGE. In an action by an adjacent lot-owner for damages occasioned by the construction of approaches to a bridge, evidence of damage caused by the bridge employes throwing dust and dirt from the bridge in baskets, is not admissible; nor is evidence of damage arising from the diversion of travel and trade. It is competent in such action to show that the diminution in value of the property arises from the general depression in trade. *E. St. L.* v. *Wiggins Ferry Co.*, 11 Bradw. 254.

901. SAME—*evidence*. Where the plaintiff claims damage by reason of the jar caused by the passage of trains over the bridge, it is competent upon cross-examination to show how the opposite approach is constructed, and that there is more vibration there, and that buildings at the opposite approach are not injured by the vibrations. *Ib.*

902. RAILWAY IN STREET—*element of damage*. In an action to recover damages caused to a house and lot by the construction of railroad tracks in a street in close proximity to the plaintiff's property, the true measure of damages is the loss sustained by the nuisance, the injury from jarring the building and the throwing of cinders and smoke upon the plaintiff's premises, and the depreciation of the value of the property by these causes may be considered; but not general depreciation in value from other causes, such as mere inconvenience in approaching or leaving the property, or the noise and confusion in the vicinity. The injury must be physical. *C., M. & St. P. R. R.* v. *Hall*, 90 Ill. 42.

903. DEPRECIATION IN VALUE—*benefits.* Damage, to property not taken for public use, to be recoverable, must be physical and real, and not speculative, and it must depreciate the value of the property or its use. The depreciation is to be determined by comparing its value before and after the structure which produced the injury, and any benefits thus conferred should be considered as well as injury inflicted by the structure, in estimating the damages. *Ib.*

904. ELEMENTS OF—*depreciation in value.* Under sec. 13, Art. 2, of the constitution, a recovery may be had in all cases, where private property has sustained a substantial damage by the making and using of an improvement that is public in its character, as a railway, and it is not required that the damage shall be caused by trespass or an actual physical invasion of the owner's real estate; but if the construction and operation of a railroad or other public improvement is the cause of the damage, though merely consequential, the party damaged may recover. Depreciation in the value of the land fronting on a highway caused by obstructing access to it, is a proper element of damage. *Ch. & W. Ind. R. R.* v. *Ayres,* 106 Ill. 511.

905. Where the usual outlet of water is obstructed so as to overflow the plaintiff's lands, he may recover for the loss of or injury to the crop of hay, &c., or the expense of securing them, in addition to the loss by the depreciation of the land. *C., R. I. & P. R. R.* v. *Casey,* 90 Ill. 514. The depreciation in the value of the land caused solely by the structure, may be considered as the measure of damages as to the real estate injured. *Ib.*

906. SAME—*instruction too broad.* In an action for obstructing water, so as to overflow the plaintiff's land, if the court instructs the jury that the depreciation in the value of the land may be considered, it will be error to further instruct that they may consider the inconvenience and damage in separating the farm, the damage caused to the land overflowed and to that not overflowed, and the expense of making roads and bridges, as these are included in the depreciation to the land. *C., R. I. & P. R. R.* v. *Carey,* 90 Ill. 514.

907. ELEMENTS—*evidence.* In an action by a lot-owner against a railway company to recover damages to his lot caused by the construction and operation of a railroad along a public street in front of the lot, it is error to allow the plaintiff to prove the difference in value of the lot and its rental value with or without the road, as such difference in part may be the result of inconveniences for which the law affords no remedy. *C., B. & Q. R. R.* v. *McGinnis,* 79 Ill. 269.

908. EXCAVATION IN STREET—*evidence as to depreciation in value.* If a railway company, in constructing its track along a public street, makes a deep excavation therein in front of the plaintiff's lots and business house, he will be entitled to recover as damages whatever diminution in value his real estate may undergo; and to show this, it is proper to prove the market value of the property before and since the injury, leaving out of view any inflated value arising from any cause. Proof of the rental value, before and since the construction of the road, will furnish some criterion by which to determine the extent of the injury. *St. L., V. & T. H. R. R.* v. *Capps,* 67 Ill. 607.

909. NON-ACTIONABLE INCONVENIENCES. The difficulty of crossing a railroad track in a public street, the detention of trains, the frightening of horses, the danger to persons crossing the track, and the like, are inconveniences which property owners on the street have to suffer, and for which they cannot recover in a suit for damages. *Stone* v. *F. P. & N. W. R. R.,* 68 Ill. 394.

910. BENEFITS. In estimating the damages done to property by the appropriation of a public street adjacent thereto, to public use other than as a street, where no part of the private property is taken, the effect on the whole property should be considered and not merely a part of it. If one part of the same property is damaged, and another part specially benefited, so that the value of the whole is not diminished, then there is no damage done; but any general benefit common to all other property affected by the work should not be considered in determining whether the property is benefited as much as injured. *Shawneetown* v. *Mason*, 82 Ill. 337.

MEASURE.

LIABILITY UNDER ORDINANCE FOR RAILROAD IN STREET.

911. Injury to Adjacent Property and Business. Where a railway company constructs its track along a public street under an ordinance requiring it to pay all damages thereby occasioned, and in so doing, makes a deep excavation in front of a person's lots and place of business, which diminishes the value of his lots and injures his business, by making his place difficult of access and dangerous for teams to approach, the company, by acting under such ordinance, will become liable to pay the lot-owner all damages caused to his property, and also to his business. *St. L., V. & T. H. R. R.* v. *Capps*, 67 Ill. 607.

912. Injury to Business. Where a party's place of business is so seriously affected by the construction of a railroad in the street in front of the same, as to make it necessary to remove to another place, he will be entitled to damages for interruption to his business during such time as would have been necessarily employed in accommodating himself to another place of business equally eligible, and his removal thereto. During such time the damage to his business should be ascertained by proof of the probable and reasonable profits which might have been made, had there been no interruption. The necessary and reasonable expenses of removal is also a proper element of damage. *Ib.* See same case, 72 Ill. 188.

913. Same—*evidence of decline in business*. In a suit by a merchant against a railway company to recover damages to his business caused by making deep excavations in the street in front of his place of business, the plaintiff proved the extent of his business in the preceding year and the decrease in the year after. The company then offered to prove the fact of a general decline in the business in which the plaintiff was engaged, which the court refused: *Held*, error to refuse the evidence. *Ib.*

914. Ordinance—*extent of damages under*. Where a railway company constructs its road in a public street under an ordinance of the town granting the privilege on condition that it shall pay all damages that may accrue to property owners by reason thereof, it will be held liable to such owners for all damages done to them during the progress of the work, as well as for such as are caused by the road when completed. *St. L., V. & T. H. R. R.* v. *Capps*, 72 Ill. 188.

915. Injury to Business—*evidence*. Where a railway company accepts a grant of the right of way over a public street upon condition that it shall pay all damages caused to property owners upon the street, a lot-owner in a suit against the company for damages, may show that his store was situated on the corner of the street along which the road ran and another street; that dirt was thrown up at the corner, so that for a time travel was entirely interrupted; that by reason of the occupation of the street, there was but a narrow passage left for travel, and there was not room enough for teams to turn in the street; that teams could not approach the store on account of the running of the cars; that there was no place to hitch teams or unload conveniently; and on account of the frequent passage of trains, it was dangerous for teams to be left standing, or to pass along the street in front of the store, as tending to show in what manner the property was injuriously affected. *Ib.*

916. Where a railway company builds its road along the street of a town under an ordinance granting the privilege upon condition that it shall pay all damages accruing to property holders on such street, by reason of the construction of the road, it will be liable to a lot-owner for the deterioration in the value of his lot in consequence of the laying of the track, and for damages for interruption to his business during a reasonable time in which to provide another equally eligible place, and remove thereto; and the damage to his business during such time should be ascertained by proof of the probable reasonable profits which might have been made. The property owner, if he chooses to remain and submit to the interruption in his business and loss of profits, may nevertheless recover from the company as damages, the necessary cost of avoiding such loss by a removal. *Ib.*

917. ORDINANCE—*liability of company under—for what injuries.* Where an ordinance of a town authorizing a railway company to build its road on a street provides that the company shall pay all damages that may accrue to property owners on such street by the construction of the road, an action will lie on the ordinance against the company in favor of any property owner who is injured by the construction of the road, either by depreciation in value of his property or loss of business sustained during the building of the road, and after its completion. *St. L., V. & T. H. R. R.* v. *Haller,* 82 Ill. 208.

918. In an action against a railway company upon an ordinance of a town permitting it to lay its track on a street, and providing for the payment of damages by the company to property owners, the parties will be governed and their rights measured by the ordinance without reference to the constitutional provision in regard to compensation for property taken or damaged for corporate purposes, or to the common law on the subject as announced in *Moses* v. *P., Ft. W. & Ch. R. R.,* 21 Ill. 516, and *Murphy* v. *Chicago,* 29 Ill. 279. *St. L., V. & T. H. R. R.* v. *Haller,* 82 Ill. 208.

919. MEASURE OF DAMAGE. In a suit under a town ordinance providing for the payment of damages to property owners occasioned by constructing a railroad track in the street, the difference in the value of property caused by the construction of the road, is the measure of damages, and this may be shown by a comparison of the sales of other property similarly situated before and after the construction of the road, or by the difference in its rental value, if held for the purpose of renting; but if not held for that purpose, then the difference in rental value would not be a criterion. *Ib.*

920. In such case, if there have been no sales of property of a character similar to that claimed to be injured, either before or after the construction of the road, from which the depreciation in value can be ascertained, it is proper to resort to evidence of the noise and jarring of the earth, and smoke and dust caused by passing trains, rendering the house, if a dwelling, uncomfortable, and injuring the furniture and walls of the house, as an aid to the jury in estimating the depreciation in value of the property. *Ib.*

921. PAST, PRESENT AND FUTURE DAMAGES. In an action brought for a deterioration in the value of real estate occasioned by a nuisance of a permanent character, or which is treated as permanent by the parties, all damages for the past and the future injury of the property may be recovered, and one recovery in such a case is a bar to all future actions for the same cause. *Ch. & E. Ill. R. R.* v. *Loeb,* 118 Ill. 203.

922. Where private lots in a city are physically damaged or injured in value by the construction and operation of a railroad in close proximity thereto along a public street, the right of action, if any exists, is vested in the owner of the lots immediately upon the construction of the road, to recover for all damages, past, present and future, and a subsequent grantee of the lots cannot maintain an action at all for the proper use and operation of the road after his purchase. *Ib.*

923. The just compensation to be made for damage to land is intended as an indemnity, not for successive constantly accruing damages as they may afterwards be suffered, but for all the land-owner may suffer from all the future consequences of the careful and prudent operation of the proposed public improvement. *Ib.*

924. JUDGMENT ON REPORT—*effect of order and payment.* § 10.
The judge or court shall, upon such report, proceed to adjudge and make such order as to right and justice shall pertain, ordering that petitioner enter upon such property, and the use of the same, upon payment of full compensation, as ascertained as aforesaid: and such order, with evidence of payment, shall constitute complete justification of the taking of such property. [R. S. 1887, p. 647, § 10; S. & C., p. 1050, § 10; Cothran, p. 649, § 10.]

925. EXECUTION. Unless the statute so provides, it is error to award an execution for the damages assessed or the costs of the proceeding. *Ch. & Mil. R. R. v. Bull,* 20 Ill. 218.

926. FORM OF, UNDER ACT OF 1852. The form of the judgment in a proceeding to condemn land under the act of 1852 should conform to that prescribed by § 15 of the act. *Wilson v. R., R. I. & St. L. R. R.,* 59 Ill. 273.

927. CONDITIONAL—*execution.* No execution can issue upon a judgment of condemnation for the damages awarded. The judgment should not be absolute for the payment of the sum found. The only mode to coerce payment is by *mandamus. Cook v. South Park Comrs.,* 61 Ill. 115.

928. EXECUTION. It is error, in a proceeding under the act of 1852, to award execution against the company for the damages assessed. *St. L. & S. E. Ry. v. Lux,* 63 Ill. 523.

929. The judgment must be an order authorizing the petitioner to enter upon the land and use the same upon payment of the compensation found by the jury, but there should be no award of execution therefor. *P., P. & J. R. R. v. P. & S. R. R.,* 66 Ill. 174.

930. Where the petitioner has not already entered upon the land, the judgment should be that it enter upon and use the property upon payment of the compensation found. But where it has given the requisite bond and has entered, such an order is unnecessary. *R., R. I. & St. L. R. R. v. Coppinger,* 66 Ill. 510.

931. REPORT AND JUDGMENT—*a part of the record.* The report of the damages assessed, and the judgment of the court thereon being a matter of record, will be taken notice of by the supreme court without a bill of exceptions. *Ch. Mil. & St. P. Ry. v. Melville,* 66 Ill. 329.

932. EFFECT COLLATERALLY. The judgment cannot be impeached collaterally, and it will be presumed conclusively that the party whose land was taken has received by the judgment and award, not only just compensation for the land taken, but for all such incidental loss, inconvenience and damages as might reasonably be expected to result

from the construction and use of the way or crossing in a legal and proper manner, and the judgment will afford a complete justification to the party exercising the right so acquired. *C. & A. R. R.* v. *S. & N. W. R. R.*, 67 Ill. 142.

933. Final judgment of condemnation and payment of the award, vest in the company exercising the right of eminent domain, the absolute right to use the land embraced in the judgment for all legitimate purposes. *Ib.*

934. AWARD OF EXECUTION. Where the corporation has not taken possession and used the land when the assessment of the compensation and damages is had, it is error to render judgment awarding an execution for its collection; but if the company has taken possession, and is in the occupancy of the land, such a judgment is proper. *St. L. & S. E. Ry.* v. *Teeters*, 68 Ill. 144.

935. PAYMENT—*that confers the right.* It is the payment of the money found by the jury, and not the order of the court, that confers the right of way. Such order, with evidence of payment, constitutes a justification for taking the property. *Ib.*

936. AWARD OF EXECUTION. It is error for the circuit court, on the trial of an appeal, to award execution on the judgment for the amount of the compensation and damages assessed in a proceeding to condemn land for a right of way. *S. & Ill. S. E. Ry.* v. *Turner*, 68 Ill. 187.

937. MUST BE CONDITIONAL. No order or judgment, of binding force, can be entered in a proceeding to condemn, so as to confer a present right to take or damage real estate before payment of compensation found. All that can be done, is to enter an order vesting the right to take or damage the property upon payment of such compensation. *Chicago* v. *Barbian*, 80 Ill. 482.

938. NO VESTED RIGHTS UNDER. The party seeking condemnation acquires no vested right until the sum found is paid or deposited, and the property owner has no vested right in the damages found until the same is paid or deposited. But if the property is taken or damaged by the owner's consent before compensation is made, the owner will then have a vested right in the compensation when ascertained. *Ib.*

939. In a proceeding by a city to condemn land for a street, it is error to render an unconditional judgment for the payment of the compensation and damages found by the jury. The order should simply fix the sum to be paid before taking the property, leaving the city free to abandon the improvement, if it so chooses. *Bloomington* v. *Miller*, 84 Ill. 621.

940. COLLATERALLY—*conclusive, if jurisdiction.* Where commissioners have been duly appointed according to law to condemn land for a right of way and assess damages, and have jurisdiction of the matters acted on by them, their action will be conclusive in all collateral proceedings. *Townsend* v. *C. & A. R. R.*, 91 Ill. 545.

941. An order affirming an assessment of damages for property taken for public use is a judgment and a final determination of the disputed facts and law of the case. Until reversed or otherwise impeached, it is conclusive on the parties as to the questions involved. *Beveridge* v. *West Ch. Park Coms.*, 100 Ill. 75.

942. COSTS—*limiting witnesses fees to be taxed.* The general cost act applies to proceedings to condemn, and under it the court may, after the conclusion of the evidence, limit the number of the witnesses whose fees are to be taxed against any party, not less than two, as may appear to have been necessary. *C., B. & Q. R. R.* v. *Bowman*, — Ill. —. Filed Nov. 11, 1887.

943. INTEREST ON. Interest is allowable on the sum awarded for land taken by a city to open or extend a street, if payment is neglected or refused for an unreasonable time. *Chicago* v. *Wheeler*, 25 Ill. 478.

944. No interest accrues upon an award before judgment, nor can a party causing or contributing to delay, have interest until the entry of the final judgment. But the judgment upon the award will bear interest. *Cook* v. *South Park Comrs.*, 61 Ill. 115.

945. The judgment of the circuit court on an appeal from the assessment of damages under the act of 1852 will draw six per cent. interest, where possession of the property is taken and retained by the applicant for condemnation. *Ill. & St. L. R. R.* v. *McClintock*, 68 Ill. 296.

946. INTEREST—*execution.* Where the property has not been taken or damaged, the order or judgment on the assessment of the jury will not bear interest, and no execution can be awarded for the collection of the sum assessed. *Chicago* v. *Barbian*, 80 Ill. 482.

947. INTEREST. Until possession is taken the compensation found should not bear interest, and it is error to order that it shall bear interest. *South Park Comrs.* v. *Dunlevy*, 91 Ill. 49.

948. Under proceedings to condemn for public use, the filing of the petition is not a taking of the property, and it would be a trespass to take possession before the damages are ascertained and paid. The owner, having the right to the use of the land until the damages are paid, is not entitled to interest on the value of the land from the commencement of the suit to the trial. *Ib.*

949. A judgment for the condemnation of property taken by a city to widen a street, and awarding the amount of the compensation to be paid the owner, will bear interest at six per cent. from the time possession is taken by the public. *Chicago* v. *Palmer*, 93 Ill. 125.

950. It being the duty of the park commissioners to pay for lands condemned by them for a boulevard within a reasonable time after confirmation of the proceedings and the title to the property is settled, they will be held liable to pay interest on the compensation awarded for the property condemned after demand made by the owner and the establishment of his title to the property, although the land is vacant and unoccupied, and possession has not been taken. *Beveridge* v. *West Ch. Park Comrs.*, 100 Ill. 75.

951. VESTED RIGHT. The rights of the land-owner and the party seeking condemnation, being correlative, and the change of title being dependent upon payment of the condemnation, money, it follows that no interest can be collected for failure to pay the condemnation money, for until payment the land-owner has no vested right therein, and can maintain no action therefor. *Beveridge* v. *W. Ch. Park Comrs.*, 7 Bradw. 460.

952. INTEREST. Where possession is acquired of land for a park or other public purpose, and payment of the compensation is withheld, it is proper to require the payment of interest thereon from the time possession is taken. *Phillips* v. *South Park Comrs.*, — Ill. —. Filed Jan. 25, 1887.

WHEN CONDEMNATION IS COMPLETE.

953. ACTION FOR CONDEMNATION MONEY. The party seeking condemnation acquires no title in the land until possession is taken and the land appropriated to the use for which it was condemned and payment of the damages, and the land-owner acquires no vested right to the condemnation money until possession is taken by the other, and hence can maintain no action therefor before that time. *Beveridge* v. *W. Ch. Park Comrs.*, 7 Bradw. 460.

954. MANDAMUS—*to compel payment.* Where a street has been laid out or extended and the damages for the land taken assessed, and the report thereof accepted and confirmed, and a warrant issued for. the collection of the assessment to pay for the property taken, and such street ordered to be opened, the parties entitled to the damages may by *mandamus* compel the city to collect and pay over the same. *Higgins* v. *Chicago*, 18 Ill. 276.

955. REMEDY TO COLLECT—*estopped.* In an action of assumpsit against a city to recover the damages awarded to the plaintiff by commissioners for lots taken for the extension of a street, the city will be estopped from denying the validity of the proceeding. *Chicago* v. *Wheeler*, 25 Ill. 478.

956. SAME—*case.* The owner of land taken by a city for a public street may maintain an action on the case against the city for a breach of duty in neglecting to collect the assessments of benefits out of which to pay him the damages assessed in his favor for the land so taken. He is not confined to the remedy afforded by *mandamus*. *Clayburg* v. *Chicago*, 25 Ill. 535.

957. RIGHT TO ABANDON PROCEEDING—*public road.* The only way to avoid the payment of the damages assessed for a county road is to vacate the order directing the road to be opened. *Sangamon Co.* v. *Brown*, 13 Ill. 207.

958. SAME—*park.* The park commissioners, in condemning land for park purposes, may abandon the proceeding at any time before taking possession of the land. The assessment of damages and confirmation by the court does not invest them with the title to the land. *Beveridge* v. *W. Ch. Park Comrs.*, 7 Bradw. 460.

959. The proceedings may be abandoned at any time after the damages are assessed, and before payment thereof or its deposit for the owner, where the property has remained unmolested; and the court will not, in such case, compel the payment of the compensation by *mandamus*. *Chicago* v. *Barbian*, 80 Ill. 482.

960. SAME—*injunction.* Where the condition of the order is not complied with in a reasonable time, by the payment of the damages and taking possession of the property, a court of equity will enjoin any attempt to proceed under it. *Ib.*

961. SAME—*street.* There is no error in refusing a village permission to discontinue a proceeding to condemn for a street, as this may be done by ordinance at any time after the assessment. *Hyde Park* v. *Dunham*, 85 Ill. 569.

962. PAYMENT NECESSARY TO COMPLETE CONDEMNATION. The damages assessed for right of way, on appeal from an order laying out a public road, must be paid before the road can be constructed. *Sangamon Co.* v. *Brown*, 13 Ill. 207.

963. SAME—*right to possession.* The damages or compensation awarded, with the costs of the condemnation, must be paid before the petitioner can take possession of the land condemned, or acquire any right to it whatever. *C. & M. R. R.* v. *Bull*, 20 Ill. 218.

964. The constitution (1848) does not require that the compensation allowed for land taken for right of way shall precede the entry upon the land. If the compensation is held until called for, and then paid or tendered, the prior entry will not be a trespass. *Johnson* v. *Joliet & Ch. R. R.*, 23 Ill. 202.

965. The constitution of 1848 does not require that compensation shall be made before the land is taken and used. It is sufficient if provision is made for its payment. *Shute* v. *C. & M. R. R.*, 26 Ill. 436.

966. SAME—*park.* Until the damages assessed for land condemned for a public park are paid, it cannot be occupied for the purposes intended. *People* v. *Williams*, 51 Ill. 63.

967. A judgment of condemnation of land for a public park, without payment of the damages assessed, confers no right to the land condemned. It is only by payment of the damages that the owner can be deprived of the title, or the use or possession of his land. *Cook* v. *South Park Comrs.*, 61 Ill. 115.

968. Under the "act to incorporate the Mississippi Railroad company," approved Feb. 15, 1865, the entry of judgment on the report of the commissioners, and payment thereof, was essential to the passing of the title. *P. & R. I. Ry.* v. *Rice*, 75 Ill. 329.

969. Until the compensation awarded is paid the petitioner has no right to enter upon the premises. *Schreiber* v. *C. & E. R. R.*, 115 Ill. 340; *St. L. & S. E. Ry.* v. *Teeters*, 68 Ill. 144; *Ch. & Iowa R. R.* v. *Hopkins*, 90 Ill. 316.

980. A condemnation of land for a right of way upon due proceedings, will not deprive the owner of his title, or right of possession, or of alienation, without payment of the compensation and damages awarded. *Ch. & Iowa R. R.* v. *Hopkins*, 90 Ill. 316.

981. No fixed rights acquired by condemnation before payment of the sum awarded. Until this is done the petitioner may abandon the location and the owner may use the property. *Schreiber* v. *Ch. & E. R. R.*, 115 Ill. 340.

982. INJUNCTION—*of use before payment.* The jurisdiction of a court of equity to afford preventive relief by injunction to restrain commissioners of highways from appropriating private lands for a highway, is undoubted. *Willett* v. *Woodhams*, 1 Bradw. 411.

983. If the compensation awarded is not paid, the company condemning may be restrained by injunction from using the right of way until it is paid. But non-payment will not render the condemnation invalid. *Shute* v. *Ch. & Mil. R. R.*, 26 Ill. 436.

984. Injunction to prevent the opening of part of a highway where the entire road cannot be opened. *Green* v. *Green*, 34 Ill. 320.

985. An attempt to open a road over improved land before the land owner's damages are adjusted and paid, may be restrained by a court of equity. *Coms. Highways* v. *Durham*, 43 Ill. 86.

986. WHEN COMPENSATION TO BE PAID—*possession.* The eminent domain act requires the payment of the compensation for land taken for a public use, or a tender or a deposit of the same with the county treasurer, before possession shall be taken. This is a condition precedent to the taking of possession. *Phillips* v. *South Park Coms.*, — Ill. —. Filed Jan. 25, 1887.

OF THE RIGHT TO POSSESSION.

987. EJECTMENT BY OWNER—*demand.* Where a railroad company had land condemned for right of way, but failed to pay the damages assessed, and the owner sued and recovered judgment for the damages upon which an execution was issued and returned no property found, the company having entered into possession by the owner's consent and built its road, and having leased the same to another company, against whom the owner brought ejectment: *Held*, that the action would not lie without notice to quit. *C. B. & Q. R. R.* v. *Knox College*, 34 Ill. 195.

988. Where possession is lawfully taken of property condemned for a right of way, the mere reversal of the judgment of condemnation without taking any further steps, will not render the possession unlawful and authorize a recovery by the land-owner in ejectment. *St. L., A. & T. H. R. R.* v. *Karnes*, 101 Ill. 402.

989. Where the condemnation is void for want of proper notice, and the company has notice that the owner claims the land and informs him that he will have to sue if he gets anything, this will obviate the necessity of any formal demand before bringing ejectment by the owner. *C. & A. R. R.* v. *Smith*, 78 Ill. 96.

990. Ejectment will lie against a railway corporation by the owner for land used by it for the purposes of its road where the land has not been legally condemned. *Smith* v. *Ch., A. & St. L. R. R.*, 67 Ill. 191.

991. Under the act of 1852, where an appeal is taken to the circuit court, to entitle the petitioner to possession pending the appeal, it must give a bond to the defendant to secure the payment of the final award and judgment. If possession is forcibly taken pending the appeal without giving such bond, it will be illegal, and may be recovered back in an action of forcible entry and detainer. *Mitchell* v. *Ill. & St. L. R. R. & Coal Co.*, 68 Ill. 286.

992. EFFECT OF GIVING POSSESSION. The general railroad law authorizing the acquisition of lands for right of way, and giving the right to take possession and use such lands, does not mean that if an owner permits a railroad company to enter pending litigation to ascertain the damages, or without litigation, he will lose not only his damages, but also the land. The owner will lose none of his rights by permitting the company to take possession without grant of condemnation. *I. C. R. R.* v. *Ind. & Ill. Central Ry.*, 85 Ill. 211.

993. LICENSE TO ENTER—*evidence of.* The mere fact that a railroad company has long been in possession of land occupied as a right of way, in the absence of all other proof, does not raise a presumption that the owner had given a license to enter and construct the road. *T., P. & W. Ry.* v. *Darst*, 61 Ill. 231.

994. SALE OF RIGHT OF WAY—*possession as evidence of.* The mere fact that a railway company has entered upon land and constructed its road over the same and occupied it about thirteen years, does not raise a presumption that the owner had sold the right of way to the company. *Ib.*

995. Where a railway corporation has taken possession of land without the owner's consent and without condemnation, and wrongfully holds the same, the law affords the owner two remedies—an action of ejectment and an action to recover the value of the land. *Smith* v. *Ch., Alton & St. L. R. R.*, 67 Ill. 191.

996. MANDAMUS—*to compel condemnation.* After a railway company has obtained the possession of land for its right of way, and is in the use of it, *mandamus* will not lie to compel it to institute proceedings to condemn the land. *Ib.*

997. EJECTMENT—*breach of condition.* Where the owner of land gave a railway company a written agreement for a conveyance of a right of way over the same, which contained an irrevocable license to enter and occupy a part thereof as a right of way; *held*, that the failure of the company to perform conditions subsequent, such as fencing, afforded no grounds for revoking the license under which the company entered and made its road, and hence the owner could not recover possession of the right of way in ejectment for a breach of such conditions. *Morris* v. *I., B. & W. Ry.*, 76 Ill. 522.

998. GRANT OF RIGHT OF WAY—*only by deed.* Where a railway

—9

company, in conveying a tract of land, reserved a strip on each side of its track, and another strip crossing the first, for railroad purposes, upon which another company, some sixteen years afterwards, laid the track of its road by permission: *Held*, that the reservation in the deed passed no title, legal or equitable, to the latter company as to any of the strip not actually occupied by it *I. C. R. .R.* v. *Ind. & Ill. Central Ry.*, 85 Ill. 211.

999. POSSESSION—*extent of.* Where a railway company constructs its track over the land of another, and erects buildings thereon without any written evidence of title, and does not inclose the same, its pos session will be limited to the ground actually occupied, *Ib.*

1000. RIGHT OF WAY—*by dedication.* The statute providing that streets, &c., designated on a town plat, when properly certified, &c., shall operate as a conveyance in fee to the public, does not apply in favor of individuals or private corporations. So, the reservation in a deed of a strip of land for railroad purposes, according to a diagram which shows the name of the railway company, will not operate as a conveyance of the strip to the company, or as a dedication. *Ib.*

1001. EJECTMENT—*by subsequent grantee.* The purchaser of land over which a railroad is constructed and operated without having acquired the right of way, may, upon receiving a conveyance of the legal title, maintain ejectment against the company for the land so tortiously taken and occupied. *Ch. & Iowa R. R.* v. *Hopkins*, 90 Ill. 316.

1002. TELEGRAPH COMPANY—*of the rights acquired.* A telegraph company, by the condemnation of land for its use, does not acquire the fee to the land or the right to use it for any other purpose than to erect poles and suspend wires on them, and maintain and repair the same, and use the structure for telegraph purposes. This includes the right at all times to enter upon the strip when necessary to construct or repair the line, doing as little damage as possible, but not the right to cultivate the ground. The only exclusive right of occupancy is the ground occupied by the poles. *Lockie* v. *Mut. Union Tel. Co.*, 103 Ill. 401.

1003. TRESPASS—*against telegraph company.* Trespass *quare clausum fregit* lies against a telegraph company by the owner of land for entering upon a highway over his land and erecting poles thereon without his assent. *Board of Trade Tel. Co.* v. *Barnett*, 107 Ill. 507.

1004. RIGHT TO TAKE POSSESSION. A railway company has no right to the possession of land for its right of way until the damages for the taking of the same have been assessed and paid, and if it takes possession before that is done, without the owner's consent, it is a trespasser, and the owner may bring ejectment or trespass, or both, and recover his property, and such damages as he may have sustained by the unlawful act. *Ch., St. L. & Western R. R.* v. *Gates*, — Ill. —. Filed March 23, 1887.

1005. EFFECT OF JUDGMENT—*passing title.* A judgment of condemnation of land for the widening of a street under the act of 1887 relating to the city of Chicago, as effectually concludes the former land owner from asserting title to the land taken, as a sale on execution or a recovery in ejectment. *Morris* v. *Chicago*, 11 Ill. 650.

1006. REVERSION—*street.* Where a deed of land for a street provides that where the same shall cease to be used as a street, or the street shall be abandoned or vacated, the land shall revert to the grantor, or his heirs or assigns, on vacation of the street, the land by virtue of such clause will revert; and also upon general principles, without such reservation. *Helm* v. *Webster*, 85 Ill. 116.

1007. PASSING TITLE. The final judgment of the circuit court approving the report of the commissioners appointed under the general law of 1859, relating to plank roads, etc., passes the title to the land condemned to the corporation. *Skinner* v. *Lake View Avenue Co.,* 57 Ill. 151.

1008. DIVESTITURE OF TITLE—*street.* Where condemnation of land is effected for a street, the damages assessed and accepted by the owners, who thereby give their assent to the proceedings, and possession is taken, the title is thereby divested from such owners, notwithstanding errors in the proceedings. *Rees* v. *Chicago,* 38 Ill, 322.

1009. The proprietor of land over which a railroad passes, after condemnation, has no right to build a fence on the right of way, or make cattle guards along the road. *A. & S. R. R.* v. *Baugh,* 14 Ill. 211.

1010. COMPENSATION—*to whom paid—persons entitled—attaching creditor.* Where a creditor of the land-owner has attached the land and obtained a judgment, payment of the money awarded in a proceeding to condemn, to such creditor, not exceeding his judgment, will be a payment to the party interested, in accordance with the statute. *C., B. & Q. R. R.* v. *Chamberlain,* 84 Ill. 333.

1011. RES ADJUDICATA —*judgment.* An adjudication upon an appeal from an award of commissioners, that the condemnation money belonged to the party appealing, is conclusive upon all the parties to the original proceeding, although they had no notice of the appeal. *Ib.*

1012. NAMING THE PARTIES TO BE PAID—*subsequent adjudication.* In a proceeding to condemn land under the act of 1852, owned by several persons as tenants in common, where there are adverse and conflicting claims by tax titles, attachment and judgment liens, it is sufficient for the commissioners, under § 6, to state in their report separately, the compensation to be paid for each lot of land, leaving it for the court to determine in regard to the rights of the respective claimants to the money awarded. *Ib.*

1013. FORECLOSURE. Where mortgaged property is condemned and appropriated to public use, and the compensation awarded to the owner or mortgagor exceeds the sum due on the mortgage and is not paid, it is not proper on bill to foreclose, to order a sale of the premises. The sum found due should be ordered paid out of the condemnation money. *Colehour* v. *State Savings Institution,* 90 Ill. 152.

1014. A railway company seeking the condemnation of a part of a lot for the purposes of its road, has no cause to complain of an order of court fixing the compensation to be paid, and directing the money to be paid to the treasurer of the county for the benefit of the owners of the property affected or those interested in it. Such an order does not determine who is entitled to the compensation awarded. *Ch. & W. Ind. R. R.* v. *Prussing,* 96 Ill. 203.

1015. MORTGAGED PROPERTY. Where the property of a mortgagor is condemned for public use and the compensation to be paid is assessed, the holder of the mortgaged debt will be entitled to be first paid out of it the amount due him, and the mortgagor the balance. *South Park Coms.* v. *Todd,* 112 Ill. 379.

1016. OWNER UNKNOWN OR NON-RESIDENT. If the owner is not known or is a non-resident, the money should be paid into the county treasury for his use. If paid to one not entitled to it, the court will compel its payment again to the rightful claimant. The commissioners awarding the compensation have no authority to determine to whom the money shall be paid. *Ib.*

1017. LANDLORD AND TENANT. A lessee is entitled to compensation for his unexpired term before he can be deprived of the use of his property. But if his term expires before the final hearing, he will have no interest to be taken, and cannot have compensation for improvements. *Schreiber* v. *Ch. & E. R. R.*, 115 Ill. 340.

1018. After service, a tenant, cannot by taking a new lease, not before secured by contract, acquire any new rights to compensation. *Ib.*

1019. Under the statute, the compensation awarded, is required to be paid, either to the person entitled to it, or to the county treasurer. It is error to direct its payment into court to await further proceedings to determine who is entitled to it. *McCormick* v. *W. Ch. Park Coms.*, 118 Ill. 655.

1020. The compensation, for which the public is liable in condemning land, must go to those who are entitled to the property itself, in proportion to their several interests. *Chicago* v. *Garrity*, 7 Bradw. 474.

1021. LANDLORD AND TENANT. As between landlord and tenant, the condemnation of land does not operate as an extinguishment in whole or in part of the lease, but the tenant remains liable to his landlord for the entire rent. *Ib.*

1022. APPORTIONMENT. A tenant is entitled to receive from the public full compensation for so much of his leasehold estate as is appropriated to the public use. The landlord's compensation should be diminished by reason of the existence of the lease-hold estate only by such an amount as the evidence shows that the actual rental value of the premises exceeds the rent reserved. *Ib.*

1023. The damages awarded the landlord and tenant respectively are the results of independent assessments; and because the aggregate assessment may exceed the entire value of the property taken, the public power seeking the condemnation, has such an interest therein, that it may insist upon a proper apportionment of damages between landlord and tenant. *Ib.*

1024. RIGHTS UNDER JUDGMENT. Until compensation is paid, there is no right of entry, and the company may abandon the location and adopt another. Until the selection by the company becomes binding, the owner may exercise all the rights of ownership not materially interfering with the condemnation proceeding, and so may remove machinery and buildings from the premises. *Schrieber* v. *Ch. & E. R. R.*, 115 Ill. 340.

1025. POSSESSION BEFORE CONDEMNATION—*trespass.* Where a railroad is located and operated over land of an estate without condemnation or otherwise acquiring the right of way, the taking and retaining of the land is a continuing trespass, and on judicial sale the whole land, including the so-called right of way, passes to the purchaser, and he will be entitled to compensation for the land taken and damages for any injury to the part not taken, on a proceeding to condemn. *Ch. & Iowa R. R.* v. *Hopkins*, 90 Ill. 316.

JUDGMENT.

1026. AWARD CONSTRUED. In a condemnation proceeding, the commissioners, after assessing the value of the estate and the improvements thereon, further awarded that if the improvement should be retained by the owner for three months, there was no damage from the interruption of his business, and if he should retain the possession for two months, then the damages to the business were fixed at $1,600, and if he should retain the possession one month, at $3,200: *Held*, that it rested with the railway company when to take possession, and that if it took possession within three months, it would have to pay the damages named, but that the owner could not force it to take possession at any time he might select, and then recover the damages provided by the award to be paid upon his having to give possession at that time, and if the company did not take possession within the three months it was not liable for any damages. *Glennon* v. *Ch M. & St. P. Ry.*, 79 Ill. 501.

1027. ARBITRATION—*enforcement of award.* A submission of all matters in dispute with regard to a right of way claimed by a railway company over a party's land is sufficiently broad to embrace an award as to the building of fences and crossings as well as the payment of a sum of money. No judgment for a sum of money can be rendered on such an award, but it may be enforced under § 8 of the statute relating to arbitration and award. *Kankakee & S. W. R. R.* v. *Alfred*, 3 Bradw, 511.

1028. BINDING FORCE—*collaterally.* The record of a condemnation proceeding where the jurisdiction appears, is competent evidence and cannot be impeached collaterally for errors or irregularities. *G. & Ch. Union R. R.* v. *Pound*, 22 Ill. 399.

1029. In trespass for removing a fence to open a road, the proceedings to establish the road cannot be attacked collaterally for mere errors not going to the jurisdiction, and parol evidence to show the jury adopted an improper basis in the assessment of damages, is inadmissible. *Hankins* v. *Calloway*, 88 Ill. 155.

1030. PRESUMPTION IN FAVOR OF. Where the court acquires jurisdiction under the act of 1852, by the proper notice, and the filing of the petition, its subsequent action in appointing commissioners will be presumed to be correct, and that they had the requisite qualifications. *C., B. & Q. R. R.* v. *Chamberlain*, 84 Ill. 333.

1031. COLLATERALLY. The report of the commissioners of the damages assessed, under the act of 1852, became final and conclusive upon the parties in all collateral proceedings without it appearing that notice had been given of the filing of the same with the clerk. It will be presumed notice was given. *Ib.*

1032. ACQUIESCENCE IN. Where the parties acquiesce in and ratify the award, it will be conclusive in respect to the interest claimed without regard to the giving of a notice. *Ib.*

1033. FRAUDULENT—*void—injunction.* Where a railway company proceeds to condemn for its own use, the road and track of another *de facto* railroad company, concealing the object and purpose, and giving no notice, and the whole proceeding shows it to be the carrying out of a scheme for the fraudulent and inequitable purpose of getting possession of the latter company's right of way and road, without making compensation, a court of equity will restrain the taking of possession under such fraudulent proceeding. *Cin., Laf. & Ch. R. R.* v. *Danville & Vin. Ry.*, 75 Ill. 113.

1034. COLLATERALLY—*error on face of proceedings.* Where the verdict of a jury, in a proceeding to condemn land for a public road, shows on its face that benefits were allowed against the value of the land taken, it will render the order establishing the road absolutely void. Such defect goes to the jurisdiction of the commissioners. *Hyslop* v. *Finch*, 99 Ill. 171.

1035. COSTS. Expenses attending an assessment of damages for right of way include costs, and they stand the same as the damages, and must be paid before possession can be taken of the land. *Ch. & Mil. R. R.* v. *Bull*, 20 Ill. 218.

1036. CROSS PETITION—*new parties by—what it may show.* § 11. Any person not made a party may become such by filing his cross petition, setting forth that he is the owner or has an interest in property, and which will be taken or damaged by the proposed work; and the rights of such last named petitioner shall thereupon be fully considered and determined. [R. S. 1887, p. 647, § 11; S. & C., p. 1051, § 11; Cothran, p. 649. See notes to § 9, *anti*, §§ 760–765.]

1037. APPEALS—*when lies and practice on.* § 12. In all cases, in either the circuit or county court, or before a circuit or county judge, an appeal shall lie to the supreme court. [R. S. 1887, p. 647, § 12; S. & C., p. 1051, § 12; Cothran, p. 650, § 12.]

1038. AS TO PUBLIC ROAD—*on question of damages.* A person whose land has been taken for a road has the right to be heard upon the question of damages upon an appeal to the supervisors, and it is error to dismiss his appeal, and the mode of appeal is not changed by the fact that the proposed road is upon a county line. *Deitrick* v. *Highway Comrs.,* 6 Bradw. 70.

1039. SAME—*joinder in by tenants in common.* Tenants in common may join in an appeal, but parties having different interests cannot. They must prosecute separate appeals. *Sangamon Co.* v. *Brown,* 13 Ill. 207.

1040. SAME—*where right exists.* No appeal is given to the owner of land from an order of the county court laying out a public road until the court orders the road to be opened, nor can land be appropriated for a road until such order has been made. *Ib.*

1041. SAME—*who has the affirmative.* The land-owner on the trial of an appeal, takes the affirmative and must prove the title to the land and show that he will sustain damage by the construction of the road. The county is defendant. *Ib.*

1042. SAME—*question involved.* If the county court had jurisdiction and proceeded regularly, the only question for review is the amount of the damages. The propriety of the road is not involved. *Ib.*

1043. SAME—*error in proceeding.* The circuit court may inquire into the regularity and validity of the proceedings, and if the county court has proceeded illegally or without lawful authority, the circuit court should reverse the order and leave the county court to proceed anew. *Ib.*

1044. SAME—*costs.* The county is liable for costs where an appeal is successfully prosecuted, and a material increase in the damages assessed is a successful prosecution. *Ib.*

1045. SAME—*refusal to lay road.* An appeal does not lie from a decision of the county court refusing to open and construct a road. *Ib.*

1046. AS TO PLANK ROAD, &c. The order of approval of the report of the commissioners appointed under the general law of 1859, relating to plank, gravel and macadamized roads, is a final judgment from which an appeal lies to this court, notwithstanding the act providing for the condemnation is silent as to an appeal or writ of error. *Skinner* v. *Lake View Avenue Co.,* 57 Ill. 151.

1047. SAME—*freehold involved.* Where, under the statute, the petition was presented to the court, commissioners were appointed who made their report, and the clerk recorded the orders and the court confirmed the report. *Held,* that this constituted a condemnation of the land by which the title passed to the corporation, and it relating to a free hold, an appeal laid to the supreme court. *Ib.*

1048. RIGHT OF WAY—*law of 1845.* Where the charter of a railway company provides for the condemnation for right of way under the act of March 3, 1845, an appeal will lie from the assessment of the commissioners. *Austin* v. *Belleville & Illinoistown R. R.*, 19 Ill. 310.

1049. SAME—*right to dismiss proceeding on.* On the removal of a proceeding to condemn land for right of way, by appeal or *certiorari* to the circuit court, the company seeking to condemn has the right to dismiss the proceeding, and it is error to refuse leave to do so. *Joliet & Ch. R. R.* v. *Barrows*, 24 Ill. 562.

1050. SAME—*by certiorari under statute.* Where an appeal is given from an assessment for a right of way, and the land-owner has not had notice of the proceeding in time to take an appeal, he may have a trial *de novo* by *certiorari* under the statute. *Ib.*

1051. SAME—*width of road fixed by report.* Upon an appeal from the assessment of the commissioners under the charter of the Peoria & Rock Island Railway company, to the circuit court, the report of the commissioners is the foundation of the appeal, and the width of land therein described, must control. *P. & R. I. Ry.* v. *Bryant*, 57 Ill. 473.

1052. COSTS ON—*law of 1852.* On appeal from the assessment of damages under the act of 1852, since the adoption of the new constitution, the land-owner should not be compelled to pay costs, if the assessment is confirmed or not increased. *People* v. *McRoberts*, 62 Ill. 38.

1053. QUESTIONS INVOLVED—*title to land.* The circuit court on appeal from the award of the commissioners can consider only the questions decided and reported by them. The question of title is not involved. *P., P. & J. R. R.* v. *Laurie*, 63 Ill. 264.

1054. CONSTITUTIONAL RIGHT. An appeal lies from the judgment of the circuit court condemning land for right of way under the act of 1852. This is a constitutional right conferred by that clause of the constitution defining the jurisdiction of the supreme court. *St. L. & S. E. Ry.* v. *Lux*, 63 Ill. 523.

1055. SERVICE OF NOTICE OF. In a proceeding to condemn land under the act of 1852, for the right of way of a railroad, notice of an appeal by the land-owner from the award of the commissioners to the circuit court, served upon the attorney of the railway company, is a nullity. *Hartman* v. *Belleville & O'Fallon R. R.*, 64 Ill. 24.

1056. WIDENING STREET. An appeal or writ of error lies to the final judgment of the circuit court in a proceeding to condemn property by a municipal corporation for the purpose of widening a street. *Hyde Park* v. *Dunham*, 85 Ill. 569.

1057. QUESTIONS INVOLVED IN. An order of the county court made in a proceeding to condemn land for a right of way after the allowance and perfecting of an appeal from the final judgment to the supreme court, authorizing the petitioner to enter on the premises pending the appeal, will not be involved in the appeal. *L. S. & M. S. R. R.* v. *Ch. & W. Ind. R. R.*, 100 Ill. 21.

1058. An appeal will lie directly from the county court to the supreme court in a proceeding to condemn land for right of way for a railroad under the eminent domain act. *Kankakee & Seneca R. R.* v. *Straut*, 101 Ill. 653.

1059. § 12 of the eminent domain act expressly gives an appeal directly to the supreme court from the judgment of condemnation, and there is nothing in the practice act that takes away this right. *P. & P. U. Ry.* v. *P. & F. Ry.*, 105 Ill. 110.

1060. SEPARATE APPEALS. Each separate owner is entitled to a separate appeal. *Johnson* v. *F. & M. R. Ry.*, 116 Ill. 521.

1061. FINAL ORDER—*dismissal of cross claim.* Where a cross petition filed in a proceeding to condemn under the eminent domain act, which brings before the court and states a claim of ownership or interest not stated in the original petition, is dismissed, the order of dismissal is final as to the rights claimed under it, and an appeal lies from the order of dismissal. *Johnson* v. *F. & M. R. Ry.*, 116 Ill. 521.

1062. SAME—*disposition of the compensation.* After judgment of condemnation awarding the payment of the compensation assessed to a party as the owner of the property, a subsequent order directing the payment of the condemnation money into court to await its further order as to whom to be paid, is such a final order as may be reviewed by appeal or writ of error. *McCormick* v. *W. Ch. Park Comrs.*, 118 Ill. 655.

1063. EXECUTOR—HEIRS. Where the land-owner dies pending a proceeding to condemn his executor cannot appeal from the judgment, unless he has some interest in the land by the will of the deceased owner. If he has no such interest the heirs alone can appeal. *Bower* v. *G. & M. R. R.*, 92 Ill. 223.

1064. RE-TAXATION OF COSTS. Under a former statute, review of taxation of costs by county court in condemnation case could be obtained in circuit court by appeal thereto, and motion for re-taxation. *Peoria & B. V. R. R.* v. *Bryant*, 15 Ill. 438.

1065. § 9 of act of 1852, requiring the execution of an appeal bond on taking an appeal from the award to the circuit court, and § 12, which permits the land to be entered upon pending the appeal, are repealed by the new constitution. *People* v. *McRoberts*, 62 Ill. 38.

1066. ABANDONMENT OF CLAIM—*failure to prosecute after reversal.* After reversal of a judgment of condemnation, if the land-owner deems the compensation awarded him insufficient, he should, within two years after the reversal, have the cause remanded and re-docketed, giving the proper notice, and have another trial. If he fails to do so and retains the sum paid him he will be regarded as having abandoned any claim for further compensation. *St. L., A. & T. H. R. R.* v. *Karnes*, 101 Ill. 402.

1067. BOND TO GIVE POSSESSION PENDING APPEAL—*Conditions and approval.* § 13. In cases in which compensation shall be ascertained as aforesaid, if the party in whose favor the same is ascertained shall appeal such proceeding, the petitioner shall, notwithstanding, have the right to enter upon the use of the property upon entering into bond, with sufficient surety, payable to the party interested in such compensation, conditioned for the payment of such compensation as may be finally adjudged in the case, and in case of appeal by petitioner, petitioner shall enter into like bond with approved surety. Said bonds shall be approved by the judge before whom such proceeding shall be had, and executed and filed within such time as shall be fixed by said judge. [R. S. 1887, p. 647, § 13; S. & C., p. 1052, § 13; Cothran, p. 650, § 13.]

1068. RIGHT TO POSSESSION—*pending appeal.* Under the act of 1852, when an appeal is taken from the award, if the party seeking the condemnation desires to enter upon and occupy the property pending the appeal, bond must be given to the person whose land is sought, to secure the payment of the judgment that may be rendered. Possession taken forcibly pending an appeal without giving such bond, is illegal, and may be recovered back by forcible entry and detainer. *Mitchell* v. *Ill. & St. L. R. R. & Coal Co.*, 68 Ill. 286.

, **1069.** COMPENSATION—*payment to county treasurer, &c.* § 14. Payment of compensation adjudged may, in all cases, be made to the county treasurer, who shall, on demand, pay the same to the party thereto entitled, taking receipt therefor, or payment may be made to the party entitled, his, her or their conservator or guardian. [R. S. 1887, p. 648, § 14; S. & C., p. 1052, § 14, Cothran, p. 650, § 14.

1070. RECORD—*of verdict and judgment.* § 15. The court or judge shall cause the verdict of the jury and the judgment of the court to be entered upon the records of said court. [R. S. 1887, p. 648, § 15; S. & C., p. 1052, § 15; Cothran, p. 650, § 15.]

1071. REPEAL. § 16. All laws and parts of laws in conflict with the provisions of this act are hereby repealed: *Provided,* that this act shall not be construed to repeal any law or part of law upon the same subject passed by this general assembly; but in all such cases this act shall be construed as providing a cumulative remedy. [R. S. 1887, p. 648, § 16; S. & C., p. 1053, §16; Cothran, p. 650, § 16.]

BENEVOLENT INSTITUTIONS.

An act for the further protection of the state institutions. Approved and in force March 9, 1867.

1072. LANDS OF STATE INSTITUTIONS NOT TAKEN. § 1. *Be it enacted by the People of the State of Illinois, represented in the general assembly,* That no part of any land heretofore or hereafter conveyed to the state of Illinois, for the use of any benevolent institutions of the state (or to any such institutions), shall be entered upon, appropriated or used by any railroad or other company for railroad or other purposes, without the previous consent of the general assembly; and no court or other tribunal shall have or entertain jurisdiction of any proceeding instituted or to be instituted for the purpose of appropriating any such land for any of the purposes aforesaid, without such previous consent. [Laws 1867, p. 165; R. S. 1887, p. 648, § 17; S. & C., p. 1053, § 17; Cothran, p. 650, § 17.

CHAPTER 82.

LIENS UPON RAILROADS,

An act to protect contractors, sub-contractors and laborers in their claims against railroad companies, or corporations, contractors or sub-contractors. Approved April 3, 1872. In force July 1, 1872.

1073. LIEN ON PROPERTY—*for material, supplies, labor, &c.* § 1. *Be it enacted by the people of the state of Illinois, represented in the general assembly,* That all persons who may have furnished, or who shall hereafter furnish to any railroad corporation now existing, or hereafter to be organized under the laws of this state, any fuel, ties, material, supplies, or any other article or thing necessary for the construction, maintenance, operation or repair of such roads, by contract with said corporation, or who shall have done and performed, or shall hereafter do and perform any work or labor for such construction, maintenance, operation or repair by like contract, shall be entitled to be paid for the same as part of the current expenses of said road; and in order to secure the same, shall have a lien upon all the property, real, personal and mixed, of said railroad corporation as against such railroad, and as against all mortgages or other liens which shall accrue after the commencement of the delivery of said articles, or the commencement of said work or labor: *Provided,* suit shall be commenced within six months after such contractor or laborer shall have completed his contract with said railroad corporation, or after such labor shall have been performed or material furnished. [Laws 1871–2, p. 279: R. S. 1887, p. 852, § 55; S. & C., p. 1533, § 52; Cothran, p. 936, § 51.]

1074. LIEN—*for what it is given—not money loaned.* This lien is given only for materials used, supplies furnished and for labor performed, in constructing, repairing, operating or maintaining the road. The loan of money or the payment of its creditors, is not embraced in the statute giving the lien. *C. & V. R. R.* v. *Fackney,* 78 Ill. 116.

1075. A party, who at the request of a railway company, takes up its certificates of indebtedness given to its laborers and others for the boarding of hands, is not entitled to any lien. *Ib.*

1076. ASSIGNMENT. The lien of a laborer upon the road for the sum due him is not assignable at law. The lien is enforceable only in equity. *Ib.*

1077. UNDER ACT OF 1861. Under the act of 1861, no one is entitled to a lien, unless his contract was directly with the railroad company, and suit is brought within three months after the action accrues. *Arbuckle* v. *Ill. Midland Ry.,* 81 Ill. 429.

1078. ACT OF 1872. This act relates only to labor and materials furnished after its passage, and gives no lien for labor, &c., furnished before its passage. *Ib.*

1079. Lien of material-man, who begins proceedings to enforce his lien within six months after last delivery of materials, is superior to lien of mortgage made after date of such last delivery and before bringing suit to enforce lien. *C. & A. R. R.* v. *Union Rolling Mill Co.*, 109 U. S. 702.

1080. Statutory lien held not waived by special contract, that the contractor shall have a lien on the rails till payment; nor to give credit beyond the time within which the statutory lien shall be enforced, when the purchaser fails to perform the conditions upon which that credit was agreed to be given. *Ib.*

1081. SUB-CONTRACTOR—LABORER—LIEN. § 2. Every person who shall hereafter, as sub-contractor, material-man, or laborer, furnish to any contractor with any such railroad corporation, any fuel, ties, materials, supplies, or any other article or thing, or who shall do and perform any work or labor for such contractor in conformity with any terms of any contract, express or implied, which such contractor may have made with any such railroad corporation, shall have a lien upon all the property, real, personal and mixed, of said railroad corporation: *Provided*, such sub-contractor, materialman or laborer, shall have complied with the provisions of this act; but the aggregate of all liens hereby authorized shall not, in any case, exceed the price agreed upon in the original contract to be paid by such corporation to the original contractor: *And, provided, further*, that no such lien shall take priority over any existing lien. [R. S. 1887, p. 852, § 56; S. & C., p. 1533, § 53; Cothran, p. 936, § 52.]

1082. LIEN DOES NOT EXTEND BEYOND SUB-CONTRACTOR. The statute giving liens on railroads does not extend beyond sub-contractors. One furnishing materials to a sub-contractor has no lien against the railroad company or its property. *Cairo & St. L. R. R.* v. *Watson*, 85 Ill. 531.

1083. PETITION—*must show the necessary steps taken*. In a proceeding by a sub-contractor to obtain a lien against a railway company for work and materials furnished according to an agreement with the original contractor, it must appear that all the steps required by the statute have been taken. *Cairo & St. Louis R. R.* v. *Cauble*, 4 Bradw. 133.

1084. RELEASE OF, BY CONTRACTOR. A release of all claims to a lien by the contractor to the owner, is a waiver of his right to a lien, and the sub-contractors taking their contracts subject to the fulfillment of the original contract, are equally bound, and not entitled to a lien. *Whitcomb* v. *Eustace*, 6 Bradw. 574.

1085. OF LABORER—*relation of sub-contractor to general contractor*. The work in a general contractor's contract was required to be performed in such manner as not to relieve him from the immediate charge and responsibility of the work, and were such that the company might forfeit the same for the neglect to put on a sufficient force to complete the work in the time stipulated, or to require him to make up balances due to the laborers or persons furnishing materials or supplies monthly. *Held*, that the relations of the sub-contractor to the general contractor were such that the work done and materials furnished under sub-contracts could be regarded as materials furnished or labor done under his contract, so as to enable those furnishing the same to enforce a lien against the road. *Solomon* v. *Nicholson*, 113 Ill. 351.

1086, Where a general contractor for building a road is held liable to the company to protect it against liens of laborers, and material-men, in a contract sub-letting a part of the work, reserved the option to retain in his own hands the amount of estimates, or such part thereof as he might deem necessary, and pay the laborers and other creditors of the sub-contractor, and charge the amount thereof as so much money paid to him, the general contractor may keep back estimates due the sub-contractor, and pay it out on debts incurred by him in attempting to perform his sub-contract; and in so doing the general contractor cannot be charged with meddling in his affairs, and such general contractor may make such payment through the sub-contractor as his agent, or by any other agent., *Ib.*

1087. NOTICE OF LIEN — *copy of contract, when to be served.* § 3. The person performing such labor, or furnishing such material, shall cause a notice, in writing, to be served on the president or secretary of such railroad corporation, substantially as follows, viz:

To ......, president, (or secretary, as the case may be) of the ......: You are hereby notified that I am (or have been) employed by ...... as a laborer (or have furnished supplies, as the case may be) on or for the ......, and that I shall hold all the property of said railroad (or railway, as the case may be) company to secure my pay.

If there shall be a contract in writing between the original contractor and sub-contractor, material-man or laborer, a copy of such contract, if the same can be obtained, shall be served with such notice and attached thereto, which notice shall be served at any time within twenty days after the completion of such sub-contract, or such labor: *Provided,* that no lien shall attach in favor of any person performing such labor or furnishing material until such notice shall have been served as above, or filed for record as hereinafter provided. [R. S. 1887, p. 852, § 57; S. & C., p. 1534, § 54; Cothran, p. 937, § 53.]

1088. NOTICE TO COMPANY ESSENTIAL—*sufficiency of petition as to notice.* A sub-contractor is not entitled to a lien, unless he complies with the statute in giving notice to the company. A petition showing the filing of a notice with the circuit clerk, without averring that the president and secretary of the company did not reside in the county, or could not be found in the county, is fatally defective as failing to show a right to the lien. *Cairo & St. Louis R. R. v. Cauble,* 85 Ill. 555.

1089. SAME—*copy of contract.* Copy of sub-contractor's contract must accompany the notice, but a copy of the contract between the original contractor and the company need not be attached thereto. *Cairo & St. Louis R. R. v. Cauble,* 4 Bradw. 133.

1090. WHEN NOTICE TO BE FILED WITH CLERK—RECORD OF SAME—*mailing copy to president, &c.* § 4. If neither the president or the secretary of such railroad corporation shall reside or can be found in the county in which the sub-contract was made, or labor performed, the laborer, or person furnishing labor or material, shall file said notice in the office of the clerk of the circuit court; and the clerk of the circuit court shall file and keep a record of said notice, and cause a

copy of the same to be mailed to the president or secretary
of said company, for which he shall receive the sum of twenty-
five cents, and said clerk shall keep a list of the names of the
persons so claiming lien, and the names of the corporation
against which such liens are claimed. [R. S. 1887, p. 853, §
58; S. & C., p. 1534, § 55; Cothran, p. 937, § 54.]

1091. ACTION FOR SUM DUE—JOINDER OF PARTIES—*filing
transcript of justice in circuit court.* § 5. If the money due
the person having given notice as aforesaid, shall not be paid
within ten days after the money shall become due and payable, then such person may commence suit therefor, in any
court having jurisdiction of the amount claimed to be due,
against the corporation with which the original contract was
made; or he may commence suit, as aforesaid, against such
railroad corporation and original contractor jointly, and execution to issue as in other cases. If execution, issued on
judgment obtained before a justice of the peace, shall be returned not satisfied, a transcript of such judgment may be
taken to the circuit court, and spread upon the records thereof,
and shall have all the force and effect of judgments obtained
in the circuit court, and execution issued thereon as in other
cases. [R. S. 1887, p. 853, § 59; S. & C., p. 1534, § 56; Cothran, p. 937, § 55.]

1092. ATTORNEY'S FEE—*to be taxed as costs.* § 6.
Whenever any suit, so brought, shall be determined in favor
of the plaintiff, the court shall allow, if before a justice, $5,
if in a court of record, $20, attorney's fees to be taxed as
costs. [R. S. 1887, p. 853, § 60; S. & C., p. 1535; § 57; Cothran, p. 938, § 56.]

1093. FAILURE OF ORIGINAL CONTRACTOR TO COMPLETE
CONTRACT; PETITION—*notice and decree.* § 7. Should the
original contractor in any case fail to complete his contract,
any person entitled to a lien, as aforesaid, may file his petition in any court of record, in any county through which
the road may be constructed, against the railroad corporation
and the contractors, setting forth the nature of his claim,
and the amount due as near as may be, [and] the fact that
the contractor has failed to complete his contract. The clerk
of said court shall thereupon cause a notice to be published
for four successive weeks in a newspaper printed in the
county, setting forth that said petition has been filed, and the
time when the writ issued on the same shall have been made
returnable, and all persons entitled to liens under this act
may enter their appearance and interplead in said cause, and
have their claims adjudicated; and it shall be the duty of the
court, in case the petitioner or claimants, or either of them,
establish their claims, to enter a decree against said corpor-

ation and original contractor, for the amount to which the persons so establishing their claims are respectively entitled, and such decree shall have the same force and effect as decrees in other cases. [R. S. 1887, p. 853, § 61; S. & C., p. 1535, § 58; Cothran, p. 938, § 57.]

1094. LIMITATION—*suit in three months.* § 8. The lien hereby created shall continue for three months from the time of the performance of the sub-contract, or doing of the work or furnishing the material as aforesaid, except when suit shall be commenced by petition as aforesaid, and in such cases all liens shall be barred by decree entered in such cause. [R. S. 1887, p. 853, § 62; S. & C., p. 1535, § 59; Cothran, p. 938, § 58.]

1095. TIME OF FILING PETITION. The statute provides that the lien shall continue for three months from the time of the performance of the work or furnishing the material; and suit to enforce such lien must be begun within the time limited. *C. & St. L. R. R.*, v. *Cauble*, 4 Bradw. 133.

1096. DECREE. Should be against the railroad and the original contractor, and the lien should only be enforced and the property of the company sold in default of payment, within a day to be fixed by the court. *Ib.* § 9, repealed. See R. S. chap. 131, § 5, and therefore omitted.

CHAPTER 110.

PRACTICE.

1097. ACTION AGAINST RAILWAY COMPANY—*in what county brought.* § 2. * * * * * Actions against a railroad or bridge company, may be brought in the county where its principal office is located, or in the county where the cause of action accrued or in any county into or through which its road or bridge may run. [Laws of 1877, p. 146; Laws of 1871-2, p. 338, § 2; R. S. 1887, p. 970, § 2; S. & C., p. 1773, § 2; Cothran, p. 1090, § 2. See Laws 1861, p. 180, § 1.]

1098. See *Bristol* v. *Ch. & Aurora R. R.*, 15 Ill. 436; *Peoria Ins. Co.* v. *Warner*, 28 Ill. 429; *Ill. Cen. R. R.* v. *Swearingen*, 33 Ill. 289; *Mineral Point R. R.* v. *Keep*, 22 Ill. 9.

1099. SERVICE ON CORPORATION—*return of—publication.* § 4. An incorporated company may be served with process by leaving a copy thereof with its president, if he can be found in the county, in which the suit is brought, if he shall not be found in the county, then by leaving a copy of the process with any clerk, secretary, superintendent, general agent, cashier, principal, director, engineer, conductor, station agent or any agent of said company found in the county, * and in case the proper officer shall make return upon such process that he cannot in his county find any clerk, secretary, superintendent, general agent, cashier, principal, director, engineer, conductor, station agent or other agent of said company, then such company may be notified by publication and mail in like manner and with like effect, as is provided in sections twelve and thirteen of an act entitled, "An act to regulate the practice in courts of chancery," approved March 15, 1872. [R. S. 1887, p. 970, § 5, as amended by Laws of 1877, p. 146, which added the portion after asterisk * ; S. & C., p. 1777, § 5; Cothran, p. 1091, § 5. For service from justice of the peace, see R. S. 1887, p. 822, § 21; S. & C., p. 1440, § 21; Cothran, p. 888, § 21.]

1100. SHOWING PARTY SERVED IS PRESIDENT—*amending.* On bill to foreclose mortgage, the summons against a bank was returned served by delivering a copy to F. M. The sheriff was allowed to amend his return out of court by adding that F. M. was president of the bank. *Held,* that the amendment was properly allowed. *Montgomery* v. *Brown,* 2 Gilm. 581.

1101. AN AGENT—*of foreign railway company.* If railroad companies having their officers and offices, do business, and have agents and property in this state, service of process may be made upon such agents in this state in the same manner as upon agents of local corporations. *Mineral Point R. R.* v. *Keep,* 22 Ill. 9.

1102. AGENCY MAY BE DENIED. If the fact of the agency is denied, the return of the officer as to that, is not conclusive. This should be put in issue by plea in abatement. *Ib.*

1103. WHAT KIND OF AGENT. The service of process upon any agent other than the law agent of a corporation, is sufficient, if properly made and returned. *Ch. & R. I. R. R.* v. *Fell,* 22 Ill. 333.

1104. ON PRESIDENT. Where a corporation is sued, the service should be on its president, if he resides in the county in which the suit is brought. *Ill. & Miss. Tel. Co.* v. *Kennedy,* 24 Ill. 319.

1105. RETURN. The return must be positive as to the service on the president, and the sheriff must take the responsibility of determining the fact. To serve the writ on A. B., as president, is not in compliance with the statute. *Ib.*

1106. WHEN ON AGENT. Process may be served upon an agent of a corporation in any county, provided the president of the company does not reside in the county where the process is issued. *Peoria Ins. Co.* v. *Warner,* 28 Ill. 429.

1107. A court has jurisdiction over a corporation of this state by service upon an agent, although its principal place of business may be in a different county from that where the agent was served. *Ib.*

1108. ON COUNTY. In suits against a county the process must be served upon the clerk of the county court, and the service must be at his office. *Kane Co.* v. *Young,* 31 Ill. 194.

1109. OUT OF COUNTY. Where the action was brought in the plaintiff's county where the cause of action accrued, against a corporation of the state, having its principal office in another county, and service of process was made upon the president in such foreign county: *Held,* that the service was insufficient to give jurisdiction. *Stephenson Ins. Co.* v. *Dunn,* 45 Ill. 211; *Ins. Co.* v. *Holzgrafe,* 46 Ill. 422.

1110. In such case the process should be sued out to the county of the plaintiff's residence, and if the president does not reside, or cannot be found therein, it may be served upon any other agent of the company found in the county, and a return of such facts will give the court jurisdiction. *Ib.*

1112. ON AGENT. In order that a return of service on an agent may be held good, it must show that the president of the company did not reside in, or was absent from the county. *St. L., A. & T. H. R. R.* v. *Dorsey*, 47 Ill. 288.

1113. MUNICIPAL CORPORATIONS. In an action against a municipal corporation, service upon the mayor and city clerk was held sufficient. The general statute has no application to such corporations. *People* v. *Cairo*, 50 Ill. 154.

1114. ABSENCE OF PRESIDENT. A return of service upon the cashier of an incorporated company showed, "the president not found in my county, he being a non-resident." *Held*, sufficient. *Reed* v. *Tyler*, 56 Ill. 288.

1115. JUSTICES' SUMMONS. A justice's garnishee summons was returned, "served the within by reading to the within named company therein Jan. 15, 1870." *Held*, that the service was a nullity and gave the court no jurisdiction. *Grand Tower M. & M. & Transp. Co.* v. *Schirmer*, 64 Ill. 106.

1116. The act of 1853 requires the service of process upon an incorporated company to be made on its president, if he is a resident of the county, and if he is absent from the county, or does not reside therein, that service shall be made by leaving a copy with any one of the several officers therein named. The service must be by copy, and the return should state the name of the person so served. *Ib.*

1117. ON AGENT.. Where the president of an insurance company does not reside in the county where suit is brought against the company, the statute authorizes service to be made upon an agent of the company resident in the county. *Bills* v. *Stanton*, 69 Ill. 51.

1118. BY COPY ONLY. Service of process on a railroad company under the practice act in force July 1, 1872, can only be by leaving a copy with the proper person, and cannot be by reading the same. *C. & V. R. R.* v. *Joiner*, 72 Ill. 520.

1119. ON AGENT. Where the return of the officer states that he read the process to a station agent, (naming him) of the defendant, the president and secretary not being residents of the county, it is defective, both, because it shows an attempted service by reading instead of by copy, and because it does not show that the president could not be found in the county. The fact that he was not a resident of the county does not exclude the idea that he might have been found therein at the time of service. *Ib.*

1120. RAILWAY COMPANY. The return on a summons was: "Served the within named railroad company by reading the same and delivering a copy thereof to C. D., cashier of said railroad company this, &c., the president of said company could not be found in my county this," &c.: *Held*, that the last date was evidently the date of the return of the writ and that the return shows that on the first named day. when the writ, was served, the president could not be found, and that the service and return was in strict conformity to the statute. *Ch. & Pac. R. R.* v. *Kœhler*, 79 Ill. 354.

1121. AGENCY DENIED. On motion to quash a return of service of a summons against a corporation, which shows service on one as agent, where the agency is denied, the defendant must disprove the agency, or the motion will be overruled. *Protection Life Ins. Co.* v. *Palmer*, 81 Ill. 88.

1122. The question whether the person served was an agent, cannot be raised by plea in abatement. Such a plea does not furnish a better writ. *1b. Contra*, see *Mineral Point R. R. v. Keep*, 22 Ill. 9.

1123. The return of a sheriff on a summons against an incorporated company, that he has "served the within named company by reading and delivering a copy thereof to A. B., president of said company," shows a sufficient service on the company. *Rock Valley Paper Co. v. Nixon*, 84 Ill. 11.

1124. AGENT OTHER THAN PRESIDENT. To authorize the service of summons against a corporation upon any officer or agent other than its president, it must appear by the return that the president cannot be found in the county; and even where this fact appears, a return of service on A. B. "as secretary" cannot be sustained. It must be stated he is secretary of the company. *Ch. Planing Mill Co. v. Merchants' Nat. Bank*, 86 Ill. 587.

1125. AGENCY OF FOREIGN CORPORATION. In a suit against a corporation created by an act of congress, not residing or doing business in this state, service of process upon an agent appointed by the land commissioner of the corporation and its trustees, whose business is merely to receive and transmit offers for lands and to assist in making sales, will not give the court jurisdiction, such person not being an agent of the corporation, in the sense of the statute. *Union Pac. R. R. v. Miller*, 87 Ill. 45.

1126. ABATEMENT TO RETURN. A corporation may put in issue the fact of the service of process upon it by plea in abatement, and thus contradict the officer's return, which is only *prima facie* evidence of the truth of the facts therein recited. *Union Nat. Bank v. First Nat. Bank*, 90 Ill. 56. See *Mineral Point R. R. v. Keep*, 22 Ill. 9; *Holloway v. Freeman, Id.* 197; *Sibert v. Thorp*, 77 Ill. 43; *Protection Life Ins. Co. v. Palmer*, 81 Ill. 88; *C. & St. L. R. R. v. Holbrook*, 92 Ill. 297.

1127. AGENTS OF FOREIGN CORPORATION. Where a foreign corporation does business and has agents in this state, with property, service may be had upon such corporation through such agents or officers doing business here, the same as upon domestic corporations. *Midland Pac. Ry. v. McDermid*, 91 Ill. 170.

1128. But where a foreign corporation does not transact its business in this state, and has no office or agents located in this state, service of process upon one of its officers or agents while temporarily in this state on private business, or passing through it, will confer no jurisdiction on the courts over such corporation. *Ib.*

1129. 'SUFFICIENCY OF RETURN. A sheriff's return of service of summons against a railway corporation was: "Sept. 4, 1872, served by reading to and delivering a true copy to C. D., a director of the defendant, the president of the defendant not residing or being found in my county:" *Held*, on bill to enjoin the collection of the judgment, that the return was sufficient and gave the court jurisdiction *Cairo & St. Louis R. R. v. Holbrook*, 92 Ill. 297.

1130. FOREIGN CORPORATION. Foreign corporations doing business in this state are liable to be sued the same as domestic corporations or citizens, and process may be served upon its agents in this state, and the word "process" in the practice act embraces process of every kind including garnishee process. *Hannibal & St. Joseph R. R. v. Crane*, 102 Ill. 249.

1131. INTERESTED DIRECTOR. On a bill by a director of a private corporation and others, stockholders and creditors of the corporation, the only service on the corporation was by leaving a copy of the summons with the complainant director, the return stating that "the president, clerk, secretary, superintendent, general agent, cashier and principal of said company not found." The bill alleged that the president and all the other directors and officers of the company were non-residents: *Held*, that the service as to the corporation was void, the director with whom the notice was left, being a party complainant in the suit, and the service being void, advantage might be taken of it on error, as well as in the trial court. *St. L. & S. Coal & Mining Co.* v. *Edwards*, 103 Ill. 472; *St. Louis & Sandoval Coal & Mining Co.* v. *Sandoval Coal & Mining Co.*, 111 Ill. 32.

1132. SUFFICIENCY OF RETURN. A return to a summons against a private corporation was as follows: "Served this writ on the within named defendant, C. S. E. U. Co., by delivering a copy thereof to E. N. K., director and treasurer of said company, the president of said company not found in my county, the 23d day of November, 1883." *Held*, that the return was good, filling the requirements of the statute. *Ch. Sectional Electric Underground Co.* v. *Congdon Brake Shoe Manuf. Co.*, 111 Ill. 309.

1133. PLEA IN ABATEMENT. A defendant corporation may plead in abatement to the service of process by contradicting the sheriff's return; and when it tenders a material issue and is properly verified, it is error to strike the plea from the files. *Ib.*

1134. SAME—*its sufficiency.* A plea in abatement by a corporation to the jurisdiction over its person, showing its organization under the laws of this state, and its representation by its president, naming him; that at the time of the issuing and service of the summons the president was a resident of the county, and not absent from the same, and that the service was not made upon him, presents an immaterial issue, and is obnoxious to demurrer, in not putting in issue the return that the sheriff was unable to find the president in the county. *Ib.*

1135. FOREIGN CORPORATION. Judgment against a foreign corporation doing business in this state upon service on their agents, is a personal one and conclusive in other states. *Penn. Co.* v. *Sloan*, 1 Bradw. 364, 373.

1136. This section applies to foreign corporations doing business in this state. *Ib.*

1137. The statute provides that service may be made upon a corporation by leaving a copy of the summons with the president, secretary, &c., if either can be found in the county; if not, then by leaving a copy of the summons with any director, clerk, &c., of such company found in the county. These constitute two classes, and service upon one class is primary to service upon the other; and before service upon persons of the second class will confer jurisdiction, it must appear affirmatively that service could not be had upon persons of the first class. *St. L., V. & T. H. R. R.* v. *Dawson*, 3 Bradw. 118.

1138. The return of the officer must show that the president of the company did not reside in, or was absent from the county, to make a service on a director, clerk, &c., a good one. *Ib.*

1139. Service upon a foreign insurance company which states that the president of the company was not found in the city of A, but fails to state that he was not found in the county where suit is brought, is insufficient. *Mich. State Ins. Co.* v. *Abens*, 3 Bradw. 488.

1140. ON AGENT—*company having ceased to do business in state.* The act relating to foreign insurance companies provides that when such company ceases to transact business in this state, the agents last designated, or acting as such, shall be deemed to continue for the purpose of serving process, &c., in such case, and service must be made upon such last designated agents of the company, and the sheriff takes upon himself the responsibility of determining whether service is actually made upon an officer of the company. *Ib.*

1141. WHO MEANT BY LAST DESIGNATED AGENT. The statute evidently refers to the agents last acting in the entire state, and not such as may have been dispensed within any particular county where the plaintiff happens to reside, provided others remain in the jurisdiction upon which service can be made. *Ib.*

1142. Foreign corporations doing business in Illinois may be sued here in the federal court though the statute has provided no specific form of service. *Wilson Packing Co.* v. *Hunter,* 8 Biss. 429.

1143. A foreign insurance company doing business in this state may be served with process under the above section. *Johnson* v. *Hanover Fire Ins. Co.,* 11 Biss. 452.

1143a. DOMICILE OR RESIDENCE OF RAILWAY COMPANY. See *S. & M. R. R.* v. *Morgan Co.,* 14 Ill. 163; *Bristol* v. *Ch. & Aurora R. R.,* 15 Ill. 436; *Mineral Point R. R.* v. *Keep,* 22 Ill. 9; *St. Clair* v. *Cox,* 106 U. S. 350; *Bank* v. *Earle,* 13 Pet. 588; *State* v. *Milw. &c., Ry.,* 45 Wis. 579; *C., D. & V. R. R.* v. *Bank,* 82 Ill. 493.

1143b. COUNTY IN WHICH TO BE SUED. *Bristol* v. *Ch. & A. R. R.,* 15 Ill. 436; *Winnesheik Ins. Co.* v. *Holzgrafe,* 46 Ill. 422; *C., D. & V. R. R.* v. *Bank,* 82 Ill. 493.

CHAPTER 114.

INCORPORATION OF RAILWAY COMPANIES.

An act to provide for the incorporation of associations that may be organized for the purpose of constructing railways, maintaining and operating the same; for prescribing and defining the duties and limiting the powers of such corporations when so organized. Approved and in force March 1, 1872. Laws 1871-2,' p. 625; R. S. 1887, p. 1000; S. & C., p. 1907; Cothran, p. 1136.

1144. INCORPORATION—PURPOSE AND POWER—*right to own and operate roads.* § 1. *Be it enacted by the people of the state of Illinois, represented in the general assembly,* That any number of persons, not less than five, may become an incorporated company for the purpose of constructing and operating any railroad in this state,* and that any and all railroads or transportation companies authorized to be incorporated and transact business in this state by virtue of this act, shall be and they are hereby authorized and empowered to purchase, own, operate and maintain any railroad sold or transferred under order or powers of sale or decree of, or sale under foreclosure of mortgage or deed of trust, and corporations heretofore organized under the provisions of the act hereby amended, their successors or assigns, shall have and possess all the powers and privileges conferred by this act. [As amended by act approved May 11, 1877. In force July 1, 1877. Laws 1877, p. 163. Amendment adds all after asterisk. * R. S. 1887, p. 1000; S. & C., p. 1907; Cothran, p. 1136.]

1145. HORSE AND DUMMY ROADS. It is doubtful whether this law has any application to "horse and dummy railroads." Chapter 66, R. S., entitled, "horse and dummy railroads," does not provide for the incorporation of this class of railroads. *Wiggins Ferry Co.* v. *E. St. L. Union Ry.,* 107 Ill. 450.

1146. INCORPORATION—*as a connecting road.* A company may organize under the general railroad law to construct a road exclusively within the limits of a city, for the purpose of transferring freights in railroad cars between the different depots, wharehouses, elevators, manufactories, &c., that are or may be on its line, or may be reached by its lateral branches. *Ib.*

1147. CONSTITUTION OF 1870—*effect on law of 1849.* The general railroad act of 1849, so far as it provided for the formation of railway companies, was not abrogated by the constitution of 1870, and a corporation organized under that law in 1871, followed by a user of corporate franchises, is a *de facto* corporation. *Cin., La Fay. & Ch. R. R.* v. *Dan. & Vin. R. R.,* 75 Ill. 113, 116.

1148. PRIOR LAWS—*repeal.* The provision of the general railroad law of 1849, prohibiting railroads from entering cities without municipal consent, if not repealed by implication by the act of 1872, is wholly so by the act of March 31, 1874. *Ch. & W. Ind. R. R.* v. *Dunbar,* 100 Ill. 110, 128.

1149. INCORPORATION—*under former laws.* Under the law of 1849, a railway company was not fully organized and entitled to exercise all its powers until the route and *termini* of its road were approved by the legislature. *Gillinwater* v. *Miss & Atlantic R. R.,* 13 Ill. 1.

1149a. The Atlantic & Mississippi Railroad Company was a valid and subsisting corporation, having full power to construct its road. *People* v. *Miss. & At. R. R.,* 14 Ill. 440.

1149b. When the prerequisites of the charter have been complied with, the powers of the corporation come into existence, and those of the commissioners cease. *Smith* v. *Bangs,* 15 Ill. 399.

1149c. Special charters will be valid notwithstanding the constitutional provision (Art. 10, § 1, Const. 1848,) requiring general laws for such purposes, without any recital or preamble. *Johnson v. Joliet & Ch. R. R.,* 23 Ill. 202.

1150. ORGANIZATION—*before being abrogated by constitution.* Acts and steps taken by corporators with a view to organize prior to the adoption of the constitution of 1870 held, sufficient to show corporate existence and prevent the abrogation of charter under § 2, Art. 11, of constitution. *McCartney* v. *Ch. & Evanston R. R.,* 112 Ill. 611.

1150a. NEW COMPANY OR RE-ORGANIZATION. An act provided that parties interested in a trust deed of a railway company, on purchase at the trustees' sale, should be incorporated by a different name from the old company, and be invested with all the corporate powers, &c., given to the old company, but gave the old stockholders no rights in the new one, and did not require the new company to pay the debts of the old one: *Held,* that the act created a new corporation and was not a re-organization of the old one, and that the new one took its purchase subject to no liens, except such as were paramount to the trust deed. *Morgan Co.* v. *Thomas,* 76 Ill. 120.

1151. DE FACTO CORPORATION—*sufficient collaterally.* An organization in fact followed by *user* of corporate franchises, is sufficient, except in a direct proceeding by *quo warranto* or *scire facias* by the state. The legality of the incorporation cannot be questioned collaterally. *Rice* v. *R. Is. & A. R. R.,* 21 Ill. 93; *Tarbell* v. *Page,* 24 Ill. 46; *Hamilton* v. *Carthage,* 24 Ill. 22; *Mendota* v. *Thompson,* 20 Ill. 197; *Jameson* v. *People,* 16 Ill. 257; *Mitchell* v. *Deeds,* 49 Ill. 416; *Marsh* v. *Astoria Lodge,* 27 Ill. 421; *Lewiston* v. *Proctor,* 27 Ill. 414; *Lawson* v. *Kolbenson,* 61 Ill. 405; *Baker* v. *Backus,* 32 Ill. 79; *McCarthy* v. *Lavasche,* 89 Ill. 270; *Cin., La. F. & Ch. R. R.* v. *Dan. & Vin. R. R.,* 75 Ill. 113, 116; *Goodrich* v. *Reynolds,* 31 Ill. 490; *Osborn* v. *People,* 103 Ill. 224; *P. & P. U. Ry.* v. *Peo. & F. Ry.,* 105 Ill. 110;

People v. *Trustees of Schools*, 111 Ill. 171. Curing defects in organization. See post 1464, 1465.

1152. ARTICLES OF INCORPORATION — RECORDING. § 2. Such persons shall organize by adopting and signing articles of incorporation, which shall be recorded in the office of the recorder of deeds in each county through or into which such railway is proposed to be run, and in the office of the secretary of state. R. S. 1887, p. 1001; S. & C., 1908; Cothran, p. 1136. See post 1173, 1206.

1153. ARTICLES—*their contents.* § 3. Such articles shall contain:

First—The name of the proposed corporation.

Second—The places from and to which it is intended to construct the proposed railway.

Third—The place at which shall be established and maintained the principal business office of such proposed corporation.

Fourth—The time of the commencement and the period of the continuance of such proposed corporation.

Fifth—The amount of the capital stock of such corporation.

Sixth—The names and places of residence of the several persons forming the association for incorporation.

Seventh—The names of the members of the first board of directors, and in what officers or persons the government of the proposed corporation and the management of its affairs shall be vested.

Eighth—The number and amount of shares in the capital stock of such proposed corporation.

[R. S. 1887, p 1001, § 3; S. & C., p. 1908, § 3; Cothran, p. 1136, § 3.]

1155. CORPORATION—*when brought into existence—general powers—evidence of incorporation.* § 4. When the articles shall have been filed and recorded as aforesaid, the persons named as corporators therein shall thereupon become and be deemed a body corporate, and shall thereupon be authorized to proceed to carry into effect the objects set forth in such articles, in accordance with the provisions of this act. (*a*) As such body corporate they shall have succession, and in their corporate name may sue and be sued, plead and be impleaded. The said corporation may have and use a common seal, which it may alter at pleasure; may declare the interests of its stockholders transferrable; establish by-laws, and make all rules and regulations deemed necessary for the mangement of its affairs in accordance with law. (*b*) A copy of any articles of incorporation filed and recorded in pursuance with this act, or of the record thereof, and certified to be a copy by the secretary of state, or

his deputy, shall be presumptive evidence of the incorporation of such company, and of the facts therein stated. (c)
[R. S. 1887, p. 1001; S. & C., p. 1908; Cothran, p. 1137.
(a) See next section. (b) See post, pp. 1157-1165. (c) See
post, pp. 1166-1171 (c).]

(*A*) WHAT CONSTITUTES A CORPORATION.

1156. WHEN INCORPORATION IS COMPLETE—*recording of arti-
cles.* The recording of the articles of incorporation seems essential to
the corporate existence. *Buff. & Alleg. R. R.* v. *Cary,* 26 N. Y. 75;
Ind. Furnace & M. Co. v. *Herkimer,* 46 Ind. 142; *Hunt* v. *Kansas & M
Bridge Co.,* 11 Kan. 412; *Oroville & V. R. R.* v. *Plumas Co.,* 37 Cal
354; *Abbott* v. *O. Smelting Co.,* 4 Neb. 416; *Baili Calvert Col. Ed.*
Soc., 47 Md. 117; *Stone* v. *Gr. Western Oil Co.,* 41 Ill. 85. See *Stowe* v.
Flagg, 72 Ill. 401; *Cresswell* v. *Oberly,* 17 Bradw. 281.

1156a. CORPORATION — *when formed.* A railway corporation,
under the general law, does not become a legal body until all the requirements of the statute have been complied with, and the articles filed
in the office of the secretary of state. While they remain in the hands
of a subscriber, before filing, he may erase his subscription or modify
it. *Burt* v. *Farrar,* 24 Barb. 518. But see *Cross* v. *Pinckneyville Mill
Co.,* 17 Ill. 54.

1156b. Where a general law provides that persons, who shall, by
articles in writing, associate themselves, and comply with the law,
shall become a body corporate, such persons will not, by merely
executing the articles to that effect, without complying with the other
provisions of the law, become a corporation. *Bigelow* v. *Gregory,* 73
Ill. 197.

1156c. Where a general law provides that persons may incorporate
by complying with its provisions, one of which is, that before commencing business, its articles shall be published in a certain way, and
a certificate of the purposes of the organization shall be filed in a certain public office, the performance of these acts is a necessary prerequisite to the existence of such corporation. *Ib.*

1156d. Until a fire insurance company has fully completed its organization by filing the auditor's certificate with the county clerk,
that, &c., the transaction of business is unauthorized. *Gent* v. *M. &.
M. Mut. Ins. Co.,* 107 Ill. 652.

1156e. A corporation must have full and complete organization
and existence as an entity before it can enter into any kind of contract, or transact any business. *Ib.*

1156f. CORPORATE EXISTENCE—*difference under charter and
general law.* There is a marked difference, as to the effect of irregularities and omissions in the organization of corporations, between a
case where the corporation is created by special charter followed by
acts of *user,* and a case where individuals seek to form themselves
into a corporation under a general law. In the latter case it is only
by a compliance with the statute that corporate existence can be acquired. *Bigelow* v. *Gregory,* 73 Ill. 197.

1156g. SAME—*where subscription of capital stock essential.* By
the filing and recording of articles of incorporation, a corporation is
created as efficient for all purposes as if its powers were conferred by
a special charter. Where the capital stock is fixed by the articles, or
by the charter, it must be all subscribed before the corporation will
have a legal existence. *Temple* v. *Lemmon,* 112 Ill. 51; *Allman* v.
Hav., Rantoul & E. R. R., 88 Ill. 521; *Stoneham Branch R. R.* v.

Gould, 2 Gray, 277; *N. Bridge* v. *Storey*, 6 Pick. 45, note; *Salem Mill Dam* v. *Ropes*, 6 Pick. 23; *Worcester & Nashua R. R.* v. *Hinds*, 8 Cush. 110; *N. H. Cent. R. R.* v. *Johnson*, 3 N. H. 390; *S. & R. R. R.* v. *Cushing*, 45 Me. 124.

1156h. SUBSTANTIAL COMPLIANCE. A substantial compliance with the statute is sufficient to make the organization valid. *People* v. *Stockton & V. R. R..* 45 Cal. 306: The omission of the names of the directors will not be fatal. *Eakright* v. *L. & N. I. R. R.*, 13 Ind. 404. But the articles, unless complete in substance, will not hold subscribers. *Duchess & C. R. R.* v. *Mabbett*, 58 N. Y. 397; *Monterey & S. V. R. R.* v. *Hildreth*, 53 Cal. 153.

1156i. CORPORATE EXISTENCE—*who may question it.* Where a company under the act of 1849 had taken all the steps to be incorporated, except to file the certificate of incorporation in the office of the secretary of state, it was *held*, that while this omission might sustain a *quo warranto* to oust the corporation of its franchises, it did not follow that it was not a corporation as to third persons. *Baker* v. *Backus*, 32 Ill. 79; *Stone* v. *Gr. Western Oil Co.*, 41 Ill. 85; *Tarbel* v. *Page*, 24 Ill. 46; *Hudson* v. *Green Hill Sem.*, 113 Ill. 618; *Baker* v. *Neff*, 73 Ind. 68; *Williamson* v. *Kokomo B. & L. Assoc.*, 89 Ind. 389; *Central Ag. Assoc.* v. *Alabama Co.*, 70 Ala. 120.

(B) GENERAL POWERS.

1157. BY-LAWS—*right to establish.* A railway company has an implied power to establish by-laws; but whether the power is conferred expressly or by implication, it is limited to such as are lawful and reasonable. *Chandler* v. *N. Cross R. R.*, 18 Ill. 190; *K. & P. R. R.* v. *Kendall*, 31 Me. 470; *Kent* v. *Quicksilver Mining Co.*, 78 N. Y. 159, 178, 182.

1157a. SAME—*as to stock and voting.* Company may make such by-laws regulating stock and the manner of voting upon it, as are consistent with its charter. *Chandler* v. *N. Cross R. R.*, 18 Ill. 190.

1157b. SAME—*binds members.* A person, by becoming a member of a corporation submits himself to the operation of all by-laws for its government, and by implication, agrees to be bound by them so far as they are within the corporate authority to enact. *People* v. *Board of Trade*, 45 Ill. 112.

1157c. SAME—*as to stranger.* The by-laws of a corporation are not evidence for it against strangers who deal with it, unless they are brought to their knowledge, and assented to by them. *Smith* v. *N. Car. R. R.*, 68 N. C. 107.

1157d. SAME—*personal liability by.* In the absence of legislative enactment or contract, a personal liability cannot be created against a stockholder by a by-law of the company. *Kennebeck, &c., R. R.* v. *Kendall*, 31 Me. 470.

1157e. SAME—*in consonance with nature and purpose of corporation.* The nature and purpose for which a corporation is created, is the controlling consideration in determining the validity of its by-laws. If they are foreign to its character, and a departure from its purposes, they are void. If otherwise, and in harmony with the general laws, they are valid. *People* v. *Board of Trade*, 45 Ill. 112.

1157f. As to reasonableness of by-laws regulating conduct of members. See *Dickenson* v. *Chamber of Commerce*, 29 Wis. 45; *State* v. *Chamber of Commerce*, 20 Wis. 63, 71.

1157g. BY-LAWS—*on whom binding.* Where the charter provides that the corporate powers of the company shall be exercised by a board of directors or managers, who may adopt by-laws for the government of the officers and affairs of the company, a by-law adopted at the first meeting of the stockholders, all of whom were present and participated therein, and who were the only persons interested in the company, either as officers, managers or stockholders, is binding, notwithstanding they may, in the adoption thereof, have designated themselves as stockholders, instead of managers. *People* v. *Sterling Burial Case, Mfg. Co.*, 82 Ill. 457.

1157h. SAME—*estopped to deny validity.* Where a stockholder participates in the adoption of by-laws, and acts, and acquires rights under them, and through his instrumentality they are held out to the public as the laws of the corporation, and outside parties acquire rights in the corporation on the faith of the validity of such by-laws, such stockholder will be estopped to deny their validity. *Ib.*

1157i. RULES AND REGULATIONS—*showing ticket before entering car.* A railway company may establish a rule requiring passengers to produce their tickets before entering the cars. *C., B. & Q. R. R.* v. *Boger*, 1 Bradw. 472.

1157j. SAME—*extra fare of one having no ticket.* Passengers neglecting to purchase tickets before embarking on cars, may be charged additional fare, if afforded proper facilities for getting tickets. If they pay from station to station without tickets, they may be compelled to pay an extra charge at each station. *C., B. & Q. R. R.* v. *Parks*, 18 Ill. 460.

1157k. SAME—*facilities for getting tickets.* The company must furnish proper facilities for procuring tickets, if it intends to charge extra fare when tickets are not obtained. If a ticket is applied for and not furnished, that fact may be shown by the station agent, and his certificate should be evidence to the conductor of such fact. *St. L., A. & Ch. R. R.* v. *Dalby*, 19 Ill. 353.

1157l. SAME—*liable for not adopting, &c.* Railway companies must adopt proper rules for the running of trains, and conform to them, or be responsible for all consequences. *C., B. & Q. R. R.* v. *George*, 19 Ill. 510.

1157m. SAME—*liability to employes for not adopting proper ones.* An employe entering the service of company with the knowledge that no provision has been made for protecting him from moving trains about the depot grounds, will have no cause of action for injuries resulting to him by reason of the neglect of the corporation to make such rules and regulations as prudence would require in that respect. *Haskin* v. *N. Y. Cent. & Hudson River R. R.*, 65 Barb. 129.

1157n. SAME—*procuring ticket before entering cars.* Where a railroad company carries passengers on freight trains, and in such case, requires tickets to be shown before entering the train, and a passenger disregards the rule, he can be expelled, but only at a regular station. *I. C. R. R.* v. *Sutton*, 42 Ill. 438.

1157o. A passenger who knowingly disregards a rule requiring tickets to be purchased before taking passage, may be expelled at any regular station, the same as one refusing to pay fare. *C. & A. R. R.* v. *Flagg*, 43, Ill. 364; *I. C. R. R.* v. *Sutton*, 53 Ill. 397.

1157p. PASSENGERS—*on what trains.* A railway company has the right to devote a portion of its trains exclusively to the carrying of freight, and to entirely exclude passengers from the same. It is not required to carry passengers on its freight trains, or freight on its passenger trains. *C. & A. R. R.* v. *Randolph*, 53 Ill. 510.

1157q. Where a passenger purchases a ticket, he only acquires the right to be carried according to the custom of the road. He has a right to go to the place for which his ticket calls, on any train that usually carries passengers to that place. *Ib.*

1157r. TRAINS NOT STOPPING AT ALL STATIONS. Railway companies furnishing reasonable means for carrying passengers to all their stations, have the right to run trains that only stop at designated, or the principal stations on their road, and it is the duty of a passenger to learn before getting on a train whether it will stop at all stations, or the principal ones. *Ib.*

1157s. RULES AND REGULATIONS — *in respect to passengers.* Whatever rules tend to the comfort, order and safety of the passengers on a railroad, the company is authorized to make and enforce. But such rules must always be reasonable and uniform in respect to persons. They must not discriminate on account of color. *Ch. & N. W. Ry.* v. *Williams,* 55 Ill. 185.

1157t. SAME—*ladies' car.* A rule setting apart a car for the exclusive use of ladies, and gentlemen with ladies, is a reasonable one and may be enforced. *Ch. & N. W. Ry.* v. *Williams,* 55 Ill. 185; *Bass* v. *Ch. & N.W. Ry.,* 36 Wis. 450.

1157u. SAME—*colored passengers.* Under some circumstances it might not be an unreasonable rule to require colored persons to occupy separate seats in a car furnished by the company, equally as comfortable and safe as those furnished for other passengers. But in the absence of any reasonable rule on the subject, the company cannot lawfully, from caprice, wantonness or prejudice, exclude a colored woman from the ladies' car, merely on account of her color. *Ch. & N. W. Ry.* v. *Williams,* 55 Ill. 185.

1157v. SAME—*as to passengers.* A railway company has the right to require of its passengers the observance of all reasonable rules, calculated to insure comfort, convenience, good order and behavior, and to secure the safety of its trains and the proper conduct of its business. *I. C. R. R.* v. *Whittemore,* 43 Ill. 420.

1157w. SAME—*surrender of ticket.* A rule requiring passengers to surrender their tickets to the conductor when called for, is a reasonable one and may be enforced. *Ib.*

1157x. SAME—*reasonableness of.* The reasonableness of a rule adopted by a company for the government of its business, is purely a question of law. *I. C. R. R.* v. *Whittemore,* 43 Ill. 420, 423.

1157y. SAME—*as to passengers on freight trains.* It is not an unreasonable rule to require that all persons desiring to ride on freight trains, shall procure tickets sold expressly for such trains. *I. C. R. R.* v. *Nelson,* 59 Ill. 110.

1158. A railway company has the clear right to make a rule that no one shall be carried as a passenger on its freight trains. But if it is accustomed to carry passengers on such trains, it will not be justified in refusing to carry a passenger, or in putting him off. *I. C. R. R.* v. *Johnson,* 67 Ill. 312.

1158a. It may require that passengers procure tickets before riding on freight trains, and conductors may expel from the cars, at regular stations, such as neglect to comply with the regulation. *T., P. & W. Ry.* v. *Patterson,* 63 Ill. 304.

1158b. SAME—*ladies' waiting room at depot.* Where separate waiting rooms are provided at a depot for ladies and gentlemen, a regulation that no gentleman without a lady shall be admitted in the ladies' room, is not only reasonable, but necessary to enable the company to discharge its duty to protect females at the depot from violence and insult. *T., W. & W. Ry.* v. *Williams,* 77 Ill. 354.

1158c. RULES AND REGULATIONS—*family ticket.* A family ticket will authorize a son residing with the holder as a member of the family to ride upon the road, although he may be over twenty-one years of age. But if the purchaser was informed when he bought the ticket that a son over that age would not be allowed to ride on it, such regulation of the company would be binding on the holder of the ticket, or any person attempting to ride on it. *Ch. & N. W. Ry.* v. *Chisholm, Jr.,* 79 Ill. 584.

1158*d.* SAME—*evidence of.* The published schedule of regulations respecting family tickets are not evidence, unless notice thereof is brought home to the party to be affected. *Ib.*

1158*e.* SAME—*passengers on freight train.* The law imposes no obligation on railway companies to carry passengers on freight trains, nor freight on passenger trains. It only requires them to carry both, leaving them to regulate the manner in which it shall be done. *Arnold* v. *I. C. R. R.,* 83 Ill. 273.

1158*f.* SAME—*As to servants and passengers.* A railway corporation has the right to make reasonable rules for the conduct of its employes and also for the conduct of its passengers. *C., B. & Q. R. R.* v. *McLallen,* 84 Ill. 109.

1158*g.* SAME—*reasonableness.* Whether a rule be reasonable or unreasonable, and therefore *ultra vires,* is a question of law for the court; but whether such rules are adequate for the safety of others, and the management of the train, is a question of fact for the jury. *C., B. & Q. R. R.* v. *McLallen,* 84 Ill. 109.

1158*h.* The reasonableness of regulations of a railway company affecting third persons, is a mixed question of law and fact. *Bass* v. *Ch. & N. W. Ry.,* 36 Wis. 450.

1158*i.* SAME—*witness may not construe.* A question asking a witness whether under a certain rule there would be any objection to doing a thing a certain way, is improper, as calling on the witness to construe the rule. *Penn. Co.* v. *Stœlke,* 104 Ill. 201.

1158*j.* SAME—*preventing a person from travelling on cars.* A railway company has no power to adopt rules and regulations prohibiting decently behaved persons from travelling on its road, who will pay their fare and conform to all reasonable requirements for the safety and comfort of passengers. *C., B. & Q. R. R.* v. *Bryan,* 90 Ill. 126.

1158*k.* SAME—*manner of entering car.* Company has the right to make all reasonable rules respecting the time, manner and place of entering cars; and these rules when known to the passenger, he is bound to conform to, or he cannot recover for an injury sustained thereby. 26 Iowa, 124.

1159. In an action for an injury from a collision, it is not sufficient for the company to show that the plaintiff was at the time acting in disobedience of a proper order to secure his safety. It should further appear that the injury was *caused by* such disobedience. *L. & Upper Miss. R. R.* v. *Montgomery,* 7 Ind. 474.

1160. TICKET—*rule requiring passengers to show, and also to surrender ticket.* See *B. & O. R. R.* v. *Blocher,* 27 Md. 277; *Davis* v. *K. C., St. J. & C. B. R. R.,* 53 Mo. 317; *Northern R. R.* v. *Page,* 22 Barb. 130.

1161. A regulation requiring passengers either to present evidence to a conductor of a right to a seat, when reasonably required so to do, or to pay fare, is reasonable; and for non-compliance therewith, a passenger may be lawfully put off the train. *Townsend* v. *N. Y. Cent. & H. River R. R.,* 56 N. Y. 295.

1162. TO TAKE AND NEGOTIATE NOTES. A railway company has the inherent power to take and negotiate promisory notes in the ordinary course of business. *Frye* v. *Tucker,* 24 Ill. 180; *Goodrich* v. *Reynolds,* 31 Ill. 490; *Foy* v. *Blackstone,* 31 Ill. 538.

1163. POWER TO LEASE, OR TAKE LEASE. Power to a railway company to lease its road to another corporation, or to receive from another corporation a lease of the road of the latter, is conferred only by special authorization in charter or other legislative action. Such power is not among the ordinary powers of railway companies. *Penn. R. R.* v. *St. L. &c. R. R.*, 118 U. S. 290.

1164. A railway company cannot transfer or lease its lines, unless authorized by statute. *Troy & Boston R. R.* v. *Boston & Hoosac Tunnel & Western Ry.*, 86 N. Y. 107; *Atty. Genl.* v. *Niagara Falls Bridge Co.*, 20 U. Canada, 34; *Abbott* v. *J. G. & K. R. R.*, 80 N. Y. 27. See *Ill. Mid. Ry.* v. *People*, 84 Ill. 426.

1165. Without enabling legislation, a railroad company possesses no power to lease its road to a foreign corporation, and surrender its road and franchises into its control. *Archer* v. *T. H. & Ind. R. R.*, 102 Ill. 493.

(C) EVIDENCE OF INCORPORATION.

1166. BOOKS—*to show exercise of corporate acts.* Where certain steps are required to be taken before a corporation has existence, such as the opening of books, subscription of the capital stock and the choice of directors, the corporation books showing the election of officers, is *prima facie* evidence to show that the prerequisites of the statute have been complied with, and that the corporation has an existence. *Ryder* v. *A. & S. R. R.*, 13 Ill. 516, 523.

1167. The books of a railway company showing its organization are competent evidence for that purpose. *Peake* v. *Wabash R. R.*, 18 Ill. 88.

1168. JUDICIAL NOTICE. This court cannot take judicial notice of the existence of a railroad in a county. *Log., Peo. & B. R. R.* v. *Caldwell*, 38 Ill. 280. See *Danv. & White Lick Pl. R. Co.* v. *State*, 16 Ind. 456.

1169. USER UNDER GENERAL LAW. To show an incorporation under a general law, except as against the state, it is sufficient to show a *user* by a professed organization under the law. *Mitchell* v. *Deeds*, 49 Ill. 416; Abbot's Trial Evid., 30.

1170. ADMISSION OF CORPORATE EXISTENCE by dealing with the body as a corporation. *Mitchell* v. *Deeds*, 49 Ill., 416; *Miami Powder Co.* v. *Hotchkiss*, 17 Bradw. 622; *Brown* v. *Scottish A. M. Co.*, 110 Ill. 235; *Hudson* v. *Green Hill Seminary*, 113 Ill. 618.

1171. THE ARTICLES OF ASSOCIATION of a corporation certified by the secretary of state, are *prima facie* evidence of the fact that the full amount of the capital stock required by the articles has been subscribed. *Jewell* v. *Rock River Paper Co.*, 101 Ill. 57.

1171a. PROOF OF INCORPORATION—*organization under general law.* The existence or the formation of the corporation under the general law, may be proved, unless the law otherwise provides, by producing the certificate of organization which the law requires to be filed, with proof of its filing. *Chamberlain* v. *Huguenot Manf. Co.*, 118 Mass. 532; *Leonardsville Bank* v. *Willard*, 25 N. Y. 574; *Augur, &c.* v. *Whittier*, 117 Mass. 451; *Hawes* v. *Anglo Saxon Petroleum Co.*, 101 Mass. 385; *Priest* v. *Essex Hat Co.*, 115 Mass. 380; see also *Mokelumne* v. *Woodbury*, 14 Cal. 424; *New Eel River Drain Assoc.* v. *Durbin*, 30 Ind. 173.

1171b. The statute makes a certified copy of the articles evidence equally with the original. In the absence of such a provision the original would be the best evidence. *Jackson* v. *Leggett*, 7 Wend. 377; *Evans* v. *Southern Turnpike*, 18 Ind. 101.

1171c. Where the corporate existence of the plaintiff is denied, the original articles of association, properly recorded, may be read in evidence, without a certificate of the clerk that it is a true copy. *Fortin* v. *U. S. Wind Engine & Pump Co.*, 48 Ill. 451.

1172. LIMIT OF CHARTER—RENEWAL. § 5. No such corporation shall be formed to continue more than fifty years in the first instance, but such corporation may be renewed from time to time, in such manner as may be provided by law, for periods not longer than fifty years: *Provided*, that three-fourths of the votes cast at any regular election for that purpose shall be in favor of such renewal, and those desiring a renewal shall purchase the stock of those opposed thereto at its current value. [R. S. 1887, p. 1001, § 5; S. & C., p. 1908, § 5; Cothran, p. 1137, § 5.]

1173. BY-LAWS RECORDED. § 6. A copy of the by-laws of the corporation, duly certified, shall be recorded as provided for the recording of the articles of association in section 2 of this act; and all amendments and additions thereto, duly certified, shall also be recorded as herein provided, within ninety days after the adoption thereof. [R. S. 1887, p. 1001, § 6; S. & C., p. 1909, § 6; Cothran, p. 1137, § 6. Cited in *Allman* v. *Havan. &c. R. R.*, 88 Ill. 521.]

1174. PUBLIC OFFICE IN THIS STATE—*books of stock—inspection of.* **§ 7.** Every such corporation organized under the provisions of this act shall have and maintain a public office or place in this state for the transaction of its business, where transfers of all its stock shall be made, and in which shall be kept for public inspection, books, wherein shall be recorded the amount of capital stock subscribed and by whom, the names of the owners of its stock, the number of shares held by each person, and the number by which each of said shares is respectively designated, and the amounts owned by them respectively, the amount of stock paid in, and by whom, the transfers of said stock, the amount of its assets and liabilities, and the names and places of residence of all its officers. [R. S. 1887, p. 1001, § 7; S. & C., p. 1909, § 7; Cothran, p. 1137, § 7.]

1175. DIRECTORS—THEIR ELECTION AND CLASSIFICATION—VACANCY. § 8. All the corporate powers of every such corporation shall be vested in and be exercised by a board of directors, who shall be stockholders of the corporation, and shall be elected at the annual meetings of stockholders at the public office of such corporation within this state. The number of such directors, the manner of their election, and the mode of filling vacancies, shall be specified in the by-laws, and shall not be changed except at the annual meetings of the stockholders. The first board of directors shall classify themselves by lot in such manner that there shall be, as nearly as practicable, three directors in each class. Those belonging to the first class shall go out of office at the end of one year, those of the second class at the end of two years, and in like manner those of each class shall go out of office at the expiration of a number of years corresponding to the number of his class; and all vacancies occurring by reason of expiration of term shall be filled by election for a term of years equal to the number of classes. [R. S. 1887, p. 1001, § 8; S. & C., p. 1909, § 8; Cothran, p. 1138, § 8. Post 1187, 1425.]

1176. DIRECTORS—*trustees.* Directors of a railway corporation are trustees of the funds and other property of the corporation for the stockholders. *Cheeney* v. *L., B. & M. Ry.,* 68 Ill. 570; *Holder* v. *L., B. & M. Ry.,* 71 Ill. 106; *Gil., Clinton & Springf. R. R.* v. *Kelley,* 77 Ill. 426; *Peterson* v. *Ill. Land & Loan Co.,* 6 Bradw. 257; *Blake* v. *Buffalo Creek R. R.,* 56 N. Y. 485.

1176a. SAME—*interest in contracts with company.* It is illegal for directors of a railway company to become members of a company contracting to build the road, so as to share in the profits. *G., C. & Sp. R. R* v. *Kelley,* 77 Ill. 426. See *European & N. Am. R. R.* v. *Poor,* 59 Me. 277.

1177. COMPENSATION. The president and directors of a railway company are not entitled to any compensation for their ordinary services as such officers, unless the amount is fixed in the by-laws, or by resolution spread upon the record, before the services are rendered. *Cheeney* v. *L., B. & M. Ry.,* 68 Ill. 570; *Am. Cent. R. R.* v. *Miles,* 52 Ill. 174; *Merrick* v. *Peru Coal Co.,* 61 Ill. 472; *R., R. I. & St. L. R. R.* v. *Sage,* 65 Ill. 328; *Holder* v. *L., B. & M. Ry.,* 71 Ill. 106; *Gridley* v. *L., B. & M. Ry.,* 71 Ill. 200; *Hall* v. *Vt. & Mass. R. R.,* 28 Vt. 401; *Barstow* v. *City R. R.,* 42 Cal. 465.

1177a. It is not sufficient to prove that the matter of allowing compensation was talked over by the board, where the record of their proceedings fails to show any allowance. *R., R. I. & St. L. R. R.* v. *Sage,* 65 Ill. 328.

1177b. Where the by-laws of a private corporation provide that the officers shall receive such compensation for their services as shall be determined at the annual meeting of the stockholders, or at any special meeting called for that purpose, and none are ever so fixed, an officer performing the ordinary duties and services pertaining to his office, will not be entitled to recover for such services, in the absence of any agreement to pay him for the same. *Ill. Linen Co.* v. *Hough,* 91 Ill. 63.

1178. COMPENSATION—*for services not incident to office.* Directors employed to perform duties or services disconnected with their office, may recover or receive compensation for such services. *Holder* v. *L., B. & M. Ry.,* 71 Ill. 106; *Gridley* v. *L., B. & M. Ry.,* 71 Ill. 200; *Ill. Linen Co.* v. *Hough,* 91 Ill. 63.

. A director appointed to perform duties not pertaining to his office, such as to solicit the subscription of stock, or to procure the right of way, may recover for such services when rendered; but he cannot recover for services performed as a member of the executive committee, nor in making efforts to contract for the construction of the road, including time and travel, as these are a part of his duties as director. *Cheeney* v. *L., B. & M. Ry.,* 68 Ill. 570.

1178b. If the finance committee of a railway company audits an account of the president for ordinary services, and draws an order for its payment, where no compensation has been provided before the services were rendered, it will be illegal, and no recovery can be had. *Gridley* v. *L., B. & M. Ry.*, 71 Ill. 200.

1179. POWERS OF DIRECTORS. Charter directors can do such acts only as are necessary to set the association in motion as a corporation; they cannot make contracts, or incur liabilities for the construction of the road. *Allman* v. *Hav., R. & E. R. R.*, 88 Ill. 521.

1179a. SAME—*increase of capital.* A special charter which in terms vested all corporate powers in the directors, *held*, not to authorize them to increase the capital stock without assent of the stockholders. *Ry. Co.* v. *Allerton*, 85 U. S. 233.

1180. ELECTION OF DIRECTORS—*by-laws.* A railway company may make such by-laws regulating stock and the manner of voting upon it as are consistent with its charter. *Chandler* v. *N. Cross R. R.*, 18 Ill. 190.

1180a. SAME—*freedom in voting.* One stockholder has no right to direct how the votes of another shall be cast, nor for whom. *Ryder* v. *A. & S. R. R.*, 13 Ill. 516.

1181. SAME—*proxy by city.* The city of Alton and non-residents had a right to become stockholders in this company; and the city might give its proxy to any one it chose. *Ib.*

1182. STOCKHOLDERS MEETINGS — *how called between annual meetings.* § 9. A meeting may be called at any time during the interval between such annual meetings, by the directors, or by the stockholders owning not less than one-fourth of the stock, by giving thirty days' public notice of the time and place of such meeting in some newspaper published in each county through or into which the said railway shall run, or be intended to run, provided there be a newspaper published in each of the counties aforesaid; and if, at any such special meeting so called, a majority in value of the stockholders equal to two-thirds of the stock of such corporation, shall not be represented in person or by proxy, such meeting shall be adjourned from day to day, not exceeding three days, without transacting any business; and if, within said three days, two-thirds in value of such stock shall not be represented at such meeting, then the meeting shall be adjourned, and a new call may be given and notified as hereinbefore provided. [R. S. 1887, p. 1002, § 9; S. & C., p. 1909, § 9; Cothran, p. 1138, § 9. Post 1206.]

1183. ANNUAL STOCKHOLDERS' MEETING—*report or statement of corporate affairs.* § 10. At the regular annual meeting of the stockholders of any corporation organized under the provisions of this act, it shall be the duty of the president and directors to exhibit a full, distinct and accurate statement of the affairs of the said corporation; and at any meeting of the stockholders, or a majority of those present (in person or by proxy), may require similar statements from the president and directors, whose duty it shall be to furnish such statements when required in manner aforesaid.

1184. POWERS OF STOCKHOLDERS—*to fix amount of loans and interest.* And at all general meetings of the stockholders, a majority in value of the stockholders of any such corporation may fix the rates of interest which shall be paid by the corporation for loans for the construction of such railway and its appendages, and the amount of such loans.

1185. SAME—*removal of officers.* At any special meeting, by a two-thirds vote in value of all the stock, such stockholders may remove any president, director or other officer of such corporation, and elect others instead of those so removed.

1186. STOCKHOLDERS—*right to examine books, &c.* All stockholders shall, at all reasonable hours, have access to and may examine all the books, records and papers of such corporation. [R. S. 1887, p. 1002, § 10; S. & C., p. 1910, § 10; Cothran, p. 1138, § 10.]

1187. ELECTION OF DIRECTORS—*on failure to elect at proper time.* § 11. In case it shall happen, at any time, that an election of directors shall not be made on the day designated by the by-laws of such corporation for that purpose, the corporation, for such cause, shall not be dissolved, if within ninety days thereafter the stockholders shall meet and hold an election for directors in such manner as shall be provided by the by-laws of such corporation: *Provided,* that it shall require a majority in value of the stock of such corporation to elect any member of such board of directors, and a majority of such board of directors shall be citizens and residents of this state. [Const., art. 11, § 11; R. S. 1887, p. 1002, § 11; S. & C., p. 1910, § 11: Cothran, p. 1139, § 11.]

1187a. DIRECTORS—*constitutional provision as to residence of, construed.* The constitutional provision (art 11, § 11) that "a majority of the directors of any railroad corporation, now incorporated, or hereafter to be incorporated by the laws of this state, shall be citizens and residents of this state," has no application to a railway corporation formed prior to the adoption of the constitution by the consolidation of a railway company of this state with one of another state, by the consent of each of such states. Such a corporation exists under the laws of the two states and cannot be said to be incorporated solely under the laws of this state. *O. & M. Ry. v. People,* — Ill. —. Filed Jan. 18, 1888.

1188. OFFICERS—*their duties.* § 12. There shall be a president of such corporation, who shall be chosen by and from the board of directors, and such other subordinate officers as such corporation, by its by-laws, may designate, who may be elected or appointed, and shall perform such duties and be required to give such security for the faithful performance thereof as such corporation, by its by-laws, shall require: *Provided,* that it shall require a majority of the directors to elect or appoint any officer. [R. S. 1887, p. 1002, § 12; S. & C., p. 1910, § 12; Cothran, p. 1139, § 12.]

1189. PRESIDENT. The president of a corporation may perform all acts which are incident to the execution of the trust reposed in him, such as custom or necessity has imposed upon the office, and this without express authority. *Mitchell* v. *Deeds*, 49 Ill. 416.

1190. POWER OF OFFICERS AND AGENT. A corporation, unless otherwise provided by its charter, may by resolution, or by-law, appoint any person agent to dispose of its property or negotiable securities. No officer of the corporation has such exclusive power, unless given by the charter. *Ib.*

1191. POWER OF PRESIDENT—*to employ counsel.* Where the by-laws of a corporation make it the duty of the president to exercise a general supervision over its entire business, and provide that its property shall be under his control, and as such president for several years before he had acted as its attorney, this will be evidence of his authority to employ an attorney. *Wetherbee* v. *Fitch*, 117 Ill. 67.

1191a. SUPERINTENDENT. The general superintendent may, in the exercise of his power as such, bind the company for the discharge of liabilities assumed by a station agent towards an injured employe. *T. W. & W. Ry.* v. *Rodrigues*, 47 Ill. 188.

1191b. STATION AGENT. Where a railroad station agent engages a surgeon to attend an employe injured in the service of the company, although the act is unauthorized, yet the company will be liable, if, upon due notice given to the general superintendent, the act is not repudiated. *T. W. & W. Ry.* v. *Prince*, 50 Ill. 26; *T. W. & W. Ry.* v. *Rodrigues*, 47 Ill. 188. See also *Ind. & St. L. R. R.* v. *Morris*, 67 Ill. 295. Admissions of agent when binding on company. *C. B. & Q. R. R.* v. *Coleman*, 18 Ill. 297.

1192. PAYMENT OF SUBSCRIPTIONS TO CAPITAL STOCK—*forfeiture of payment.* § 13. The directors of such corporation may require the subscribers to the capital stock of such corporation to pay the amount by them respectively subscribed, in such manner and in such installments as they may deem proper. If any stockholder shall neglect to pay any installment as required by a resolution or order of such board of directors, the said board shall be authorized to declare such stock and all previous payments thereon forfeited for the use of the corporation; but the said board of directors shall not declare such stock so forfeited until they shall have caused a notice in writing to be served on such stockholder personally, or by depositing the same in a postoffice, properly directed to the postoffice address of such stockholder, or if he be dead, to his legal representatives, with necessary postage for its transmittal properly prepaid, stating therein that in accordance with such resolution, or order, he is requested to make such payment, at a time and place and in the manner to be specified in such notice, and that if he fails to make the same in the manner requested, his stock and all previous payments thereon will be forfeited for the use of such corporation; and thereafter such corporation, should default in payment be made, may sell the same and issue new certificates of stock therefor: *Provided*, that the notice as aforesaid shall be personally served or duly deposited, as above required, at least sixty days previous to the day on which such payment is required to be made. [R. S. 1887, p. 1002, § 13; S. & C., p. 1910, § 13; Cothran, p. 1139, § 13.]

RELEASE OF SUBSCRIPTION.

1192a. ALTERATION OF CHARTER—*authorizing consolidation.* An amendment of the charter authorizing the consolidation of the road to be built, with any other intersecting road and there terminating the same, is not such an alteration of the original project as to excuse the payment of a subscription for stock. *Sprague* v. *Ill. River R. R.*, 19 Ill. 174.

1192b. An act of incorporation may be amended, and if the amendment is accepted by the directors, the stockholders under the original act, unless otherwise stated, will be held liable. *Ill. River R. R.* v. *Zimmer*, 20 Ill. 654.

1192c. It is no defense to an action to collect an installment of a subscription, that the company has accepted an amendment to its charter after the defendant had subscribed, authorizing it to extend its road, and otherwise to assume new and increased responsibilities. *Rice* v. *R. I. & Alton R. R.*, 21 Ill. 93; *Hays* v. *O. O. & F. R. V. R. R.*, 61 Ill. 424.

1192d. It is no defense that the charter has been so changed as to authorize the company to purchase stock in other railroad companies, even though the *terminus* of the road is thereby changed. *T. H. & Alton R. R.* v. *Earp*, 21 Ill. 291.

1192e. MATERIAL CHANGE IN ENTERPRISE—*releases subscribers.* Where a charter to build a railroad across the state as a continuous project under one management, with a common interest, is, after subscription, so amended as to divide the project into three parts, to be under separate control, and no proper acceptance of the change of the charter is manifested, subscribers to the stock will thereby be released. *Fulton Co.* v. *Miss. & Wab. R. R.*, 21 Ill. 338, 370. See *Ross* v. *C., B. & Q. R. R.*, 77 Ill. 127.

1192f. A subscriber who agrees to be subject to the rules and regulations of the directors which they may adopt, cannot avoid payment, because the charter has been amended, reducing the number of days' notice to be given of calls, if the amendment of the charter has been accepted. *Ill. River R. R.* v. *Beers.* 27 Ill. 185.

1192g. A subscriber will be liable on his subscription, although the legislature may have authorized, and the directors may have adopted a change of route from that originally fixed, provided the change does not make an improvement of a different character, and his interest is not materially affected by the alteration. *Banet* v. *Alton & Sang. R. R.*, 13 Ill. 504, 511.

1192h. A subscription to stock may be collected, although amendatory acts have been subsequently passed, affecting the original charter, by extending its powers. *P. & O. R. R.* v. *Elting*, 17 Ill. 429.

1192i. INJUNCTION—*of collection of subscription.* If a railway company ceases to prosecute work, attempts to misapply its means, or attempts any radical change in the character of the enterprise, it may be enjoined from collecting the obligations given to support the original undertaking. *Ill. Grand Trunk R. R.* v. *Cook*, 29 Ill. 237.

1192j. When subscriber who is also a director is estopped by his acts from alleging that the corporation has ceased to be what it was when he subscribed. *Ross* v. *C., B. & Q. R. R.*, 77 Ill. 127.

—11

1192k. Fraud as a defense to a suit on a subscription. *Hays* v. *Ot., Os. & Fox River Valley R. R.*, 61 Ill. 422. Failure of consideration. *O. O. & F. R. V. R. R.* v. *Black*, 79 Ill. 262. Mismanagement of corporate affairs. *Chetlain* v. *Repub. Life Ins. Co.*, 86 Ill. 220.

1193. STOCKHOLDER—*who is one.* An agreement to subscribe a certain amount of stock when books shall be opened, does not make the party a stockholder and as such liable for calls. *Thrasher* v. *Pike Co. R. R.*, 25 Ill. 393, 405.

1194. SUBSCRIPTION—*must be to corporation seeking to enforce.* One corporation cannot recover on subscriptions made to another, however identical the object sought by the two companies, or the parties composing them. *Ib.*

1195. RELEASE OF—*void as to creditors.* As against creditors, the release or surrender of the obligation of a subscriber of stock, by the directors, is void. *Union Mutual Life Ins. Co.* v. *Frear Stone Manfg. Co.*, 97 Ill. 537, 549; *Upton* v. *Tribilcock*, 1 Otto, 45; *Sawyer* v. *Hoag,* 17 Wall. 610; *Burke* v. *Smith,* 16 Wall. 390; *New Albany* v. *Burke*, 11 Wall. 96; *Zirkel* v. *Joliet Opera House*, 79 Ill. 334; *Melvin* v. *Lamar Ins. Co.*, 80 Ill. 446. Release as against other stockholders. See *Chandler* v. *Brown,* 77 Ill. 333.

1196. Any device by which members of a corporation seek to avoid liability which the law imposes on them, is void as to creditors. *Union Mut. L. Ins. Co.* v. *Frear Stone Manfg. Co.*, 97 Ill. 537.

1197. CAPITAL STOCK—*trust fund for creditors—release of subscriber.* The capital stock subscribed is a trust fund for creditors which the directors cannot give away to their prejudice. Any agreement releasing stockholders from payment of their subscriptions, is void. *Ib. Putnam* v. *New Albany*, 4 Biss. 365.

1198. CAPITAL STOCK—*must be all subscribed before any subscription is collectable.* Until the whole amount of the capital stock fixed has been subscribed, the corporation has no existence, and the directors cannot make any calls, or assessments on the shares of those who have subscribed. *Allman* v. *Hav., Rantoul & Eastern R. R.*, 88 Ill. 521; *Temple* v. *Lemon*, 112 Ill. 51. See cases ante 1156g.

1199. WHEN STRICT COMPLIANCE REQUIRED—*rights depending on.* In actions on contracts, like subscriptions for stock, where the very consideration is the legal organization of a corporation having a right to existence, the inquiry may extend to the due compliance with all of the requirements of the law. Abbott's Trial Evid. 19; *Railway Co.* v. *Allerton*, 18 Wall. 233; 1 Morawetz on Corporations, §§ 29, 137, 408; *Bray* v. *Farwell*, 81 N. Y. 607; *Peoria, &c., R. R.* v. *Preston*, 35 Iowa 118, 121; *Hoagland* v. *Cinn. &c., R. R.*, 18 Ind. 452; *Selma, &c,. R. R.* v. *Anderson*, 51 Miss. 829; *Swartwout* v. *Mich. Air Line R. R.*, 24 Mich. 390; *Santa Cruz R. R.* v. *Schwartz*, 53 Cal. 106.

1200. STOCK — *personalty — transfers of—purchase of prohibited—use of corporate funds.* § 14. The stock of such corporation shall be deemed personal estate, and shall be transferable in the manner prescribed by the by-laws of such corporation. But no shares shall be transferable until all previous calls thereon shall have been paid; and it shall not be lawful for such corporation to use any of the funds thereof in the purchase of its own stock, or that of any other corporation, or to loan any of its funds to any director or other officer thereof, or to permit them or any of them to use the same for other than the legitimate purposes of such corporation. [R. S. 1887, p. 1003, § 14; S., & C., p. 1911, § 14: Cothran, p. 1139, § 14.]

1200a. PERSONAL ESTATE. Statute making stock personal property is but declaratory of the common law. *Mohawk, &c. R. R. v. Clute,* 4 Paige, 384, 393; *Hutchins* v. *State Bank,* 12 Met. 426; *Johns* v. *Johns,* 1 Ohio St., 350.

1200b. TRANSFERS OF STOCK—*by-laws of company.* Certificates of stock in a railway company, unlike negotiable paper, can only be assigned by an act of the company, or in pursuance of a by-law. *Hall* v. *Rose Hill & Evanston Road Co.,* 70 Ill. 673.

1200c. SAME—*by issue of new certificate.* If the purchaser of stock of a railway company applies to procure a transfer of the same to him, and the directors order the transfer to him, and new certificates to be issued to him, he will become an innocent holder, if he acts in good faith, and the company will be estopped to deny that the stock thus issued is valid. *Ib.*

1200d. If the secretary issues new certificates of stock to one claiming to have purchased shares therein, without taking up or cancelling the original, the new certificates will be invalid. *Ib.*

1200e. CERTIFICATE OF STOCK—*presumption of its proper issue.* The certificate of stock in a railway company, issued by its secretary, is *prima facie* evidence that it was regularly issued; but this presumption may be overcome by other evidence, as by showing no order was passed for its issue. If the order was passed and not entered of record, that may be shown by the holder. *Ib.*

1200f. ASSIGNMENT—*relief against equitable assignee by assignor.* A court of equity will not give the assignor of stock relief against a *bona fide* purchaser, merely because the latter may have failed to have the stock transferred to him upon the books of the corporation, as required by law. It is no concern of the assignor whether the assignee ever becomes invested with the legal title, or the right to membership in the corporation. Such stock may be regarded as a *chose* in action, the equitable title of which, as between the parties, may be transferred without observing the requirements of the charter or by-laws of the company as to the mode of transfer so as to pass the legal title. *Otis* v. *Gardner,* 105 Ill. 436.

1201. ASSIGNMENT OF STOCK—*neglect to enter on books.* Where a charter requires all sales and transfers of stock to be made upon the books of the corporation in order to be valid, this provision will be regarded as designed for the protection of the company, and perhaps a purchaser without notice; but as between the assignor and purchaser, a sale and transfer will be good without being entered upon the books, and will be enforced in equity. *Kellogg* v. *Stockwell,* 75 Ill. 68.

1201a. EQUITABLE ASSIGNMENT—*rights and liabilities of assignee.* The equitable assignee or owner of stock in an incorporated company can use it as his own property, control it and receive dividends thereon, the same as though he had the legal title; and therefore as between himself and his assignor, he is bound to assume the burdens imposed upon the owner of the legal title arising out of assessments made upon the stock. *Ib.*

1201b. SAME—*protection of assignor in equity.* Where shares in the capital stock of an incorporated company have been sold and transferred, but not in accordance with the charter or by-laws of the company, so as to pass the legal title, and the assignor is compelled to make payment of assessments, or is liable to be called upon for payment, a court of equity at the suit of the assignor, will require the assignee to pay or indemnify him, as the case may require. *Ib.*

1201c. TRANSFER OF CERTIFICATES—*assignee not protected*. Certificates of stock are not securities for money, nor do they possess the qualities of commercial obligations, so as to protect a *bona fide* purchaser or holder from equities of the corporation against them; and when stock of a corporation is fraudulently issued by one of its officers and transferred to a third person as collateral security for a debt, it is not error upon a bill filed for that purpose, to order the certificates of such stock returned and cancelled. *Campbell* v. *Morgan,* 4 Bradw. 100.

1201d. TRANSFER OF SHARES—*as against creditors of assignor*. Where the board of directors of a corporation are expressly empowered by the charter to provide for the mode of transfer of shares of· stock, and the board does, by a by-law, provide that such transfer shall only be made upon the books of the secretary on the presentation of the stock certificates properly indorsed, a transfer by indorsement and delivery only, will not be valid as against a creditor of the assignor who levies his execution upon such shares without notice of the transfer. *People's Bank* v. *Gridley,* 91 Ill. 457.

1201e. SAME—*as between the parties*. As between the vendor and vendee of shares of stock in a corporation whose charter or by-laws require transfers of stock upon its books, a sale and transfer will be good without being entered upon the company's books, and will be enforced in equity, and the vendee required to pay subsequent assessments, or indemnify the vendor against their payment. *Ib.*

1201f. TRANSFER OF STOCK—*as against execution creditors*. The provision of the statute making shares of stock in a private corporation subject to levy and sale on execution, contemplates that, as against a judgment creditor, the title to stock in such corporation can only pass by transfer on the books of the company. *Ib.*

1202. CORPORATION—*liability for refusing to transfer stock*. A corporation will be liable in case for refusing to transfer on its books shares of its capital stock which it has issued, to a purchaser of the same, unless such stock is absolutely void for fraud or want of consideration, in which latter event no action will lie against the corporation for such refusal. *Protection Life Ins. Co.* v. *Osgood,* 93 Ill. 69.

1202a. EQUITABLE TRANSFER—*passes only equitable title*. The charter of a private corporation provided that the stock should be transferred in such manner as the directors might determine, and the by-laws of the company provided that the secretary should keep a book upon which all transfers of stock should be made by the holder or holders, or by his or their attorney, duly constituted in writing. A holder of certificates of stock delivered the same with a blank assignment and power of attorney indorsed thereon, to a borrower of the same, which power authorized the assignee to have the stock transferred on the books of the company, but no such transfer was ever made upon the books, and such holder, being the borrower, transferred the certificates as collateral security for a loan. *Held,* that the legal title never passed by the transfer for want of an assignment on the books of the company, but that the pledgee took an equitable title as security for his money, of which he could not be divested by the real original owner. *Otis* v. *Gardner,* 105 Ill. 436.

1202b. Where certificates of stock are assigned in blank with a power of attorney for a transfer on the books of the company, with no limitation as to their use by the assignee, he will, as to persons dealing with him without notice of any defect of power in him, be authorized to make any legitimate use of them, and he may transfer them as security for a loan. *Ib.*

1202c. ASSIGNMENT—*in the absence of any by-law, &c., on subject*.

In the absence of any by-law or other regulation to the contrary, an assignment of the certificate of stock by indorsement and delivery, will be sufficient to authorize the assignee to vote. *People* v. *Devin*, 17 Ill. 84.

1202*d*. TRANSFER—*new certificate not necessary*. A transfer of shares upon the books makes the transferee a shareholder, although no new certificate is issued. The certificate is merely the evidence of the holder's rights. *First Nat. Bank* v. *Gifford*, 47 Iowa, 575, 583; *Hawley* v. *Upton*, 102 U. S. 314.

1203. RAILWAY COMPANY—*of the right to purchase its own or other stock*. The weight of authority in this country is in favor of the power of a corporation to purchase its own capital stock, except where the circumstances are such as to show that the purchase was fraudulent in fact, or that the corporation was insolvent at the time of such purchase. *Fraser* v. *Ritchie*, 8 Bradw. 554.

1203*a*. PURCHASE OF ITS OWN STOCK—*as against creditors of company*. A corporation has not the power as against creditors to extinguish its capital stock. So, where a corporation conveyed to one of its shareholders a large amount of real estate and other property, and in return received the surrender of the shares of stock held by him, which were then cancelled: *Held*, that a judgment creditor of the corporation could maintain a bill to subject the property so conveyed, to the payment of his judgment; and that it made no difference that there might be enough property remaining with the corporation to satisfy his judgment. The lien attached to the whole stock, and the creditor could not be remitted to his remedy against the remaining shares. *Peterson* v. *Ill. L. & L. Co.*, 6 Bradw. 257.

1203*b*. Although a corporation has the power to purchase its own stock, yet in equity the transaction may be impeached, if it operates to the injury of creditors. *Clapp* v. *Peterson*, 104 Ill. 26.

1203*c*. The shareholders of a corporation are conclusively charged with notice of the trust character which attaches to its capital stock. *Ib.*

1203*d*. Private corporations may purchase their own stock in exchange for money or other property, and hold, re-issue or retire the same, if it is done in entire good faith, and the exchange is of equal value, and is free from all fraud, actual or constructive, and if the corporation is not insolvent or in process of dissolution, and the rights of creditors are not affected thereby. *Ib.*

1203*e*. The purchase of its own stock by a corporation by the exchange of its property of equal value, though made in good faith without any element of fraud about it, there not being anything in the apparent condition of the company to interfere with the making of the exchange, will not be allowed where it injuriously affects a creditor of the company, even though the fact of indebtedness was not at the time established or known to the stockholder. *Ib.*

1204. EQUITABLE LIEN OF CREDITORS ON CAPITAL STOCK. The capital stock of a private corporation is a fund set apart for the payment of its debts, and its creditors have a lien in equity. If diverted they may follow it as far as it can be traced, and subject it to their claims, except as against holders who have taken it *bona fide* for a valuable consideration and without notice. *Ib.*

1204*a*. CAPITAL STOCK, A TRUST FUND—*notice thereof to stockholders*. The shareholders of a corporation are conclusively charged with notice of the trust character which attaches to the capital stock. As to it, they cannot occupy the *status* of innocent purchasers, but they are to all intents and purposes privies to the trust. When, therefore, they have in their hands any of this trust fund, they hold it *cum onere*, subject to all the equities which attach to it. *Ib.*

1205. Purchase of its own stock. The directors of a railway company, when not prohibited by the charter, have the lawful power to purchase shares of its own stock issued to others. *C. P. & S. W. R. R.* v. *Marseilles*, 84 Ill. 145, 643; *Fraser* v. *Ritchie*, 8 Bradw. 554. See also *Peterson* v. *Ill. Land & Loan Co.*, 6 Bradw. 257; *Chetlain* v. *Republic L. Ins. Co.*, 86 Ill. 220; *Dupee* v. *Boston Water Power Co.*, 114 Mass. 37; *State* v. *Building Assoc.*, 35 Ohio St. 258.

INCREASE OF CAPITAL STOCK.

CALL OF SPECIAL MEETINGS FOR — OTHER BUSINESS.

1206. Of the notice of such meeting — *record of proceedings.* § 15. In case the capital stock of any such corporation shall be found insufficient for constructing and operating its road, such corporation may, with the concurrence of two-thirds in value of all its stock, increase its capital stock, from time to time, to any amount required for the purpose aforesaid. Such increase shall be sanctioned by a vote, in person or by proxy, of two-thirds in amount of all the stock of such corporation, at a meeting of such stockholders called by the directors of the corporation for such purpose, by giving notice in writing to each stockholder, to be served personally or by depositing the same in a postoffice, directed to the postoffice address of each of said stockholders severally, with necessary postage for the transmittal of the same, prepaid, at least sixty days prior to the day appointed for such meeting, and by advertising the same in some newspaper published in each county through or into which the said road shall run or be intended to run (if any newspaper shall be published therein), at least sixty days prior to the day appointed for such meeting. Such notice shall state the time and place of the meeting, the object thereof, and the amount to which it is proposed to increase such capital stock; and at such meeting the corporate stock of such corporation may be so increased, by a vote of two-thirds in amount of the corporate stock of such corporation, to an amount not exceeding the amount mentioned in the notices so given. Should the directors of any such corporation desire at any time to call a special meeting of the stockholders, for any other necessary purpose, the same may be done in the manner in this section provided, and if such meeting be attended by the owners of two-thirds in amount of the stock, in person or by proxy, any other necessary business of such corporation may be then transacted, except the altering, amending or adding to the by-laws of such corporation : *Provided*, such business shall have been specified in the notices given. And the proceedings of any such meeting shall be entered on the journal of the proceedings of such corporation. Every order or resolution increasing the capital stock of any such corporation shall be duly recorded as required in section 2 of this act. [R. S. 1887, p. 1003, § 15; S. & C., p. 1911, § 15; Cothran, p. 1140, § 15.]

1207. SPECIAL CHARTER—*power under, to increase capital stock.* A special charter which in terms vested all corporate powers in the directors, *held,* not to authorize them to increase the capital stock without the assent of stockholders. *Ry. Co.* v. *Allerton,* 85 U. S. 233.

1207a. INCREASE OF CAPITAL—*power of directory.* A charter authorized an increase of the capital stock, but failed to provide by whom the power might be exercised: *Held,* that the directors did not merely by virtue of their position as such, have authority to increase the capital stock without the assent of the shareholders. *Eidman* v. *Bowman,* 58 Ill. 444.

1207b. It seems the management and transaction of all business for which a corporation is created, and its general affairs, are within the usual powers of the board of directors, but a power given to a corporation to increase its capital stock, cannot be exercised by the directors, except they be specially authorized so to do, either by the charter or by the shareholders. *Eidman* v. *Bowman,* 58 Ill. 444.

1207c. INCREASE OF STOCK—*who entitled to shares.* If the capital stock of a corporation be increased by proper authority, the right to such additional stock vests in the original stockholders, each one to take in proportion to the amount held by him of the original stock, if he will pay for it. This right may be waived, but if it is not, the party entitled cannot be deprived of it by the board of directors of the corporation or otherwise. *Eidman* v. *Bowman,* 58 Ill. 444.

1208. STOCKHOLDERS' LIABILITY—*holder in representative capacity exempted.* § 15½. No person holding stock in any such corporation as executor, administrator, guardian or trustee, and no person holding such stock as collateral security, shall be personally subject to any liability as stockholders of such corporation; but the person pledging the stock shall be considered as holding the same, and shall be liable as a stockholder accordingly. [R. S. 1887, p. 1003, § 16; S. & C. p. 1912, § 16; Cothran, p. 1140, § 16.]

1209. STOCKHOLDERS' INDIVIDUAL LIABILITY—*for debts to extent of unpaid subscriptions.* § 16. Each stockholder of any corporation formed under the provisions of this act, shall be held individually liable to the creditors of such corporation to an amount not exceeding the amount unpaid on the stock held by him, for any and all debts and liabilities of such corporation, until the whole amount of the capital stock of such corporation so held by him shall have been paid. [R. S. 1887, p. 1003, § 17; S. & C., p. 1912, § 17; Cothran, p. 1140, § 17.]

1210. POWER OF LEGISLATURE—*to impose liability on shareholders in existing corporations.* Although no power of amendment may be reserved in a charter, the legislature may, after its grant, impose an individual liability on stockholders and officers of a corporation by subsequent legislation, without infringing upon any constitutional rights of the stockholder. *Shufeldt* v. *Carver,* 8 Bradw. 545; *Fogg* v. *Sidwell, Id.,* 551.

1210a. RESERVATION OF POWER—*to regulate by general laws.* A reservation in a charter, or an amendment thereto, of the right of the legislature to bring the corporation under general laws, does not bind the legislature to enact any specific law, and does not operate as a contract with the stockholders that they shall be subjected to any specific additional primary liability on their contracts of subscription. But the legislature may enact such general laws as it thinks best and such laws may be even penal in their character. *Diversy* v. *Smith*, 103 Ill. 378.

1210b. CONSTITUTION OF 1848—*as providing for reservation of power over corporations.* § 2, art. 10, of the constitution of 1848 was designed to express the reservation of power in the legislature, in granting charters, to provide from time to time by proper laws for securing dues and debts from corporations by individual liability of the corporators, or otherwise. *Weidenger* v. *Spruance*, 101 Ill. 278; *Diversy* v. *Smith*, 103 Ill. 378.

1210c. Where a special charter of an insurance company contains a provision that it may be altered, amended or repealed at any time, there can be no doubt of the power of the legislature to amend such charter in such manner as it may see proper, in reference to the rights, duties and liabilities of the company and its stockholders. *Butler* v. *Walker*, 80 Ill. 345; *Diversy* v. *Smith*, 103 Ill. 378.

1210d. A general law, making trustees and corporators of insurance companies, including those already acting under special charters, severally liable for all debts of their companies, to the amount by them subscribed, *until the whole amount* of the capital shall be paid in, is not a law impairing the obligation of any contract. *Weidenger* v. *Spruance*, 101 Ill. 278.

1211. The real obligation of the contract of such subscriber to the capital stock of a corporation, is that he will pay for his stock. A mere expectation on his part that the law will not be enforced, requiring all the capital stock to be paid in, is not a vested right. If the stockholders and the corporation fail to have the stock paid in, it is competent for the legislature to impose a reasonable penalty, such as that prescribed by the insurance law of 1869. *Ib.*

1211a. The legislature had the right to repeal so much of the act of 1857, relating to private corporations, as made the stockholders personally liable to creditors, to the amount of their stock, there being no vested right in such provision. A law changing the remedy for the collection of a debt is not liable to any constitutional objection. *Richardson* v. *Akin*, 87 Ill. 138.

1212. As to the *individual liability* of stockholders under similar laws and special acts, as to the *evidence* of being stockholders, and *remedies* to enforce such liability. See post 2812.

1213. EMINENT DOMAIN—*acquisition of land by condemnation.* § 17. If any such corporation shall be unable to agree with the owner for the purchase of any real estate required for the purposes of its incorporation, or the transaction of its business, or for its depots, station buildings, machine and repair shops, or for right of way or any other lawful purpose connected with or necessary to the building, operating or running of said road, such corporation may acquire such title in the manner that may be now or hereafter provided for by any law of eminent domain. [R. S. 1887, p. 1003, § 18; S. & C., p. 1912, § 18; Cothran, p. 1141, § 18; ante 303 et seq.]

1214. EMINENT DOMAIN—*material for road by condemnation.* § 18. Any such corporation may, by their agents and employes, enter upon and take from any land adjacent to its road, earth, gravel, stone, or other materials, except fuel and wood, necessary for the construction of such railway, paying, if the owner of such land and the said corporation can agree thereto, the value of such material taken and the amount of damage occasioned thereby to any such land or its appurtenances; and if such owner and corporation cannot agree, then the value of such material, and the damage occasioned to such real estate, may be ascertained, determined and paid in the manner that may now or hereafter be provided by any law of eminent domain, but the value of such materials, and the damages to such real estate, shall be ascertained, determined and paid for before such corporation can enter upon or take the same. [R. S. 1887, p. 1004, § 19; S. & C., p. 1912, § 19; Cothran, p. 1141, § 19; § 179 ante.

1215. If the contractors who are bound to furnish all materials by their contract take materials for the construction of their road, the corporation will be liable to make compensation therefor. *Lesher* v. *Wabash Nav. Co.,* 14 Ill. 85; *Hinde* v. *Wabash Nav. Co.,* 15 Ill. 72: cited and distinguished in *Scammon* v. *Chicago,* 25 Ill. 424.

1216. A railway corporation is liable to third persons for the tortious acts of its contractors while constructing the road. *Ch., St. Paul & Fond Du Lac R. R.* v. *McCarthy,* 20 Ill. 385; *West* v. *St. L., V. & T. H. R. R.,* 63 Ill. 545.

1217. So, it is liable for the acts of its lessees, or contractors in operating and using the road under its authority. *O. & M. R. R.* v. *Dunbar,* 20 Ill. 623; *Ch. & R. I. R. R.* v. *Whipple,* 22 Ill. 105; *Ill. Central R. R.* v. *Read,* 37 Ill. 484; *I. C. R. R.* v. *Finnigan,* 21 Ill. 646; *P. & R. I. R. R.* v. *Lane,* 83 Ill. 448; *P., C. & St. L. Ry.* v. *Campbell,* 86 Ill. 443; *Balsley* v. *St. L., A. & T. H. R. R.,* 119 Ill. 68.

1218. A railway company allowing another to operate its unfenced road will be liable for stock killed through neglect to fence. *Ill. Central R. R.* v. *Kanouse,* 39 Ill. 272; *T., P. & W. Ry.* v. *Rumbold,* 40 Ill. 143; *Wab., St. L. & P. Ry.* v. *Peyton,* 106 Ill. 534.

1219. Consolidated company liable for the acts of the companies consolidated. *C., R. I. & P. R. R.* v. *Moffitt,* 75 Ill. 524.

1219a. The liability of a railway company for injuries by the wrongful acts of any lessee, contractor or other person, done in the exercise of any of its franchises, is limited to "wrongs done by them while in the performance of acts which they would have had no right to perform, except under the charter of the company" sought to be made liable. *St. L., A. & T. H. R. R.* v. *Balsley,* 18 Bradw. 79.

1219b. As to liability of a railway company for acts or torts of a receiver operating the road. See *Wyatt* v. *O. & M. .R R.,* 10 Bradw. 289; *Brown* v. *Wabash Ry.,* 96 Ill. 297; *Metz* v. *B., C. & P. R. R.,* 58 N. Y. 61; *O. & M. R. R.* v. *Davis,* 23 Ind. 553; *Bell* v. *I. C. & L. R. R.,* 53 Ind. 57; *Turner* v. *H. & St. Jo. R. R.,* 74 Mo. 602; *O. & M. R. R.* v. *Anderson,* 10 Bradw. 313; High on Receivers, §§ 396, 397: *contra, Cent. Trust Co.* v. *Wab., St. L. & P. R. R.,* 26 Fed. Rep. 12.

1219c. If the trustees of a railway company do business in the name of the company, they are liable to suit in that name, and their property is responsible for liabilities incurred while transacting business in that name. *Wilkinson* v. *Fleming*, 30 Ill. 353.

1219*d*. COMPANY LIABLE—*for acts of trustees.* Trustees selected by the corporation as well as the bondholders, while in possession operating the road to earn money to pay debts of the corporation, will be regarded as the agents of the corporation so far as relates to the transaction of business with third persons, and such persons may sue the corporation and recover damages in respect to transactions with such trustees, and will not be compelled to sue the trustees. *Gr. T. Manf. & Transp. Co.* v. *Ullman*, 89 Ill. 244.

1220. ADDITIONAL POWERS. § 19. Every corporation formed under this act shall, in addition to the powers hereinbefore conferred, have power :—

ENTRY UPON LANDS — *to examine, survey and lay its road. First* — To cause such examination and survey for its proposed railway to be made as may be necessary to the selection of the most advantageous route; and for such purpose, by its officers, agents or servants, may enter upon the lands or waters of any person or corporation, but subject to responsibility for all damages which shall be occasioned thereby. [R. S. 1887, p. 1004, § 20; S. & C., p. 1912, § 20; Cothran, p. 1141, § 20.]

1221. Where a railway in the exercise of the powers conferred upon it, commits an injury to the land of another by entering upon it in order to make preliminary surveys, or by taking materials therefrom, or the like, and the law under which such acts are done, prescribes a mode for assessing damages for such injuries, an action of tort will not lie theref r, but the statutory remedy must be pursued — it being in general, exclusive. *Smith* v. *C., A. & St. L. R. R.* 67 Ill. 191.

1222. On an assessment of damages for right of way, it is error to admit evidence of a violent entry upon the land going to show a willful trespass. *L. B. & M. R. R.* v. *Winslow*, 66 Ill. 219.

1223. LOCATION. The Illinois Central Railroad company had the right under its charter, to locate its road in the waters of Lake Michigan. *I. C. R. R.* v. *Rucker*, 14 Ill. 353.

1224. The grant of a right to extend to and unite with any other railroad in this state gives the right to extend to any other railroad within the prescribed limits. *Bellville & Ill. R. R.* v. *Gregory*, 15 Ill. 20.

1225. Railroad crossing another has the right to select the point and manner of intersection. *L., S. & M. S. Ry* v. *Ch. & W. Ind. R. R.*, 97 Ill. 506.

1225*a*. CHANGE OF LOCATION. After having once fixed the terminal points of its road, and located its depot in a town or city, a railway company has no power afterwards to change the same without legislative authority, but it will be held to its election. *People* v. *L. & N. R. R.*, 120 Ill. 48.

1226. ACQUISITION OF PROPERTY BY VOLUNTARY GRANT. *Second.* To take and hold such voluntary grants of real estate and other property as shall be made to it, in aid of the construction and use of its railway, and to convey the same when no longer required for the uses of such railway, not incompatible with the terms of the original grant. [R. S. 1887,

p. 1004, § 20; S. &. C., p. 1913, § 20; Cothran, p. 1141, § 20.]

1227. A deed to a railway company "of the right of way" of the railroad, with nothing to define its extent in width, when the charter does not define the extent of the right of way, is too indefinite to constitute color of title. *Wray* v. *C., B. & Q. R. R.*, 86 Ill. 424.

1228. EFFECT OF RELEASE. A contract of a party by which he agrees to release and convey a right of way to a railway company over any lands he may own, as soon as the road is located, will preclude him from claiming damages from the construction of the road over his lands. *Conwell* v. *S. & N. W. R. R.*, 81 Ill 232.

1229. ACQUISITON OF PROPERTY—*by purchase—disposition of same.* Third—To purchase, hold and use all such real estate and other property as may be necessary for the construction and use of its railway, and the stations and other accommodations necessary to accomplish the object of its incorporation, and to convey the same when no longer required for the use of such railway. [R. S. 1887, p. 1004, § 20; S. & C., p. 1913, § 20; Cothran, p. 1142, § 20.]

1230. CONVEYANCE—*estate granted.* A deed to a railway company conveying no land, but only the right to construct, maintain and use in, through, upon and over certain lands, all such railroad tracks, depots, warehouses, &c., as the company should find necessary or convenient for transacting its business, and to keep thereon without disturbance, all property belonging to or in the possession of the company, to have and to hold the said rights and easements so long as the same should be used for such purposes, and for no other, even forever, passes only an easement which is a freehold of inheritance, though only a base or qualified fee, which may be defeated. *Wiggins Ferry Co.* v *O & M. Ry.*, 94 Ill. 83.

1231. POWERS—*to lay out and construct road 100 feet wide—when may take more.* Fourth—To lay out its road, not exceeding one hundred feet in width, and to construct the same; and for the purpose of cuttings and embankments, to take as much more land as may be necessary for the proper construction and security of the railway; and to cut down any standing trees that may be in danger of falling upon or obstructing the railway, making compensation therefor in manner provided by law. [R. S. 1887, p. 1004, § 20; S. & C., p. 1913, § 20; Cothran, p. 1142, § 20.]

1232. WIDTH OF RIGHT OF WAY. Company not bound to take and pay for all the lands described in the petition, if less will answer its purposes. *Peoria & R. I. Ry.* v. *Bryant*, 57 Ill. 473.

1233. Width of the land as described in the report of the commissioners was *held*, to control, and where acquiesced in by the company, with knowledge, it is concluded. *Ib.*

1234. *Alteration* of the route subsequent to the assessment of damages gives the land-owner a right to recover for damages resulting therefrom. *Peoria & R. I. R. R.* v. *Birkett*, 62 Ill. 332.

1235. POWERS—*to build road across or upon streams, highways, streets, &c.—consent to, or condemnation.* Fifth—

To construct its railway across, along or upon any stream of
water, watercourse, street, highway, plank road, turnpike or
canal, which the route of such railway shall intersect or
touch; but such corporation shall restore the stream, water-
course, street, highway, plank road and turnpike thus intersected or touched, to its former state, or to such state as not
unnecessarily to have impaired its usefulness, and keep such
crossing in repair: *Provided,* that in no case shall any railroad company construct a road-bed without first constructing
the necessary culverts or sluices, as the natural lay of the
land requires for the necessary drainage thereof. Nothing
in this act contained shall be construed to authorize the erection of any bridge, or any other obstruction, across or over
any stream navigated by steamboats, at the place where any
bridge or other obstructions may be proposed to be placed,
so as to prevent the navigation of such stream; nor to authorize the construction of any railroad upon or across any
street in any city, or incorporated town or village, without
the assent of the corporation of such city, town or village:
Provided, that in case of the constructing of said railway
along highways, plank roads, turnpikes or canals, such railway shall either first obtain the consent of the lawful authorities having control or jurisdiction of the same, or condemn
the same under the provisions of any eminent domain law
now or hereafter in force in this state. [R. S. 1887, p. 1004,
§ 20; S. & C., p. 1913, § 20; Cothran, p. 1142, § 20. See
ante 60.]

1236. RAILWAY OVER STREAMS—*duty as to culverts.* Duty of
railway company in constructing its road under legislative authority
over water courses on private land, to make suitable bridges, culverts,
or other provisions for carrying off the water effectually, and to keep
them in suitable repair. *I. C. R. R.* v. *Bethel,* 11 Bradw. 17.

1236a. In constructing culverts for the passage of water the company must exercise ordinary care and skill, and bring to bear on the
work such engineering knowledge, care and skill ordinarily applied
to works of that kind, as may be reasonably deemed sufficient to avoid
damages from the stream, in connection with the work, in all ordinary
floods or freshets. *Ib.*

1236b. If the construction of a railroad over a water course was
not improperly done, and is washed out by an extraordinary flood,
leaving *debris* upon the land of an adjacent owner, beyond the company's right of way, the company is not bound to remove such material; and if by reason of it being so lodged, the waters of the stream are
diverted in a subsequent freshet, it will not give to such adjacent
owner any right of action. *Ib.*

1236c. BRIDGES AND WATER COURSES—*obstruction of flow of
water.* A railway company is only required to construct its bridges
across water courses with such care and skill as to make them sufficient to pass the water in all ordinary floods and freshets. *P., Ft. W.
& C. R. R.* v. *Gilleland,* 56 Pa. St 445; *Town of China* v. *Southwick,*
12 Me. 238: *Lawler* v. *Baring Boom Co.,* 56 Me. 443; *Norris* v. *Vt.
Cent. R. R.,* 28 Vt. 99; *Henry* v. *Vt. Cent. R. R.,* 30 Vt. 638; *Sprague*

v. *Worcester.* 13 Gray. 193; *Smith* v. *Agawam Canal Co.*, 2 Allen, 355.

1237 OBSTRUCTION OF NATURAL FLOW OF WATER—*liability for.* A railway company is liable for any injury that may result to the owner of lands from an obstruction created by it in the natural flow of surface water. *K. & S. R. R.* v.*Horan,* 22 Bradw. 145.

1238. The fact that a railway owns a right of way over the plaintiff's land, does not authorize it to make such a change thereon, by structures or otherwise, as to flow water back upon the land of the plaintiff, or others, and thereby inflict an injury. *C., R. I.& P. R. R.* v. *Carey,* 90 Ill. 514. See also, *J. N. W. & S. W. R. R.* v. *Cox,* 91 Ill. 500.

1239. As to measure of damages in case of obstructing the free passage and flow of water. See *K. & S. R. R.* v. *Horan,* 22 Bradw. 145; *C., R. I. & P. R. R.* v. *Carey,* 90 Ill. 514.

1240. A railway company has no right, by an embankment or other artificial means, to obstruct the natural flow of the surface water, and thereby force it in an increased quantity upon the lands of another, and if it does so, it is liable for any injury that the owner of the land may sustain by reason thereof. *T. W. & W. Ry.* v. *Morrison,* 71 Ill. 616. See also *Gillham* v. *Madison Co. R. R.* 49 Ill. 484; *Laney* v. *Jasper,* 39 Ill. 46; *Gormley* v. *Sanford,* 52 Ill. 158; *C., B. & Q. R. R.* v. *Schaffer.* — Ill.—. Filed March 28, 1888.

1241. A railway company by obstructing the flow of a water course will not be liable to the owner of cattle, who has no interest in the grounds overflowed, but who made a contract with the owner of the lands so overflowed to feed the same, after the obstruction was made. *T., W. & W. Ry.* v. *Hunter,* 50 Ill. 325.

1242. RESTORING FORMER USEFULNESS. The statute requiring the restoration of the stream crossed by a railroad to its former usefulness, applies to streams not navigable as well as to those navigable. *C., R. 1. & P. R. R.* v. *Moffitt,* 75 Ill. 524.

1243. Where a railroad crosses a stream not navigable under legislative authority, which imposes a duty to leave the stream in such condition as not to materially destroy its usefulness, the company will be under substantially the same obligation as would be upon a private owner of the land and stream who had undertaken to interfere with the water course in the same way; and if it so constructs its bridge as to obstruct the stream by the accumulation of drift, &c., and thus overflow the lands of others, it will be liable for the damages. *Ib.*

1244. BRIDGE—*when built by city.* A bridge built by a railway company over a navigable stream within the limits of a city, for the use of the railroad, under an ordinance of the city granting permission and providing the manner in which it should be built, may be regarded as having been constructed by the city, and as falling fairly within the power given to it to construct and repair bridges and regulate the use of them. *McCartney* v. *Ch. & E. R. R.,* 112 Ill. 611.

1244a. As to obstruction of navigable stream by a bridge, and an action in respect thereto, see *Ill. Packet Co.* v. *Peoria Bridge Assoc.,* 38 Ill. 467; *Miss. River Bridge Co.* v. *Lonergan,* 91 Ill. 508, 516.

1244b. Town responsible, if it makes such a bridge as will obstruct the free navigation of the stream. *Town of Harlem* v. *Emmert,* 41 Ill. 319.

OF THE USE OF HIGHWAYS.

1245. GRANT OF USE—*whether exclusive, or joint use.* A grant of power to a railway company to construct its road upon, or across any highway its route may intersect, the corporation to restore the same to its former state, or so as not to impair its usefulness, is equivalent to allowing a joint use of the highway by the company with the public, protecting its use as an ordinary highway against any impairment. It does not authorize a use to the exclusion of ordinary travel. *P., Ft. W. & Ch. R. R.* v. *Reich*, 101 Ill. 157.

1246. GRANT OF USE—*statute authorizing construed.* § 26 of the act of 1849, authorizing county or town officers having charge of lands belonging to their county or town, to grant the right of way over the same to railroad corporations, has application only to lands which belong to counties or towns as owners thereof, and not to lands in which they hold the nominal title only for a prescribed public use, such as for a street or a highway. *Ib.*

1247. The commissioners of highways of a town, having no title to an avenue or public highway, are powerless to grant the same to a railway company, by deed, so as to pass an exclusive right to its use, and a deed by them attempting to grant such right, is void. *Ib.*

1247a. RIGHT TO USE OF HIGHWAY—*duty as to public travel.* A railway corporation may take possession of such part of any public road as may be within the limits of the right of way, and may construct its railway across any established road, whenever it is necessary to do so; but the railway must be so constructed that it will not impede the passage or transportation of persons or property along the road. If the corporation finds it necessary to appropriate a public road, or any portion of it, in such a manner, or to such an extent, that it is no longer fully available for its original use, a duty arises for the corporation forthwith, at its own expense, to change its site and to reconstruct the road on the most favorable location in as perfect a manner as the original road, for the public. *Commonwealth* v. *Penn. R. R.* Opinion of Sup. Court of Pa. Filed Jan. 3, 1888. 20 Ch. Legal News, p. 284.

USE OF STREETS.

1248. TITLE TO STREETS—*vested in corporation.* Where a city or town is laid out by plat under the statute, the fee or legal title to the streets, indicated on the plat, is vested in the corporation for the use of the public. *Canal Trustees* v. *Havens*, 11 Ill. 554; *Moses* v. *P., Ft. W. & Ch. R. R.*, 21 Ill. 516; *Belleville* v. *Stookey*, 23 Ill. 441; *I., B. & W. R. R.* v. *Hartley*, 67 Ill. 439; *C. & V. R. R.* v. *People*, 92 Ill. 170; *People* v. *Walsh*, 96 Ill. 232.

1249 DEDICATION OF—*acceptance necessary.* To make a complete dedication of streets and alleys by a town plat, so as to pass the title to the corporation, there must be some act showing an acceptance. Until acceptance the fee remains with the original proprietor. *Hamilton* v. *C., B. & Q. R. R.*, — Ill. —. Filed March 28, 1888.

1250. RIGHT TO USE STREETS FOR. The use of steam as a motive power along the streets of a city may be granted. *Moses* v. *P., F. W. & Ch. R. R.*, 21 Ill. 516. It is a legitimate use of a street or highway to allow a railroad track to be laid in it. *Murphy* v. *Chicago*, 29 Ill. 279. As to power of cities over their streets and liability for injury from change of grade, see *Roberts* v. *Chicago*, 26 Ill. 249; *Nevins* v. *Peoria*, 41 Ill. 502; *Quincy* v. *Jones*, 76 Ill. 231; *Stack* v. *E. St. Louis*, 85 Ill. 377; *Chicago* v. *Brophy*, 79 Ill. 277; *Shawneetown* v. *Mason*, 82

Ill, 337; *Aurora* v. *Gillett,* 56 Ill. 132; *Aurora* v. *Reed,* 57 Ill. 29; *Dixon* v. *Baker,* 65 Ill. 518; *Alton* v. *Hope,* 68 Ill. 167.

1251. SAME—*charter construed.* Authority by a charter to construct a railroad from V. to, and into the city of C., with the general power to cross any road or highway on the route, only gives such power outside of the corporate limits of C. It cannot by any fair intendment be held as a grant of the use of the streets of the city for railroad tracks. *C., D. & V. R. R.* v. *Chicago,* 121 Ill. 176.

1252. GRANT OF RIGHT TO LAY TRACK IN STREET—*who may question.* Those having the control of public roads may authorize travel on them by means of railways, and where a railway company has constructed its track upon and along a public highway, such use and possession is a matter between the road authorities and the company, and the right cannot be questioned in an action of ejectment by the owner of the land over which the public road has been established. *Edwardsville R. R.* v. *Sawyer,* 92 Ill. 377.

1253. A city has the power to allow the construction of a railroad upon or over its streets, and the public will be bound by whatever may be lawfully done in regard to the streets by the city. *C. & N. W Ry.* v. *People,* 91 Ill. 251.

1254. GRANT OF USE OF STREET—*how made—binds city.* Although a city charter may give power to make all *ordinances* necessary and proper to carry out the express powers, the action of a city council, though in the form of a resolution, in connection with its deed, granting the use of streets for railroad tracks, will be a sufficient grant of permission to so use the streets. *Quincy* v. *C., B. & Q. R. R.,* 92 Ill. 21.

1255. Where a city under a resolution adopted, conveys a street absolutely to a railway company, the resolution and the deed will give the company the right to construct, maintain and operate its track upon the street, even if invalid to pass the entire dominion over the street; and when such right is exercised, the city cannot resume the grant to the exclusion of the company. *Ib.*

1256. GRANT OF RIGHT—*passes to successor of grantee.* Where a city, under special authority of law, grants to a railway company the right to use certain parts of its streets for railroad tracks, the grant containing no clause restricting the use of the streets to the grantee, the right to such use of the streets may be transferred to another railway company, which is authorized by law to acquire and succeed to all the property, &c., of the former company. *Quincy* v. *C., B. & Q. R. R.,* 94 Ill. 537.

1257. LOCATION IN CITIES—*limited to assent of city authorities.* The fourth clause of this section gives the company authority to select its own route and fix its *termini;* but this is limited by the fifth clause, providing that a railroad shall not be laid on or across any street without the assent of the municipality. This clause excludes railways from cities, except with the assent of their councils. *Hickey* v. *Ch. & W. Ind. R. R.,* 6 Bradw. 172.

1258. POWER OF CITY TO REGULATE—*delegation of power.* Cities have full power to regulate the location and use of railroad tracks within their limits, and this power cannot be delegated. Ordinances granting permission to construct tracks in streets must definitely fix the location and *termini.* Ordinance held void for uncertainty in this respect. *Ib.* Overruled. See *Ch. & W. Ind. R. R.* v. *Dunbar,* 100 Ill. 110, and *Chicago* v. *Ch. & W. Ind. R. R.,* 105 Ill. 73.

1258a. An ordinance authorizing the corporation to allow other companies to use its tracks upon such terms as they may agree, is void as an attempted delegation of power. *Ib.* Overruled. *Chicago* *v. Ch. & W. Ind. R. R.*, 105 Ill. 73; *Ch. & W. Ind. R. R.* v. *Dunbar*, 100 Ill. 110.

1259. POWER TO BUILD ROAD IN CITY—*legislative recognition of power.* A provision of an act amendatory of a charter of a railway company, that the rate of speed at which its trains, &c., may be run in the city, shall be under the control of the common council, is a legislative recognition of its right to construct its road within the city limits. *McAuley* v. *Col., Ch. & Ind. Cent. Ry.*, 83 Ill. 348.

1260. GRANT BINDING ON CITY. A city is bound by its grant of the right to lay railroad tracks in streets, so as to bar its recovery in ejectment. *Quincy* v. *C., B. & Q. R. R.*, 92 Ill. 21.

1261. ASSENT OF CITY—*necessary only for use of streets.* A railway company organized under the general law of 1872, has authority to select its own route, to lay out its road, and to construct the same; and this power by necessary implication, carries with it the power of fixing the terminal points of the proposed road, subject only to the limitation that the *construction* upon or across any street in any city must be with the assent of such city. *Ch. & W. Ind. R. R.* v. *Dunbar*, 100 Ill. 110.

1261a. The lines selected for a proposed railroad may, without the assent of the city, cross streets, and the company may, without such assent, acquire the right of way and construct its road upon every part of such line, except the parts to be upon or across streets. *Ib.*

1261b. SUFFICIENCY OF ORDINANCE—*to give use of streets.* A city ordinance granting permission to a railway company to construct and operate a railroad within the city limits, is not void because it fails to designate the precise line upon which the road may be constructed, and omits to designate the precise points at which the road may be constructed across and upon the several streets to be intersected by it. *Ib.*

1262. DELEGATION OF AUTHORITY. Permission granted by a city to a railway company to construct its road across streets at any points to be selected by the company within a given district, is not a delegation to the company of powers which can only be exercised by the council, as the power to locate the line of the road is given by statute to the railway company alone, and not to the city authorities. The city of Chicago has power to make provision for the location of a railroad within its limits, but no power to locate. *Ib.*

1262a. The mere existence of a power in the city council "to provide for the location, grade and crossings" of railroads within the city, and "to change the location, grade and crossings" of railroads, until exercised, is no limitation upon the power of a railway company to select the route and locate its road within the city. *Ib.*

1263. ASSENT OF ADJACENT LOT-OWNERS. The clause in the city act that "the city council shall have no power to grant the use of, or the right to lay down any railroad tracks in any street" of the city, "except upon petition of the owners of the lands representing more than one-half of the frontage of the street, or so much thereof as is sought to be used for railroad purposes," has reference only to cases where the city may propose to grant the privilege to a railroad company to run along a street for a given distance, and not to a case where the road merely crosses a street. *Ib.*

1263a. CONDITION TO GRANT OF PERMISSION. A provision in a city ordinance that the permission to construct a railroad within the city is upon condition that the railroad company shall permit any other railroad companies, not exceeding two in number, which have not then the right of entrance into the city, to use the main track of the road therein authorized to be laid, *jointly* with such road so authorized, does not render the ordinance invalid, as it confers upon the railroad company no power not given it by law, nor does it deprive the city of any power whatever. *Ib.*

1264. LIMITATION ON RIGHT GRANTED—*right to lease track.* An ordinance giving a railroad company license to construct its track along or across the streets and alleys of a city, upon the condition it shall permit any other companies, not exceeding two in number, to use its main track upon such fair and equitable terms as may be agreed upon, will not be construed as prohibiting the company from leasing the use of its track to more than two other companies. Such provision is a limitation, not upon the right of the company to admit other companies to a joint use of its track, but upon the exclusive enjoyment of the estate granted by the city. *Chicago* v. *Ch. & W. Ind. R. R.,* 105 Ill. 73.

1265. § 20 of the charter of East St. Louis, authorizing the council to make contracts with any street or horse railroad company for the use of any street, &c., upon the consent of the owners of three-fourths of the property per foot fronting on such street, &c., applies exclusively to street or horse railroads, strictly so called, and has no application to railroads contemplated in the general railroad law. *Wiggins Ferry Co.* v. *East St. Louis Union Ry.,* 107 Ill. 450.

1265a. Under the general law relating to railroads, it is only necessary to procure the assent of the municipal authorities of a city to authorize a railroad company to construct its track or tracks over or along a public street therein. That act, as revised in 1874, does not require the assent of the abutting lot-owners, and in the absence of any special statutory provision requiring such assent, it will not be necessary. *Wiggins Ferry Co.* v. *E. St. L. Union Ry.,* 107 Ill. 450.

1266. In cities, towns or villages organized under the general incorporation law, which requires such assent, or under special charters containing a similar provision to that in the general law, this rule does not apply, and the assent of the requisite number of the abutting property owners will be required as well as that of the municipality. *Ib.*

1267. RIGHT TO MAKE ROAD IN CITY—*charter construed.* The words "to" and "from" a place or city, are construed to mean to or from a point within the place to or from which a corporation is authorized to construct a railroad. Authority to construct and operate a railroad from the city of C. to any point in the town of E., is held to authorize the location and operation of the road from any point within the city of C. *McCartney* v. *Ch. & E. R. R.,* 112 Ill. 611.

1268. Where a railway company is authorized to build a railroad from a city to another place, the fact it is also empowered to contract with a horse railroad company for the joint or separate operation of either or both companies' roads, as may be agreed on, will not operate as a limitation upon the railway company in respect to its entrance into the city. *Ib.*

1269. LEGISLATIVE RECOGNITION OF RIGHT. An act of the legislature confirming a city ordinance granting the right to lay track in the city will remove any doubt as to the company's right to construct its road in the city, and be regarded as a recognition of the right. *Ib.*

1270. Under the 9th and 25th clauses of § 1, art. 5, of the general incorporation law, the common council in cities incorporated under that law, is vested with the exclusive control and regulation of the streets, and with the power to direct and control the location of railroad tracks within the limits of their cities; and being inconsistent with the 9th clause of § 1, art. 5 of the amended charter of the city

—12

of Chicago, adopted in 1867, must prevail over the latter. *Ch. Dock & Canal Co.* v. *Garrity*, 115 Ill. 155.

1271. LIMITATION ON CITY. The power of the city council of Chicago to direct and control the location of railroad tracks, is subject to the limitation imposed by the 90th clause of § 1, art. 5, of the act, making a petition of the owners of the land representing more than one-half the frontage of the street, necessary to the grant of the right to lay a railway track in any street of the city. *Ib.*

1272. That clause, is to be construed as including both corporations and individuals. The word "company" in the clause must be held to embrace natural persons as well as corporations. *Ib.*

1273. A city council under the general incorporation act, may grant to private individuals, or to a private corporation, the right to lay railroad tracks in the streets, connecting with public railway tracks, previously laid, and extending to the manufacturing establishments or warehouses, of those laying the tracks. In such case the tracks so laid, become, in legal contemplation, part of the railway with which they connect, and are open to the public, and subject to public control in all respects as other railway tracks. *Ch. Dock & Canal Co.* v. *Garrity*, 115 Ill. 155.

1274. The use of the streets of a city, however, whether for vehicles drawn by animals, for riding upon animals, for foot-men, or for the passage of railway cars, must be for the public. No corporation or individual can acquire an exclusive right to their use, or for merely private purposes. *Ib.*

1275. Railroad tracks laid on streets of a city, connected with existing railroads, and extending to public warehouses, malt houses or manufacturing establishments, or to public wharves and landings, are in their nature public, and for the public good, and all railroad companies are required by law to permit such connections to be made with their tracks. *Ch. Dock & Canal Co.* v. *Garrity*, 115 Ill. 155.

1276. CONSENT OF CITY — *to construction of railway in city — repeal.* The general railroad act of 1849 prohibiting railroads from entering cities without municipal consent, is wholly repealed. *Ch. & W. Ind. R. R.* v. *Dunbar*, 100 Ill. 110.

1277. TRACK IN STREET — *when not a nuisance.* A railroad track laid upon a street of a city by authority of law, properly constructed, and operated in a skillful and careful manner, is not, in law, a nuisance. *Ch. & E. Ill. R. R.* v. *Loeb*, 118 Ill. 203.

1278. RIGHT TO CROSS STREETS, &c. The fifth paragraph of § 20 of the railroad act of 1872, is an absolute grant of power by the state to railway companies to construct their roads across any public highway. It is only where the railroad is to be constructed *along* or lengthwise of a highway, that the consent of the local authorities is necessary. *Cook Co.* v. *Gr. Western R. R.*, 119 Ill. 218; *Ch. & W. Ind. R. R.* v. *Dunbar*, 100 Ill. 110.

1279. SAME — *who may question right.* The only authority that can call in question the right of a railway company to construct its track across or along a street or highway within an incorporated city or village, is such city or village. The county authorities cannot even question the validity of an ordinance of a city or village for the construction of a railroad within such city or village. *Ib.*

1280. REGULATING USE OF STREETS — *by whom.* The act of 1872, relating to cities and villages, confers upon them full authority to regulate the use of streets, to provide for and change the location, grade and crossings of railroads, to require railroad companies to fence their roads, to construct cattle guards and crossings of streets, to keep the same in repair, to maintain flagmen at such crossings, to compel the roads to raise or lower their tracks, &c. This invests incorporated cities and villages with exclusive authority over the matter of railroad crossings over streets and highways within their limits, and excludes the jurisdiction of the county or town authorities. *Ib.; Ch. Dock & Canal Co.* v. *Garrity,* 115 Ill. 155, 163.

1281. GRANT OF RIGHT TO USE STREET—*must clearly appear.* A permission to a railway company to occupy a public street with railway tracks, must plainly appear, and not be left to be derived by doubtful implication from the generality of language used, which does not unmistakably manifest the intention to give such permission. *Ch., Dan. & Vin. R. R.* v. *Chicago,* 121 Ill. 176.

1281a. A city ordinance, after a careful mention and specification of what streets, might be used by a railway company in which to lay down its tracks and side tracks, contained a general clause giving authority, also, to lay down all such tracks "as may be necessary to the convenient use of any depot ground said company may now own, or hereafter acquire in the vicinity of or adjoining said line of road," without the specification of any streets: *Held,* that such general clause gave no authority in respect to the use of the streets, additional to those which had been specifically named in the preceding part of the ordinance. *Ib.*

1281b. RIGHT TO USE STREETS—*charter construed.* Authority in the charter of a railway company to construct a railroad from V., in the state of Indiana, to, and into the city of C., with the general power to cross any road or highway on the route, is to be held only as giving such power outside of the corporate limits of the city of C. By no fair intendment can it be held as a grant of the use of the streets of the city for tracks of the road. *Ib.*

1282. The grant in a charter to a railway company of the right to run its road through a town, cannot, by any reasonable or fair intendment, operate as a grant of the use of the streets, or either of them, to the company. *St. L., V. & T. H. R. R.* v. *Haller,* 82 Ill. 208.

1283. GRANT OF USE CONSTRUED. An ordinance or resolution of a city appropriated certain streets to a railway company, "so far as the said company may require to appropriate the same in crossing them, in the construction of their railroad track, switches, turn-tables, &c., and other machinery and fixtures to be used or employed by them in operating their said road, subject, however, to this proviso: that the same shall be occupied with as little detriment and inconvenience to the public as possible," and requiring the crossing to be so graded as to make any embankment that should be made, no obstruction: *Held,* that this was but a provision for a joint use with the general public. *St. L., A. & T. H. R. R,* v. *Belleville,* 122 Ill. 376.

1284. MUNICIPAL ASSENT—*petition of lot-owners.* Clause 90 of § 1 of art. 5 of chap. 24, R. S., provides that city and village authorities shall have no power to grant the right to lay a railway track in a street except upon petition of the owners of the land, representing more than one-half of the frontage of the street, or so much thereof as is sought to be used for railroad purposes. (Ante 151.) A compliance with this clause is a prerequisite to the validity of corporate consent. *Hickey* v. *Ch. & W. Ind. R. R.,* 6 Bradw. 172. See R. S., chap. 24, art. 5, § 1, clauses 7, 9, 10, 24, 25 and 90.

1285. RIGHT UNDER CONSENT OF CITY. Where the city has duly granted such right to a railway company, the latter may build and operate its road without interference, subject to its liability to respond to abutting lot-owners for all legal damages. *Ch. & W. Ind.*

R. R. v. *Berg,* 10 Bradw. 607; *Same* v. *George, Id.* 646; *Same* v. *Phillips, Id.* 648.

1286. DUMMY RAILROADS—*consent of property holders.* Under the act of 1874 in relation to horse and dummy railroads, no petition of the adjoining property-owners is necessary. *Hunt* v. *Ch. & D. Ry.,* 20 Bradw. 282.

1287. RAILROAD IN STREETS. A city or town cannot confer upon any one an exclusive right to the use of a public street, thereby depriving it of its character of a public highway. *St. L., A. & T. H. R. R.* v. *Belleville,* 20 Bradw. 580.

1288. LIMITATION ON RIGHT TO LOCATE ROAD. The power of the city council to provide for the location of railroads within its streets is no limitation on a railway company to locate its road in a city, until it is exercised. *Ch. & W. Ind. R. R.* v. *Dunbar,* 100 Ill. 110.

1289. NEW USE OF ROAD IN STREET. Where a railway company lays its track in a street of a city, having the right to construct a track for passenger cars only, the city, under chap. 24, art. 5, § 1, clause 90, has no power afterwards to grant the use of the track for the operation of freight cars upon it, except upon a petition of property-owners upon the street, as required in the statute, and a grant of the use of such track for freight purposes, without such petition, being void, such use is unlawful and a public nuisance, which the state may cause to be abated. *McCartney* v. *C. & E. R. R.,* 112 Ill. 611.

REMEDY FOR IMPROPER USE OF STREETS.

1290. INJUNCTION. Where the municipality granting the right to lay a railroad track in a street, owns the fee of the streets, the owners of lots fronting on such streets, cannot enjoin the laying of the track. *I., B. & W. R. R.* v. *Hartley,* 67 Ill. 439; *Stetson* v. *Ch. & Evanston R. R.,* 75 Ill. 74; *Patterson* v. *Ch. Danv. & Vin., R. R.,* 75 Ill. 588; *C., B. & Q. R. R.* v. *McGinnis,* 79 Ill. 269; *P. & R. I. Ry.* v. *Schertz,* 84 Ill. 135; *Truesdale* v. *Peoria Grape Sugar Co.* 101 Ill. 561.

1291. A court of equity will not assume jurisdiction to control the use of a street in an incorporated city by a railway company, or the manner in which the track is laid, or in which the business of the road is operated, for the reason that this power is conferred by law upon the corporate authorities of the city, and the court cannot supervise the exercise of such power at the suit of the people. *Cairo & Vincennes R. R.* v. *People,* 92 Ill. 170; *Ch. & Pac. R. R.* v. *Francis,* 70 Ill. 238.

1292. INJUNCTION will issue to restrain the laying of track when this condition has not been performed, but the complainant must show some special injury to his property. *Hickey* v. *Ch. & W. Ind. R. R.,* 6 Bradw. 172, 186, 187.

1293. BILL TO ENJOIN—*sufficiency.* A bill, to enjoin the laying of a railroad track in a street, averring that there was no petition of the the property owners representing more than one-half of the frontage of the street, is too broad. It should be confined to so much of the street as was sought for railroad purposes. An averment that no such petition was presented as the statute requires, is of a conclusion of law, and is not traversable. *Schuchert* v. *Wabash Ry.,* 10 Bradw. 397.

HIGHWAY CROSSINGS.

1294. CHANGE IN—*equitable interference.* Where a railway company is authorized to change highways intersected by it, so as to afford a more convenient crossing, &c., the option to change the crossing will be vested in the company, and the exercise of such option cannot be controlled by a court of equity, if proper care and skill are observed. *I. C. R. R.* v. *Bentley*, 64 Ill. 438.

1294a. Where a highway has been changed under competent legal authority, there being no want of proper care and skill, a court of equity will have no jurisdiction to order the same to be restored to its former location, on the ground of its being a private nuisance. . *Ib.*

1295. HIGHWAY CROSSINGS—*duty to put and keep in safe condition*. In the absence of express provision in its charter, a railway company is under obligation to leave every highway that it crosses in a safe condition for the use of the public, and where this duty is imposed by the charter, the same duty will rest upon its successor. *People ex rel* v. *C. & A. R. R.*, 67 Ill. 118.

1295a. SAME—*change in place of intersection*. Where municipal authorities with the assent of a railway company, discontinue a road crossing considered dangerous, and substitute another a short distance from the old one, the change will not exonerate the company from keeping up such new crossing. · *Ib.*

1295b. A railway company will not be relieved of its duty to keep in proper condition its intersection with highways, merely because of a slight deflection of a highway by the proper authorities so as to . change the precise place of crossing. *Ib.*

1296. APPROACHES AND CROSSINGS OVER NEW STREETS. Where, long after the construction of a railroad, a street was extended so as to cross the same, and the city passed an ordinance requiring the railway company to make a safe and proper crossing by grading the approaches of the street at the crossing, such duty not being imposed by its charter or any general law in force when the company was created: *Held*, that such burden could not be imposed on the company even by the legislature without compensation. *I. C. R. R.*, v. *Bloomington*, 76 Ill. 447.

1297. RESTORING USE OF STREET CROSSED. It is the duty of a railway company in constructing its track across a street, to restore the street to its former usefulness, or to such a state as not necessarily to impair its usefulness, and to keep such crossing in repair. *P., D. & E. R. R.* v. *Lyons*, 9 Bradw. 350.

1298. A railway company is under the statutory duty, in the construction of its road across a public highway, to restore the highway to its former state, or in a sufficient manner not to impair its usefulness. And if a highway can be restored in a manner not to impair its usefulness only by constructing the highway over the railway, it is the duty of the company to so restore it, and the omission is a breach of duty. *C., B. & Q. R. R* v. *Payne*, 59 Ill. 534.

1299, GRANT TO LAY TRACK IN STREET—*where the city has only an easement*. Where the fee of a street is in the land-owner who dedicated the same, the state or city may grant permission to a railway company to lay its track across or along the same; but the owner will be entitled to compensation for the additional burden placed upon the land. *I., B. & W. R. R.* v. *Hartley*, 67 Ill. 439; *Stetson* v. *Ch. & Evanston R. R.*, 75 Ill. 74.

1300. *Liability of city* for acts done in street by its permission. See *Murphy* v. *Chicago*, 29 Ill. 279; *City of Pekin* v. *Winkel*, 77 Ill. 56; *Stack* v. *East St. Louis*, 85 Ill. 377; *Pekin* v. *Brereton*, 67 Ill. 477; *Quincy* v. *Jones*, 76 Ill. 231. See Eminent Domain.

1300a. If the municipal authorities of a city or town authorize a structure upon a public street, or other obstruction, that causes injury to adjacent lot-owners, it will be liable for the damages sustained.

What it does by another it does by itself. *Stack* v. *East St. Louis*, 85 Ill. 377.

1301. Liability of railway company for injury to lot-owners for building and operating railroad on adjoining street under license or permission of city or village authorities. *St. L., V. & T. H. R. R.* v. *Capps*, 67 Ill. 607; *Stone* v. *F., P. & N. W. R. R.*, 68 Ill. 394; *C., B. & Q. R. R.* v. *McGinnis*, 79 Ill. 269; *St. L., V. & T. H. R. R.* v. *Haller*, 82 Ill. 208; *Ch. & Pac. R. R.* v. *Francis*, 70 Ill. 238; *Eberhart* v. *Ch., Mil. & St. P. Ry.*, 70 Ill. 347; *Ch., Mil. & St. P. Ry.* v. *Hall*, 90 Ill. 42; *P., Ft. W. & Ch. R. R.* v. *Reich*, 101 Ill. 157.

1301a. OBSTRUCTING STREETS—*at crossings.* An ordinance of a city prohibiting railway companies from allowing their engines, cars, &c., to stand or remain on a traveled railroad crossing used by teams and travel passing and repassing, to the hindrance and detention of the same, is valid, and a conviction thereunder may be had on proof that cars were left standing on a switch, leaving a space between them of only ten or twelve feet at the crossing, and also another car on another switch opposite such open space and about twelve feet therefrom, and that it was dangerous to attempt to pass through with a scary team. *Great Western R. R.* v. *Decatur*, 33 Ill. 381.

1302. If a railway company unnecessarily obstructs the streets of a town with its cars, contrary to an ordinance of the town, it will be liable for the penalty prescribed for so doing. *I. C. R. R.* v. *Galena*, 40 Ill. 344.

1303. A town ordinance that "no person shall put or cause to be put in any street, sidewalk or other public place within the city limits, any dust, dirt, filth, shavings or other rubbish or obstructions of any kind," is broad enough to embrace the obstruction of a street by a railroad company with its cars. *Ib.*

1304. INTERSECTION AND CONNECTIONS WITH OTHER RAILWAYS. *Sixth*—To cross, intersect, join and unite its railways with any other railway before constructed, at any point in its route, and upon the grounds of such other railway company, with the necessary turnouts, sidings and switches, and other conveniences, in furtherance of the objects of its connections; and every corporation whose railway is or shall be hereafter intersected by any new railway, shall unite with the corporation owning such new railway in forming such intersections and connections, and grant the facilities aforesaid; and if the two corporations cannot agree upon the amount of compensation to be made therefor, or the points and manner of such crossings and connections, the same shall be ascertained and determined in manner prescribed by law. [R. S. 1887, p. 1004, § 20; S. & C., p. 1914, § 20; Cothran, p. 1142, § 20. See 1213 *supra.*]

1305. Railway company may condemn right of way for itself across right of way of another railway company. *St. L., J. & C. R. R.* v. *S. & N. W. R. R.*, 96 Ill. 274.

1306. Has no right to take property already devoted to public use for same public use. *L. S. & M. S. R. R.* v. *Chicago & W. Ind. R. R.*, 97 Ill. 506.

1307. If the legislature does not prescribe in what manner one railway shall cross another, chancery has jurisdiction in a proper case to control the matter. *Chicago & N. W. R. R.* v. *Chicago & Pacific R. R.*, 6 Biss. 219.

1308. Where a railway company built a side track from its road, connecting with a side track of another company leading to a public warehouse, whereby it could reach such warehouse over a part of the track of such other company, and the circuit court enjoined it from removing such connecting track, it was *held*, that in view of the statutory provision that every railroad corporation shall permit connections to be made, there was no error in enjoining the removal of the track at the suit of the owners of the warehouse. *Hoyt* v. *C., B & Q. R. R.*, 93 Ill. 601.

1309. RIGHT TO CONNECT. By the rules of the common law, railway companies cannot be compelled to permit individuals to connect side tracks of their own with the tracks of such companies, in order to enable the latter to carry grain to warehouses or elevators which have been erected off their lines of road. *People ex rel* v. *Ch. & N. W. Ry.*, 57 Ill. 436.

1310. The owner of a lot of ground in Chicago, having erected a grain elevator thereon, was permitted by contract with a railway company, to connect a side track, extending from his elevator to the company's line, with its track. So far as appeared the contract was purely personal. *Held*, that a subsequent lessee of the elevator did not succeed to any of the rights of his lessor in respect to such contract. *Ib.*

1311. Where the city had, by ordinance, granted to the lessor the privilege of laying down a track along one of its streets, in order that he might connect his elevator with the line of a railroad, such grant of authority being made especially to the lessor, the mere leasing of his elevator to a third person would not operate to pass to the lessee any of the rights secured to the lessor by the ordinance. *Ib.*

1312. A private switch from a railroad to coal lands which is not owned by the railway company, but by individuals for their own private use, is not a public highway within the meaning of § 12, art. 11, of the constitution, and therefore is not free to all persons for the transportation of their persons and property thereon. *Kœlle* v. *Knecht*, 99 Ill. 396.

1313. Contract between two railway corporations held not a lease or consolidation, but a contract of connection of the two roads leaving the domestic corporation, the owner of the road, its property, franchise, &c. *Archer* v. *T. H. & I. R. R.*, 102 Ill. 493.

1314. Railroad tracks laid on the streets of a city connected with existing railroads, and extending to public warehouses, malt houses or manufactories, or to public wharves and landings, are in their nature public, and for the public good, and all railroad companies are required by law to permit such connections to be made with their tracks. *Ch. Dock & Canal Co.* v. *Garrity*, 115 Ill. 155.

1315. CONNECTIONS MEANT. The provision in the constitution of Colorado "that every railroad company shall have the right with its road, to intersect, connect with or cross any other railroad," only implies a mechanical union of the tracks of the roads so as to admit of the convenient passage of cars from one to the other, and does not of itself imply the right of connecting business with business. *A., T. & S. F. R. R.* v. *D. & N. O. R. R.*, 110 U. S. 667.

1316. TRANSPORTATION OF PERSONS AND PROPERTY— *motive power used. Seventh*—To receive and convey persons and property on its railway, by the power and force of steam or animals, or by any mechanical power. [R. S. 1887, p. 1004, § 20; S. & C., p. 1914, § 20; Cothran, p. 1143, § 20.

1817. The passage of an ordinance by a city granting permission to a railway company to lay down tracks in certain streets, &c., which is accepted, with a resolution of such company that the proper construction of the ordinance is that the permission granted thereby, was to operate in the city, cars with animal power only, and that the company should not connect with any other railroad on which other power is used, does not create a contract between the people of the city and the railway company, to abandon for all time to come the use of steam within the city as a motive power; but such company may afterwards, on permission of the city, use steam to move its cars within the city. *McCartney* v. *C. & E. R. R.*, 112 Ill. 611.

1818. Authority in the charter of a railway company to build either a horse railroad or a steam railroad within a city, confers a continuing option to use either steam or animal power, or both, upon its road, or any part of it, which may be exercised from time to time. Under it, the use of either motive power may be changed and the other substituted as the company may see fit. *Ib.*

1319. BUILDINGS, MACHINERY AND FIXTURES—*necessary to accommodate public. Eighth*—To erect and maintain all necessary and convenient buildings and stations, fixtures and machinery, for the construction, accommodation and use of passengers, freights and business interests, or which may be necessary for the construction or operation of said railway. [R. S. 1887, p. 1005, § 20; S. & C., p. 1914, § 20; Cothran, p. 1143, § 20.

1320. RULES AND REGULATIONS, AS TO TRANSPORTATION— *compensation or charges. Ninth*—To regulate the time and manner in.which passengers and property shall be transported, and the compensation to be paid therefor, subject, nevertheless, to the provisions of any law that may now or hereafter be enacted. [R. S. 1887, p. 1005, § 20; S. & C., p. 1914, § 20; Cothran, p. 1143, § 20. See post 1427, 1460.]

1321. RULES AND REGULATIONS. Railway companies must adopt proper rules for the running of trains, and conform to them, or be responsible for the consequences of running out of time, resulting in collision. *C., B. & Q. R. R.* v. *George*, 19 Ill. 510.

1322. Regulation requiring a passenger on freight train to buy a ticket before entering the train proper. *I. C. R. R.* v. *Sutton*, 42 Ill. 438; *C. & A. R. R.* v. *Flagg*, 43 Ill. 364.

1323. Company to keep office open for sale of tickets for a reasonable time before and up to the time fixed for departure of trains—not up to time of actual departure, and if tickets not then procured, to charge extra price on the train. *St. L., A. & T. H. R. R.* v. *South*, 43 Ill. 176.

1324. Railway companies have the power to make all reasonable rules for the government of their trains; and as to certain classes of trains, they may require tickets to be purchased before entering the train. *C. & A. R. R.* v. *Flagg*, 43 Ill. 364.

1325. A railway company may expel a passenger from its train at a place other than a station, for the violation of any reasonable rule other than that of non-payment of fare. Refusal to surrender ticket justifies an expulsion at a place other than a station. *I. C. R. R.* v. *Whittemore*, 43 Ill. 420.

1326. A rule adopted by a railroad company, requiring passengers to surrender their tickets to the conductor when called for, is a reasonable one, and may be enforced. *Ib.*

1327. A railway company has the right to exact of its passengers the observance of all reasonable rules, calculated to insure comfort, convenience, good order and behavior, and to secure the safety of its trains, and the proper conduct of its business as a common carrier. *Ib.*

1328. Where a passenger wantonly disregards any reasonable rule, the obligation to transport him ceases, and the company may expel him from the train, using no unnecessary force and not at a dangerous or inconvenient place. This is a common law right, and has been restricted by statute only to cases of non-payment of fare. *Ib.*

1329. Whatever rules tend to the comfort, order and safety of the passengers on a railroad, the company is authorized to make and enforce. But such rules must always be reasonable and uniform in respect to persons. *Ch. & N. W. Ry.* v. *Williams*, 55 Ill. 185.

1330. A rule setting apart a car for the exclusive use of ladies, and gentlemen accompanied by ladies, is a reasonable rule, and it may be enforced. *Ib.*

1331. The mere fact that under the rules and regulations of the company, a certain car in the passenger train has been designated for the exclusive use of ladies and gentlemen accompanied by ladies, will not justify the exclusion of a colored woman from the privileges of such car, upon no other ground than that of her color. *Ib.*

1332. Under some circumstances it might not be an unreasonable rule to require colored persons to occupy separate seats in a car furnished by the company, equally as comfortable and as safe as those furnished for other passengers. But in the absence of a reasonable rule on the subject, the company cannot lawfully, from caprice, wantonness or prejudice, exclude a colored woman from the ladies' car, merely on account of her color. *Ib.*

1333. A railroad company may require that passengers procure tickets before riding on freight trains, and conductors may expel from the cars, at regular stations, such as neglect to comply with the regulation. *T., P. & W. R. R.* v. *Patterson*, 63 Ill. 304.

1334. Where a railway company adopts a rule prohibiting passengers from being carried on its trains, or on its freight trains, without the purchase of tickets, it must furnish convenient facilities to the public by keeping open its ticket office a reasonable time in advance of the hour fixed by its time table for the departure of the train. *I. C. R. R.* v. *Johnson*, 67 Ill. 312.

1335. It is the duty of a railway company to make all reasonable and proper regulations for the safety of its employes, and it devolves on the company when sued by a servant for an injury received while in its service, and negligence is shown, to show an observance of this duty. *Pittsburgh, Ft. Wayne & Ch. Ry.* v. *Powers*, 74 Ill. 341.

1336. Where several railway companies have provided in their depot building, in a large city, separate waiting rooms for ladies and gentlemen, a regulation that no gentleman without a lady, shall be allowed to enter and remain in the ladies' room, is not only reasonable, but absolutely necessary to enable the companies to discharge a duty they owe the public of protecting females while at the depot, from violence and insult. *T., W. & W. Ry.* v. *Williams*, 77 Ill. 354.

1337. A railway has no power to adopt rules and regulations prohibiting decently behaved persons from traveling on its road who will pay their fare and conform to all reasonable requirements for the safety and comfort of passengers. *C., B. & Q. R. R.* v. *Bryan*, 90 Ill. 126.

1338. BORROWING MONEY—*issue of bonds secured by mortgage—conditions to validity of mortgage—conversion of bonds into stock.* *Tenth*—From time to time, to borrow such sums of money as may be necessary for completing, finishing, improving or operating any such railway, and to issue and dispose of its bonds for any amount so borrowed, and to mortgage its corporate property and franchises to secure the payment of any debt contracted by such corporation for the purposes aforesaid; but the concurrence of the holders of two-thirds in amount of the stock of such corporation, to be expressed in the manner and under all the conditions provided in the fifteenth section of this act, shall be necessary to the validity of any such mortgage; and the order or resolution for such mortgage shall be recorded as provided in the second section of this act; and the directors of such corporation shall be empowered, in pursuance to any such order or resolution, to confer on any holder of any bond for money so borrowed, as aforesaid, the right to convert the principal due or owing thereon into stock of such corporation, at any time not exceeding ten years after the date of such bond, under such regulations as may be provided in the by-laws of such corporation. [R. S. 1887, p. 1005, § 20; S. & C., p. 1914, § 20; Cothran, p. 1143, § 20. See post, 1467, 1468.]

1339. Where a corporation receives money and gives a mortgage to secure its re-payment, it cannot defeat a foreclosure by denying the authority of the directors to procure the loan, nor from the fact that the bond given for the loan may have been the individual obligation of the directors executing the same, and not that of the corporation. *Ottawa Northern Plank Road Co.* v. *Murray*, 15 Ill. 336.

1340. CHANGE OF POSSESSION. A railway company gave a deed of possession of its road and other property to the trustees, who, without taking possession personally, hired the former superintendent and other employes of the road, to carry on the business as their agents and servants, and put up notices all along the road of the change of management: *Held*, that such change of possession was sufficient. *Palmer* v. *Forbes*, 23 Ill. 301, 314.

1341. Where a mortgage or deed of trust gives to the trustees therein, the power to use and run the railroad, as the agents or attorneys of the company mortgaging it, that fact does not give character to the title or possession of the property, but only to the mode and manner of using it, and a proper transfer under the power will cut off all liens, not acquired prior to that transfer. *Ib.*

1342. Railroad companies can not give mortgages except in pursuance of a power conferred upon them by their charters or some general statute. *Palmer* v. *Forbes*, 23 Ill. 301, 311.

1343. Power in a railway company to mortgage its road, franchises and property, will not be construed to authorize the mortgagee to take up and sell the material of which the road is made, so as to interfere with its beneficial use by the public. *Palmer* v. *Forbes*, 23 Ill. 301.

1344. The authority to mortgage a railroad, &c., implies the authority to sell the thing mortgaged, and to convey to the purchaser all needful powers to use the thing purchased, in a proper and beneficial manner. *Ib.*

1345. The personal property of a railroad in possession of the mortgagor or trustee, is no longer subject to be taken in execution for the general debts of the company. *Palmer* v. *Forbes*, 23 Ill. 301.

1346. Money of a corporation which has been, in advance of its being earned, set apart by its board of directors to the payment of interest on its bonds secured by mortgage or trust deed on its road and franchises, and to raise a sinking fund for their redemption, is not subject to garnishee process issued by a judgment creditor of such corporation. *Galena & Ch. Union R. R.* v. *Menzies*, 26 Ill. 121, 148.

1347. Where a corporation has given a mortgage or deed of trust of all its property, tolls, incomes, franchises, &c., to secure the principal and accrued interest on its bonds, its revenues so pledged are not liable to garnishment by its judgment creditors, after the execution of such mortgage or deed of trust. *Ib.*

1348. The trustees of a railway company, if they do business in the name of the company, are liable to be sued in that name, and their property is liable for responsibilities incurred, while transacting the business under that name. *Wilkinson* v. *Fleming*, 30 Ill. 353.

1349. In a suit to foreclose a railway mortgage, the question as to the validity of a contract of consolidation, cannot be raised by the mortgagor company. Having mortgaged the property, it will not be permitted to deny its own title. *R. & M. R. R.* v. *F. L. & T. Co.*, 49 Ill. 331.

1350. BY CONSOLIDATED COMPANY—*effect on property of either company consolidated.* Where corporations created, one under the laws of Wisconsin and one under the laws of this state, consolidate, but in so doing fail to pursue the terms of their charters, and afterwards, by the legislature of this state, the contract of consolidation is confirmed, and the consolidated company is recognized as a corporation of this state, a mortgage made in the corporate name of both companies, they being the same in both states and managed by a common board of directors, upon the property of the corporation of this state, is a valid mortgage of the latter corporation. *Ib.*

1351. Where a person takes the entire management and control of the corporations so consolidating, managing the same as one company, for the better security and protection of the mortgagee, such person thereby becomes a trustee, not only for the mortgagee, but also for the mortgage corporation. *Ib.*

1352. Where a railway company executed its deed of trust on its franchise and railroad, and all property connected therewith, present and prospective, to secure the payments of its bonds, but the deed did not mention corporate subscriptions made to its capital stock, it was *held,* that the purchasers under the same, acquired no claim to county bonds issued under a subscription made by a county. *Morgan Co.* v. *Thomas*, 76 Ill. 120.

1353. The earnings of a railway company from the operation of its road, though mortgaged to secure the payment of certain bonds, before foreclosure or possession taken by the trustee, may be reached by other creditors of the company, and are liable to garnishment, when the mortgage provides that, until default, the company may possess and use the road, &c., and receive the rents, profits and increase arising therefrom. *M. V. & W. Ry.* v. *U. S. Express Co.*, 81 Ill. 534.

1354. FORECLOSURE—*sale of part of road.* If a mortgage is given by a railway company upon its entire road to secure bonds issued by it, and it procures the grading of only a part of the road in the middle, and then abandons the work, leaving each end of the road unfinished and another company organizes and completes the road, on bill to foreclose the mortgage given by the first company, it is erroneous to decree a sale of the middle portion of the road, leaving the two ends worthless. If any foreclosure can be had the entire road must be sold, and the proceeds distributed as between the bondholders of the original company and the new company in the proportion which the work done by the first company bears to the cost or value of the entire road as completed. *C., D. & V. Ry.* v. *Lœwenthal*, 93 Ill. 433.

1355. TITLE NECESSARY TO SUPPORT. Where a railway company executing a mortgage upon its road as contemplated, has no legal title to any of the right of way, but only contracts for a small portion thereof, to be conveyed upon conditions, which it never performs or has agreed to perform, and a new company is organized which builds the road and acquires the legal title to most of the right of way and is equitably entitled to the balance, no decree can be sustained under the mortgage as against the new company for the sale of its property. The mortgage creditors of such original company, have no rights superior to those of that company, and it has no such interest or title in the road as can be subjected to sale under the mortgage. *Ib.*

1356. Where a railway is, by its charter, authorized to borrow money and mortgage the whole or any part of its road, property or income then existing, or thereafter to be acquired, the company may not only mortgage its present property and rights, but such as it may thereafter acquire, and such after-acquired property will be subject to be sold on foreclosure; and this seems to be the rule independent of the authority given in the charter. *Quincy* v. *C., B. & Q. R. R.*, 94 Ill. 537.

1357. PROOF OF AUTHORITY TO MAKE. The execution of a mortgage under the seal of a corporation, regular on its face, and by the properly constituted officers, is *prima facie* evidence that the mortgage was executed by the authority of the corporation, and parties objecting, take on themselves the burden of proving it was not so executed. *Wood* v. *Whelen*, 93 Ill. 153.

1358. BONDS—*convertible into stock—validity.* The fact that bonds of a corporation secured by mortgage contain a provision that they are convertible into stock at the option of the holder, whereby the capital stock may be increased without the assent of the stockholders, will not excuse the company from paying the money it obtained on the faith of such bonds, and will not affect their validity as binding obligations on the company. *Wood* v. *Whelen*, 93 Ill. 153.

1359. IMPLIED POWER TO BORROW MONEY. A private corporation without any express authority in its charter, may borrow money for the promotion of the objects of its creation, and may secure the same on its property, by mortgage or otherwise, as being within its implied or incidental powers; and such corporation will be estopped to deny its authority to pledge its property real or personal, or both, for the payment of the money. *Ib.*

1360. Under the statute authorizing a railway company to borrow money for certain purposes, to dispose of its bonds for the sum so borrowed, and to mortgage its property and franchises to secure the same, upon the concurrence of the holders of two-thirds in amount of stock of such corporation, to be expressed at a meeting of stockholders, to be called by the directors, who are to give notice, &c., a resolution of the directors at a directors' meeting authorizing and directing the execution of a mortgage for a loan, it being shown that they were the only stockholders, except one, and that all the stockholders assented to the making of the mortgage, while not a literal compliance with the law, is a substantial compliance with its spirit, and the mortgage will be held good. *Thomas* v. *Citizens' Horse Ry.*, 104 Ill. 462.

1361. The statute requires the concurrence of the holders of two-thirds in amount of the stock of the corporation to the proposition to borrow money and mortgage the corporate property. In such case whether the stockholders received such a notice of the meeting as the statute requires, is a matter of no importance, if they met and acted upon the question. This action is as binding as if they had the proper notice. *Thomas* v. *Citizens' Horse Ry.*, 104 Ill. 462.

1362. Under a statute requiring the concurrence of the holders of two-thirds of the stock of a corporation to the mortgage of the corporate property for a loan of money, to be expressed at a meeting of the stockholders called by the directors for that purpose, a meeting of the directors, who are the only stockholders, except one, at which all assent to the proposition, is in effect a meeting of the stockholders, and the act of the directors that of the stockholders. The requirement of the concurrence of the holders of two-thirds of the stock is intended for the protection of the stockholders,' and is a matter in which the public have no interest. *Ib.*

1363. ESTOPPEL. Where power is conferred upon a corporation to borrow money, and secure the same by mortgage on its property, such corporation after having received the loan on the security of its mortgage, will not be allowed to avoid liability by questioning its power to make the mortgage, or showing a defective execution of the power conferred upon it. *Thomas* v. *Citizens' Horse Ry.*, 104 Ill. 462.

1364. Even if the directors of a private corporation have no authority to borrow money and mortgage its real estate for its repayment, yet if the stockholders ratify their action by approving the minutes of their proceedings before the loan is effected, and afterwards receive the benefit of the loan and pay interest thereon, the stockholders will be estopped from questioning the authority of the directors on bill to foreclose the mortgage. *Aurora Agricultural and Hort. Soc.* v. *Paddock*, 80 Ill. 263.

1365. IMPLIED POWER—*to mortgage.* The power to mortgage, when not expressly given or denied to a private corporation, will be regarded as an incident to the power to acquire and hold real estate and to make contracts. *Aurora Agl. & Hort. Soc.* v. *Paddock*, 80 Ill. 263.

1366. OF CHATTELS. A mortgage or deed of trust by a railway corporation, embracing all its real and personal property, with its franchise, made in pursuance of express authority in its charter, and recorded in each county through which the road passes, will create a valid and binding lien on its personal as well as its real property, notwithstanding it has not been acknowledged in accordance with the requirements of the chattel mortgage act. That act has no application to railroad mortgages. *Cooper* v. *Corbin*, 105 Ill. 224. On this point see *Palmer* v. *Forbes*, 23 Ill. 301; *Hunt* v. *Bullock*, 23 Ill. 320; *Titus* v. *Mabee*, 25 Ill. 257; *Titus* v. *Ginheimer*, 27 Ill. 462; *Peoria & Springfield R. R.* v. *Thompson*, 103 Ill. 187.

1367. As to retiring first issue of bonds, with a new series, and exchanging bonds for lots mortgaged, released of the mortgage, &c., see *Chicago & Great Western R. R. Land Co.* v. *Peck*, 112 Ill. 408.

1368. The power of a railroad corporation to sell and transfer promissory notes and choses in action, does not include the power to mortgage them. *Morris* v. *Cheney*, 51 Ill. 451. See also *Hatcher* v. *T., W. & W. R. R.*, 62 Ill. 477.

1369. ROLLING STOCK AND MOVABLE PROPERTY—*personalty.* § 20. The rolling stock and all other movable property belonging to any such corporation, shall be considered personal property, and shall be liable to execution and sale, in the same manner as the personal property of individuals. [R. S. 1887, p. 1005, § 21; S. & C., p. 1915, § 21; Cothran, p. 1143, § 21. See ante 62–66.]

1370. Prior to the adoption of the constitution of 1870 railway, cars or rolling stock of a railway company were held to be real estate. *Palmer* v. *Forbes*, 23 Ill. 301; *Titus* v. *Mabee*, 25 Ill. 257; *Titus* v. *Ginheimer*, 27 Ill. 462; *Fahs* v. *Roberts*, 54 Ill. 192, 194; *Minnesota Co.* v. *St. Paul Co.*, 2 Wall. 645.

1371. A railroad track is real estate. *S. & M. R. R.* v. *Morgan Co.*, 14 Ill. 163. See ante 62.

1372. Fuel, office furniture, material for lights and all other detached property of that kind, not like road equipments, designed for the continued use of the road, is personalty. *Hunt* v. *Bullock*, 23 Ill. 320.

1373. The act of Feb. 14, 1855, directing that the track and superstructure of a railroad be known as "fixed and stationary personal property," has reference only to the collection of the revenue, and did not change the nature of this property for other purposes. *Maus* v. *L., P. & B. R. R.*, 27 Ill. 77.

1374. The land constituting the right of way of a railroad, with the ties, rails, &c., in place on the track, turnouts, depot grounds and the buildings on the same, are real estate, but the rolling stock is made by statute for the purposes of taxation, personal property. *Union Trust Co.* v. *Weber*, 96 Ill. 346; revenue act of 1872, § 44, p. 14; S. & C., 2043.

1375. *Held*, in 1870, that cord-wood of a railway company was subject to levy and sale on execution, while rails and bridge timber for repairing the road were not. *Fahs* v. *Roberts*, 54 Ill. 192.

1375a. ROLLING STOCK—*real estate for purpose of mortgage and conveyance.* The rolling stock of a railroad is a part of the realty, so as to pass by a mortgage or conveyance of the road. *Mich. C. R. R.* v. *Ch. & Mich. L. S. R. R.*, 1 Bradw. 399, 409.

1375b. Before the constitution of 1870 a freight car on the road, side track or turntable of the company was realty, and like timber, fruit trees and buildings, only became personalty, when detached from the realty by the owner. *Titus* v. *Mabee*, 25 Ill. 257, 261.

1376. CAPITAL STOCK AND BONDS—*limitation on issue—fictitious increase of.* § 21. No such corporation shall issue any stock or bonds, except for money, labor or property actually received and applied to the purposes for which such corporation was organized. All stock dividends, and other fictitious increase of the capital stock or indebtedness of any such corporation, shall be void. [R. S. 1887, p. 1005, § 22; S. & C., p. 1915, § 22; Cothran, p. 1143, § 22.]

1377. An agreement to subscribe a certain amount of stock when books are opened does not make the party so agreeing a stockholder and as such liable to calls. His failure to subscribe does not make him liable for the value of the stock he agreed to take. *Thrasher* v. *Pike County R. R.*, 25 Ill. 393.

1378. FRAUDULENT ISSUE OF. If the directors of a railway company gratuitously give away certificates of stock, being a major part thereof, to contractors building the road, for the purpose of giving them a controlling influence in the election of officers and the management of the road, a court of equity will declare the same void, especially where a part of the directors are interested in the contract with the contractors. *G., C. & S. R. R.* v. *Kelly*, 77 Ill. 426; *People ex rel.* v. *Logan County*, 63 Ill. 374.

1379. The fact that a railway company has fraudulently encumbered its road or given a lease thereon, so as to lessen the value of the stock, or issued a large amount of stock to the lessee of the road, is no defense against an application to compel a county to subscribe a sum voted by the people at a legal election. *People ex rel.* v. *Logan County*, 63 Ill. 374.

1380. If the directors of a railway company gratuitously gives another company to whom it has leased its road a large amount of stock, or gives such stock for the fraudulent purpose of depriving the stockholders of dividends, or of destroying the value of their shares, or to prevent them from exercising their legal power of control over the road in the election of directors, or otherwise, a court of equity will afford relief. *Ib.*

1381. STOCK WHEN VOID. Stock issued in violation of the law under which the company is incorporated, is illegal and void, and the corporation cannot be required to transfer the same upon its books, notwithstanding it may have been issued with the consent of all the stockholders of the company at the time. *People ex rel.* v. *Sterling Burial Case Mfg. Co.*, 82 Ill. 457.

1382. FRAUDULENT ISSUE OF STOCK. Where a stock-yard company was organized by the officers of a railway company and others, and the only means put into the same was by the railroad company, through its officers, who also controlled the stock-yard company, and the latter company issued stock to the extent of its charter, a portion of which was used as a corruption fund, and the balance divided between certain members of the company, they paying nothing therefor, it was *held*, that the issue of the stock was in violation of law and in fraud of the rights of the stockholders of the railway company, and vested in the recipients of the same no rights which a court of equity would enforce or protect. *Tobey* v. *Robinson*, 99 Ill. 222.

1383. The object of § 13, art. 11, of the present constitution in providing that, "no railroad shall issue any stock or bonds, except for money, labor or property actually received and applied to the purposes for which such corporation was created," and that "all stocks, dividends and other fictitious increase of the capital stock or indebtedness of such corporation shall be void," was to prevent reckless and unscrupulous speculators, under the guise or pretense of building a railroad, or of accomplishing some other legitimate corporate purpose, from fraudulently issuing and putting upon the market, bonds or stocks that do not, and are not intended to represent money or property of any kind, either in possession or in expectancy, the stock or bonds in such cases being entirely fictitious. *Peoria & Springfield R. R.* v. *Thompson*, 103 Ill. 187.

1384. Where one for a present consideration, in good faith, purchases bonds or stocks in the regular course of business from a railway company, and such consideration is accepted by the proper officer of the company, and nothing appears to show that it is to be used or applied to other than legitimate corporate purposes, such bonds or stocks when thus issued will be regarded as having been issued for money, labor or property, "actually received and applied" within the meaning of the constitutional provision. *Ib.*

1385. ESTOPPEL. Although a contract entered into by the agents or officers of a private corporation, is *ultra vires,* and not binding while executory, yet if the company permits the other party without objection to go on and perform the contract, and it thereby obtains and appropriates to its own use, money, property or labor in furtherance of some legitimate corporate purpose, it will be estopped from denying its liability on such contract. *Ib.*

1385a. DIVIDENDS — *to be general.* The dividends declared must be general on all the stock, so that each shareholder may receive his proportionate share. *Ryder* v. *Alton & Sang. R. R.,* 13 Ill. 516, 520.

1385b. WHO MAY INCREASE. The charter of a bank fixed its capital stock at $100,000, with power to increase it to $500,000, without providing by whom this power should be exercised: *Held,* that the board of directors, as such, had no authority to increase the capital stock without the assent of the shareholders. *Eidman* v. *Bowman,* 58 Ill. 444.

1385c. INCREASE OF CAPITAL — *who entitled to shares.* Where capital stock is increased legally, the right to such additional stock vests in the original stockholders, each one to take in proportion to the amount held by him of the original stock, if he will pay for it. This right may be waived, but if it is not, the party entitled cannot be deprived of it by the board of directors or otherwise. *Ib.*

1386. CONSOLIDATION—*limitation on right.* § 22. No such corporation shall consolidate its capital stock with any other railway owning a parallel or competing line. And in no case shall any consolidation take place, except upon sixty days' notice thereof given, which notice shall be given in manner and form as prescribed in the fifteenth section of this act. [R. S. 1887, p. 1005, § 23; S. & C., p. 1915, § 23; Cothran, p. 1143, § 23. See ante, 67.]

1387. All the rights and powers vested in railway corporations, will, upon their consolidation, be conferred upon and united in the consolidated company. The right of one to a municipal subscription passes to the new company. *Robertson* v. *City of Rockford,* 21 Ill. 451.

1388. WITH FOREIGN CORPORATION. Under the act of 1853, incorporating the Rockton & Freeport Railway Company, that company was authorized to consolidate with a corporation outside of this state, and to place the control of the consolidated stock under the control of the board of directors of the foreign company. *Racine & Miss. R. R.* v. *Farmers' Loan & Trust Co.,* 49 Ill. 331.

1389. EFFECT OF CONSOLIDATION. The consolidation of the stock of a railway company created by the laws of Wisconsin, with one created by the laws of this state, does not constitute the corporations thus consolidating, one corporation of both states, or of either, but the corporation of each state continues a corporation of the state of its creation, although the same persons as officers and directors manage and control both corporations—as one body. *Ib.*

1390. Defect in consolidating may be cured by subsequent legislation recognizing consolidated company. *Ib.,* 345; *Mitchell* v. *Deeds,* 49 Ill. 416.

1391. LIABILITY FOR DEBTS OF COMPANIES UNITED—*act not retrospective.* The act of 1867, which provides that in case of consolidation of two or more railroad companies, the consolidated company shall be liable for all debts of each company entering into the arrangement, is not retrospective, but was designed to apply to companies which might consolidate after its passage. *Hatcher* v. *T. W. & W. Ry.,* 62 Ill. 477.

1392. A railway company gave a deed of trust under which its road, property and income was sold by the trustee to parties who organized a new company under the old name. Afterwards under an act of the legislature, the president of the old company transferred the corporate franchise to the purchasers, and the old company ceased to exist. *Held,* that the purchasers having acquired a valid title to the property of the old corporation without liability for any of its debts which were not prior liens, their rights could not be taken away or impaired by subsequent legislation. *Ib.*

1393. DUTY OF NEW CORPORATION FORMED BY. After consolidation the new company becomes liable to perform the duties required of the companies so consolidated, and if no part of the franchise is reserved to either of the old companies, they will not be liable to the public for the performance of duties devolving upon the new company. *Peoria & Rock Island Ry.* v. *Coal Valley Mining Co.,* 68 Ill. 489.

1394. EVIDENCE OF CONSOLIDATION. In a suit against a consolidated railway company upon promissory notes given by one of the original companies forming the new one, copies of the articles of consolidation on file in the office of the secretary of state, duly certified by the secretary of state and authenticated by his seal of office, are competent evidence to prove the consolidation the same as the original articles would be. *C.. C. & I. C, Ry.* v. *Skidmore,* 69 Ill. 566.

1395. CONSOLIDATED COMPANY—*in what name to be sued.* Where a railroad company, after the execution of promissory notes, is consolidated with another company, and the company thus formed assumes a new name, it may be sued by such new name, and it will be estopped from denying the name by which it is sued. *Ib.*

1396. EFFECT ON PRIOR LIABILITIES. Where the articles of consolidation of two railway companies provided that the new company should assume the debts and liabilities of the old companies, and should assume and carry out all their unexecuted contracts, and the act of the legislature ratifying and confirming the consolidation, saved the rights and remedies of creditors, it was *held,* that a person performing labor under a contract with one of the old companies, might maintain an action against the new company to recover whatever sum was due him under his contract. *Western Union R. R.* v. *Smith,* 75 Ill. 496.

1397. Where a new corporation is formed by amalgamation of two or more distinct corporations into one, it succeeds to the rights and faculties of the several components, and must as a necessary consequence, be subject to all the conditions and duties imposed by the law of their creation, except in so far as the act allowing the consolidation may otherwise provide. *C., R. I. & P. R. R.* v. *Moffitt,* 75 Ill. 524.

1398. Where a railway company constructed a bridge across a stream not navigable, but affording a large volume of water, by driving piles with spans of only seventeen feet, and leased its road to another company, which, while operating the same as lessee, built a new bridge at the same place, constructed in the same manner, except that the spans were enlarged to fifty feet, but left the piles of the old bridge standing, a portion of the tops being cut off, after which these two companies consolidated, forming a new one, with a different name, the new company continuing to operate the road and use the bridge in such condition: *Held*, that the new company was liable in damages to a riparian owner above, whose land was overflowed and injured in consequence of an obstruction by drift caused by the manner in which the bridge was built and used. *Ib.*

1399. A consolidated railroad company, formed under legislative sanction, succeeds to all the rights conferred upon the several companies thus united, by their respective charters, but it is not invested with any greater or other rights than were possessed by the constituent companies forming the consolidation. *Ruggles* v. *People*, 91 Ill. 256.

1400. WHETHER LIABLE TO TAXES IN THIS STATE. A railway corporation formed under our laws by the consolidation of other companies, one of which was incorporated under the laws of this state, and the others in other states, the new or consolidated company is to be considered as incorporated under the laws of this state, within the meaning of § 1 of the revenue act of 1872, and the capital stock of such corporation in this state is liable to taxation here. *O. & M. R. R.* v. *Weber*, 96 Ill. 443.

1401. REMEDY — *at law.* Where a consolidated company by virtue of its consolidation, becomes liable for the debts of the companies composing it, the creditor's remedy is complete and adequate at law, and a court of equity will have no jurisdiction. *Arbuckle* v. *Ill. Midland Ry.*, 81 Ill. 429.

1402. Amendatory act held a legislative recognition of the existence of a consolidated company, and of the name it adopted by the articles of consolidation, amounting to a legislative ratification of the consolidation, &c. *McAuley* v. *C. C. & I. C. Ry.*, 83 Ill. 348, 352.

1403. POWER CONSTRUED AS TO LEASING OR CONSOLIDATING. Under an amendment to a railroad charter providing that the company shall have power to consolidate and construct its road with any other continuous line of railroad, either in this state or the state of Indiana, upon such terms as may be agreed upon between the companies uniting or connecting, and for that purpose giving full power to the company to make and execute such contracts with any other company as will secure the object of such consolidation, or connection, the domestic corporation can only do one of two things: — either consolidate its road with another railroad in this or the state of Indiana, or make an agreement for connection with such road, so as to secure a continuous line. Under such law it has no power to lease its road to a foreign company. *Archer* v. *T. H. & I. R. R.*, 102 Ill. 493.

1404. CONTRACT—*whether a consolidation or agreement for connection.* A contract between a railway company of this state and one of the state of Indiana, provided that on the completion of the two roads to the state line, the latter company should take charge of and operate the road in this state, with its equipments for 999 years, for which it was to be allowed 65 per cent. of the gross receipts from all traffic moved on the line, or business done thereon as a consideration for working and maintenance expenses, the remaining 35 per cent. to be appropriated: first to the payment of interest on the first and second mortgage bonds of the Illinois company, according to their priority; and second, all the surplus to be paid over to the Illinois company semi-annually, to be disposed of by it for the benefit of its stockholders; also that if the 35 per cent. should not, for any cause be sufficient in amount to protect the interest on the mortgage bonds and the sinking fund therefor as they matured, together with the payment of taxes and proper cost of maintaining organization, so that the rights of stockholders might be preserved, then, in that event, the lessees should advance for the company whatever might be needed, to be accounted for under yearly averages of the lease during the contract: *Held*, that the agreement was not a lease of the Illinois road, nor a contract of consolidation, but one of connection between the two roads only, leaving the Illinois corporation the owner of the road, though in the use and under the control of the Indiana company. *Archer* v. *T. H. & I. R. R.*, 102 Ill. 493.

1405. Where a railway corporation of this state is consolidated with a similar corporation of another state, in conformity with the laws of this state, the new company so created will be clothed with all the rights, privileges and powers conferred by the laws of this state upon the old corporation of this state. *Cooper* v. *Corbin*, 105 Ill. 224.

1405a. Under the laws of this state permitting the consolidation of railway companies, all the powers, rights, franchises and immunities to which the several companies were entitled, pass to the new or consolidated company. Such consolidation will not defeat a donation made by a town to one of the companies. *Niantic Savings Bank* v. *Town of Douglas*, 5 Bradw. 579.

1406. ANTECEDENT DEBTS AND LIABILITIES. The absolute consolidation of two or more railway companies, without any provision being made for the old debts of the former companies, renders the new company liable for them. *Tysen* v. *Wabash Ry.*, 11 Biss. 510: 15 Fed. Rep. 763.

1407. Persons who purchase bonds of a railway company while a statute is in force authorizing the consolidation of railways, must be held to have contemplated at the time of their purchase, that the company issuing them might consolidate with other companies. *Ib.*

1408. When a director may be estopped from objecting to a consolidation. See *Mowrey* v. *Indianapolis & Cincinnati R. R.*, 4 Biss. 78.

1409. EFFECT OF. The consolidation of two companies does not necessarily work a dissolution of both and the creation of a new corporation. Whether such will be the effect, depends upon the legislative intent manifested in the statute under which the consolidation takes place. *Central R. R. & Banking Co.* v. *Georgia*, 92 U. S. 665.

1410. In view of the legislation in Illinois, great liberality should be exercised in regard to contracts for consolidations between different railroad companies. By the general language of the statute relating to the union and consolidation of different lines of road, the means by which the result is to be, or has been obtained, have not been clearly designated, but that has been left to be adjusted by contracts between the parties. *Dimpfel* v. *O. & M. Ry.*, 9 Biss. 127.

1411. CONSENT OF STOCKHOLDERS. To effect a consolidation of railway companies subsisting under special charters not providing therefor, the consent of every stockholder must be given, and any one dissenting stockholder is entitled to an injunction against such consolidation. *Mowrey* v. *I. & C. R. R.*, 4 Biss. 78.

1412. As to consolidation of parallel or competing lines see *State* v. *Vanderbilt*, 37 Ohio St. 590.

1413. A consolidated railroad corporation, formed by the union of two corporations, each created by a different state, is, so far as property and business within each state are concerned, subject, unless otherwise provided in the act of consolidation, to the same control therein, as before, and in each state is to be treated as a domestic corporation. *Peik* v. *Chicago & N. W., R. R.*, 94 U. S. 164; *Muller* v. *Dows*, 94 U. S. 444.

1414. Without enabling legislation a railway company has no

power to lease its road to a foreign corporation. *Archer* v. *T. H. & I. R. R.*, 102 Ill. 493.

1415. The general railroad law of 1865 prohibited consolidation with, or lease to a foreign corporation, without written consent of the resident stockholders. *Ib.*

1416. Consolidation of railway companies under act of Feb. 28, 1854, *held*, to confer on consolidated company same capacity to receive municipal subscription as any one of the merged companies had. *Empire* v. *Darlington*, 101 U. S. 87.

1417. RIGHT TO CONSOLIDATE. Under the act of Feb. 28, 1854, railroad companies organized under the laws of this state, and whose lines of road are so intersected as to constitute a continuous line within this state, might consolidate their property, stocks, rights and franchises, and thereby constitute a new corporation under a new name, possessing the property, rights, powers and franchises of the constituent companies as given by their charters; and thereupon the constituent companies, as independent legal entities, would cease to exist, and all their duties and obligations, whether to the public, or to private persons, would be cast upon, and must be assumed and discharged by the new consolidated company. *O. & M. Ry.* v. *People*, — Ill. —. Filed Jan. 18, 1888.

1418. ROADS OF SEVERAL STATES. A corporation *de jure*, as well as *de facto*, may be created with the consent and under the authority of two or more states, by the voluntary consolidation of corporations, created and existing by virtue of the laws of such states respectively. On the consummation of such consolidation of railway companies in two or more states, authorized by the laws of the states creating them, a new corporation will be created, having in each state all the powers, rights and franchises that the constituent companies had in the same state, but not in one state the powers, &c., of the constituent company in the other state. The new corporation will stand in each state as the original corporation had previously stood in the same state. *Ib.*

1419. A consolidated railroad company, formed by the consolidation of two or more companies of different states, having a capital stock which is a unit, and only one set of stockholders, who have an interest, as such, in all its property everywhere, and a single board of directors, will have its domicile in each state, and its stockholders, directors and officers may, in the absence of any statutory provision to the contrary, hold meetings and transact corporate business in either of the states, though in relation to either state, the consolidated company will be a separate corporation, governed by the laws of that state as to its property therein, and subject to taxation in conformity with the laws of such state, and to all the police power of the state in respect to its property and franchise within such state. *Ib.*

1420. EFFECT OF. Upon the creation of the consolidated corporation the constituent corporations of the different states do not necessarily cease to exist, although they lie dormant, and their property, rights, powers and franchises are possessed and exercised by the new consolidated corporation. *Ib.*

1421. On consolidation the original companies become extinct and the new company succeeds to the ownership of the two roads together with all their other property and rights, and becomes subject to all the liabilities and burdens of each of the old companies. *People* v. *L. & N. R. R.*, 120 Ill. 48.

1421a. POWERS OF NEW COMPANY. A consolidated company formed under legislative sanction, succeeds to all the rights conferred upon the several companies thus united, by their respective charters, but not with greater or other rights than were possessed by the constituent companies. *Ruggles* v. *People*, 91 Ill. 256.

RAILROADS—CONSOLIDATION.

An act to provide for the consolidation of certain railroad corporations. Approved June 14, 1883. In force July 1, 1883. L. 1883, p. 124.

1422. WHAT RAILROADS MAY CONSOLIDATE, AND HOW. §
1. *Be it enacted by the People of the State of Illinois, represented in the general assembly:* Whenever any railroad which is situated partly in this state, and partly in one or more other states, and heretofore owned by a corporation formed by consolidation of railroad corporations of this and other states, has been sold pursuant to the decree of any court or courts of competent jurisdiction, and the same has been purchased as an entirety, and is now, or hereafter may be, held in the name or as the property of two or more corporations incorporated respectively under the laws of two or more of the states in which said railroad is situated, it shall be lawful for the corporation so created in this state to consolidate its property, franchises and capital stock with the property, franchises and capital stock of the corporation or corporations of such other state or states in which the remainder of such railroad is situated, and upon such terms as may be agreed upon between the directors, and approved by the stockholders owning not less than two-thirds in amount of the capital stock of such corporations. Such approval may be given by the stockholders of such corporation of this state at any time, in writing or by vote, at any annual or special meeting, upon sixty days' notice given by publication in any newspaper published in the county where the general office of such company is situated, and such meeting is to be held: *Provided,* that no consolidation shall take place with any railroad owning a parallel or competing line; and a majority of the directors of such consolidated company shall be citizens and residents of this state; and where the line of the road of the original company has been located in this state, and aid in the construction thereof voted by any municipality by way of subscription or donation, and received by the company, and the road as so located not yet completed, then the consolidated company shall have no power or right to change such line as so located so as to make the same substantially different from the line so located at the time the aid was voted. [R. S. 1887, p. 1008, § 39; S. & C., p. 1916, § 24; Cothran, p. 1189*b*, § 178. See ante, 1386–1421.]

1423. RESIDENCE OF DIRECTORS. The constitutional provision that "a majority of the directors of any railroad corporation now incorporated, or hereafter to be incorporated by the laws of this state, shall be citizens and residents of this state," has no application to a railway corporation formed prior to the adoption of the constitution, by the consolidation of a railway company in this state with one of another state, by the consent of each of such states. Such a corporation exists under the laws of the two states, and cannot be said to be incorporated solely under the laws of this state. *O. & M. Ry.* v. *People*, — Ill. —. Filed Jan. 18, 1888.

1424. SAME — *when it takes effect – recording articles of incorporation — evidence of existence of new company.* § 2. Such consolidation shall take effect upon the filing and recording of such articles of consolidation in the office of 'the secretary of state of the state of Illinois, and a certified copy thereof in the office of the recorder of the various counties in which said railroad is situated. A certified copy of such articles of consolidation, under seal of the secretary of state, shall be deemed and taken to be *prima facie* evidence of the existence of such consolidated corporation. [R. S. 1887, p. 1008, § 40; S. & C., p. 1916, § 25; Cothran. p. 1189b, § 179. See ante, 1386.]

1425. . GENERAL OFFICE TO BE KEPT IN THIS STATE — *list of stockholders, their residences, &c.— registry of stock and transfers.* § 3. Such consolidated corporation shall, at all times, keep a general office within this state, at which shall be kept a complete list of all stockholders of such corporation, their places of residence, the amount of stock owned by each, and where the stock of such corporation may be registered and transferred: *Provided,* that nothing contained in this bill shall be construed to impair or affect the rights of any party holding unsettled claims against any of the corporations to be consolidated. [R. S. 1887, p. 1008, § 41; S. & C., p. 1917, § 26; Cothran, p. 1189c, § 180. See ante 67 and 1187.]

An act relating to lessees in this state of railroads in adjoining states. Approved March 30, 1875. In force July 1, 1875. Laws 1875, p. 96

1426. PURCHASE OF LEASED ROADS—*operated by companies created under laws of this state. Roads out of state connecting with roads in state.* § 1. *Be it enacted by the people of the state of Illinois, represented in the general assembly,* That all railroad companies incorporated or organized, or which may be incorporated or organized under the laws of this state, or of this and any adjoining state, which now are, or at any time hereafter may be, in possession of and operating connecting railroads in·states adjoining this state under lease in perpetuity, or for a period of not less than twenty years, shall have power to purchase the remaining interests, property and franchises of the lessors of such railroads situate in such adjoining states, on such terms and conditions as may be agreed upon by the parties, or their assigns, to such lease: *Provided,* that nothing in this act shall be so construed as to authorize any corporation acting by, or organizing under the laws of any other state to purchase or otherwise become the owners of any railroad in this state. [R. S. 1887, p. 1009, § 45; S. & C., p. 1917, § 27: Cothran, p. 1147, § 37.]

Act of March 1, 1872, resumed.

1427. DIRECTORY—*annual sworn reports by, to auditor.* § 23. The directors of every such corporation shall annually make a report, under oath, to the auditor of public accounts, and to such other officers as may be designated by law, of all its actings and doings, which, in part, shall include such matters relating to such corporations as may be now or hereafter prescribed by law. [R. S. 1887, p. 1005, § 24; S. & C., p. 1917, § 28; Cothran, p. 1144, § 24.]

1228. LEGISLATIVE CONTROL OVER—*prevention of unjust discrimination and extortions in charges.* § 24. The general assembly shall have power to enact, from time to time, laws to prevent and correct abuses, and to prevent unjust discriminations and extortions in the rates of freight and passenger tariff, and to establish reasonable maximum rates of charges for the transportation of persons or property on any railway that may be constructed under the provisions of this act, and to enforce such laws by adequate penalties to the extent, if necessary for that purpose, of forfeiture of the property and franchises of any such corporation. [Const. art. 11, §§ 12 and 15. See ante, 68-86. R. S. 1887, p. 1005, § 25; S & C., p. 1917, § 29; Cothran, p. 1144, § 25.]

LEGISLATIVE CONTROL IN GENERAL.

(a) POLICE REGULATIONS.

1429. The legislature has the power by the enactment of general laws, from time to time, as the public exigency may require, to regulate corporations in the exercise of their franchises, so as to provide for the public safety. *Galena & Ch. Union R. R. v. Loomis*, 13 Ill. 548.

1430. In the exercise of privileges a corporation is as much subject to the general police laws of the state, as is any individual pursuing his lawful business. *Ib.*

1431. Corporations are artificial persons, created with limited powers and capacities, and subject to the general laws and legislation of the state, the same as natural persons are. They cannot be deprived of rights secured to them by contract, without just compensation; but like natural persons, in the exercise of their rights of oganization and existence, they are subject to the control of the legislature by general laws. *Bank of the Republic* v. *Hamilton County*, 21 Ill. 53.

1432. The general rights and powers of a private corporation and, which are not intended to be secured to it as its property, are subject to legislative control in the same manner as the general rights of individuals. *Ib.*

1433. Corporations are subordinate to and under the control of the government to the same extent as individuals. They are subject to the general police regulations of the state. *O. & M. R. R.* v. *McClelland*, 25 Ill. 140.

1434. The act to regulate the duties and liabilities of railroad companies passed in 1855, applies to companies previously incorporated. *Galena & Ch. Union R. R.* v. *Crawford*, 25 Ill. 529.

1435. A railroad company takes its charter upon the implied condition that its franchises shall be exercised subject to the power of the state to impose reasonable regulations upon it as the comfort, safety or welfare of society may require. *C. & A. R. R.* v. *People, use*, 105 Ill. 657. See *I. C. R. R.* v. *People*, 95 Ill. 313.

1436. Corporations created within the state are amenable to the police power of the state to the same extent as are natural persons, but to no greater extent. The legislature may require of these bodies the performance of any and all acts, which they are capable of performing, that it may require of natural persons. *Ruggles* v. *People*, 91 Ill. 256.

1437. The police power of the state when exercised by the legislature in the passage of laws for the protection of life, liberty and property, or laws for the general welfare, has no limitations or restrictions, except such as are found in the constitution. *Hawthorn* v. *People*, 109 Ill. 302.

1438. A state has all power necessary for the protection of the property, health and comfort of the public, and it may delegate this power to local municipalities in such measure as may be deemed desirable for the best interests of the public and the state may resume it again when deemed expedient. *Harmon* v. *Chicago*, 110 Ill. 400.

1439. PUBLIC BURDENS. The police power of a state, comprehensive as it is, has its limitations. A purely public burden cannot be laid upon a private individual, except as authorized in cases to exercise the right of eminent domain, or by proceedings to enforce special assessments or special taxation. *Chicago* v. *O'Brien*, 111 Ill. 532.

1440. By the grant of corporate franchises to railway companies to procure the right of way and operate their trains by the power of steam, the state does not deprive itself of its inherent power to enact all police laws necessary and proper to protect the life and property of its citizens. *T., P. & W. Ry.* v. *Deacon*, 63 Ill. 91.

1441. Railway corporations are subject to police regulations the same as private citizens. The legislature, where the public exigencies require it, has power to regulate corporations in the exercise of their franchises, so far as to provide for the public safety. The exercise of this right in no manner interferes with or impairs the powers conferred by their acts of incorporation. *T., W. & W. Ry.* v. *Jacksonville*, 67 Ill. 37.

1442. Under this power it has been held that the legislature may require railroad corporations, notwithstanding no such right has been reserved in their charters, to fence their tracks, to put in cattle guards, to place upon their engines a bell, and to do many other things for the protection of life and property. *Ib.*

1443. This police power is inherent in the state and it can not part irrevocably with its control over that which is for the health, safety and welfare of society. *Ib.*

1444. What are reasonable regulations, and what are subjects of police powers, must, necessarily, be judicial questions. The law making power is the sole judge when the necessity exists, and when, if at all, it will exercise the right to enact such laws. *Ib.*

1445. Like other powers of government, there are constitutional limitations to the exercise of the police power. The legislature cannot under the pretense of exercising this power, enact laws not necessary to the preservation of the health and safety of the community, that will be oppressive and burdensome upon the citizens. If it should prohibit that which is harmless in itself, or command that to be done, which does not tend to promote the health, safety or welfare of society, it would be an unauthorized exercise of power, and it would be the duty of the court to declare such legislation void. *Ib.*

1446. In matters pertaining to the internal peace and well being of the state, its police powers are plenary and inalienable. It is a power coextensive with self protection. Everything necessary for the protection, safety and best interests of the people of the state may be done under this power. Persons and property may be subjected to all reasonable restraints and burdens for the common good. *Dunne* v. *People*, 94 Ill. 120.

1447. Where mere property interests are involved, this power like other powers of government, is subject to constitutional limitations; but where the internal peace and health of the people are concerned, the only limitations imposed are, that "such regulations must have reference to the comfort, safety and welfare of society." What will endanger the public security must, as a general rule, be left to the wisdom of the legislative department. *Ib.*

1448. With certain constitutional limitations, the rights of all persons, whether natural or artificial, are subject to such legislative control as the legislature may deem necessary for the general welfare, and in this respect there is no difference between the rights of natural and artificial persons. *Ward* v. *Farwell*, 97 Ill. 593.

1449. Laws imposing police regulations are usually construed as applying to existing corporations as well as those afterwards created. *Galena & Ch. Union R. R.* v. *Loomis*, 13 Ill. 548; *Western Union R. R.* v. *Fulton*, 64 Ill. 271; *Indianapolis & St. Louis R. R.* v. *Blackman*, 63 Ill. 117.

1450. The state in the exercise of this power may require reports. See ante, § 23, Railroad Law; *State* v. *Southern Pacific R. R.*, 24 Tex. 80.

1451. The numbering of cars. *F. & P. Pass. R. R.* v. *Philadelphia*, 58 Pa. St. 119.

1452. The fixing and posting of rates. *Ch. & N. W. R. R.* v. *Fuller*, 17 Wall. 560.

1453. A slow rate of movement at certain places. *T., P. & W. Ry.* v. *Deacon*, 63 Ill. 91; *C., R. I. & P. R. R.* v. *Reidy*, 66 Ill. 43; *C., B. & Q. R. R.* v. *Haggerty*, 67 Ill. 113; *L., S. & M. S. R. R.* v. *Berlink*, 2 Bradw 427; *C. & N. W. Ry.* v. *Schumilowsky*, 8 Bradw. 613; *Garland* v. *C. & N. W. Ry.*, 8 *Id.* 571; *C. & A. R. R.* v. *Robinson*, 9 Bradw. 89; *P., D. & E. Ry.* v. *Miller*, 11 Bradw. 375; *C., B. & Q. R. R.* v. *Dougherty*, 12 Bradw. 181.

1454. The disuse of steam in cities. *R. R.* v. *Richmond*, 96 U. S. 521; 26 Gratt. 83.

1455. The stationing of a flagman at highway crossing. *T., W. & W. Ry.* v. *Jacksonville*, 67 Ill. 37; *L., S. & M. S. R. R.* v. *Kaste*, 11 Bradw. 536, 114 Ill. 79; *Ch., R. I. & P. Ry.* v. *Eininger*, 114 Ill. 79.

1456. The stopping of trains at county seats or certain stations. *Davidson* v. *State*, 4 Tex. App. 545; *R. R.* v. *LeGeirre*, 51 Tex. 189; *C. & A. R. R.* v. *People*, 105 Ill. 657.

1457. To keep ticket office open. *St. L., A. & T. H. R. R.* v. *South*, 43 Ill. 176.

1458. Police regulation must not interfere with foreign or interstate commerce. If it does it is void. *R. R.* v. *Husen*, 5 Otto, 465; *Salzenstein* v. *Mavis*, 91 Ill. 391.

1459. ELECTION OF DIRECTORS—*minority representation—cumulative voting—proxy.* § 25. In all elections for directors or managers of such railway corporations every stockholder shall have the right to vote, in person or by proxy, for the number of shares of stock owned by him, for as many persons as there are directors or managers to be elected, or to cumulate said shares, and give one candidate as many votes as the number of directors, multiplied by the number of his shares of stock, shall equal; or to distribute them, on the same principle, among as many candidates as he shall think fit; and such directors or managers shall not be elected in any other manner. [R. S. 1887, p. 1005, § 26; S. & C., p. 1917, § 30; Cothran, p. 1144, § 26. See ante, 59.]

1460. RATES OF CHARGES—*to induce aid by donation or subscription, binding on corporation and its successor.* § 25½. In all cases when any corporation organized under this act to induce aid in its construction, either by donation or subscription to its capital stock, shall desire to fix the rates for any period of time for the transportation of passengers or freight, such corporation may adopt a resolution fixing such rates, and the time for which the same is to be fixed, and have the same recorded in the office of the recorder of deeds in the several counties through which said road is proposed to be run; and during the time for which they are fixed, said rates shall in no case be amended by said corporation or its successors: *Provided,* that said rates shall not exceed the rates allowed by law. [R. S. 1887, p. 1005, § 27; S. & C., p. 1918, § 31; Cothran, p. 1144, § 27.]

1461. As to the binding effect of contract respecting rates of charges for freight, upon successor, see *C. & A. R. R. v. Ch., V. & W. Coal Co.,* 79 Ill. 121; *People* v. *L. & N. R. R.,* 120 Ill. 48.

1462. LIMITATION—*for commencement of and completion of road.* § 26. If any railway corporation organized under this act, shall not, within two years after its articles of association shall be filed and recorded as provided in the second section of this act, begin the construction of its road, and expend thereon twenty-five per cent. on the amount of its capital, within five years after the date of its organization, or shall not finish the road and put it in operation within ten years from the time of filing its articles of association, as aforesaid, its corporate existence and powers shall cease. [R. S. 1887, p. 1006, § 28; S. & C., p. 1918, § 32; Cothran, p. 1144, § 28.]

1463. REPEAL OF PRIOR ACTS — *saving of rights acquired under acts repealed — adoption of this act — validates prior organization — no release from prior obligations and liabilities—action against corporation.* § 27. That an act entitled "An act to amend 'an act to provide for a general system of railroad incorporations,' approved November 5, 1849," approved February 13, 1857, and also all of an act entitled "An act to provide for a general system of railroad incorporations," approved November 5, 1849, except the sections of the last named act numbered 34, 35, 36, 37, 38, 39, 40, 41, 42 and 45, and all laws in conflict with the provisions of this act, be and the same are hereby repealed: *Provided, however*, that all general laws of this state in relation to railroad corporations, and the powers and duties thereof, so far as the same are not inconsistent with the provisions of this act, shall remain in force and be applicable to railroad incorporations organized under this act. The repeal of the acts and parts of acts mentioned in this section shall not be construed so as to effect any rights acquired thereunder; but all corporations formed or attempted to be formed under such acts or parts of acts, notwithstanding any defects or omissions in their articles of association, may, if they will adopt or have adopted this act, be entitled to proceed thereunder, and have all the benefits of this act; and all such corporations that have adopted or that will adopt this act, are hereby declared legal and valid corporations, within the provisions of this act, from the date of the filing of their respective articles of association. And the fixing of the *termini* by any such corporation shall have the same effect as if fixed by the general assembly: *Provided*, that all corporations to which this act shall apply shall be held liable for, and shall carry out and fulfill all contracts made by them, or for, or on their behalf, or of which they have received the benefit, whether such corporation, at the time of the making of such contract or contracts, was organized, or had attempted to organize, under the general laws of the state of Illinois, or not; whether said contract was for right of way, work and labor done, or materials furnished, or for the running of trains or carrying passengers or freight upon such road, or upon any other road in connection therewith. And if such corporation has or does take possession of or use such right of way, labor or material so furnished by other persons or corporations, it shall be evidence of its acceptance of such contract so entered into by such person or corporation with said persons or corporations for its benefit. And upon said corporation failing to pay said sum as it ought equitably to pay for such right of way, labor or materials, or fail to carry out such contracts as aforesaid, so made with persons or corporations, it shall be held liable in an action at law or in chancery for the recovery of the value of said right of way, labor or materials, and for damages for non-fulfillment of such contract, in any court of competent jurisdiction in any county through which the road of such corporation may be located: *And, provided, further,* that this act shall not in any manner legalize the subscription of any township, county or city to the capital stock of any railroad company, nor authorize the issuing of any bonds by any township, city or county in payment of any subscription or donation. [R. S. 1887, p. 1006, § 29; S. & C., p. 1918, § 33; Cothran, p. 1144, § 29.]

1464. This section applies only to such corporations as adopt the act, and has no application to contracts to carry coal at reduced rates. *C. & A. R. R.* v. *Chi., Verm. & Wilm. Coal Co.,* 79 Ill. 121, 127.

1465. CURATIVE LEGISLATION—*validation of defective organization.* Any defect in the organization of a railway company, under the general railroad law of 1849, may be cured by subsequent legislation. *Illinois Grand Trunk R. R.* v. *Cook,* 29 Ill. 237; *Goodrich* v. *Reynolds, Wilder & Co.,* 31 Ill. 490; *Mitchell* v. *Deeds,* 49 Ill. 416; *Hatcher* v. *T., W. & W. R. R.,* 62 Ill. 477.

1466. Legislation held to create a new and distinct corporation and not a reorganization of the old company. *Morgan Co.* v. *Thomas,* 76 Ill. 120.

An act to enable railway companies to borrow money and to mortgage their property, and franchises therefor. Approved May 7, 1873. In force July 1, 1873. Laws of 1873, p. 141.

1467. BORROWING MONEY BY COMPANIES FORMED PRIOR TO MARCH 1, 1872. § 1. *Be it enacted by the people of the state of Illinois, represented in the general assembly,* That every railroad company organized under any law or laws of this state, in force before the first day of March, A. D 1872, is hereby empowered from time to time to borrow such sums of money as may be necessary for completing, furnishing, improving or operating any such railroad, and to issue and dispose of its bonds for any amount· so borrowed, and to mortgage its corporate property and franchises to secure the payment of any debt contracted by such corporation for the purposes aforesaid; but the concurrence of the holders of two-thirds in amount of the stock of such corporation—to be expressed in the manner hereinafter provided—shall be necessary to the validity of any such mortgage; and the order or resolution for such mortgage shall be recorded as provided in this act; and the directors of such corporation shall be empowered, in pursuance of any such order or resolution, to confer on any holder of any bond, for money so borrowed as aforesaid, the right to convert the principal due or owing thereon into stock of such corporation at any time not exceeding ten years after the date of such bond, under such regulation as may be provided in the by-laws of such corporation. [R. S. 1887, p. 1007, § 35; S. & C., p. 1919, § 34; Cothran, p. 1145; § 30, See ante, 1338, 1368.]

1468. BORROWING MONEY, MORTGAGE—*concurrence of stockholders, how shown.* § 2. The concurrence of the holders of at least two-thirds in amount of the capital stock of such corporation in the creation of any such debt and the execution of any such mortgages, shall be made manifest by the votes cast by such stockholders in person or by proxy, on the passage of appropriate orders or resolutions at a meeting of the stockholders of such corporation, called by the directors thereof for such purpose. [R. S. 1887, p. 1007, § 36; S. & C., p. 1919, § 35; Cothran, p, 1146, § 31. See ante, 1338, et seq.]

1469. NOTICE OF MEETING—*contents, and how given.* § 3. The directors of such corporation shall give notice of such meeting by causing written or printed notices thereof to be either personally served upon or duly mailed (postage prepaid) to such stockholders whose names and address shall be known to said directors, such notice to be so mailed at least sixty days before the time fixed for such meeting. The said notices shall state the time and place of such meeting and the purpose thereof, as well as the amount of the proposed indebtedness. The said directors shall also cause like notices to be inserted in some newspaper published in each county through which said road shall run, (if any newspaper shall be published therein) at least sixty days prior to the day appointed for such meeting. [R. S. 1887, p. 1008, § 37; S. & C., p. 1919, § 36; Cothran, p. 1146, § 32.]

1470. RESOLUTION TO MORTGAGE—*record of, where.* § 4. When such meeting shall be held, the resolution or order authorizing the creation of such indebtedness, and the execution of the mortgage to secure the same, together with the result of the vote thereon, shall be recorded in the office of the recorder of deeds of each county through which said road shall run, and shall also be recorded in the office of the secretary of state. [R. S. 1887, p. 1008, § 38; S. & C., p. 1920, § 37; Cothran, p. 1146, § 33.]

STOCK TRANSFER OFFICES.

An act to require railroad corporations to have and maintain a public office, or place in the state of Illinois where transfers of stock may be made, and to enforce the provisions of section nine (9), article eleven (11), of the constitution of Illinois. Approved June 18, 1883. In force July 1, 1883. [L. 1883, p. 128. R. S. 1887, p. 1006; S. & C., p. 1920; Cothran, p. 1189c. See ante, 61.]

1471. SHALL HAVE PUBLIC OFFICE—BOOK WITH TRANSFERS OF STOCK REGISTERED. § 1. *Be it enacted by the people of the state of Illinois, represented in the general assembly:* Each and every railroad corporation, organized or doing business in this state, under the laws or authority thereof, shall have and maintain a public office, or place in this state for the transaction of its business, where transfers of shares of its stock shall be made by such railroad corporation, upon the request of the owner of shares thereof, presenting the certificate thereof. Every such railroad corporation shall

keep a book in which the transfers of shares of its stock shall
be registered, and another book containing the names of its
stockholders, which book shall be open to the examination of
the stockholders.

1472. FINES FOR FAILURE TO COMPLY. § 2. Any railroad corporation, organized or doing business in this state
under the laws or authority thereof, failing to comply with
the provisions of section one (1), of this act, within ninety
(90) days after the taking effect of this act, shall upon conviction thereof, be fined in any sum not less than one thous-
and dollars ($1,000), nor more than two thousand dollars
($2,000). In case any such railroad corporation shall fail to
comply with the provisions of said section one (1) within six
months after the taking effect of this act it shall, upon conviction thereof, be fined in any sum, not less than two thous-
and dollars ($2,000), nor more than four thousand dollars
($4,000); and for every year after the taking effect of this
act, any such railroad corporation shall fail to comply with
the provisions of said section one (1), it shall, upon conviction, be fined not less than four thousand dollars ($4,000):
Provided, that in all cases under this act either party shall
have the right of trial by jury.

1473. FINES RECOVERED IN ACTION OF DEBT. § 3. The
fines hereinbefore provided for, may be recovered in an action
of debt in the name of the people of the state of Illinois.

1474. RAILROAD AND WAREHOUSE COMMISSIONERS — *duty
of, to enforce act.* § 4. It shall be the duty of the railroad
and warehouse commissioners to personally investigate and
ascertain whether the provisions of this act are violated by
any railroad corporation in this state; and whenever the facts
in any manner ascertained by said commissioners shall, in
their judgment, warrant such prosecution, it shall be the
duty of said commissioners to immediately cause suits to be
commenced and prosecuted against any railroad corporation
which may violate the provisions of this act. Said suits and
prosecutions may be instituted in any county in this state,
through or into which the line of the railroad corporation
sued for violating this act may extend. And such railroad
and warehouse commissioners are hereby authorized to employ counsel to assist the attorney general in conducting
such suit on behalf of the state. No such suits commenced
by said commissioners shall be dismissed, except said railroad and warehouse commissioners and the attorney general
shall consent thereto.

1475. FINES TO BE USED FOR COUNTY PURPOSES. § 5. All
fines recovered under the provisions of this act shall be paid
into the county treasury of the county in which the suit is

tried, by the person collecting the same in the manner now provided by law, to be used for county purposes.

An act to enable railroad companies to enter into operative contracts and to borrow money. Approved Feb. 12, 1855. Private laws 1855, p. 304.

1476. DOMESTIC CORPORATIONS—*power to contract with each other and foreign corporations, and to purchase real and personal property.* § 1. *Be it enacted by the people of the state of Illinois, represented in the general assembly,* That all railroad companies incorporated or organized under, or which may be incorporated or organized under the authority of the laws of this state, shall have power to make such contracts and arrangements with each other, and with railroad corporations of other states, for leasing or running their roads, or any part thereof; and also to contract for and hold in fee simple or otherwise, lands or buildings in this or other states for depot purposes; and also to purchase and hold such personal property as shall be necessary and convenient for carrying into effect the object of this act. [R. S. 1887, p. 1009, § 42; S. & C., p. 1921, § 43; Cothran, p. 1146, § 34]

1477. This act held applicable to horse railways. *Chicago* v. *Evans*, 24 Ill. 52.

1478. One company using the road of another, must conform to the charter of the road used, or leased. *Ib.*

1479. A railway company selling tickets over its own and other roads, is liable for the safety of passengers and baggage to the place of destination. *I. C. R. R.* v. *Copeland*, 24 Ill. 332.

1480. It would seem that a liability for not delivering goods over connecting roads, exists. *Ib.*

1481. Liability of company for acts of its lessees or contractors. *West* v. *St. L., V. & T. H. R. R.*, 63 Ill. 545.

1482. An unanthorized lease of the road will not release a sub-scriber from his subscription. *O. O. & F. R. V. R. R.* v. *Black*, 79 Ill. 262.

1483. Relief of stockholders against an unauthorized lease is in equity and not at law. *Ib.*

1484. Under this act railway companies have power to make contracts and arragements with each other for leasing and running their respective roads, or any part thereof. *Ill. Midland Ry.* v. *People ex rel*, 84 Ill. 426.

1484a. PARTNERSHIP—*joint contracts of.* As a general rule corporations are not capable of forming partnerships, but they may make joint contracts by which two or more may become liable. *Marine Bank* v. *Ogden*, 29 Ill. 248.

1484b. POWER TO MAKE CONTRACTS. While it is true that railway corporations can only make such contracts as the legislature may authorize, yet when made within their powers, their contracts, in legal effect, are the same as like contracts made by natural persons under similar circumstances. *People* v. *L. & N. R. R.*, 120 Ill. 48.

1484c. CONTRACT TO STOP TRAINS AT A PARTICULAR POINT. Where a railway company accepts a municipal subscription or donation under conditions imposed by a vote of the people, and by its contract with a county board, to maintain a depot for passengers and freights within the limits of a town, and to stop all its passenger trains at such depot, and afterwards consolidates with another company which owes no such duty, the new company, thereby formed, will become bound by such contract. But a purchaser of the property of the original railway company at judicial sale, will take the property without assuming any liability to perform its personal contracts. *People* v. *L. & N. R. R.*, 120 Ill. 48.

1485. CONNECTIONS. § 2. All railroad companies incorporated or organized, or which may be incorporated or organized as aforesaid, shall have the right of connecting with each other, and with the railroads of other states, on such terms as shall be mutually agreed upon by the companies interested in such connection. [R. S. 1887, p. 1009, § 43; S. & C., p. 1921, § 44; Cothran, p. 1147, § 35.

1486. § 3 of this act, giving power to borrow money, repealed by statutes, chap. 131, § 5.

An act to facilitate travel and transportation. Approved and in force Feb. 25, 1867. Laws of 1867, p. 174.

1487. RAILWAY BRIDGES—*duty to make and allow connections with by railway companies.* § 1. *Be it enacted by the people of the state of Illinois, represented in the general assembly:* Railroads terminating, or to terminate at any point on any line of continuous railroad thoroughfare where there now is or shall be a railroad bridge for crossing of passengers and freight in cars over the same as part of such thoroughfare, shall make convenient connections of such railroads, by rail, with the rail of such bridge; and such bridge shall permit and cause such connections of the rail of the same with the rail of such railroads, so that by reason of such railroads and bridge, there shall be uninterrupted communication over such railroads and bridge as public thoroughfares. But by such connections no corporate rights shall be impaired. [R. S. 1887, p. 1009, § 44; S. & C., p. 1921, § 45; Cothran, p. 1147, § 36.

An act to facilitate the carriage and transfer of passengers and property by railroad companies. Approved May 24, 1877. In force July 1, 1877. L. 1877, p. 167.

1488. POWER TO OWN AND USE WATER CRAFT. § 1. *Be it enacted by the people of the state of Illinois, represented in the general assembly,* That all railroad companies incorporated under the laws of this state, having a *terminus* upon any navigable river bordering on this state, shall have power to own for their own use any water craft necessary in carrying across such river any cars, property or passengers transported over their lines, or transported over any railroad terminating on the opposite side of such river to be transported over their lines.

1489. No CONDEMNATION—*for landing.* Provided, that no right shall exist under this act to condemn any real estate for landing for such water craft, or for any other purpose.

And this act only apply to such railroad companies as shall own the landing for such water craft.

1490. FERRY PRIVILEGES—*subject to rights of others, and laws regulating.* *Provided, also,* that nothing in this act shall be held to impair or affect any right or privilege granted any ferry company incorporated under the laws of this state; and that all the powers and rights herein granted said railroad companies shall be subject to whatever rights and privileges may have heretofore been granted to any ferry company in this state, and that nothing in this act shall prevent said railroad companies from being subject, in the use of such water craft, to all laws of the state regulating ferries now in force or hereafter to be in force:

1491. CONSOLIDATION—*right of state protected.* *And, provided, further,* that nothing in this act shall be held or construed to authorize any railroad or railway company doing business under any charter granted by this state, to consolidate with any railroad or railway company out of this state, so as to form one continuous line of railroad, or otherwise to alter, modify or repeal any provision of any such charter granted by this state; or to impair the rights of this state as now reserved to it in any such charter. [R. S. 1887, p. 1009, § 47; S. & C., p. 1922, § 46; Cothran, p. 1148, § 39.]

An act to empower township trustees to sell and convey right of way and depot grounds for the use of railroads crossing school lands. Approved April 13, 1875. In force July 1, 1875. Laws of 1875, p. 96.

1492. RIGHT OF WAY—*over school lands—depot grounds.* § 1. *Be it enacted by the people of the state of Illinois, represented in the general assembly,* That the trustees of schools of any township concerned are hereby authorized and empowered, in their corporate capacity, to sell and convey to any railroad company which may construct a railroad across any of the public school lands of such township, the right of way and necessary depot grounds. All money received by such trustees for any right of way or depot grounds so sold, to be turned over by such trustees to the treasurer of the township, for school purposes. [R. S. 1887, p. 1009, § 46; S. & C., p. 1922, § 47; Cothran, p. 1148, § 38.]

CONDITIONAL SALES OF RAILWAY ROLLING STOCK.

An act to render valid leases, bailments and conditional sales of railway rolling stock. Approved May 30, 1881. In force July 1, 1881. [Laws 1881, p. 126; R. S. 1887, pp. 1010, 1011, §§ 50, 51, 52, 53, 54, 55; S. & C., p. 1923, §§ 48, 49, 50, 51, 52, 53; Cothran, p. 1189a, §§ 172, 173, 174, 175, 176, 177.]

1493. CONDITIONAL SALE OF ROLLING STOCK—*sale, lease or contract reserving title in vendor or lessor until payment of price.* § 1. *Be it enacted by the people of the state of Illinois, represented in the general assembly.* In all cases where any cars, carriages, locomotives or vehicles used upon
—14

railways shall be delivered to any person or persons, or corporation, by the manufacturer or builder thereof, under lease, bailment, conditional sale, or other contract, providing that the title to the same shall remain in, or not pass from, the lessor, bailor or conditional vendor, until conditions fulfilled according to the terms of such contract, such contract shall be held and considered to be good, valid and effectual, according to the terms, tenor and effect thereof, both in law and in equity, as against all persons whatsoever, when the same shall be reduced to writing, acknowledged, and filed for record, as hereinafter provided. The provisions of this act shall apply only to sales made by manufacturers to purchasers, and no contract made in pursuance hereof shall be good for a longer period than four (4) years, nor shall any such contract be renewed. And it shall be the duty of the managers of all such corporations to list and return such property for taxation the same as is done by all other railroads owning their own rolling stock in this state.

1494. INSTRUMENT—HOW SIGNED AND EXECUTED. § 2. The instrument of writing evidencing such contract shall be signed by the lessor, bailor or conditional vendor, and by the lessee, bailee or conditional vendee, or their agents, and acknowledged by one or other of them or their agents, in the same manner as provided by law for the acknowledgment of conveyances of real estate, and shall be filed for record in the recorder's office of each county through or into which the railroad availing itself of such additional purchase is operated.

1495. INSTRUMENT TO BE ADMITTED TO RECORD. § 3. Such instruments, when properly acknowledged, shall be admitted to record at the request of any person interested, upon the payment of the legal fees, without regard to the residence of the parties.

1496. INSTRUMENT AS EVIDENCE. § 4. Every such instrument executed, acknowledged and recorded in pursuance of this act, may be read in evidence without any further proof of the execution thereof; and when it shall appear, by affidavit, or otherwise, that the original thereof is lost or cannot be produced, a copy of the record thereof, certified by the recorder, may be read in evidence in the like manner and to the same effect as the original thereof.

1497. TO WHAT THIS ACT APPLIES. § 5. This act shall not apply to railway rolling stock leased in the ordinary way without condition regarding purchase and sale, nor shall it effect the legality of any instrument of sale or lease existing at the time of the passing of this act.

1498. CONTRACT, NOTICE OF, TO CREDITORS—STATEMENT, RECORDING. §. 6. Any and all contracts mentioned in section one (1) of this act, which shall be executed, acknowledged and recorded, in pursuance of the provisions hereof, shall be held and considered to be full and sufficient notice to all persons whatsoever, but shall cease to be notice as against third persons after the expiration of the day the last payment thereunder, or other conditions thereof, shall become due, or to be performed by the terms thereof: *Provided*, that the lessor, bailor or conditional vendor shall, within ten (10) days from the first of January in each year, file a sworn statement with the recorder of each county where the lease or sale bill provided for in section one (1) of this act is recorded, and pay the recorder for putting the same on record, which statement shall show the names and dates and description of the said contract, and the amount due and unpaid thereon; and upon failure to make such statement, or if such statement is false, or made with the intent to deceive and mislead any creditor of said railroad, in any way, then such lessor, bailor or conditional vendor shall thereby lose all benefits which he or they would otherwise have under the provisions of this act, and any person or creditor may treat the property described in such conditional contract for sale, as though the sale had been unconditional, and not subject to any lien for purchase money whatever, and may levy execution or attachment thereon, or purchase the same, freed from any lien of such lessor, bailor or conditional vendor.

An act compelling railroad companies in this state to build and maintain depots for the comfort of passengers, and for the protection of shippers of freight at towns and villages on the line of their roads. Approved May 23, 1877. In force July 1, 1877. L. 1877, p. 165.

1499. REQUIRED TO BUILD AND MAINTAIN DEPOTS. § 1. *Be it enacted by the people of the state of Illinois, represented in the general assembly,* That all railroad companies in this state carrying passengers or freight shall, and they are hereby required to build and maintain depots for the comfort of passengers and for the protection of shippers of freight, where such railroad companies are in the practice of receiving and delivering passengers and freight, at all towns and villages on the line of their roads having a population of five hundred or more. [R. S. 1887, p. 1010, § 48; S. & C., p. 1924, § 54; Cothran, p. 1148, § 40.]

1500. Liability of company for an injury caused by defect in floor of platform. *T., W. & W. Ry.* v. *Grush*, 67 Ill. 262.

1500a. Company not liable to proprietor of an eating house for obstructing passage from depot to such house when dangerous to pass over the tracks. *Disbrow* v. *Ch. & N. W. Ry.*, 70 Ill. 246.

1501. Railroad depot grounds and passenger houses are *quasi* public, and a person in going to such houses and passing over depot grounds in a proper manner, is not a trespasser. *I. C. R. R.* v. *Hammer*, 72 Ill. 347.

1502. What is such negligence in persons passing over such grounds as to preclude a recovery. *Ib.*.

1503. It is the duty of a railway company, before the departure of its passenger train to clear the way by the removal of freight trains between it and the depot buildings, so that passengers can approach the passenger trains with safety. *Ch. & N. W. Ry.* v. *Coss*, 73 Ill. 394.

1504. Regulation that men unaccompanied by women, shall not be allowed to remain in the ladies' room is a reasonable one. *T., W. & W. Ry.* v. *Williams*, 77 Ill. 354.

1505. It is made the duty of railway companies to establish depots, and so operate their roads as to afford the public reasonable safety and dispatch in the transaction of business; and to effect this, it is necessary that they should at all reasonable times provide a ready and convenient means of access to their stations and depots. *C., B. & Q. R. R.* v. *Hans*, 111 Ill. 114.

1506. PENALTY. § 2. Any railroad company in this state failing to comply with the provisions of the preceding section after this act shall go into effect, and within ninety days after notice in writing of its failure to comply with the provisions of said section shall have been served upon any agent of said railroad by the authorized agent of any town or village aggrieved, shall pay for each and every day it shall neglect, the sum of fifty dollars ($50.00) to be recovered in an action of debt before any justice of [the] peace, in the name of the people of the state of Illinois, in any town or village aggrieved. Said penalty to be paid to the said town or village for the school fund. [R. S. 1887, p. 1010, § 49; S. & C., p. 1924, § 55; Cothran, p. 1149, § 41.]

UNION DEPOTS.

An act authorizing the formation of union depots and stations for railroads in this state. Approved April 7, 1875. In force July 1, 1875. [L. 1875, p. 97. R. S. 1887, pp. 1011, 1012, 1013, §§ 56, 57, 58, 59, 60, 61; S. & C., pp. 1924, 1925. 1926, §§ 56, 57, 58, 59, 60, 61; Cothran, pp. 1149, 1150, 1151, §§ 42, 43, 44, 45, 46, 47.]

1507. CORPORATION — HOW FORMED — ARTICLES OF INCORPORATION — CONTENTS. § 1. *Be it enacted by the People of the State of Illinois, represented in the General Assembly,* That in order to facilitate the public convenience and safety in the transmission of goods and passengers from one railroad to another, and to prevent the unnecessary expense, inconvenience and loss attending the accumulation of a number of stations, any number of persons, not less than five, are hereby authorized to form themselves, or any two or more railroad companies may themselves form or join individuals in forming a corporation for the purpose of constructing, establishing and maintaining a union station for passenger or freight depots, or for both, in any city, town or place in this state, with the necessary offices and rooms convenient for the same, and appurtenances thereto, and for that purpose may make and sign articles, in which shall be stated the number of years the same is to continue, the city, town or place in which the same is to be located, the amount of the capital stock of said company, which shall not exceed three millions of dollars, the amount of each share of stock, the names and places of residence of its directors, which shall not be less than five nor exceed fifteen, who shall manage its affairs for the first year, and until others are chosen in their place, and shall also state the amount of stock taken by each subscriber.

1508. ARTICLES OF ASSOCIATION—PRESENTATION OF WITH PETITION TO CIRCUIT COURT. § 2. Any association of persons or corporations, desiring to become incorporated under the provisions of this act, shall present their articles of association to the circuit court of the county in which such city or place is, or to the judge thereof in vacation, with the petition from such members for a certificate of incorporation under the provisions of this act, to which petition shall be added or appended a certificate of at least two railroad companies who have tracks leading into said city, town or place, stating its public utility, and that they expect to make arrangements for its use when it shall be constructed, signed by the presidents of their respective companies.

1509. CERTIFICATE OF INCORPORATION. § 3. If the circuit court, or any judge thereof, in vacation, shall be satisfied that said certificate has been signed by such companies, then the said court or judge, upon filing the said petition, articles and certificate aforesaid, with the clerk of the court, shall grant to the said association a certificate of incorporation, which may be in the following form, to-wit:

WHEREAS, A, B and C, etc., (stating the names) have filed in the office of the clerk of the circuit court their articles of association, in compliance with the provisions of an act entitled "An act authorizing the formation of union depots and stations for railroads in this state," approved (stating day of approval,) with their petition of incorporation, under the name and style of ....; they are therefore hereby declared a body politic and corporate, by the name and style aforesaid, with all the powers, privileges and immunities granted in the act above named. By order of circuit court (or judge thereof) ......, attest:........., clerk of the circuit court of county ....

And thereupon, upon filing the same, or a certified copy thereof, in the office of the secretary of state, the said association, from the time of such filing, shall be a corporation under the laws of this state.

1510. CORPORATE POWERS DEFINED—PROVISOS AND LIMITATIONS. § 4. Every corporation formed under this act, in addition to the general powers conferred by the laws of this state in relation to corporations, shall have power—

1511. ACQUISITION OF PROPERTY—BY CONVEYANCE AND BY CONDEMNATION. *First*—To take and hold such real estate as it may acquire either by conveyance to said corporation, or such as it may acquire under the provisions of this act by condemnation, and which shall be necessary for the transaction of its business.

1512. EXERCISE OF EMINENT DOMAIN—FOR WHAT USES AND

PURPOSES. *Second*—To take, occupy and condemn any land and real estate, or any interest therein, needed for the establishment of such union station or depot and necessary approaches thereto, and the same proceedings shall be had therefor as are now or may hereafter be provided by law, concerning the condemnation of lands for or by railroad companies in the state, so far as such laws are applicable to the purposes of this act; and when so condemned, the said land and any interest therein shall belong to such corporation for the purposes of this act: *Provided*, that nothing in this act shall be construed to authorize the condemnation as depot grounds of any railroad which is not of the same guage of those joining in the petition: *Provided further*, that none of the provisions of this act relating to the condemnation of land shall extend to any land, or lands to which any municipal corporation has a title. [See eminent domain, ante 179.]

1513. LAYING TRACKS—MAKING CONNECTIONS—USE OF STREETS—DAMAGES. *Third*—With the consent of the corporate authorities of the city, town or place in which said station or depot is to be constructed, to have the right to lay the necessary track or tracks over, upon or under such streets or roads of said city, town or place as may be necessary to make the necessary connections with railroads proposing to use said union depot, and may, with such consent, also construct such station or depot under, over or upon any such streets or roads: *Provided*, that all injury, if any, that may be occasioned to the property fronting on any streets or roads, by the laying of any railroad tracks, or the location of any depot upon such streets or roads, under the provisions of this act, shall be assessed and the assessment paid into the city treasury, to the use of the owners of the property s o injured by the corporation so appropriating such streets or roads, before such corporation shall have the right to lay any track or locate any depot over, under or upon such streets or roads.

1514. BORROWING MONEY—MORTGAGE OF ITS PROPERTY. *Fourth*—From time to time to borrow such sums of money as may be necessary for the construction, completion and furnishing or repairing of such station or depot, and to issue or dispose of their bonds for such amounts, at such prices as they shall think proper, and to mortgage their corporate property and franchises for the purpose of securing the same.

1515. RECEIVING SUBSCRIPTIONS—LEGISLATIVE CONTROL RESERVED. *Fifth*—To open, from time to time, books of subscription to the remainder of the capital stock not taken by the subscribers to the articles of association. The general assembly shall have power to enact, from time to time, laws to prevent and correct abuses, and to prevent unjust discrimination and extortions in the management and prosecution of the business of any corporation formed under this act, and to enforce such laws by adequate penalties.

1516. DIRECTORS—ELECTION AND TERM OF OFFICE—NOTICE OF ELECTION. § 5. After the directors named in the articles of corporation shall have served for one year, there shall be an annual election of directors, to be conducted in the manner prescribed in the constitution of this state. The directors so elected shall serve for the ensuing year, and notices of such election, appointing a time and place, shall be given by the directors as originally constituted for the first annual election, and thereafter by their successors in office, which notice shall be published not less than twenty days previous thereto, in some newspaper published in the English language, in the city, town or place in which said station or depot is located.

·1517. USE OF UNION DEPOT—NO DISCRIMINATION. § 6. There shall be no discrimination against or in favor of any railroad company using or desiring to use the said union depot, but the terms, conditions and regulations adopted for the use of the same, shall be, so far as practicable, uniform, and apply alike to all railroads using or desiring to use said union depot.

FENCING AND OPERATING RAILROADS.

An act in relation to fencing and operating railroads. Approved March 31, 1874. In force July 1, 1874.

1518. FENCING TRACK — CATTLE GUARDS — DAMAGES TO STOCK—ATTORNEY'S FEES. § 1. *Be it enacted by the people of the state of Illinois, represented in the general assembly,* That every railroad corporation, shall, within six months after any part of its line is open for use, erect and thereafter maintain fences on both sides of its road or so much thereof as is open for use, suitable and sufficient to prevent cattle, horses, sheep, hogs or other stock from getting on such railroad, except at the crossings of public roads and highways,* and within such portion of cities and incorporated towns and villages as are or may be hereafter laid out and platted into lots and blocks*, with gates or bars, at the farm crossings of such railroad, which farm crossings shall be constructed by such corporation when and where the same may become necessary, for the use of the proprietors of the lands adjoining such railroad; and shall also construct, where the same has not already been done, and thereafter maintain at all road crossings now existing or hereafter established, cattle-guards, suitable and sufficient to prevent cattle, horses, sheep, hogs and other stock from getting on such railroad; and when such fences or cattle-guards are not made as aforesaid, or when such fences or cattle-guards are not kept in good repair, such railroad corporations shall be liable for all damages which may be done by the agents, engines or cars of such corporation, to such cattle,'horses, sheep, hogs or other stock thereon†, and reasonable attorney's fees in any court wherein suit is / brought for such damages, or to which the same may be appealed†; but where such fences and guards have been duly made ;¦and¦ kept in good repair, such railroad corporation shall not be liable for any such damages, unless negligently or willfully done. [R. S. 1887, p. 1013, § 62; S. & C., p. 1927, § 62; Cothran, p. 1151, § 48.]

Amendment of 1877. May 23, 1877, substituted single for double damages.

Amendment of 1879 substituted words between asterisks (*——*) for "and within such portion of cities and incorporated towns and villages, as are or may be hereafter laid out and platted into lots and blocks," and inserted clause as to attorney's fees between daggers (†——†).

CONSTITUTIONALITY OF STATUTE.

1519. POLICE REGULATION — *imposing additional duties.* The act of 1855 requiring roads open for use, to be fenced, and making them liable for a failure to do so to pay all damages to stock, is a proper police regulation, and not an *ex post facto* law or law impairing the obligations of a contract. *O. & M. R. R.* v. *McClelland*, 25 Ill. 140; *C., M. & St. P. R.*.R. v. *Dumser*, 109 Ill. 402.

1520. Under the police power, the legislature may require existing railway corporations to fence their roads and put in cattle-guards, although no such right may be reserved in their charters. *T., W. & W. Ry.* v. *Jacksonville*, 67 Ill. 37.

1521. Act of 1874, making company neglecting to fence liable for double damages, is not unconstitutional. *C. & St. L. R. R.* v. *Peoples*, 92 Ill. 97; *C. & St. L. R. R.* v. *Warrington*, 92 Ill. 157.

1522. Railway corporations may be compelled to fence their tracks by the imposition of fines, penalties or forfeitures, and a law providing a forfeiture or penalty for a neglect to fence, which is given to the owners of stock killed, is not open to the constitutional objection of depriving one of property "without due process of law." *C. & St. L. R. R.* v. *Warrington*, 92 Ill. 157.

1523. In what manner and to what extent railway corporations shall be required to inclose their tracks, and when it shall .be done, would seem to be ordinarily within the legislative discretion. *C., M. & St. P. R. R.* v. *Dumser*, 109 Ill. 402.

1524. As a police regulation for the protection of the public safety in travel by railroads, the legislature may well require the fencing of such roads, and provide penalties for securing the performance of such requirements. *P., D. & E. Ry.* v. *Duggan*, 109 Ill. 537.

1525. The act of 1879 making railway companies liable for attorney's fees in addition to the damages sustained by the owners of stock through a neglect to fence their track and keep the same in repair, is not open to the objection of special legislation. It may be upheld as being in the nature of a penalty for a neglect to comply with a proper police regulation. *P., D. & E. Ry.* v. *Duggan*, 109 Ill. 537.

CONSTRUCTION.

1526. APPLICATION. The act of 1855 to regulate the duties and liabilities of railroad companies, applies to companies previously incorporated. *Galena & Ch. Union R. R.* v. *Crawford*, 25 Ill. 529.

1527. REMEDIAL, NOT PENAL. This statute is not a penal statute but is remedial and will receive a liberal construction. *O. & M. R. R.* v. *Brubaker*, 47 Ill. 462.

1528. FARM CROSSINGS. The word "necessary" in the statute requiring railway corporations to construct farm crossings "when and where the same may become necessary for the use of proprietors of lands, adjoining such railroads," was used in its more popular sense, and is equivalent to the words "reasonably convenient." *Chalcraft* v. *L. E. & St. L. R. R.*, 113 Ill. 86.

1529. Where the erection and maintenance of a proposed farm crossing over a railroad track will directly affect the operation of the road as a means of public transportation, by seriously increasing the danger of collisions, this will be a sufficient reason why such crossing should not be made, and if attempted to be made by the landowner, he may be restrained from doing so by injunction. *Ib.*

1529a. Railroads must be fenced or enclosed with gates or bars at all road crossings which are not used and treated by the people and road authorities as public highways. *T. H. & I. R. R.*, v. *Elam*, 20 Bradw. 603.

EFFECT OF OTHER LAWS ON THIS PROVISION.

1530. The act of 1867, to prevent domestic animals from running at large in certain counties, is not so far repugnant to the general railroad law requiring the fencing of railroads, as to repeal the same by implication. *O. & M. Ry.* v. *Jones*, 63 Ill. 472.

1531. The act of 1869, giving the land-owner a right to build a fence along the track on his own premises, and hold the company liable therefor upon its failure to fence on notice, does not release railway companies from their liability under the act of 1855 for stock killed. *T., P. & W. Ry.* v. *Pence*, 68 Ill. 524.

1532. The law prohibiting domestic animals from running at large in force October 1, 1872, does not by implication, repeal or modify any of the provisions of the act of 1855 requiring railway companies to fence their roads, and the same is true in regard to the law preventing male animals from running at large. *R., R. I. & St. L R. R.* v. *Irish*, 72 Ill. 404.

1533. WITHIN WHAT TIME TO FENCE—*burden of proof.* The plaintiff must show that the road has been open to use six months prior to the injury of his stock. *W., St. L. & P. Ry.* v. *Neikirk*, 13 Bradw. 387; *O. & M. R. R.* v. *Brown*, 23 Ill. 94; *O. & M. R. R.* v. *Meisenheimer*, 27 Ill. 30; *O. & M. R. R.* v. *Jones*, 27 Ill. 41; *T., P. & W. Ry.* v. *Wickery*, 44 Ill. 76.

1534. The declaration must show that the road has been open to use six months prior to the accident and the neglect to fence. *Galena & Ch. Union R. R.* v. *Sumner*, 24 Ill. 631.

1535. An averment in a declaration: "nevertheless more than six months after said railroad was in use, to-wit: on the 1st day of May,

1864, the said defendant neglected to erect." &c.: *Held,* a sufficient breach of the statutory duty. *Great Western R. R.* v. *Hanks,* 36 Ill. 281.

1536. An averment that the road has been in use more than six months prior to the accident and still remains unfenced, by reason of which neglect of duty, the injury occurred, is material and must be proved. *C. & A. R. R.* v. *Taylor,* 40 Ill. 280.

1537. An omission to state in an instruction that it must be proved that the road had been operated for six months prior to the accident, is a harmless error, where it clearly appears that the road had been in use for a much longer period. *Ch. & N. W. Ry.* v. *Dement,* 44 Ill. 74.

1538. The obligation of a railway company to fence its line of road does not attach until it has been in operation for six months. *T., P. & W. Ry.* v. *Miller,* 45 Ill. 42.

1539. If a company which has not been in operation six months has built a fence, it will be under no obligation to keep it in repair until the duty to fence has attached. *Ib.*

1540. CHANGE OF OWNERSHIP. A purchaser of a railroad which has been open for six months before its sale, will be liable for injury to stock resulting from the want of a fence, before six months after the change of ownership. *T., P. & W. Ry.* v. *Arnold,* 51 Ill. 241.

1541. In such case the new owner is not entitled to a period of six months after the change of ownership in which to comply with the law, but takes possession-subject to all consequences resulting from a want of compliance with the law. *Ib.*

1542. An averment that the road at the place where the injury occurred has been opened for six months is not sufficient, where it does not appear but that the stock strayed upon the track at another place where the road had not been opened for six months. *T. P. & W. Ry.* v. *Darst,* 51 Ill. 365.

1543. If an instruction for the plaintiff undertaking to enumerate the facts upon which a recovery may be had, omits the essential facts that the road has been opened six months, a judgment for the plaintiff will be reversed, unless such omitted fact is shown by the evidence. *Ch. & N. W. Ry.* v. *Diehl,* 52 Ill. 441.

1544. Declaration held insufficient on special demurrer to show with sufficient certainty that the company had failed to erect a proper fence for six months after the road had been opened. *T., P. & W. Ry.* v. *Bookless,* 55 Ill. 230.

1545. A railway company is liable under the statute if it fails to fence within six months after it begins to run trains on the track for construction purposes. Being under the control of contractors will not change this liability. *R. R. I. & St. L. R. R.* v. *Heflin,* 65 Ill. 366.

1546. EVIDENCE. Proof that plaintiff's steers were killed by the trains of the company in the fall of 1870, and his horses and hogs in the summer of 1871: *Held,* as showing inferentially that the road had been open for use six months before the horses and hogs were killed. *R., R. I. & St. L. R. R.* v. *Spillers,* 67 Ill. 167.

EXCEPTED PLACES.

1547. DEPOT GROUND. A railway company is not bound to fence the grounds about a station. This section is not to be construed to embrace depots and stations. *T. H. & Ind. R. R.* v. *Bowles,* 16 Bradw. 261.

1548. PLEADING. In an action under the statute the plaintiff should aver in his declaration that the animals were not killed within the limits of a village, &c. *C., B. & Q. R. R.* v. *Carter*, 20 Ill. 390.

1549. The declaration must not only show the duty of the company to fence, and its failure to do so, but must also negative the exceptions in the act, and aver that the animals were not injured at any point on the road within those exceptions, &c. *Galena & Ch. Union R. R.* v. *Sumner*, 24 Ill. 631.

1560. To recover, plaintiff must prove every material allegation in his declaration, and that the injury did not occur at any of the excepted places, and this though the declaration is defective, in not negativing the exceptions in the statute. *O. & M. R. R.* v. *Brown*, 23 Ill. 94.

1561. The declaration should show that the injury to the stock did not happen at a place where the company is not bound to maintain a fence. *Ill. C. R. R.* v. *Williams*, 27 Ill. 48.

1562. If a horse gets upon the track within a city and is driven by a train beyond the city, where he is killed without negligence on the part of the company, it will not be liable. *Great Western R. R.* v. *Morthland*, 30 Ill. 451.

1563. In an action under the statute, the declaration must negative all the exceptions in the statute. *Great Western R. R.* v. *Bacon*, 30 Ill. 347.

1564. In an action before a justice of the peace, the plaintiff must show by proof that there was no public crossing where the killing or injury occurred, and that the company was bound to fence at that point. *O. & M. R. R.* v. *Taylor*, 27 Ill. 207.

1565. The declaration need not negative the possibility that the animal may have been killed at a farm crossing. If road not properly fenced at such crossing the company is liable, and if properly fenced that is a matter of defence. *Great Western R. R.* v. *Helm*, 27 Ill. 198.

1566. An averment that the animal killed got on the track "without the limits of towns, cities and villages, and not at the road crossings or public highways," is sufficient. The important point, is where the animal got upon the track, and not where it was killed. *Great Western R. R.* v. *Hanks*, 36 Ill. 281.

1567. It is sufficient if the declaration negatives the killing in the excepted places named in the statute. *Ib.*

1567a. DEPOTS. A railway company is not required to fence its track upon its depot grounds in a town. *Galena & Ch. Union R. R.* v. *Griffin*, 31 Ill. 303.

1568. The question of the obligation of a railway company to fence its road at a particular place is one of law and not of fact, and should not be left to a jury to decide. *I. C. R. R.* v. *Whalen*, 42 Ill. 396.

1569. The want of an averment that a fence was necessary at the place where stock is injured, is cured by proof on the trial, and a verdict for the plaintiff. *T., P. & W. Ry.* v. *McClannon*, 41 Ill. 238.

1570. The necessity of fencing a railroad at a given point is not obviated by there being an embankment at that place from twelve to twenty feet in height, it not appearing it was sufficient to prevent stock from getting upon the track. *T., P. & W. Ry.* v. *Sweeney*, 41 Ill. 226.

1571. Where testimony is admitted without objection, showing an injury to animals, happened at a place where the company was bound to fence its road, an instruction will not be erroneous, merely because it fails to exclude, *all* the places excepted in the statute. *T., P. & W. Ry.* v. *Parker*, 49 Ill. 385.

1572. It is not enough to aver in the declaration that the road was not fenced at the place where the injury occurred, as the stock may have got upon the track at another place where the road was fenced. *T., P. & W. Ry.* v. *Darst,* 51 Ill. 365.

1573. Where the evidence shows that the stock was not killed within a corporation or near a crossing, the jury may infer that it was not killed within the limits of a city, town or village. *St. L. & S. E. Ry.* v. *Casner,* 72 Ill. 384.

1574. Where a railway company had a switch outside the platted limits of an incorporated village, but adjacent to the same, and in this locality there was a ware house and a store, and it was used as much by the public as any part of the village, and the switch was so located that it could not be reached by teams for loading and unloading if a fence was erected there: *Held,* that these facts were sufficient to justify the inference that the place was ground open to the public, where a fence was not required. *T. W. & W. Ry.* v. *Chapin,* 66 Ill. 504.

1575. A railway company is not bound to fence its track, or make cattle-guards within the limits of a village; and a place where there is a station house, a ware house, a store, a blacksmith shop, a post office and five or six dwelling houses, whether they are situate upon regularly laid out streets or not, comes fully up to the requirements of a village for the purposes of excusing a railroad company from fencing its track therein. *T. W. & W. Ry.* v. *Spangler,* 71 Ill. 568.

1576. A town or village, within the meaning of the statute, may exist, although there is no plat of the same dedicating the streets, &c., in the manner pointed out in the statute. *I. C. R. R.* v. *Williams,* 27 Ill. 48.

1576a. A place composed of a few houses with a population of two hundred persons is a village within the limits of which a railway company is not bound to fence its road, *Ewing* v. *C. & A. R. R.,* 72 Ill. 25.

1577. It will be presumed that houses compose a village. Where the proof is that an animal was killed beyond the houses, it will be presumed it was killed outside of the village. If the village extends beyond the houses and includes the place were the killing occurred, the company must show that fact. *O. & M. R. R.* v. *Irvin,* 27 Ill. 178; *Ewing* v. *C. & A. R. R.,* 72 Ill. 25.

1578. The court is not disposed, if it had the power, to extend the exception in the statute to cases not therein named, without proof of facts showing a necessity for relieving railway companies from the duty to fence their tracks. *C., M. & St. P. R. R.* v. *Dumser,* 109 Ill. 402.

1579. Where there is no public necessity for keeping a railroad track open at any point, whether in or out of the limits of a city, town or village, the company must fence the same, or respond in damages for an injury to stock resulting from an omission to do so. *Ib.*

1580. Where a railway company has a station at a place on its road where trains stop to receive and discharge passengers and freights, which is not in a city, town or village laid out into lots and blocks, and has side tracks at such station, this court cannot as a matter of law hold that the company is exempted from fencing its track at such place. *C., M. & St. P. R. R.* v. *Dumser,* 109 Ill. 402.

1581. This statute is not intended to apply to public stations or depot grounds, although they may not be within the limits of a city, town or village, or at a highway crossing. But side tracks not at stations or depots, and such parts of side tracks as do not constitute a part of the depot yard, may well be held to be within the statute. *C., B. & Q. R. R.* v. *Hans,* 111 Ill. 114.

1582. CATTLE GUARDS IN STREET. If a railway company constructs cattle-guards within the limits of a town it must keep the same in repair. *Ch. & R. Island R. R.* v. *Reid,* 24 Ill. 144.

1583. What is not a public road crossing, but is a farm crossing. *P., P. & J. R. R.* v. *Barton,* 80 Ill. 72.

EXTENT OF LIABILITY FOR NEGLECT TO FENCE.

1584. INJURY TO EMPLOYE. The liability of a railway company to an employe or servant for a personal injury growing out of its neglect to fence its track, doubted. Liability seems to be limited to owners of stock injured. *Wabash Ry.* v. *Brown,* 2 Bradw. 516.

1585. Railway companies not bound to maintain a fence as a protection to its employes. The duty to fence is imposed as a protection of the owners of cattle. *Wabash Ry.* v. *Brown,* 5 Bradw. 590.

1586. The liability for injury to animals from a neglect to fence is limited to such damages as may be done by the agents, engines or cars of the company, and not to injury resulting from fright. *I., B. & W. Ry.* v. *Schertz,* 12 Bradw. 304; *Ch. & N. W. Ry.* v. *Taylor,* 8 Bradw. 108.

1587. The injury for which the statute gives, an action, must be caused by actual collision with the engine or cars of the company. Consequential damages resulting from fright to animals, not caused by actual collision, or any negligence or willful misconduct on the part of the company, are not embraced in the statute. *Schertz* v. *I., B. & W. Ry.,* 107 Ill. 577.

1588. Where a horse gets on the track for want of a fence and is frightened by an approaching train and in its flight is injured by jumping a cattle-guard or by running into a wire fence, without negligence in those having charge of the train, the company will not be liable. *Ib.*

1589. DAMAGE TO CROPS. The statute is not intended for the protection of land-owners from damage to their crops resulting from a neglect to fence. The liability for a neglect to fence extends only to the owners of cattle injured thereby. *P., D. & E. Ry.* v. *Schiller,* 12 Bradw. 443.

1590. FOR WHAT ANIMALS. The statutory liability extends to the killing or injury of mules and asses, these animals being included in the terms "horses and cattle." *O. & M. R. R.* v. *Brubaker,* 47 Ill. 462; *T., W. & W. Ry.* v. *Cole,* 50 Ill. 184.

1591. INJURY MUST BE CAUSED BY WANT OF FENCE — *or defects therein.* No liability for stock killed which break over a sufficient fence. The bad condition of the fence at other places cannot be shown. *C., B. & Q. R. R.* v. *Farrelly,* 3 Bradw. 60; 39 Ill. 433.

1591a. The statutory duty of a railroad company to maintain suitable and sufficient cattle-guards to prevent stock from getting on its track is not complied with, where, for 'an unreasonable time, it permits its guards to remain filled up with snow, ice or any other substance which destroys their usefulness. *I., B. & W. Ry.* v. *Drum,* 21 App. Rep. 331.

1591b. ATTORNEY'S FEES. Reasonable attorney's fees may be recovered for the second as well as the first trial, although the new trial was granted by consent. Not allowed for services in appellate court. *I., B. & W. R. R.* v. *Buckles,* 21 App. R. 181.

NEGLECT TO FENCE ROAD.

1592. CONNECTION OF OMISSION WITH INJURY. Where a railway company neglects to build a fence as it had agreed to do, its liability for sheep killed will not depend upon the fact whether such fence would have made a perfect inclosure of the sheep, but entirely upon the fact whether the neglect of duty contributed to the injury. *Joliet & Northern Ind. R. R.* v. *Jones,* 20 Ill. 221.

1593. Although a railway company may fail to make a fence according to its contract, it will not be liable to the other party for sheep killed, where it does not appear that they got upon the track because the fence was not built, and it appears that his negligence in respect to the animals was the direct and proximate cause of the injury. *Ib.*

1594. In an action under the statute to recover for stock killed, it is sufficient to prove the neglect to fence and the killing. No other negligence need be proved. *T. H., A. & St. L. R. R.* v. *Augustus,* 21 Ill. 186.

1595. Where the declaration counts only upon a common law liability, negligence in the management of the train must be shown, and no recovery can be had for a mere neglect to fence the track. *T. H., A. & St. L. R. R.* v. *Augustus,* 21 Ill. 186.

1596. Liability for suffering a cattle-guard in a public street to get out of repair. *C. & R. I. R. R.* v. *Reid,* 24 Ill. 144.

1597. Since the act of 1855 railway companies are liable for injuries to cattle that may stray upon their track through the want of the required fences. *Galena & Ch. Union R. R.* v. *Crawford,* 25 Ill. 529.

1598. Declaration need not show that the injury did not occur at a farm crossing. If the road is not properly fenced at such crossing, the company will be liable, and if it is, that is a matter of defense. *Great Western R. R.* v. *Helm,* 27 Ill. 198.

1599. EVIDENCE. Proof of the killing of an animal by a railway company upon its track does not show its liability. It must appear that the company has been guilty of negligence, or that the case comes within the statute of 1855. *Great Western R. R.* v. *Morthland,* 30 Ill. 451.

1600. EVIDENCE OF INSUFFICIENT FENCE. While the fact that a horse was killed upon a railroad track does not of itself prove negligence in the company, yet if killed at a point where it was its duty to fence the track, this is a circumstance which may be considered in determining the question, whether the fences and cattle-guards were good and sufficient. *C. & A. R. R.* v. *Utley,* 38 Ill. 410.

1601. INSUFFICIENCY—*frightened animal.* If a horse takes fright and runs away and gets upon a railroad at a point required to be fenced, and is killed upon the track, the insufficiency of the fence or cattle-guard at that point will alone render the company liable. *C. & A. R. R.* v. *Utley,* 38 Ill. 410.

1602. WHO LIABLE UNDER STATUTE. Where a railway company by contract allows another company to run trains over its unfenced road, by one of which trains stock are injured by reason of the omission to fence, both companies will be liable to the owner of the stock, and he may sue either company. *I. C. R. R.* v. *Kanouse,* 39 Ill. 272.

1603. LIABILITY WITHOUT PROOF OF NEGLIGENCE. Where such company fails to fence its track as required by law, or erects an insufficient one, or fails to maintain a good and sufficient fence, it will be liable for all damages resulting from such omission of duty, without reference to the manner in which its engines may have been controlled. *St. L., A. & T. H. R. R.* v. *Linder,* 39 Ill. 433.

1604. WHERE TWO COMPANIES LIABLE. A railway company allowing another company to use its road will be liable for injuries done by the trains of the latter company, to stock happening from the road being unfenced, the same as if done by its own trains, and it seems the other company is also liable therefor. *T., P. & W. Ry.* v. *Rumbold,* 40 Ill. 143.

1605. NEGLIGENCE INFERRED. Where the killing of stock is attributable to a defective fence, which it was the duty of the company to provide, but which it failed to do, negligence is inferred; but if it has performed this duty, then negligence must be proved as in ordinary cases. *I. C. R. R.* v. *Whalen,* 42 Ill. 396.

1606. If a railway company neglects to fence its road, and an injury to an animal occurs, which is fairly attributable to such neglect, the mere fact that the animal is at large where it is not in violation of any general or local law, will not relieve the company of its liability, even though the animal may go upon the track from uninclosed lots adjacent to the crossing, and is not standing when injured on the actual intersection of the railway and the highway. *T., W. & W. Ry.* v. *Furgusson,* 42 Ill. 449.

1607. Where cattle are injured upon a railroad at a place where the company is required to fence its road, and it has been in operation several years without that being done, the company will be liable for the damages resulting from such neglect of duty. *T., P. & W. Ry.* v. *Wickery,* 44 Ill. 76.

1608. FAILURE TO KEEP IN REPAIR. Where the company suffers bars at a farm crossing to be left down for the period of three months, it will be guilty of negligence, and liable for injury to stock getting on the track in consequence of the bars being down, unless they are left down by the owner of the stock. *I. C. R. R.* v. *Arnold,* 47 Ill. 173.

1609. Railway companies are required to fence their roads with fences sufficient to turn stock, and to keep the same in repair. They are required to put in gates at farm crossings, which are a part of the fence; and their duty to keep the fences in repair, includes the duty of keeping their gates safely and securely closed, so as to afford equal protection from stock getting upon their road at such places as at other places. *Ch. & N. W. Ry.* v. *Harris,* 54 Ill. 528.

1610. SUFFICIENCY OF GATES. A railway company is not required to fasten gates to fences so as to make it impossible for stock to open them under any and all circumstances. It will be sufficient if it uses the fastening commonly adopted by persons reasonably prudent and careful, and which are regarded by them as safe. *C. & A. Ry.* v. *Buck,* 14 Bradw. 394.

1611. CATTLE-GUARDS. The requirement to build cattle-guards at road crossings is not different from that to build fences along the track, and a failure to build such cattle-guards imposes no greater or other liability than a failure to fence. *P., D. & E. Ry.* v. *Schiller,* 12 Bradw. 443.

1612. ELECTION OF REMEDY. Where stock is killed by a railway company through a neglect to fence its road within the time required, the owner may, at his election, sue under the statute, or upon the common law ground of negligence, or both. *R., R. I. & St. L. R. R.* v. *Phillips,* 66 Ill. 548.

1613. Where the plaintiff proceeds under the statute, he need only prove the killing of his cattle by the defendant's train and its neglect to fence. He is not required to show any other negligence. *R., R. I. & St. L. R. R.* v. *Lynch,* 67 Ill. 149.

1614. Where the plaintiff declares upon the statutory liability growing out of a neglect to fence the road within six months after the same is opened and used, no recovery can be had unless the company was bound to fence its road. *R., R. I. & St. L. R. R.* v. *Lynch,* 67 Ill. 149.

1615. Where a railway company has been operating trains over its road for more than six months, and has failed to fence its track, and while passing through plaintiff's farm with its train, kills the plaintiff's stock upon the track, the company will be liable for the damages. *T., P. & W. Ry.* v. *Crane*, 68 Ill. 355.

1616. NEGLIGENCE PRESUMED. The design of this section was to afford some protection from hazard, of trains running at a high rate of speed, by fencing; and if this is omitted by a railway company, it will be presumed to be guilty of negligence, without any other proof than that of the omission to fence. *T., P. & W. Ry.* v. *Pence*, 68 Ill. 524.

1617. Where a railway company kills stock upon its track at a place not a public crossing, or where not required to fence, and it has been in operation more than six months before and has not fenced its track at such place, and the owner of the land has not agreed to fence the road, the company will be liable, without proof of any actual negligence, even though the owner may not prove that the stock got upon the track at the point not fenced. *T., P. & W. Ry.* v. *Pence*, 68 Ill. 524.

1618. SUFFICIENCY OF FENCE. Where the proof shows that the company's fence at the place where a person's mare got upon the track and was killed, was not sufficient to prevent domestic animals from getting upon the road, the company will be liable. *C. & A. R. R.* v. *Umphenour*, 69 Ill. 198.

1619. Railway companies are responsible to the owners of stock killed by their trains where they have not fenced their roads; and the party injured can recover without proof of actual negligence in running their trains. *T., P. & W. Ry.* v. *Logan*, 71 Ill. 191.

1620. Where a railway company fails to fence its track as required, it must see that its servants so conduct its trains, that injury shall not result to stock that may get upon its track, if it can be avoided by care and caution. In failing to fence it takes the hazard, and where injury results therefrom, it must be required to respond in damages. *T., P. & W. Ry.* v. *Lavery*, 71 Ill. 522.

1621. Where a mule escapes from an enclosure without the fault of the owner, and gets upon a railroad track at a point not fenced, but where it is the duty of the company to have had a fence, and is injured by a train, the company will be liable. *T., P. & W. Ry.* v. *Delehanty*, 71 Ill. 615.

1622. Where a railway company fails to fence its track as required by law, it is sufficient to fix its liability, if the plaintiff's stock in consequence thereof, and without any contributing negligence on his part, goes upon the track of the railroad, where it is killed or injured by its locomotive or train. *Ewing* v. *C. & A. R. R.*, 72 Ill. 25.

1623. In a suit against a railway company to recover for the killing of the plaintiff's cow, where the evidence tended to show that the cow got upon the track through the negligence of its servants in failing to keep a gate at a farm crossing in repair, it was held that a verdict finding the company liable would not be disturbed. *T. W. & W. Ry.* v. *Nelson*, 77 Ill. 160.

1624. A railroad ran through a common field of several square miles, owned by different parties, some of whom resided therein, which was fenced only on the outside. The road had been in operation for more than six months, and the company had not fenced its track entirely through the enclosure: *Held*, that the company was liable for stock killed by its trains inside of the inclosure. *P., P. & J. R. R.* v. *Barton*, 80 Ill. 72.

1625. A railway company is required to put in cattle-guards at public road crossings to keep cattle from getting on its track. *P., P. & J. R. R.* v. *Barton,* 80 Ill. 72.

1626. Where a railroad crosses a private farm crossing inside of an inclosure, it is its duty to place bars or gates there, and it will be liable for stock killed through its failure to do so. *Ib.*

1627. Where stock is killed or injured by reason of the insufficiency of the fences of a railway company along its track, and the fences have been out of repair so long that the company must have known it, and the owner of the stock is guilty of no negligence, the company will be liable for the injury. *O. & M. Ry.* v. *Clutter,* 82 Ill. 123.

1628. A railway company which fails to fence its track as required by law, is liable for any damage resulting from such failure whether caused by its own trains or those of another company using its tracks. *E. St. L. & C. Ry.* v. *Gerber,* 82 Ill. 632.

1629. A railway company will be liable for any damage done by its trains resulting from a failure to fence the track on which the damage is done, although the track may belong to another company. *E. St. L. & C. Ry.* v. *Gerber,* 82 Ill. 632.

1629a. The words "on both sides of its road" as used in the act, mean the margin or border of the entire ground used as a right of way. *People* v. *O. & M. Ry.,* 21 App. R. 21; *O. & M. Ry.* v. *People,* 121 Ill. 483.

1629b. Mandamus lies to compel railway company to fence its right of way on the margin thereof. *O. & M. Ry.* v. *People,* 121 Ill. 483.

1630. NEGLIGENCE IN KEEPING GATES CLOSED. The company is not required to patrol the line of its road to see that the gates at the farm crossings are not left open; nor to keep a guard upon the road to discover and counteract such carlessness immediately upon its occasion. It is only negligent where it has had a reasonable time to discover such breach, or has been notified and failed to take proper action. *C., B. & Q. R. R.* v. *Sierer,* 13 Bradw. 261.

1631. A railway company is not required to keep a patrol on the line of its road to see that the gates at farm crossings are kept closed; but if its employes seeing such a gate open, do not close it, when not left open by a person to whom an injury afterwards results, the company will be liable for such injury. *I. C. R. R.* v. *McKee,* 43 Ill. 119.

1632. If a horse gets upon a railroad track through an open gate in the fence of the company, where it is killed, the company will not be liable, unless the gate has been so long open, as to raise the presumption that the servants of the company knew it, or to charge them with negligence. *C. B. & Q. R. R.* v. *Magee,* 60 Ill. 529; 47 Ill. 206.

1633. Where the evidence tends to show that a horse killed upon a railroad, got upon the track through an open gate at a farm crossing, it is error to instruct the jury, that if the road was not so fenced as to prevent the horse from getting on it under any circumstances, to find for the plaintiff. *Ib.*

1634. LEAVING GATE OPEN. Where it appeared that two horses got upon the track of a railway company, through an open gate at a farm crossing where they were killed by a train, the company having permitted the gate to remain open for a week previous to the accident: *Held,* that the company was guilty of such negligence as to render it liable. The fact that the plaintiff's horses entered the close of another through an insufficient fence upon the highway, and passed from thence upon the plaintiff's road, will not effect his right to recover. *Ch. & N. W. Ry.* v. *Harris,* 54 Ill. 528.

1635. NEGLIGENCE—*failure to discover breaches, &c.* Where a
—15

sufficient fence has been made, and from accident or wrong over which the company has no control, it becomes insufficient to turn stock, it will have a reasonable time in which to discover and repair the same. The company need not have a patrol at all times, night and day, passing along the road to see the condition of the fence. If this is done daily, and the company when informed of the defect at once makes the necessary repairs, it will not be liable for an injury resulting from the temporary insufficient condition. *I. C. R. R.* v. *Swearingen*, 33 Ill. 289.

1636. Where an employe whose duty it was to keep fences in repair, passed over the road at 4 o'clock P. M. on Saturday and found the fences in repair, and again on Monday morning passed over the road, and found the fence had recently been broken, and that through such breach stock got upon the track and was injured: *Held*, that the company had used reasonable diligence in keeping the fence in repair. *I. C. R. R.* v. *Swearingen*, 47 Ill. 206.

1637. TIME TO MAKE REPAIRS. Where a casual breach occurs in the fence without the knowledge or fault of the company, through which stock get upon the track and are injured, the company will not be liable unless it has had a reasonable time in which to discover such breach, or has been notified, and failed to repair before the injury. *Ib.*

1638. DILIGENCE TO DISCOVER BREACH IN OR DEFECTS. While railway companies are not required to keep a guard on their roads to discover a breach in their fence as soon as it occurs, and repair it at once, still the law requires them to keep such a force as may discover breaches and openings in their fences and close them in a reasonable time. To neglect repairing for a week or more, is a neglect of duty that will ordinarily render them liable for an injury ensuing therefrom. *Ch., & N. W. Ry.* v. *Harris*, 54 Ill. 528.

1639. While railway companies will be held to a high degree of diligence in keeping their fences in good repair, they are not required to do impossible things, nor are they required to keep a constant patrol, night and day. *Ch. & N. W. Ry.* v. *Barrie*, 55 Ill. 226.

1640. If a breach occurs in the fence by the unlawful act of a stranger through which stock get upon the track and are injured, in the absence of negligence on its part, the company will not be liable, unless the accident happened after the lapse of sufficient time for it, in the exercise of reasonable diligence, to have discovered and repaired the breach. *Ib.*

1641. A railway company will not be liable for the temporary insufficient condition of its fence, unless it has notice thereof, and neglects, thereafter to repair. *C. & A. R. R.* v. *Umphenour*, 69 Ill. 198.

1642. Where a railroad is inclosed by a sufficient fence, and a casual breach occurs therein, without the knowledge or fault of the company, and through such breach, stock get upon the track and are injured, the company is not liable unless it has had a reasonable time to discover such breach, or has been notified, and fails to repair before the injury occurs. *I. & St. L. R. R.* v. *Hall*, 88 Ill. 368.

1643. Where a railway company is required to keep its track fenced, and a breach is made in the fence by parties not in the employ or under the control of the company, and the company has no knowledge of such breach, and there are no circumstances showing that it was authorized to anticipate the breach being made, and by reason of such breach stock gets upon the track and is killed before the company has had a reasonable time to discover the breach, the company will not be liable; and a covenant or condition in a deed conveying the right of way, to fence the same, will not add to the defendant's liability under the statute. *C. & A. R. R.* v. *Saunders*, 85 Ill. 288.

1644. NOTICE OF DEFECTS. Where defects in its fence are known to the company, the failure of the adjoining land-owner to use reasonable efforts to notify the company of such defects, will not justify it in failing to repair the same. *C., B. & Q, R. R.* v. *Seirer*, 60 Ill. 295.

1645. ESTOPPEL TO DENY LIABILITY TO MAINTAIN. A railway company which erects a fence and gate along its right of way, a few feet beyond the same, and upon the land of the adjoining owner, and keeps the same in repair for some time, and then suffers it to get out of repair, whereby stock escapes through the same and is killed upon the track, cannot escape liability to the owner of such stock, on the ground that such fence and gate are not on the right of way, when it has given no prior notice that it will not keep up such repairs any longer. *C. & E. Ill. Ry.* v. *Guertin*, 115 Ill. 466.

1646. Where a railway company after erecting and maintaining for many years a fence along the side of and near its right of way, near a station, suffers it to become defective, so that stock gets over it and upon its track and are killed, it cannot exonerate itself from liability on the ground of its higher duty to the public of keeping its depot grounds open. So long as it permits such a fence to stand as a fence required by statute, it will be estopped from denying its duty to keep it in proper repair. *Ib.*

1647. WHERE RAILWAY NOT IN DEFAULT AS TO FENCING— *burden of proof.* Where the injury is not the result of a neglect to fence, the burden of proof to show negligence, rests on the plaintiff. *Ch. & N. W. Ry.* v. *Taylor*, 8 Bradw. 108.

1648. If a horse takes fright and gets upon the track by breaking a fence or leaping a guard, which would be sufficient under all ordinary circumstances, it will not devolve upon the company to prove an absence of negligence in running its trains, and it will not be liable, except on proof of its carelessness or willful injury. *C. & A. R. R.* v. *Utley*, 38 Ill. 410.

1649. To make a railway company liable for killing or injuring stock, except for neglect to fence, the plaintiff must show that the injury resulted from a want of ordinary care on the part of the company. *P. D. & E. R. R.* v. *Dugan*, 10 Bradw. 233.

1650. In an action at common law against a railway company for killing cattle, negligence of the company must be averred and proved. It is otherwise if the action is brought under the statute for neglect to fence. *T. H., A. & St. L. R. R.* v. *Augustus*, 21 Ill. 186.

1651. Where stock gets upon the track of a railway company without its fault, the law requires evidence beyond the mere proof that they were injured or killed by the engine or cars of the company, to establish its liability. It is necessary to show negligence on the part of the servants of the company having charge of the train at the time the injury occurred. *Ch. & N. W. Ry.* v. *Barrie*, 55 Ill. 226.

1652. EVIDENCE OF NEGLIGENCE. A case of negligence is not made out by simply showing the killing of stock upon the road by the agents or cars of the company. *Ch. & Miss. R. R.* v. *Patchin*, 16 Ill. 198.

1653. The mere fact of killing an animal by a railway company, does not render the company liable, unless it has been guilty of negligence, or the case comes within the statute of 1855. *Great Western R. R.* v. *Morthland*, 30 Ill. 451.

FENCING NOT INVOLVED.

1654. LIABLE ONLY FOR GROSS NEGLIGENCE. Railway companies are not liable for injuries to cattle, unless they be willfully or maliciously done, or done under circumstances showing gross negligence. They are not bound to use the highest possible degree of care towards animals coming upon their tracks. *Great Western R. R.* v. *Thompson*, 17 Ill. 131.

1655. A railway has a right to run its cars upon its track without obstruction, and an animal has no right upon the track without the consent of the company; and if suffered to stray there, it is at the risk of the owner. *Central Military Tract R. R.* v. *Rockafellow*, 17 Ill. 541.

1656. Where an animal is allowed to stray upon the track of a railroad company, the company will not be liable for an injury to it, except gross negligence of the company is shown. *Ib.*

1657. Animals straying upon the track of an uninclosed railroad, are strictly trespassers, and the company is not liable for their destruction, unless its servants are guilty of willful negligence, evincing reckless misconduct. *I. C. R. R.* v. *Reedy*, 17 Ill. 580.

1658. While a train was running through a town upon the depot grounds at the usual rate of speed, the bell being rung, a colt ran upon the track from behind a building so near the road that it could not be seen by the engineer in time to check the train, but as soon as he saw it he blew the whistle and the brakes were put down, but the colt was killed. The track at that point was not fenced. *Held*, that the company being guilty of no negligence, was not liable. *Galena & Ch. Union R. R.* v. *Griffin*, 31 Ill. 303.

1659. If an animal is suddenly driven on the track by a dog and is killed without there being any fault on the part of the engineer, the company will not be liable. *I. C. R. R.* v. *Wren*, 43 Ill. 77.

1660. Gross or willful negligence on the part of a railway company will make it liable for an injury to an animal, even though the animal be improperly on the track. *I. C. R. R.* v. *Wren*, 43 Ill. 77.

1661. If injury could have been prevented by ordinary care—*liable*. If by the use of ordinary care and diligence, animals on a railroad track can be saved from injury, it is the duty of the company to employ that degree of care. No other rule will afford sufficient protection to animals which are lawfully on the track, as they are, if they get upon it from the range or commons. *I. C. R. R.* v. *Baker*, 47 Ill. 295.

1662. Where stock are upon the track and a train is approaching, though down a slight grade, and the engine-driver, instead of stopping his train to drive off the stock, pursues them to a point where by means of ditches filled with water on each side of a high embankment, there is but little probability the animals will leave the track, and they are overtaken and killed, the company is guilty of gross negligence, and will be liable, even if it appears that the animals got upon the track within the limits of a town. *I. C. R. R.* v. *Baker*, 47 Ill. 295.

1663. Where the engineer saw a lot of mules on the track and sounded the whistle to frighten them off, but they ran along the track into a cut, where they were killed, and it appearing that he might have stopped the train: *Held*, culpable negligence on the part of the engineer for which the company was liable. *I. C. R. R.* v. *Middlesworth*, 46 Ill. 494.

1664. Where cattle killed upon the track could have been seen by the engineer in charge of the train for a distance of more than half a mile, and there was nothing to obstruct his view, and without any effort to stop the train or giving any signals of alarm, rushed upon them at a rapid rate: *Held*, such gross negligence as to authorize a recovery, even though the cattle were upon the track without the fault of the company. *Ch. & N. W. Ry.* v. *Barrie*, 55 Ill. 226.

1665. Where stock is killed on a railroad track, and the engineer in charge of the train, could by the use of ordinary care and skill, without danger, have stopped the train in time to have avoided the collision, although the animals were wrongfully upon the track, the company will be liable. *T., P. & W. Ry.* v. *Bray*, 57 Ill. 514.

1666. Where a cow is killed by an engine at a place where the company is not bound to fence its track, there being no wanton or willful neglect of the company, yet if by the exercise of ordinary care and skill upon its part, the injury could have been prevented, the company will be liable. *R., R. I. & St. L. R. R.* v. *Lewis*, 58 Ill. 49.

1667. Where by the use of ordinary care and diligence on the part of the servants of a railway company, animals straying on its track can be saved from injury, it is their duty to exercise such care, and a failure to do so, will make the company liable. *T., P. & W. Ry.* v. *Ingraham*, 58 Ill. 120.

1668. If an engineer sees cattle on the track and can by ordinary care, caution and diligence, avoid injury to them, he should do so, and failing to do so, the company will be liable to the owner, even though he was negligent in allowing the cattle to get on the track. *C., M. & St. P. R. R.* v. *Phillips*, 14 Bradw. 265.

1669. BURDEN OF PROOF. Where the company has erected and maintains sufficient fences and cattle-guards, the *onus* is in the owner of the cattle to show a negligent or willful act by the company or its servants, before he can recover for an injury thereto. *Galena & Ch. Union R. R.* v. *Crawford*, 25 Ill. 529.

1670. Where a company is not bound to fence its road, it is only liable for injury to animals resulting from wantonness or gross negligence. *I. C. R. R.* v. *Phelps*, 29 Ill 447.

1671. A railway company is liable for injuries to persons and property where willfully done, or resulting from gross neglect of duty. To free such company from liability for injury, it must discharge every duty imposed by law. It must use all reasonable means to prevent injury, and an omission to do so, will create liability, unless the injured party by his negligence, has contributed in some degree to the injury. *Great Western R. R.* v. *Geddis*, 33 Ill. 304.

1672. Where cattle get upon a track at a point where the company is not bound to fence, or where others are bound to fence, and stray along the track and are killed by a train at a place where the company is bound to fence, but has failed to do so, the company will not be liable, since the injury in such case will have no connection with the failure to fence the road at the place where the animals were killed. *St. L., A. & T. H. R. R.* v. *Linder*, 39 Ill. 433.

1673. Where stock gets upon the track at a place where others than the company are bound to erect and maintain the fence, and is killed at that place or another place, the company will be liable only in case of gross negligence. *Ib.*

1674. Where the company has performed its duty in respect to fencing its road, to render it liable for stock killed, negligence must be proved as in ordinary cases. *I. C. R. R.* v. *Whalen*, 42 Ill. 396.

1675. A railway company is liable for gross negligence resulting in the destruction of property, irrespective of the question of the erection of fences. *R., R. I. & St. L. R. R.* v. *Phillips*, 66 Ill. 548.

1676. Where stock is killed by an engine of a railway company on its track within six months after it being opened for use, it is incumbent on the owner of the stock to show negligence on the part of the company, to entitle him to recover. *R., R. I. & St. L. R. R.* v. *Connell,* 67 Ill. 216.

1677. Where the evidence showed the entire sufficiency of the fences and that the plaintiff's horse was killed at the crossing of a public road, where the company had constructed and maintained suitable cattle-guards, and that he got upon the track from the road: *Held,* that the company could not be held liable, except upon the ground that the killing was willful, or the result of negligence. *C. & A. R. R.* v. *McMorrow,* 67 Ill. 218.

1678. While railway companies are not required, or permitted to fence their tracks in incorporated towns, &c., still they are bound to use all due and proper diligence to avoid injury to property, and they must exercise such diligence as to stock wrongfully running at large, or trespassing on their track or right of way. *T., W. & W. Ry.* v. *McGinnis,* 71 Ill. 346.

1679. Where a team ran away and got upon a railroad track in an incorporated town where the company was not required to fence its track, and ran along the track, until they fell into an old cattle-guard, where they were injured by a freight train, the engineer having done all he could to stop the train: *Held,* that the company was not liable. *C. & A. R. R.* v. *Rice,* 71 Ill. 567.

1680. Where stock is killed by a railroad company at a place where it is not required to fence its road, the party seeking a recovery must prove that the killing was caused through the negligence of the company; and where the proof shows that the stock was killed within the limits of a city, and there is no evidence of negligence on the part of the company, no recovery can be had. *I. C. R. R.* v. *Bull,* 72 Ill, 537.

1681. An instruction holding a railway company liable for the failure of its servants in charge of a train to use ordinary care to prevent the killing of hogs, is erroneous, if it excludes the necessary element that the injury might have been avoided by such care, and makes the liability depend upon a failure to attempt to prevent the injury, whether it could have availed or not. *G., C. & S. R. R.* v. *Spencer,* 76 Ill. 192.

1682. Where a railway company is under no statutory liability for injury to stock by its trains, by reason of its road not having been fenced, as where the road has not been open for six months, the only ground of liability will be that the injury might have been avoided by the exercise of ordinary care and prudence, and that its servants in charge, failed to exercise such care and prudence. *G., C. & S. R. R.* v. *Spencer,* 76 Ill. 192.

1683. Where stock are killed or injured within a city, town, or village, there can be no recovery had by the owner, without an averment in the declaration and proof, that the servants of the company were guilty of negligence in running its trains through such city, town or village. *P., P. & J. R. R.* v. *Barton,* 80 Ill. 72.

1684. TRESPASSING ANIMALS. Where a domestic animal running at large by the sufferance of the owner, gets upon a railroad track at the crossing of a highway, where the company is not required to fence, and is injured by a passing train, the company will not, in general, be liable, unless its servants, after they discover the animal, might by the exercise of proper care and prudence, have prevented the injury. *T., W. & W. Ry* v *Barlow,* 71 Ill. 640.

1685. In such a case it is not sufficient to entitle the owner to recover, to show that the train was running at an unusual rate of speed, or without proper care in other respects. *Ib.*

1686. Prior decisions made under the law making it lawful for cattle to run at large, do not fully apply under the present law, where no neglect in fencing is involved. *Ib.*

1687. WHAT IS NEGLIGENCE. It is negligence on the part of a railway company to permit grass or weeds to grow on its grounds so as to obstruct the view of stock by the engineer. *O. & M. R. R.* v. *Clutter*, 82 Ill. 123.

1688. FAILURE TO STOP TRAIN. The law imposes no obligation upon those in charge of a train to stop the same upon discovering an animal grazing near the track, in anticipation that it may get upon the track and be injured, and a failure to do so, is not negligence. *P., P. & J. R. R.* v. *Champ*, 75 Ill. 577.

1689. ICE AND WATER IN DITCHES. The law does not require a railway company to keep the excavations along the sides of its track free from water and ice, and it will not be liable for stock killed in consequence of ice therein so as to prevent escape from the track over the same. *P. & R. I. Ry.* v. *McClenahan*, 74 Ill. 435.

PLEADING.

1690. DECLARATION. In an action under the statute for injury to animals, the plaintiff should aver in his declaration that the animals were not within the limits of a village, &c. *C., B. & Q. R. R.* v. *Carter*, 20 Ill. 390.

1691. Gross negligence need not be averred, the degree of negligence being a matter of proof. *C., B. & Q. R. R.* v. *Carter*, 20 Ill. 390.

1692. Unless action is brought under the statute an averment of a neglect to fence may be treated as surplusage. *Ib.*

1693. UNDER STATUTE. The declaration must show not only that the company was required to fence its track and had failed to do so, but must also negative the exceptions in the act, and aver that the animals were not injured at a point on the road within these exceptions, and also that the road has been in use six months prior to the accident. *Galena & Ch. Union R. R.* v. *Sumner*, 24 Ill. 631.

1694. The declaration must show that the accident did not happen at a place where the company is not bound to maintain a fence. *I. C. R. R.* v. *Williams*, 27 Ill. 48.

1695. It need not negative the possibility that the animals may have been killed at a farm crossing. *Great Western R. R.* v. *Helm* 27 Ill. 198.

1696. In suit before a justice of the peace the plaintiff should negative by proof that there was no public crossing where the killing occurred, and show that the company was bound to fence at that point. *O. & M. R. R.* v. *Taylor*, 27 Ill. 207.

1697. The declaration must negative all the exceptions in the statute. *Great Western R. R.* v. *Bacon*, 30 Ill. 347.

1698. A declaration averred: "nevertheless, more than six months after said railroad was in use, to-wit, on &c., the said defendant neglected to erect," &c.: *Held*, a sufficient averment of a breach of the statutory duty. *Great Western R. R.* v. *Hanks*, 36 Ill. 281.

1699. An averment that the steer which was killed "strayed and got on such railroad without the limits of towns, cities and villages, and not at the road crossings or public highways," is sufficient. *Ib.*

1700. It is sufficient if the declaration negatives the killing in the excepted places named in the statute. *Ib.*

1701. The want of an averment that a fence was necessary at the place.of the accident is cured by the proof on the trial after verdict for the plaintiff. *T., P. & W. Ry.* v. *McClannon*, 41 Ill 238.

1702. Neglect to maintain a fence whereby stock is injured, is a ground of action distinct from that of negligence in leaving open a gate along the line of the fence; and where an action is predicated on the latter ground, it must be so averred in the declaration. *I. C. R. R.* v *McKee*, 43 Ill. 119.

1703. The declaration in every case must contain a full and explicit statement of all the material facts upon which a recovery is sought, so that the defendant may be prepared to meet them. *Ib.*

1704. A defect in a declaration in failing to show which of two animals was killed, and which crippled, is cured by a subsequent averment, that by the act of the defendant in running its trains upon them, they were lost to the plaintiff. *T., W. & W. Ry.* v. *Cole*, 50 Ill. 184.

1705. It is not enough to aver that the road was not fenced at the place where the injury occurred; but it should be shown that the road was not fenced at the place where the animal got upon the track. It should be shown that the stock did not get upon the track at some other place where the road was fenced. *T., P. & W. Ry.* v. *Darst*, 51 Ill. 365.

1706. An averment that the road has been opened for six months, is not sufficient, if such averment relates only to the place where the injury occurred, it not being shown but that the stock strayed upon the track at another place, where the road had not been opened for six months before the injury. *Ib.*

1707. A declaration averring that a company had failed to fence the road at the place where the animal was killed, or where it got upon the track, and that it was not killed, nor did it get upon the track at any of the excepted places, is sufficient, at least after verdict, or on motion in arrest, to show that the injury resulted from the neglect to fence. *T., P. & W. Ry.* v. *Darst*, 52 Ill. 89.

1708. The declaration should show that the road is located at some place in this state, or it will be obnoxious to a special demurrer. *T., P. & W. Ry.* v. *Bookless*, 55 Ill. 230.

1709. In a suit brought, Oct. 30, 1868, under the statute, the declaration averred, "that the defendant on the first day of January, 1867, and from thence forward, to the commencement of this suit, were possessed of and had the entire control of the "road, and had the right to run upon the same, locomotives, and trains;" "and that the defendant more than six months after the said railroad was in use, and continuously to the time of the committing of the grievances, &c., neglected to comply with the before mentioned requirements as by the statute in such case made and provided, it was their duty to do:" *Held*, on special demurrer, that it was not alleged with sufficient certainty, that the company for the period of six months after the road was "opened for use," had failed to erect proper fences. *T., P. & W. Ry.* v. *Bookless*, 55 Ill. 230.

1710. An objection to a declaration that it fails to aver that the railroad of the defendant, used by it, is in the county and state in which the action is brought, comes too late after verdict. *T., P .& W. Ry.* v. *Webster*, 55 Ill. 338.

1711. Where a count averred a neglect of the company to fence its road, and that a train was run, conducted and directed carelessly, whereby plaintiff's horse was killed: *Held*, that the plaintiff might recover on proving either ground; but that it was subject to demurrer for duplicity. The general issue is a traverse of both grounds. *C., B. & Q. R. R.* v. *Magee*, 60 Ill. 529.

1712. A declaration held not obnoxious to a demurrer for want of an allegation of time and place, when and where the injury was committed. *St. L., J. & C. R. R.* v. *Kilpatrick*, 61 Ill. 457.

1713. To recover under the statute, the declaration must state facts which bring the case substantially within the statute, and the plaintiff is not bound to show that there was negligence in the management of the locomotive or train, which was the immediate cause of the injury. *R., R. I. & St. L. R. R.* v. *Phillips*, 66 Ill. 548.

1714. COMMON LAW. A declaration charging that the defendant was the owner of the railroad and operating the same by running locomotives and trains thereon; that plaintiff's horse strayed and got upon defendant's road, and that defendant by its servants, so carelessly, negligently and improperly run, conducted and directed the locomotive and train of defendant as that said locomotive struck plaintiff's horse with great force and killed it, shows a good cause of action at common law. *R., R. I. & St. L. R. R.* v. *Phillips*, 66 Ill. 548.

1715. In pleading, the averment of negligence is sufficient to admit proof of gross negligence; and on demurrer, an averment of negligence is equivalent to whatever degree of negligence is necessary to sustain the pleading. *R. R., I. & St. L. R. R.* v. *Phillips*, 66 Ill. 548.

1716. Where the declaration charges negligence as at common law, all allegations respecting the want of sufficient fences may be rejected as surplusage. *Ib.*

1717. To recover for an injury to stock within a city, &c., the declaration must aver that the servants of the company were guilty of negligence in running its train through such city, town or village. *P., P. & J. R. R.* v. *Barton*, 80 Ill. 72.

1718. Where the value of the stock killed is laid under a *videlicit* at $200, an averment that the cattle were of the value of $19.50 each, may be regarded as surplusage. *O. & M. R. R.* v. *Clutter*, 82 Ill. 123.

DEFENSES.

1719. DEFENSE—*cost of fencing paid in compensation for right of way.* A railway company when sued for injury to animals, may show in defense, the proceedings to condemn the right of way, in which the cost of fencing the road is included in the damages, and their payment to the plaintiff. *R., R. I. & St. L. R. R.* v. *Lynch*, 67 Ill. 149.

1720. Where damages are assessed against a railway company for fencing the road in a proceeding to condemn, and the proceedings are formal and made a matter of record, then the land will thereafter be charged with the duty to fence, and the company and its successor discharged from that duty. *Ib.*

1721. In a suit to recover for the killing of stock on the ground of a neglect to fence, if the land-owner has agreed to fence the track, or has received compensation in damages for so doing, the burden is on the company to show that fact, and not upon the plaintiff to negative it. *T., P. & W. Ry.* v. *Pence*, 68 Ill. 524.

1722. If the owner of the land where the animals got upon the track, received compensation for fencing when the right of way was obtained, the burden of proof is upon the company to show that fact. *T., P. & W. Ry.* v. *Pence*, 71 Ill. 174; *T., P. & W. Ry.* v. *Lavery*, 71 Ill 522.

1723. The duty imposed by the act of 1855 upon railway companies to maintain fences along their roads, is not transferred to the owner of the land over which the road may run, by the simple employment of such owner as its agent and servant, and his performance of the contract, to erect the required fence along the road located over his land. *I. C. R. R.* v. *Swearingen*, 33 Ill. 289.

1724. The statute only contemplates the release of the company where the duty is assumed by the land-owner. *Ib.*

1725. Where the owner of land adjoining the right of way of a railway company, under an agreement with the company, erected a fence along the line between his land and the right of way, and took upon himself to maintain it, it was held as between such owner and those holding under him, with knowledge of his duty, and the company, that the duty of maintaining and repairing the fence did not rest on the company. *St. L., V. & T. H. R. R.* v. *Washburn*, 97 Ill. 253.

1726. A tenant of such land-owner having knowledge of the undertaking of his landlord, will not be allowed to allege any want of sufficiency in the fence as a ground of recovery for stock getting through the same and being killed. *Ib.*

1727. BURDEN OF PROOF. The burden of proof is not upon the plaintiff to prove the averment that there was no contract between the company and the owner of the ground, that the latter should build the fence where the accident occurred. *Great Western R. R.* v. *Bacon*, 30 Ill. 347.

CONTRIBUTORY NEGLIGENCE.

1728. CONTRIBUTORY NEGLIGENCE—*leaving gate open.* If the land-owner opens a gate at a farm crossing and negligently leaves it open, so that his stock get upon the track and are injured, his own act and neglect will defeat any recovery by him for such injury. *I. C. R. R.* v. *McKee*, 43 Ill. 119.

1729. If bars at a farm crossing are taken down by the owner or occupant of the farm, and he neglects to put them up, his own act will preclude him from a recovery in a suit by him against the company for animals injured. *I. C. R. R.* v. *Arnold*, 47 Ill. 173.

1730. Where a person repairs a break in the fence with defective materials, so that it appears sufficient when it is not so in fact, the company will not be liable to him for an injury to his animals breaking through such part of the fence without notice of its defective condition. *C., B. & Q. R. R.* v. *Seirer*, 60 Ill. 295.

1731. Where cattle break through a railroad fence and the owner of the cattle repairs it with defective material in a temporary manner, but it is apparently sufficient, and his cattle again break through the same place and are killed, and it appears that he knew the fence thus repaired, was defective and that he failed to notify the company: *Held*, that he was guilty of negligence and could not recover for his cattle. *Ib.*

1732. The owner of land adjoining a railroad has no right to remain inactive and let his cattle get upon the track, through the known deficiency of the fence along the road. When he undertakes to repair such fence and does it negligently and fails to notify the company, he will become liable for the natural consequences of his negligence. *Ib.*

1733. If by ordinary care, caution and diligence, injury to cattle on the track may be avoided, the company will be liable to the owner for an injury to them, even though he may be negligent in allowing them to get on the track. *C., M. & St. P. R. R.* v. *Phillips*, 14 Bradw. 265.

1734. An animal has no right upon a railway track, and if suffered to go there it will be at the owner's risk. If allowed to stray upon the track the company will not be liable for an injury to it, except gross negligence of the company is shown. *C., M. Tract R. R.* v. *Rockafellow*, 17 Ill. 541.

1735. A person is guilty of negligence who permits his animals to go upon a railroad track at a place where the company is not bound by law to fence. *I. C. R. R.* v. *Phelps*, 29 Ill. 447.

1736. If negligence on the part of the plaintiff is clearly proven then the defendant will be responsible only for such negligence which implies willful injury. *I. C. R. R.* v. *Goodwin*, 30 Ill. 117.

1737. The failure of a railway company to fence its track is negligence. It is also negligence in the owner of horses to put them in a field through which an unfenced railroad passes, having on blind bridles. He has the right to turn them into the field, but not so blinded as to render them incapable of avoiding danger. *St. L., A. & T. H. R. R.* v. *Todd*, 36 Ill. 409.

1738. Where the owner of horses is guilty of negligence in putting them, blindfolded, into a field through which passed an unfenced railroad, whereby they are injured, the company will only be liable for negligence which implies willful injury. But a failure of the company to fence its road, and the killing of horses thereon, which might have been avoided by reasonable efforts will amount to such injury. *Ib.*

1739. Although stock may lawfully run at large in the highways, &c., the rule is so modified in respect to railroads, that where they are not bound to fence, they will have the right to run their trains, and stock on their tracks are trespassers. *Headen* v. *Rust*, 39 Ill. 186.

1740. Animals getting on a railway track not required to be fenced, being wrongfully there, the company will be liable only for gross negligence resulting in injury to them. *Ib.*

1741. In actions against railway companies for injuries inflicted by negligence, the company will not be liable, if the plaintiff has been guilty of negligence which has contributed to the injury, unless it appears that the company has been guilty of negligence more gross than that of the plaintiff. *I. C. R. R.* v. *Middlesworth*, 43 Ill. 64.

1742. Negligence and carelessness on the part of the owner of stock, by which they get upon a fenced railroad track, where they are killed, will not lessen the railway company's liability, where the exercise of ordinary care and skill on its part would have prevented the injury. *Ib.*

1743. Railway company liable for injury to stock when wrongfully upon its track, which might have been prevented by ordinary care and without danger. *T., P. & W. Ry.* v. *Bray*, 57 Ill. 514.

1744. Such companies must exercise all due and proper diligence and care to avoid injury to stock wrongfully running at large, or trespassing on their track or right of way. *T., W. & W. Ry.* v. *McGinnis*, 71 Ill. 346; *T., W. & W. Ry.* v. *Barlow*, 71 Ill. 640.

1745. In an action against a railway company for killing stock, it is a question of fact for the jury, to be determined from all the circumstances in evidence, whether the act of the owner in permitting his animals to run at large in violation of law, is contributory negligence. *R., R. I. & St. L. R. R.* v. *Irish*, 72 Ill. 404.

1746. It is not sufficient to charge a plaintiff with contributory negligence simply to show that he permitted his stock to run at large in violation of law; but it must appear that he did so under such circumstances that the natural and probable consequence of so doing was, that the stock would go upon the railroad track and be injured. *Ewing* v. *C. & A. R. R.*, 72 Ill. 25.

1747. Whether permitting male animals to run at large is contributory negligence, depends, *first*, upon whether permitting them to run at large, was a proximate or only a remote cause of their being injured, and if it was a proximate cause, then, *secondly*, whether such negligence on the part of the owner was slight, and that of the company gross in comparison with each other. *R., R. I. & St. L. R. R.*, v. *Irish*, 72 Ill. 404.

1748. The fact that the owner of stock permits them to run at large in violation of the act prohibiting domestic animals from running at large, does not relieve railway companies from their duty to fence their roads, or their liability for stock injured in consequence of their failure to do so. *Ewing* v. *C. & A. R. R.*, 72 Ill. 25.

1749. Where animals in counties are prohibited by law from running at large, if they escape from their inclosure without the fault or knowledge of their owner and stray upon a railroad at a point where the company have failed to fence as required, and are killed, the company will be responsible. *O. & M. Ry.* v. *Jones*, 63 Ill. 472.

1750. Although a plaintiff may be guilty of negligence in permitting his animals to get upon a railroad track, it is still the duty of the company to use ordinary skill and prudence to avoid doing them injury, and failing in this, it will become liable. *R., R. I. & St. L. R. R.* v. *Irish*, 72 Ill. 404.

1751. No contributory negligence is chargeable to the owner of stock in letting them run at large, where it breaks out of his pasture without his fault. *T., P. & W. Ry.* v. *Johnston*, 74 Ill. 83.

1752. The owner of a horse who voluntarily permits the same to run at large contrary to the law in force in the county, cannot recover of a railway company for killing the same by one of its trains, upon the ground that such company has failed to fence its track at the the place where the animal was killed. *Peo., Pekin & Jack. R. R.* v. *Champ*, 75 Ill. 577.

1753. In such a case, where the plaintiff is guilty of contributory negligence, the company will not be relieved from its duty to observe all reasonable precautions to prevent injury to the property of the plaintiff. *Ib.*

1754. The mere fact that stock is running at large in violation of the statute, does not relieve railway companies from liability for an injury to them, from a neglect to fence their road, and no other negligence need be shown. *C. & St. L. R. R.* v. *Murray*, 82 Ill. 76.

1755. Where a railway company fails to fence its road and stock is killed by its trains in a county where it is lawful for stock to run at large, the question of contributory negligence in the owner in permitting his stock to run at large, cannot arise, and the company will be liable. *O. & M. R. R.* v. *Fowler*, 85 Ill. 21.

1756. BURDEN OF PROOF. In an action by the owner of stock which were allowed to go at large contrary to law, to recover of a railway company for an injury to them resulting from its track being unfenced, the burden of showing contributory negligence on the part of the plaintiff, where it does not otherwise appear, is on the company. *C. & St. L. R. R.* v. *Woosley*, 85 Ill. 370.

1757. Permitting stock to run at large in violation of the statute, does not relieve railway companies from their duty to fence their roads, or their liability for stock injured in consequence of their failure to do so. *Ib.*

1758. CONTRIBUTORY—*letting stock run at large.* In a suit against a railway company for stock killed in consequence of its neglect to fence its road, where it appears that such stock were permitted to run at large in violation of the law, the question whether the owner has been guilty of contributory negligence in permitting them to run at large, is one of fact to be determined by the jury from the circumstances of the case. *C. & St. L. R. R.* v. *Woosley*, 85 Ill. 370.

1759. To charge the owner of stock with contributory negligence in allowing them to run at large contrary to law, it must appear that he did so under such circumstances that the natural and probable consequence of doing so, was that the stock would go upon the road and be killed or injured. *Ib.*

EVIDENCE.

1760. DUTY TO FENCE. Plaintiff must show that road has been open six months prior to injury. *W., St. L. & P. Ry.* v. *Neikirk,* 13 Bradw. 387; *O. & M. R. R.* v. *Meisenheimer,* 27 Ill. 30; *O. & M. R. R.* v. *Jones,* 27 Ill. 41; *C. & A. R. R.* v. *Taylor,* 40 Ill. 280; *R., R. I. & St. L. R. R.* v. *Lynch,* 67 Ill. 149.

1761. Plaintiff must prove there was no public crossing where his animals were killed, and that company was bound to fence at that point. *O. & M. R. R.* v. *Taylor,* 27 Ill. 207.

1762. Proof that plaintiff's steers were killed in the fall of 1870, and his horses and hogs in the summer of 1871, shows inferentially that the road had been open for use six months before the horses and hogs were killed. *R., R. I. & St. L. R. R.* v. *Spillers,* 67 Ill. 167.

1763. PLACE OF INJURY. Proof that stock was not killed within a *corporation,* nor near a crossing, will justify the jury in finding that it was not killed within the limits of a city, &c. *St. L. & S. E. Ry.* v. *Casner,* 72 Ill. 384.

1764. Proof that a cow was found killed within a mile and a quarter of plaintiff's house, is sufficient to show she was killed within five miles of a settlement; and evidence that a colt, which was killed, was kept up, and only ran out to water, is sufficient to authorize the jury to infer that it was killed within five miles of a settlement. *St. L. & S. E. Ry.* v. *Casner,* 72 Ill. 384.

1765. WHERE STOCK GOT ON TRACK. In a suit against a railway company for killing stock, where the evidence is that the road was not fenced at the place where the stock was killed, it is but a fair inference that the stock got upon the road at the place where it was killed. *Ib.*

1766. Negligence in the management and running of a train is not made out by proof of the killing of stock by it. *Ch. & Miss. R. R.* v. *Patchin,* 16 Ill. 198; *Great Western R. R.* v. *Morthland,* 30 Ill. 451; *C. & A. R. R.* v. *Utley,* 38 Ill. 410; *I. C. R. R.* v. *Whalen,* 42 Ill. 396; *Ch. & N. W. Ry.* v. *Barrie,* 55 Ill. 226; *R., R. I. & St. L. R. R.* v. *Lynch,* 67 Ill. 149; *T., P. & W. Ry.* v. *Pence,* 68 Ill. 524.

1767. To recover for cattle killed by reason of neglect to fence the road, the plaintiff must prove every material allegation in his declaration, and that the injury did not occur at any of the excepted places; and this though the declaration is defective in not negativing the exceptions in the statute. *O. & M. R. R.* v. *Brown,* 23 Ill. 94.

1768. OWNERSHIP. Where parties sue in case for damages for killing cattle, claiming as joint owners, they should be held to reasonably strict proof of ownership. *I. C. R. R.* v. *Finnigan,* 21 Ill. 646.

1769. In an action for injury to animals, it is necessary to show that the plaintiff was the owner, or had possession of the same. *O. & M. R. R.* v. *Saxton,* 27 Ill. 426.

1770. CONNECTING DEFENDANT WITH INJURY. The proof must show that the injury was done by the road of the defendant sued. *O. & M. R. R.* v. *Taylor,* 27 Ill. 207.

1771. Proof that the stock was found by the side of the railroad "badly smashed up," will justify a finding that the injury was done by the cars or locomotives of the defendant. *I. C. R. R.* v. *Whalen*, 42 Ill. 396.

1772. Evidence that plaintiff's cow, when found, was lying on her back in the railway ditch, between two or three feet from the track, bloated and the blood oozing from her nose. The jury found that the cow was killed by a passing train, and the court, though doubtful of the correctness of the finding, refused to disturb it. *Ch. & N. W. Ry.* v. *Dement*, 44 Ill. 74.

1773. Where it is shown that the defendant company was incorporated by the name it bears at the session of the legislature next preceding the injury complained of, and there is no proof or suggestion that any other railroad was operated in that part of the country, where the injury was done, it may be fairly inferred that the injury was done by the defendant's road. *T., P. & W. Ry.* v. *Arnold*, 49 Ill. 178.

1774. The evidence must connect the defendant with the injury complained of; but it is not required that such fact be proved beyond a reasonable doubt. A preponderance of the evidence is sufficient. *T., P. & W. Ry.* v. *Eastburn*, 54 Ill. 381.

1775. Evidence held sufficient to connect the defendant with the injury. *R., R. I. & St. L. R. R.* v. *Lewis*, 58 Ill. 49.

1776. OF CONDITION OF FENCES. If stock is killed at a point where it is the duty of the company to maintain a fence, this is a circumstance which may be considered in determining the question whether the fences and cattle-guards were good and sufficient. *C. & A. R. R.* v. *Utley*, 38 Ill. 410.

1777. VENUE. It is not essential to a recovery to prove that the injury complained of was done within the jurisdiction of the court. *T., P. & W. Ry.* v. *Webster*, 55 Ill. 338.

1778. TIME OF INJURY. No recovery can be had for stock killed after the action is brought. *T., P. & W. Ry.* v. *Arnold*, 49 Ill. 178,

VARIANCE.

1779. Where the declaration counts on a common law liability for animals killed or injured, no recovery can be had under the statute by proving a neglect to fence. *T. H., A. & St. L. R. R.* v. *Augustus*, 21 Ill. 186; *I. C. R. R.* v. *Middlesworth*, 43 Ill. 64.

1780. Proof of the injury on the day alleged is not required. It may be shown to have taken place at any time within the statute of limitations. *T., P. & W. Ry.* v. *McClannon*, 41 Ill. 238.

1781. Under a declaration showing an injury to a horse resulting from the failure of the railway company to maintain and keep in repair its fences on its roadway, evidence that the animal strayed upon the track through a gate at a farm crossing which had been left open and was killed, is inadmissible. *I. C. R. R.* v. *McKee*, 43 Ill. 119.

1782. Where the declaration avers that the defendant carelessly "ran, conducted and directed" its trains, whereby, &c., it is error to instruct the jury that they may consider the condition of the brakes employed. In such case the action is for carelessness and not for a failure to properly equip the road. *C. B. & Q. R. R.* v. *Magee*, 60 Ill. 529.

BURDEN OF PROOF.

1783. To show negligence other than neglect to fence. *Ch. & N. W. R. R.* v. *Taylor*, 8 Bradw. 108; *Galena & Ch. Union R. R.* v. *Crawford*, 25 Ill. 529; *C. & A. R. R.* v. *Utley*, 38 Ill. 410.

1784. To show contributing negligence in plaintiff. *C. & St. L. R. R.* v. *Woosley*, 85 Ill. 370.

1785. To show no contract for owner to fence railroad track. *Great Western R. R.* v. Bacon, 30 Ill. 347.

1786. To show that land-owner has agreed to fence road or has received compensation in damages for fencing. *T., P. & W. Ry.* v. *Pence*, 68 Ill. 524; *T., P. & W. Ry.* v. *Pence*, 71 Ill. 174; *T., P. & W. Ry.* v. *Lavery*, 71 Ill. 522.

MEASURE OF DAMAGES.

1787. INJURY—*use for beef.* Where an animal is not so seriously injured but that it is of value for food, it is the duty of the owner to dispose of it to the best advantage. The measure of damage in such case, is the difference in its value as injured, from its value before the injury. *I. C. R. R.* v. *Finnigan*, 21 Ill. 646; *T., P. & W. Ry.* v. *Parker*, 49 Ill. 385.

1788. Where the weather is warm and the cattle, when found, are swollen and unfit for beef, the plaintiff is entitled to recover their full value. *T., P. & W. Ry.* v. *Sweeney*, 41 Ill. 226.

1789. Where the cattle killed, when found, are mangled, bruised and swollen, the plaintiff will not be required to use any diligence to dispose of their dead bodies to entitle him to recover their full value. *R., R. I. & St. L. R. R.* v. *Lynch*, 67 Ill. 149.

1790. Where the stock killed is in good condition, it is the duty of the owner to dispose of it to the best advantage possible, by converting it into beef or otherwise, and he is entitled to reasonable time in which to do so. *T., P. & W. Ry.* v. *Parker*, 49 Ill. 385.

1791. But where the company on the same evening of the accident takes possession of and buries the animal, it cannot be urged that the owner failed to perform his duty by not disposing of the animal for beef. *Ib.*

1792. COMPENSATORY ONLY. The damages for stock killed by a railroad company through negligence merely, as a neglect to fence its track, is compensatory only. To authorize more, circumstances of aggravation must be shown. *T., P. & W. Ry.* v. *Johnston*, 74 Ill. 83.

1793. The owner of stock killed by a railway company for want of a fence, is not entitled to interest on its value from the time of the killing. *Ib.*

MEASURE OF RECOVERY.

1794. ATTORNEY'S FEES. The statute gives an attorney's fee only in actions to recover damages for neglect to erect and maintain fences. If the suit embraces other matters, the fee should be limited to the cause of action growing out of a failure to fence. *W., St. L. & P. Ry.* v. *Neikirk*, 13 Bradw. 387.

1795. In an action against a railway company to recover for killing a colt, an attorney's fee is recoverable, if the loss is chargeable to the statutory negligence in not fencing, but not if chargeable to common law negligence. *C., M. & St. P. R. R.* v. *Phillips*, 14 Bradw. 265.

1796. Under this section the attorney's fee may be recovered in the suit for damages for the stock killed, and the law giving such fee in such a case, is not special legislation. *W., St. L. & P. Ry.* v. *Lavieux*, 14 Bradw. 469.

1797. The attorney's fee is allowable only where the railway company has failed to comply with the requirements of the statute, and such failure must appear from the evidence. *Ib.*

1798. SAME—*notice of.* The statute making a railway corporation liable, in an action for stock killed, for a reasonable attorney's fee, is notice to such corporation when sued for injury to stock, that such fee will be claimed, and it is not necessary it should have any other notice. *P., D. & E. Ry.* v. *Duggan,* 109 Ill. 537.

1799. SAME—*how recovered.* The liability of a railway company for an attorney's fee in an action to recover for an injury to animals growing out of its neglect to fence its track, under the act of 1879, arises at the same instant with its liability for damages; and such fee may be assessed in the same suit with the damages, the law not favoring a multiplicity of actions. *Ib.* As to fencing in cities see ante, 144.

1800. RIGHT OF WAY CLEAR OF COMBUSTIBLES. § 1½. It shall be the duty of all railroad corporations to keep their right of way clear from all dead grass, dry weeds, or other dangerous combustible material, and for neglect shall be liable to the penalties named in section 1. [R. S. 1887, p. 1013, § 63; S. & C., p. 1933, § 63; Cothran, p. 1152, § 49.]

1801. If the company suffers grass to accumulate on its right of way by means of which fire from an engine is communicated to the fences and grass of another, which are destroyed, the company will be liable on the ground of negligence. *R., R. I. & St. L. R. R.* v. *Rogers,* 62 Ill. 346.

1802. It is negligence to allow weeds or anything else to grow upon the right of way to such a height as to obstruct the view of a highway crossing, and if injury results to stock at such crossing that might have been avoided but for such obstruction, the company will be liable. *I. & St. L. R. R.* v. *Smith,* 78 Ill. 112.

1803. It is negligence to permit brush or other obstructions on right of way so as to prevent the view of approaching trains by persons attempting to cross the road at a highway intersection. *Dimick* v. *Ch. & N. W. Ry.,* 80 Ill. 338.

1804. Lessee of a railroad is guilty of negligence, if it fails to keep the right of way clear of all dead grass, weeds, &c., and will be liable for injury from the escape and transmission of fire from it engines. *P., C. & St. L. Ry.* v. *Campbell,* 86 Ill. 443.

1805. A railway company should not permit obstructions upon its right of way near a crossing, which will prevent the public from observing the approach of trains. *R., R. I. & St. L. R. R.* v. *Hillmer,* 72 Ill. 235.

1806. It is negligence to permit vegetation to grow upon right of way so that cattle may be concealed from view. *Bass* v. *C., B. & Q. R. R.,* 28 Ill. 9. See post, 2484.

1807. ALLOWING, ETC., ANIMAL ON RIGHT OF WAY—BREAKING FENCE, ETC. § 2. If any person shall ride, lead or drive any horse or other animal upon the track or lands of such railroad corporation, and within such fences or guards (except to cross at farm or road crossings), without the consent of the corporation; or shall tear down, or otherwise render insufficient to exclude stock, any part of such fence, guards, gates or bars—or shall leave the gates or bars at farm crossings open or down—or shall leave horses or other animals standing upon farm or road crossings, he shall be liable to a penalty of not less than $10, nor more than $100, to be recovered in an action of debt, before any court having competent jurisdiction thereof, in the name of such railroad corporation, and for the use of the school fund in the county, and shall pay all damages which shall be sustained thereby to the party aggrieved. [In lieu of L. 1855, p. 174, § 3. R. S. 1887, p. 1013, § 64; S. & C., p. 1933, § 64; Cothran, p. 1152, § 50.]

1808. The right of way is the exclusive property of the railway company upon which no unauthorized person has a right to be for any purpose, and any person traveling over it, is a wrong-doer and a trespasser. *I. C. R. R.* v. *Godfrey*, 71 Ill. 500.

1809. The mere acquiescence of the company in the use of its track or right of way by persons passing along it for a footway, does not give them a right of way, nor will the company be bound to protect or provide safeguards for such persons. *Ib.*

1810. It is negligence to walk along a railroad track. *Ib.*

1811. Party wrongfully upon the railway track held to a greater degree of care than if there lawfully. *Aurora Branch R. R.* v. *Grimes*, 13 Ill. 585.

1812. Duty of company where a team is stalled with a loaded wagon on track. *C. & A. R. R.* v. *Hogarth*, 38 Ill. 370.

1813. It is negligence for a person to walk upon the track of a railroad, whether laid in a street or an open field, and he who deliberately does so, must assume the risk of the peril he may encounter. *I. C. R. R.* v. *Hall*, 72 Ill. 222. See, also, *I. C. R. R.* v. *Hammer*, 72 Ill. 347.

1814. If the conduct of one killed while walking upon a railroad amounts to gross negligence, the company will not be liable except for willful or criminal negligence. *I. C. R. R.* v. *Hetherington*, 83 Ill. 510.

1815. Where a railroad passes over ground not used by any except employes of the company, the engineer having no reason to apprehend that any one will be on the track, his failure to take precaution to discover some one on·it, is not negligence in the company. *I. C. R. R.* v. *Frelka*, 9 Bradw. 605.

1816. WHEN COMPANY NEGLECTS TO BUILD — NOTICE. § 3. Whenever a railroad corporation shall neglect or refuse to build or repair such fence, gates, bars or farm crossings, as provided in this act, the owner or occupant of the lands adjoining such railroad, or over or through which the railroad track is or may be laid, may give notice, in writing, to such corporation, or the lessees thereof, or the persons operating such railroad, to build such fence, gate, bars or farm crossings within thirty days (or repair said fence, gate, bars or farm crossings, as the case may be, within ten days,) after the service of said notice. Such notice shall describe the lands on which said fence, gates, bars or farm crossings are required to be built or repaired. Service of such notice may be made by delivering the same to any station agent of said railroad corporation or the persons operating such railroad. [This is § 1, Laws 1869, p. 315, extended to gates,
—16

bars and farm crossings. R. S. 1887, p. 1014, § 65; S & C., p. 1934, § 65; Cothran, p. 1152, § 51.]

1817. ADJOINING OWNER MAY BUILD AND RECOVER. § 4. If the party so notified shall refuse to build or repair such fence, gates, bars or farm crossings, in accordance with the provisions of this act, the owner or occupant of the land required to be fenced shall have the right to enter upon the land and track of said railroad company, and may build or repair such fence, gates, bars or farm crossings, as the case may be, and the person so building or repairing such fence, gates, bars or farm crossings, shall be entitled to double the value thereof from such corporation, or party actually occupying or using such railroad, to be recovered, with interest at one per cent. per month, as damages, from the time such fence, gates, bars or farm crossings were built or repaired, in any court of competent jurisdiction, together with costs, to be taxed by the court. [This is § 2 of act of 1869 (L. 1869, p. 315), extended to gates, bars and farm crossings and provision for entry on right of way inserted, and double value substituted for single value. R. S. 1887, p. 1014, § 66; S. & C., p. 1934, § 66; Cothran, p. 1153, § 52.]

1818. After notice to build a certain line of fence and neglect of the company to make any part of such fence, the land-owner built half of the line and sued to recover for the part built by him: *Held,* that he was entitled to recover without first making the entire fence. *T., P. & W. Ry.* v. *Sieberns,* 63 Ill. 217.

1819. This act authorizing the land-owner to fence and hold the company liable, does not release railway companies from their liability under the act of 1855 for stock killed. The act creates no new duty upon the land-owner to fence, but merely gives him the privilege to do so, and the fence when built by the owner, will be the property of the company. *T., P. & W. Ry.* v. *Pence,* 68 Ill. 524.

1820. Where a railway neglects and refuses to fence its right of way, after notice by the owner of adjoining land, the latter may build the fence and recover double the value thereof in an action against either the corporation owning the road, or any other party occupying or using such railroad, at his election. *O. & M. R. R.* v. *Russell,* 115 Ill. 52.

1821. It is no defence to such action against the corporation that its property, &c., is in the hands of a receiver, and such corporation is enjoined from interfering with the property or disturbing the possession of the receiver. *Ib.;* *C. & St. L. R. R.* v. *Peoples,* 92 Ill. 97; *C. & St. L. R. R.* v. *Warrington,* 92 Ill. 157.

1822. To entitle the owner of land over which a railroad is operated to recover of the company double the value of any fence built by him upon its neglect to do so on proper notice, the statute must be strictly followed, and the fence must be such as the statute requires, and be built in the mode the statute contemplates. The fence must be built on the sides of the railroad. If built two feet inside of right of away, the penalty cannot be recovered. *W., St. L. & P. Ry.* v. *Zeigler,* 108 Ill. 304.

1823. FENCE IN RIGHT OF WAY. The statute is not complied with by erecting a fence several feet within the right of way. The statute contemplates that the fence shall embrace the entire right of way. *O. & M. Ry.* v. *People*, 121 Ill. 483.

1824. REMEDY TO COMPEL BUILDING OF. Where the company, on notice, erects a fence several feet inside its right of way, and refuses to let the land-owner join his fences with the same, so as to enclose his land, *mandamus* will lie to compel the company to erect a fence along the line of its right of way. *Ib.*

1825. BOARDS AT CROSSINGS. § 5. Every railroad corporation shall cause boards, well supported by posts or otherwise, to be placed and constantly maintained upon each public road or street, where the same is crossed by its railroad on the same level. Said boards shall be elevated so as not to obstruct the travel, and to be easily seen by travelers. On each side of said board shall be painted in capital letters, of at least the size of nine inches each, the words "railroad crossing," or "look out for the cars." This section shall not apply to streets in cities or incorporated towns or villages, unless such railroad corporation shall be required to put up such boards by the corporate authorities of such cities, towns or villages: *Provided,* that when warning boards have already been erected, under existing laws, the maintenance of the same shall be a sufficient compliance with the requirements of this section. [2d L. 1849, p. 32; R. S. 1887, p. 1014, § 67; S. & C., p. 1934, § 67; Cothran, p. 1153, § 53.]

1826. The failure to maintain such signal boards raises a liability only for injuries caused by such failure. When a party with full knowledge of there being a railroad crossing before him, drives upon the track and is injured by a passing train, he cannot recover for the want of such board merely. *C. & A. R. R.* v. *Robinson*, 8 Bradw. 140, 142. See notes to next section.

1827. SIGNALS AT ROAD CROSSINGS — BELL OR WHISTLE TO BE SOUNDED. § 6. Every railroad corporation shall cause a bell of at least thirty pounds weight, and a steam whistle placed and kept on each locomotive engine, and shall cause the same to be rung or whistled by the engineer or fireman, at the distance of at least eighty rods from the place where the railroad crosses or intersects any public highway, and shall be kept ringing or whistling until such highway is reached. [L. 1869, p. 308 ; (re-written). R. S. 1887, p. 1014, § 68; S. & C., p. 1935, § 68; Cothran, p. 1153, § 54.]

DECISIONS ON STATUTE.

IN GENERAL.

1828. WHAT COMPANIES BOUND BY. The law (1849) is binding on corporations created before its passage. *Galena & Ch. Union R. R.* v. *Loomis*, 13 Ill. 548: The law of 1849 was a general law, and its provisions apply to all railway companies either before or thereafter chartered. *I. & St. L. R. R.* v. *Blackman*, 63 Ill. 117. The police regulations of the act of 1849 requiring railway companies to ring a bell or sound a whistle before reaching a public road-crossing, apply to all railroads in the state, not specially exempted by their charters—as well to those chartered since the passage of the act, as to those chartered before that time. *Western Union R. R.* v. *Fulton*, 64 Ill. 271.

1829. EXEMPTION—*from duty.* An act which exempts a railway company from ringing a bell or sounding a whistle at a road-crossing, is not unconstitutional. *G. & Ch. U. R. R.* v. *Dill*, 22 Ill. 264.

1830. SIGNAL—*may be either by bell or whistle.* The statute does not require a company to both ring a bell and sound a whistle. If it does either it has discharged its duty in this respect. *C., B. & Q. R. R.* v. *Damerell*, 81 Ill. 450; *St. L., A. & T. H. R. R.* v. *Pflugmacher*, 9 Bradw. 300.

DUTY OF RAILWAY COMPANY AT HIGHWAY CROSSING.

(*a*) TO GIVE SIGNAL OF APPROACH.

1831. ONLY AT HIGHWAY CROSSING. A railway company is not required by the statute to ring a bell or blow a whistle at a farm crossing. This is required only at the intersection or crossing of a public highway. *W., St. L. & P. Ry.* v. *Neikirk*, 13 Bradw. 387.

1832. OBJECT OF SIGNALS—*for whose benefit.* The statute is designed for the protection of travelers using the highway, and not for the benefit of persons walking upon the track without right, or those crossing at a distance from the public road over a private crossing. *W., St. L. & P. Ry.* v. *Neikirk*, 15 Bradw. 172; *Harty* v. *Central R. R.*, 42 N. Y. 468; *Voak* v. *Northern Cent. R. R.*, 75 N. Y. 320: See also *I. C. R. R.* v. *Hall*, 72 Ill. 222; *I. C. R. R.* v. *Hetherington*, 83 Ill. 510.

1833. Nor was it designed for the protection of passengers leaving their seats not at a regular station. *R., R. I. & St. L. R. R.* v. *Coultas*, 67 Ill. 398.

1834. Yard master not required to give signal before uncoupling cars standing on track in yard. *C. & A. R. R.* v. *McLaughlin*, 47 Ill. 265.

1835. Nor is it necessary to give such signal to one who otherwise is informed of the approach of the train, or one who sees it approaching and attempts to cross. *C., R. I. & P. R. R.* v. *Bell*, 70 Ill. 102; *O. & M. Ry.* v. *Eaves*, 42 Ill. 288; *L., S. & M. S. R. R.* v. *Clemens*, 5 Bradw. 77.

(*b*) LIABILITY FOR NEGLECT TO GIVE SIGNAL.

1836. NEGLECT OF DUTY, MUST CAUSE THE INJURY. Unless the injury complained of is the result of the neglect to give the statutory signal of warning, there can be no recovery for it on that ground. *C., B. & Q. R. R.* v. *Doorak*, 7 Bradw. 555; *P., D. & E. Ry.* v. *Foltz*, 13 Bradw. 535.

1837. CONNECTION OF NEGLECT WITH THE INJURY, TO BE SHOWN BY THE PLAINTIFF. The company is not liable for any and all damages a party may sustain where it has omitted to give the signal. To make the company liable, it must be shown that the injury was the result of the failure to give the signal. *G. & Ch. U. R. R.* v. *Loomis*, 13 Ill. 548; *P., D. & E. Ry.* v. *Foltz*, 13 Bradw. 535.

1838. THE BURDEN OF PROOF is upon the plaintiff to show that the injury resulted from the failure to ring the bell or sound the whistle, and not upon the defendant to show the injury was not the result of such neglect. *P., D. & E. Ry.* v. *Foltz*, 13 Bradw. 535.

1839. Until some proof is given tending to show that the injury resulted from a failure to ring a bell or sound the whistle, the burden of proving a negative—that it did not arise from such failure, should not be thrown upon the company. *G. & Ch. U. R. R.* v. *Loomis*, 13 Ill. 548.

1840. The omission to ring a bell or sound a whistle at a road crossing does not render the company liable for an injury to animals, unless it is made to appear that such signal would have prevented the injury. *I. C. R. R.* v. *Phelps*, 29 Ill. 447.

1841. The omission to ring a bell or sound a whistle for the required distance on approaching a road crossing, renders the company liable for "all damage which shall be sustained by any person by reason of such neglect." But it will not *per se* render the company liable for injuries. The injury must be shown, by circumstances at least, to have been the consequence of, or caused by such neglect. *Ch. & R. I. R. R.* v. *McKean*, 40 Ill. 218.

1842. The neglect to give such warning of approach, is not of itself such negligence as will justify a recovery for the killing of an animal upon the track. The injury must be shown to be the result of the omission or neglect of duty imposed, and this the jury must determine. *I. & St. L. R. R.* v. *Blackman*, 63 Ill. 117.

1843. A recovery against a railway company for killing stock, will be sustained, if there is evidence from which the jury may fairly infer that the killing was caused by the failure to ring the bell or sound the whistle and the rapid speed of the train. *I. & St. L. R. R.* v. *Holloway*, 63 Ill. 121.

1844. The omission to ring a bell or sound a whistle for the whole distance required by the statute at the crossing of a public highway, being eighty rods, will subject the company to the penalty given; but will not subject it to liability for damages, unless they were caused by reason of such neglect. *C. & A. R. R.* v. *McDaniels*, 63 Ill. 122.

1845. While it is negligence to omit giving the signal on approaching a public crossing, yet the company is not necessarily liable for every accident that may occur where this duty is omitted. It is only where the injury happens by reason of such neglect that the company is liable. The plaintiff must show, not only this omission of duty, but also from facts and circumstances at least, that the injury was occasioned by such neglect. *C., B. & Q. R. R.* v. *Van Patten*, 64 Ill. 510.

1846. Where it reasonably appears that if the statutory signal had been given, an animal on the track would have been frightened off and been saved, the company will be liable. *C. & A. R. R.* v. *Henderson*, 66 Ill. 494.

1847. While the statute imposes a penalty for an omission to comply with its requirements, more is required to create a liability for an injury to person or property. In the latter case, where no other negligence is proved, the injury must be "by reason of the neglect" to ring the bell or sound the whistle, and the proof must show that it was the probable result of the omission. *R., R. I. & St. L. R. R.* v. *Linn*, 67 Ill. 109.

1848. The omission to give the statutory signal as the train approaches a public crossing, will not *per se* render it liable. To make the company liable, it must be a just inference from the evidence, that the injury was caused by such neglect of duty. *C., B. & Q. R. R.* v. *Lee*, 68 Ill. 576.

1849. Where it is proved that a person injured by a collision at a railroad crossing of a highway, was in the exercise of due care and caution, it may be a reasonable inference that the accident was produced by reason of such neglect to ring a bell or sound a whistle. It may be shown by circumstantial, as well as by direct evidence.. *C., B. & Q. R. R.*, v. *Lee*, 68 Ill. 576.

1850. The statute only imposes a liability upon a railway company for neglecting to give the signal as its train approaches a highway crossing, for injury resulting from that neglect of duty. Where it appears that the non-compliance with the statute did not result in injury, no cause of action will arise. The injury complained of must be the result of that neglect, either in whole or part. *I. C. R. R.* v. *Benton*, 69 Ill. 174.

1851. If the company is also guilty of other negligence, and it is doubtful which produced the injury, or if both combined produced it, then the company will be liable, if the injured party is not also in default to such an extent as to relieve the company from liability. *Ib.*

1852. It is not enough to create a liability for stock killed by a railway train, to prove that the bell was not rung or the whistle sounded. It must be made to appear by facts and circumstances proved that the accident was caused by reason of such neglect. *Q., A. & St. L. R. R.* v. *Wellhæner*, 72 Ill. 60; 40 Ill. 218.

1853. The omission to ring a bell or sound a whistle at a road crossing does not render a railway company liable for injury to animals, or to a person, unless it is made to appear the warning might have prevented the injury. *T., W. & W. Ry.* v. *Jones*, 76 Ill. 311.

1854. Where the omission appears not to have contributed in the slightest degree to an injury or accident on a train of cars, the railway company operating the same, will not be subjected to liability on that ground in a suit for damages. *T., W. & W. Ry.* v. *Durkin*, 76 Ill. 395.

1855. *Whether failure to give signal is negligence, a question of fact.* Whether or not a failure to sound a whistle, is negligence is a question of fact for the jury, and it is error in an instruction to assume that it is negligence. *T. H. & I. R. R.* v. *Jones*, 11 Bradw. 322.

1856. Whether a failure to ring a bell or sound a whistle, when not required by statute, is negligence, is a question of fact, and cannot be regarded unless its omission occasions a collision producing injury. Where such acts are not required by statute, their omission does not raise a legal inference that the injury resulted from a want of their performance. *G. & Ch. U. R. R.* v. *Dill*, 22 Ill. 264.

1857. In an action against a railway company for damages resulting from a failure to comply with the requirements of the statute relative to sounding a bell or whistle, at public road crossings, an instruction that such omission is *prima facie* negligence, is proper. In *G. & Ch. U. R. R.* v. *Dill* the statute did not apply. *St. L., J. & Ch. R. R.* v. *Terhune*, 50 Ill. 151.

1858. Whether an injury was the result of the omission of duty is a question of fact for the jury. *C. & A. R. R.* v. *McDaniels*, 63 Ill. 122.

1859. Whether the failure to ring a bell or sound a whistle on approaching a highway crossing by a train as required by the statute, is the cause of an injury sustained, is a question of fact for the jury. *I. C. R. R.* v. *Benton*, 69 Ill. 174.

1860. The failure to give the statutory signals on approaching a highway crossing, constitutes a *prima facie* case of negligence, if the injury is caused by it. *P., D. & E. Ry.* v. *Foltz*, 13 Bradw. 535.

1861. Where the statute does not require it, an omission to give a signal by sounding a bell or whistle, is not of itself evidence of negligence. *G. & C. U. R. R.* v. *Dill*, 22 Ill. 264.

1862. An instruction to the effect that if the defendants, their servants or agents, omitted to ring a bell or sound a whistle in the manner required by law, such omission constitutes a *prima facie* case of negligence, and defendants are liable to the plaintiff for the loss and damage proved to have been sustained by reason of such negligence. *Held*, proper. *C. & A. R. R.* v. *Elmore*, 67 Ill. 176.

1863. In an action for killing a cow at a road crossing, an omission to ring a bell or sound a whistle while at a distance of at least eighty rods from the crossing, constitutes a *prima facie* case of negligence in the company. *I. C. R. R.* v. *Gillis*, 68 Ill. 317.

1864. The mere omission of a railway company to ring a bell or sound a whistle on a train approaching a highway crossing where a collision occurs with a team while crossing the railroad track, cannot be said as a matter of law, to be evidence of gross negligence, so as to fix the liability of the company for the injury. To have that effect it must be a just inference from the evidence that the injury was caused by such neglect. *C., B. & Q. R. R.* v. *Harwood*, 90 Ill. 425.

1865. It is error to give an instruction which authorizes a recovery against a railway company upon the ground of negligence in omitting to sound a whistle or ring the bell, without containing a requirement of any care or caution on the part of the person injured. *C., B. & Q. R. R.* v. *Harwood*, 80 Ill. 88.

1866. Where the injury is alleged to be the result of negligence in failing to ring a bell or sound a whistle on approaching a crossing, and in running at a prohibited rate of speed, an instruction excluding from the consideration of the jury the fact whether the plaintiff received his injuries in consequence of such neglect, is erroneous. *C., B. & Q. R. R.* v. *Dvorak*, 7 Bradw. 555.

1867. The bell or whistle of the locomotive should be sounded at a reasonable distance before reaching a road crossing. If stock is killed at such crossing in consequence of a failure to give such warning the company will be liable. *Ch. & R. I. R. R.* v. *Reid*, 24 Ill. 144.

1868. An animal was run over and killed by an engine at a road crossing, a place where the statute required a bell to be rung or a whistle to be sounded, which was not done, and the jury found the injury was the result of this omission of duty: *Held*, that the company was liable to the owner of the animal for its value. *Gr. Western R. R.* v. *Geddes*, 33 Ill. 304.

1869. In an action against a railway company, the court instructed the jury for the plaintiff, that if he was injured by one of defendant's engines at a street crossing in a city, and at the time there was no bell ringing or whistle sounding upon such engine, they should find for the plaintiff, unless he by his own negligence materially contributed to the injury: *Held*, erroneous, in failing to leave it to the jury to find whether the injury was caused by such omission. *C., B. & Q. R. R.* v. *Notzki*, 66 Ill. 455.

1870. In an action against a railway company for injuries received at a road crossing by a collision with plaintiff's team, it is error to instruct the jury to find the defendant guilty of negligence from the mere fact that a bell was not rung or whistle sounded as required by law, regardless of the consideration whether the failure contributed to the accident or not. *T., W. & W. Ry.* v. *Jones*, 76 Ill. 311.

1871. Where a railway company in running a wild train on approaching a highway crossing fails to give the statutory signals at a place where the view of an approaching train is obstructed by timber and heavy foliage, this will establish a right of recovery against the company for an injury received by one while attempting to cross the railroad with his team, in favor of the party injured. *P., P. & J. R. R.* v. *Siltman*, 88 Ill. 529.

1872. Where a person was killed in attempting to cross the railroad, and it appeared the company allowed the view along its track to be obstructed by a house, brush and weeds upon its right of way, and failed to give the statutory signal on the approaching train which did the killing, until it was too late to avail, and the train was running at an unusual rate of speed to make up time : *Held*, that the negligence of the company' was gross, and even if the deceased was guilty of negligence in failing to listen or look for a train out of its time, it was slight and the company was liable. *C., B. & Q. R. R.* v. *Lee,* 87 Ill. 454.

1873. Where a railroad is so constructed that the place where it crosses a public highway is unusually dangerous to the traveling public, as where its track intersects the highway in a cut, and is approached on the road by a descending hill, and persons approaching the crossing cannot see the track owing to brush, bushes, &c. : *Held,* that a neglect to sound a bell or whistle under such circumstances, was gross negligence. *I. & St. L. R. R.* v. *Stables,* 62 Ill. 313.

1874. *Not necessary, signals should apprise persons of danger.* The statute requires every railway corporation to cause a bell of at least thirty pounds weight to be rung, or a steam whistle to be sounded the distance of at least eighty rods before a public highway is reached by a train or locomotive, and kept ringing or being sounded until the highway is reached; and where this is done, the company has discharged its duty imposed by the statute, whether such signal is heard or not. The statute does not require the giving of such signal of the approach of a train as to enable others absolutely to ascertain its approach and avoid being injured. *C., B. & Q. R. R.* v. *Dougherty,* 110 Ill. 521.

1875. If a railway company has such a bell on an engine attached to a train as the statute requires, and it is rung in the manner required, then so far as giving signals before the train reaches a public highway is concerned, the company will be without blame, whether the signal so given is observed, or heeded, or not by one attempting to cross the railroad track on the public highway. *Ib.*

1876. In an action against a railway company for a personal injury, the court instructed that it was the duty of the railway company to ring a bell or sound a whistle at a distance of at least eighty rods from the crossing and until the crossing was reached, "*so as to apprise persons of*" the approach of the train: *Held,* erroneous, as requiring a higher duty than that imposed by the statute. *P., P. & J. R. R.* v. *Siltman,* 67 Ill. 72.

1877. In a similar case the court instructed that it was the duty of the company on approaching a highway on a common level, to give "due warning," so that a person traveling on the highway with a team and carriage might stop and allow the train to pass: *Held,* erroneous, as likely to induce the jury to believe the company was bound to do more than ring a bell or sound a whistle. *C. & A. R. R.* v. *Robinson,* 106 Ill. 142.

1878. A railway company in crossing public highway must so regulate the speed of its trains and give such signals to persons passing as to apprise them of the danger of crossing the track. This, it seems, is independent of the statute. *C. & R. I. R. R.* v. *Still,* 19 Ill. 499.

1879. DUTY OF RAILWAY TO AVOID COLLISION. Railway companies in crossing public highways are bound to so regulate the speed of their trains, and to give such signals as to apprise persons of their approach. It is also the duty of those having charge of the trains to keep a lookout so as to avoid injury as far as possible to persons exercising their legal rights in traveling upon the highway. *C., B. & Q. R. R.* v. *Cauffman*, 38 Ill. 424.

1880. Railway companies in crossing public highways must so regulate the speed of their trains, and give such signals to persons passing as. to apprise them of the danger of crossing the track; and a failure in any of these duties, will render them liable for injuries inflicted and for wrongs resulting from such omissions. *R., R. I. & St. L. R. R.* v. *Hillmer*, 72 Ill. 235.

1881. Where the view of an approaching train is obstructed by brush it is the duty of the company to give the warning of its approach, and if it does not, it will be liable for an injury to one attempting to cross, who is not guilty of negligence. *Dimick* v. *Ch. & N. W. Ry.*, 80 Ill. 338.

1882. Circumstances and case stated in which it was held that a neglect to give the signals on approaching a street crossing was gross negligence, and such as to relieve the injured party from the charge of negligence. *St. L., V. & T. H. R. R.* v. *Dunn*, 78 Ill. 197.

1883. Although the law does not require a company to regulate its speed or sound a bell or whistle at a place where a collision occurs, yet if those in charge of a train have reasonable ground to believe there is danger of such a collision, and that sounding the whistle or slackening of the speed of the train may prevent it and avoid injury, the company will be *held* guilty of negligence in not giving such signal, &c. *I. C. R. R.* v. *Modglin*, 85 Ill. 481.

1884. The court is not disposed to relax the rule as to the duty to give warning of approaching trains by a continuous ringing of the ·bell or sounding of a whistle for the distance of eighty rods before arriving at a crossing, and the willful disregard of it is gross negligence. *O. & M. Ry.* v. *Eaves*, 42 Ill. 288.

1885. A railway company is liable for killing a cow on its track near the crossing where it fails to give the warning. *T., W. & W. Ry.* v. *Furgusson*, 42 Ill. 449.

1886. INJURY NEAR CROSSING. Where an animal is killed near a street crossing by a train running through an incorporated town, if the injury occurs before the train reaches the street, and the bell or whistle is not sounded, the company will be liable under the statute. *T., P. & W. Ry.* v. *Foster*, 43 Ill. 415.

1887. If the injury occurs after the locomotive has passed the street and at a place where the statute does not require the signal to be given, in that case it is a question for the jury to determine whether or not an omission to give the signal amounts to such negligence as will render the company liable. *Ib.*

1888. An ordinance requiring the stationing of flagmen and erection of bell towers at street crossings of railroads, has reference to the duties of the owners or lessees of the railroad tracks, and cannot be made the basis of liability against a railway company not owning or leasing the track, but having only a license from the owner. *L., S. & M. S. R. R.* v. *Kaste*, 11 Bradw. 536.

DUTIES IN GENERAL AT HIGHWAY CROSSINGS.

1889. It is due to the public, that all, either persons or stock, at or near a road crossing, shall be warned of the approach of a train of cars by the bell or whistle. *C., R. I. & P. R. R.* v. *Reid*, 24 Ill. 144.

1890. A railway company must use all reasonable means to prevent injury, and an omission to do so, will create liability unless the injured party has by his negligence contributed in some degree to the injury. *Great Western R. R.* v. *Geddis,* 33 Ill. 304.

1891. While a railway company is held to a very high degree of care and diligence in operating its road through the streets of a city, yet the care and caution in this respect are required to be exercised in reference to the proper uses of the streets as a thoroughfare for travel, rather than to the safety of persons in wrongfully getting on their cars. *C., B. & Q. R. R.* v. *Stumps,* 55 Ill. 367.

1892. STOPPING TRAIN. It is not the duty of an engineer on nearing a public road crossing to stop his train for the purpose of avoiding a collision with a wagon and team he may see approaching the crossing. He has the right to presume the team will stop, if he gives the proper signal. *St. L., A. & T. H. R. R.* v. *Manly,* 58 Ill. 300.

1893. But should he see a team on the track where it would not be likely to get across in time, he should use every means in his power to check his train and prevent the collision. *Ib.*

1894. It is not the duty of the engine-driver on nearing a road crossing to stop his train for the purpose of avoiding a collision with a team he may see approaching the crossing. *C., B. & Q. R. R.* v. *Damerell,* 81 Ill. 450.

1895. The law has not made it the duty of a railway company to check up its trains on discovering a person approaching a crossing from the highway with a team. The engine-driver has a right to expect he will stop until the train passes. *C., B. & Q. R. R.* v. *Lee,* 68 Ill. 576; *T., W. & W. Ry.* v. *Jones,* 76 Ill. 311.

1896. Where an employe of a railway company while in charge of a hand-car on the track was injured by a collision with a construction train, and it appeared he knew of the approach of the train in time to have got off the track : *Held,* that his negligence was such as to preclude a recovery, and that the company was not negligent in not sounding a whistle and slackening the speed of the train, as its servants had a right to expect the hand-car would be taken from the track before it was reached. *I. C. R. R.* v. *Modglin,* 85 Ill. 481.

1897. Not bound to stop a train because a person is ahead walking near the track nearly parallel with it, or standing near the track, having reason to believe such person will keep off the track. *C., R. I. & P. R. R.* v. *Austin,* 69 Ill. 426.

1898. MUTUAL RIGHTS AT CROSSING. Railway companies have the same right to use that portion of the public highway over which their track passes as other people have to use the same. This right and those of the public, as to the use of the highways at such points of intersection, are mutual, co-extensive and reciprocal; and in the exercise of such rights, all parties will be held to a due regard to the safety of others, and to the use of every reasonable effort to avoid injury to others." *I. & St. L. R. R.* v. *Stables,* 62 Ill. 313.

1899. The degree of diligence and care required of railroad companies, is not fixed by any definite and precise rule, but depends rather upon the facts and circumstances of the case, so that what would be an unnecessary act in one case, would be imperatively demanded in another. *Ib.*

1900. A railway company has no better right to cross a public highway with its trains at the intersection with its road, than individuals have to cross its road at the same place. This right is mutual, co-extensive, and in all respects reciprocal; and "in the exercise of these rights all parties must be held to a due regard for the safety of others. *G. & Ch. U. R. R.* v. *Dill,* 22 Ill. 264.

1901. While it is true the traveler has the same right to cross a railroad at its intersection with a highway that the railway company has to cross the highway, yet each in so crossing, is bound to use reasonable care and effort to avoid a collision or inflicting an injury on the other, or in receiving injury from the other. If a team can be checked on seeing the approach of a train more readily than the train, it should be so checked up. *I. C. R. R.* v. *Benton*, 69 Ill. 174.

1902. Persons traveling along a highway which crosses a railroad track, and the trains of the railroad company, have an equal right to pass over the crossing, and it is the duty of both to use reasonable and prudent precaution to avoid accident and danger; the one to look out for the approach of trains, and the other to give the required signals and warning of its approach. *Ch. & N. W. Ry.* v. *Hatch*, 79 Ill. 137.

PROOF THAT CROSSING IS OF A HIGHWAY.

1903. Proof that a road intersected by a railroad had been traveled by the public and worked and repaired by the proper authorities, is sufficient *prima facie* evidence that it is a public highway to require a railway company, when sued for neglect to sound a bell or whistle on approaching the same, to show that it is not a legal highway. *I. C. R. R.* v. *Benton*, 69 Ill. 174. See also *C. & A. R. R.* v. *Adler*, 56 Ill. 344. See *C. & A. R. R.* v. *Dillon*, — Ill. —. Filed Jan. 20, 1888.

EVIDENCE—AS TO OMISSION OF DUTY.

1904. WEIGHT—*negative and affirmative.* Positive evidence that a headlight was burning, or that a bell or whistle was sounding, is entitled to more weight than negative evidence in regard to such facts. *C. & R. I. R. R.* v. *Still*, 19 Ill. 499.

1905. As to what is negative evidence, and its comparative weight in value see *Coughlin* v. *People*, 18 Ill. 266; *Rockwood* v. *Poundstone*, 38 Ill. 199; *C., B. & Q. R. R.* v. *Cauffman*, 38 Ill. 424; *C., B. & Q. R. R.* v. *Triplett*, 38 Ill. 482; *Frizell* v. *Cole*, 42 Ill. 362; *C. & A. R. R.* v. *Gretzner*, 46 Ill. 74; *C., B. & Q. R. R.* v. *Stumps*, 55 Ill. 367; *I. C. R. R.* v. *Gillis*, 68 Ill. 317; *R., R. I. & St. L. R. R.* v. *Hillmer*, 72 Ill. 235; *Ch., D. & V. R R* v. *Coyer*, 79 Ill. 373; *C., B. & Q. R. R.* v. *Lee*, 87 Ill. 454; *C., B. & Q. R. R.* v. *Dickson*, 88 Ill. 431; *C. & A. R. R.* v. *Robinson*, 106 Ill. 142.

1906. NEGLIGENCE OF PLFF. AS A DEFENSE—*neglect to look.* It is the duty of a person about to cross a railroad track to look out and listen for an approaching train. If he fails to do so and rushes into danger that he might have seen and avoided by ordinary care, he cannot recover for any injury he thereby receives. *L. S. & M. S. R. R.* v. *Clemens*, 5 Bradw. 77; *C. & A. R. R.* v. *Robinson*, 9 Bradw. 89; *C., B. & Q. R. R.* v. *Cauffman*, 38 Ill. 424; *C. & A. R. R.* v. *Gretzner*, 46 Ill. 74; *C., R. I. & P. R. R.* v. *Bell*, 70 Ill. 102; *C., B. & Q. R. R.* v. *Damerell*, 81 Ill. 450; *L. S. & M. S. R. R.* v. *Hart*, 87 Ill. 529; *Austin* v. *C., R. I. & P. R. R.*, 91 Ill. 35.

1907. DUTY TO LOOK IN BOTH DIRECTIONS. It is culpable or gross negligence to cross the track of a railroad without looking in every direction in which the rails run to make sure that the road is clear. *Garland* v. *Ch. & N. W. Ry*, 8 Bradw. 571; *C., R. I. & P. R. R.* v. *Bell*, 70 Ill. 102.

1908. This rule applies with increased force to one who was not lawfully using the railroad track, but passing laterally along it, not at a highway crossing. *I. C. R. R.* v. *Godfrey*, 71 Ill. 500.

1909. It is incumbent on a person approaching a railway crossing to exercise care and caution by looking and listening for approaching trains. A failure to do so is gross negligence and bars a right of recovery. *St. L., A. & T. H. R. R.* v. *Pflugmacher,* 9 Bradw. 300.

1910. It is the duty of a person about to cross a railroad track to look and listen for approaching trains; and the neglect of this duty is such gross negligence as to preclude all right of recovery for an injury by a collision. *W., St. L. & P. Ry.* v. *Neikirk,* 15 Bradw. 172; *W. St. L. & P. Ry.* v. *Hicks,* 13 Bradw. 407; *G. & Ch. U. R. R.* v. *Dill,* 22 Ill. 264; *C. & A. R. R.* v. *Gretzner,* 46 Ill. 74; *T., P. & W. Ry.* v. *Riley,* 47 Ill. 514; *C. & A. R. R.* v. *Jacobs,* 63 Ill. 178; *C., B. & Q. R. R.* v. *Harwood,* 80 Ill. 88.

1911. It is negligence for a deaf person to drive an unmanageable horse across a railroad track where a train is approaching. It is his duty to keep a good look out and avoid the danger. *I. C. R. R.* v. *Buckner,* 28 Ill. 299.

1912. CARE OF ONE COGNIZANT OF THE DANGER. It is the duty of a person about to go upon a railroad track to do so cautiously, and to ascertain whether there is danger, especially, if from long employment upon the road at the particular place, he is familiar with its peculiar dangers from the numerous tracks there and their constant use in the switching of cars. *Ch. & N. W. Ry.* v. *Sweeney,* 52 Ill. 325.

1913. Every one is bound to know that a railroad crossing is a dangerous place, and he is guilty of negligence, unless he approaches it as if it were dangerous. *C. & N. W. Ry.* v. *Hatch,* 79 Ill. 137.

1914. If a person drives upon a crossing which he knows to be dangerous, without looking out or listening to ascertain whether a train is approaching, and is struck by one, he is guilty of such negligence as will prevent a recovery against the company, unless it is guilty of gross negligence. *Ib.*

1915. It is not the exercise of ordinary care and prudence for a person to drive on a railroad crossing, known to him to be dangerous, without making an effort to ascertain whether a train is approaching, or whether it is safe to drive on the track with his team. *Ib.*

1916. EXCUSE FOR WANT OF CARE. It is the duty of a person approaching a railroad crossing, to carefully look out for approaching trains. although the signals required by law are not given; and it is gross negligence to omit this precaution. *C. & A. R. R.* v. *Robinson,* 8 Bradw. 140.

1917. The failure to ring a bell or sound a whistle, or to clear the track of obstructions upon it, does not exempt the traveler on the highway from the exercise of proper care on his part. *C. & A. R. R.* v. *Robinson,* 9 Bradw. 89.

1918. The neglect of a railway company to give proper warning of an approaching train at a highway crossing, will not justify a person at such crossing, from omitting any proper act of vigilance to avoid a collision. *Ch. & R. I. R. R.* v. *Still,* 19 Ill. 499.

1919. There is nothing that can relieve a person from the duty of using due care and caution at a railroad crossing of a public highway. It is error to instruct that if the train was behind time, this excused the plaintiff from using the same care and caution required of him, had the train been on time. *T., W. & W. Ry.* v. *Jones,* 76 Ill. 311.

1920. Notwithstanding the neglect to give the statutory signal before approaching a road crossing with a train, the traveler must exercise prudence and caution; but without such warning of danger his care would necessarily be less, and any injury to him under such circumstances, must naturally be attributed in a great degree to the negligence of the company. *C. & A. R. R.* v. *Elmore,* 67 Ill. 176.

1921. The fact that the view of the track may be obstructed by other cars left standing on the side track, does not lessen the caution required of a person attempting to cross the same, but imposes upon him the duty of exercising a higher degree of diligence. *Garland* v: *Ch. & N. W. Ry.*, 8 Bradw. 571.

1922. Where a person on approaching a railroad crossing with a wagon and team, does not avail himself of his sight and hearing, when by the proper exercise thereof, he could have avoided a collision with a train at the crossing, he will be regarded as grossly negligent on his part, and cannot recover for the injury resulting, where the only neglect of the company, was the failure to give the required signal on approaching the crossing. *St. L., A. & T. H. R. R.* v. *Manley*, 58 Ill. 300.

1923. Where a person knows he is approaching a railroad crossing, whether in a city or elsewhere, it is his duty, if possible, to observe the usual and proper precautions, by looking in either direction, and watching for the usual signals of danger, before attempting to cross; and when it appears from direct testimony, or from facts and circumstances that the party was injured from a want of these precautions, he cannot recover, however serious the injury he may receive. *C., B. & Q. R. R.* v. *Van Patten*, 64 Ill. 510.

1924. It is the duty of a person coming upon a railroad crossing of the highway, to use care and caution to avoid a collision with any passing train, and to use precaution before going thereon, to ascertain whether there is a train approaching; and the failure to ring a bell or sound the whistle does not exempt travelers on highways from this duty. *C., B. & Q. R. R.* v. *Harwood*, 80 Ill. 88.

1925. STOPPING BEFORE TRYING TO CROSS. A person about to cross a railroad is not as a *matter of law* required to *stop* as well as to look and listen before attempting to cross, but he must exercise a degree of care proportioned to the danger, and whether he is bound to stop, is a question of fact. *Garland* v. *Ch. & N. W. Ry.*, 8 Bradw. 571.

1926. A person approaching a railroad track is not required to get out of his buggy and go to the track, or stand up in order to get a better view. This would be to require extraordinary care. *C. B. & Q. R. R.* v. *McGaha*, 19 Bradw. 342.

1927. Where the plaintiff carelessly walked upon the track of a railroad a few steps south of an approaching train, without looking north to see if there was danger, and paid so little heed as not to hear the bell or whistle when sounded, or notice the calls of persons warning him of danger, and was run over by the engine not moving at a high rate of speed, and there was no proof that the servants of the company wantonly or willfully caused the injury, it was *held* that the plaintiff's negligence was so gross as to preclude a recovery. *L. S. & M. S. R. R.* v. *Hart*, 87 Ill. 529.

1928. Where a person got in close proximity to a side track of a railroad and was walking along the same, where he was struck by a yard engine and killed, and it appeared that he was well acquainted with the locality, and placed himself in this dangerous position when the approaching engine was very near him, without looking back to see if any engine was on the track, and that the engine was too close to him when he got near the track to be stopped: *Held*, that his negligence was so great as to preclude any recovery against the company. *Austin* v. *C., R. & I. P. R. R.*, 91 Ill. 35.

1929. If a traveler on the highway had notice of an approaching train in time to avoid a collision, the object of giving the signal is subserved, and the failure to give them, or either of them, cannot be held to be the cause of an injury resulting from a collision under such circumstances. *C., R. I. & P. R. R.* v. *Bell*, 70 Ill. 102.

1930. WHERE TRAVELER IS WITHOUT NEGLIGENCE. Where a person in a buggy stops before attempting to cross a railroad, and looks and listens for an approaching train, and no warning is given him by bell or whistle, and the view is obstructed by brush, &c., and he is injured by a passing train, he will be guilty of no negligence, and the company being guilty of gross negligence will be liable. *C., B. & Q. R. R.* v. *McGaha*, 19 Bradw. 342.

1931. Where a person on approaching a railroad crossing looks and listens for an approaching train before passing a cornfield which obstructed the view, and after passing the same again looks and listens, and no warning is given him by bell or whistle, he will be guilty of no negligence on his part in going upon the track, and the fact that he is told to stop, that the cars are coming, which he fails to hear, will not change the rule. *Dimick* v. *Ch. & N. W. Ry.*, 80 Ill. 338.

1932. WALKING ON TRACK. It is negligence for a person to walk upon the track of a railroad, whether laid in a street, or an open field, and he who deliberately does so, will be presumed to assume the risk of the peril he may encounter. *I. C. R. R.* v. *Hall*, 72 Ill. 222.

1933. A higher degree of care and caution will be required of a person who is without right traveling on foot along a railroad track than of a traveler crossing the track upon a highway. *L., S. & M. S. R. R.* v. *Hart*, 87 Ill. 529.

1934. A person crossing a railroad track who could have seen the cars approaching, but turned his back to that direction, and had his ears so bandaged that he could not hear, is guilty of such negligence as will prevent his recovery for injuries, unless he can prove a greater degree of negligence on the part of the railway company. *Ch. & R. I. R. R.* v. *Still*, 19 Ill. 499.

1935. The plaintiff to recover for an injury resulting from a failure to give the statutory signal, must have exercised such care as might be expected of prudent men generally under like circumstances. Overruling *C. & A. R. R.* v. *Elmore*, 67 Ill. 176; *W., St. L. & P. Ry.* v. *Wallace*, 110 Ill. 114.

CONTRIBUTORY NEGLIGENCE GENERALLY.

1936. WHEN NEGLIGENCE OF A PLAINTIFF IS A BAR — *gross negligence of plaintiff.* If the plaintiff has been guilty of gross negligence contributing to his injury he cannot recover. *C., R. I. & P. R. R.* v. *Dingman*, 1 B. 162; *L. S. & M. S. R. R.* v. *Sunderland*, 2 B. 307; *L. S. & M. S. R. R.* v. *Roy*, 5 B. 82; *President, &c.*, v. *Carter*, 6 B. 421; *C., B. & Q. R. R.* v. *Olson*, 12 Brw. 245; *P., C. & St. L. Ry.* v. *Goss*, 13 Brad. 619; *L. S. & M. S. R. R.* v. *Hunt*, 18 B. 288.

1937. If the plaintiff fails to use ordinary care and this contributes to his injury, he cannot recover for mere negligence on the part of the defendant. *President, &c.*, v. *Carter*, 2 B. 34.

1938. To recover for an injury from a defective sidewalk, the plaintiff must have exercised ordinary care to avoid the injury. *Chicago* v. *Watson*, 6 B. 344; *Macomb* v. *Smithers*, 6 B. 470.

1939. One who knowingly exposes himself to danger which could readily have been avoided must attribute his injury to his own negligence. *Bloomington* v. *Read*, 2 B. 542.

1940. A plaintiff injured by his own negligence and not that of the defendant, cannot recover. *U. Ry. & T. Co.* v. *Leahey*, 9 B. 353; *Armour* v. *McFadden*, 9 B. 508.

1941. Plaintiff must show that his own negligence or misconduct has not concurred in producing the injury. *Aurora Branch R. R.* v. *Grimes*, 13 Ill. 585.

1942. Where the plaintiff is in the wrong, or not in the exercise of a legal right, or is enjoying a favor or privilege without compensation, he must use extraordinary care, before he can complain of negligence in another. *C. & A. R. R.* v. *McKenna*, 14 B. 472; *I. C. R. R.* v. *Godfrey*, 71 Ill. 500.

1943. One may go upon a sidewalk known to be out of repair and dangerous, and if injured, may recover, if ordinary and reasonable care is used. *Joliet* v. *Conway*, 17 Bradw. 577.

1944. Where the gravamen of the action is mere negligence there can be no recovery, where there is a want of ordinary care by the plaintiff to avoid the injury. *C., B. & Q. R. R.* v. *Rogers*, 17 B. 638.

1945. To entitle a plaintiff to recover for injury from negligence, there must have been no want of ordinary care on his part. *W., St. L. & P. Ry.* v. *Moran*, 13 B. 72; *C., B. & Q. R. R.* v. *Rogers*, 17 B. 638; *C., B. & Q. R. R.* v. *Dougherty*, 12 B. 181; *C., B. & Q. R. R.* v. *Colwell*, 3 B. 545; *Garfield Manf. Co.* v. *McLean*, 18 B. 447; *Gardner* v. *C., R. I. & P. R. R.*, 17 B. 262; *Dyer* v. *Talcott*, 16 Ill. 300; *G. & C. U. R. R.* v. *Fay*, 16 Ill. 558; *C., B. & Q. R. R.* v. *Van Patten*, 64 Ill. 516, 517.

1946. If the injured party alone is in fault and the accident is the result of his own negligence, he cannot recover. *St. L., A. & T. H. R. R.* v. *Manly*, 58 Ill. 300, 306.

1947. If the plaintiff's negligence is the primary cause of the injury, and the defendant is guilty of no want of ordinary care, no recovery can be had. *R., R. I. & St. L. R. R.* v. *Coultas*, 67 Ill. 398, 401.

1948. Where both parties are equally in the position of right, the plaintiff is only bound to show that his injury was produced by the negligence of the defendant, and that he exercised ordinary care and diligence to avoid it. *I. C. R. R.* v. *Godfrey*, 71 Ill. 500.

1949. Although the defendant's negligence may have been the prime cause of the injury, yet if the plaintiff by the exercise of due care, might have avoided the injury, and his negligence is slight and that of the defendant gross, when compared, the plaintiff cannot recover. *St. L. & S. E. Ry.* v. *Britz*, 72 Ill. 256.

1950. There must be fault on the part of the defendant and no want of ordinary care on the part of the plaintiff, to entitle him to recover. *G. T., M. & T. Co.* v. *Hawkins*, 72 Ill. 386, 388.

1951. Although there may be negligence on the part of the defendant, yet if there is also negligence on the part of the plaintiff, but for which the injury would not have been received, or if the plaintiff by the exercise of ordinary care and caution, could have avoided the injury and he failed to exercise it, he cannot recover. Exceptions to rule stated. *C. & A. R. R.* v. *Becker*, 76 Ill. 25.

1952. The negligence of a parent will not excuse the carrier by rail from using all the means in its power to prevent injury to the child. But where the negligence of the parent is the proximate cause of the injury, the carrier will not be responsible, unless it omits duties which might have averted the *injury*. *O. & M. Ry.* v. *Stratton*, 78 Ill. 88.

1953. A party driving upon a railroad track without looking out for an approaching train, is guilty of such gross negligence as to bar his action, unless the company is guilty of gross negligence. *Ch. & N. W. Ry.* v. *Hatch*, 79 Ill. 137.

1954. Except where the injury has been willfully or wantonly inflicted, it is an essential element to a right of recovery that the plaintiff or person injured, must have exercised ordinary care to avert the injury. *Litchfield Coal Co.* v. *Taylor*, 81 Ill. 590. •

1955. It is a requisite to the liability of a railway company as a passenger carrier, that the passenger shall not have been guilty of any want of ordinary care and prudence which directly contributed to the injury. *I. C. R. R.* v. *Green*, 81 Ill. 19.

1956. The negligence of a plaintiff which will prevent a recovery for an injury resulting from the defendant's negligence, must be such as contributes to the injury. *I. & St. L. R. R.* v. *Herndon*, 81 Ill. 143.

1957. Before a recovery can be had by a party falling into an excavation in a sidewalk, not properly protected, he must show he used due care for his safety. *Kepperly* v. *Ramsden*, 83 Ill. 354.

1958. If a passenger on a train without the direction of the company's servant leaves his seat in a passenger coach and goes into a baggage car, where he is killed, he will be guilty of such a high degree of negligence as to defeat a recovery by his personal representative against the company, unless the latter is guilty of wanton or reckless misconduct. *P. & R. I. R. R.* v. *Lane*, 83 Ill. 448.

1959. If the conduct of one killed while walking upon a railroad track amounts to gross negligence, no recovery can be had of the company, unless it was guilty of willful or criminal negligence. *I. C. R. R.* v. *Hetherington*, 83 Ill. 510.

1960. Where the plaintiff is guilty of gross negligence, he cannot recover unless the injury was wantonly or willfully caused. *L. S. & M. S. R. R.* v. *Hart*, 87 Ill. 529.

1961. Walking upon a railroad track without looking in both directions to see if a train is approaching, when such precaution would have discovered the same, is such negligence as to bar a recovery, unless the injury be willfully or wantonly inflicted. *Austin* v. *C., R. I. & P. R. R.*, 91 Ill. 35.

1962. If it appears that the plaintiff was himself guilty of gross negligence in respect to the injury complained of, he cannot recover. *C. & N. W. Ry.* v. *Dimick*, 96 Ill. 42.

1963. Gross negligence of the plaintiff contributing to the injury, is a bar to a recovery. *C. & N. Ry.* v. *Scates*, 90 Ill., 586; *I. C. R. R.* v. *Patterson*, 93 Ill. 290; *C., B. & Q. R. R.* v. *Warner*, 108 Ill. 538; *Simmons* v. *Ch. & Tomah R. R.*, 110 Ill. 340; *Abend* v. *T. H. & I. R. R.*, 111 Ill. 202; *C. & N. W. Ry.* v. *Snyder*, 117 Ill. 376; *Penn* v. *Hankey*, 93 Ill. 580.

1964. In the absence of ordinary care on the part of the plaintiff, he cannot recover for an injury caused by mere negligence, as distinguished from the willful tort of the defendant. *C., B. & Q. R. R.* v. *Johnson*, 103 Ill. 512.

1965. A plaintiff cannot recover for an injury caused even in part by his own fault in failing to use ordinary care in being treated and cured of his injuries. *Pullman Palace Car Co.* v. *Bluhm*, 109 Ill. 20.

1966. A person who voluntarily and unnecessarily places himself in a well known place of danger, but for which he would not have been injured, and he is injured or killed in consequence of such exposure, even through the gross negligence of the defendant, if the act of the latter was not wanton or willful, is guilty of such contributory negligence as to preclude a recovery. *Abend* v. *T. H. & I. R. R.*, 111 Ill. 202.

1967. If a party's negligence materially contributes to the injury, whether it contributes to the injury or to the force causing the injury, or not, it will bar a recovery. *Ib.*

1968. Where the person killed, by the use of ordinary care, could have avoided the injury, and he failed to do so, no recovery can be had. *Myers* v. *I. & St. L. R.R.*, 113 Ill. 386.

1969. In order to recover for injury from negligence, it must be shown that the injured party was at the time he was injured, observing due or ordinary care for his safety, and that while exercising such care, he was injured by the negligence of the defendant. *Calumet Iron & Steel Co.* v. *Martin*, 115 Ill. 358.

1970. Allowing a child three years old to go upon the streets, is not such negligence as to bar a recovery for an injury to the child. *Stafford* v. *Rubens*, 115 Ill. 196; *Chicago* v. *Hesing*, 83 Ill. 204.

OF MUTUAL AND COMPARATIVE NEGLIGENCE.

1971. It has never been held by our courts that the negligence of the parties can be weighed in a scale, and if inclined in favor of the plaintiff, that he may recover. *President, &c.* v. *Carter*, 2 B. 34.

1972. Where the plaintiff's slight negligence has contributed to the injury, he cannot recover, unless the defendant's negligence was gross in comparison with his own. *C. & A. R. R.* v. *Langley*, 2 B. 505; *L. S. & M. S. R. R.* v. *Berlink*, 2 B. 427.

1973. Where the plaintiff's own act contributed to the injury, he cannot recover, unless his negligence was slight and that of defendant gross in comparison. *I. C. R. R.* v. *Brookshire*, 3 B. 225.

1974. If the negligence of the injured party was only slight, and that of the defendant in comparison amounts to gross carelessness, the plaintiff may recover; but if the person injured was guilty of gross negligence, no recovery can be had, unless the negligence of the defendant was so gross as to amount to a wanton or willful wrong. *C., B. & Q. R. R.* v. *Colwell*, 3 B. 545.

1975. A plaintiff guilty of slight negligence may recover of a defendant guilty of gross negligence; but it is not enough that the negligence of the defendant should be greater than that of the plaintiff, or that any degree of disparity between the two should exist less than that which is expressed by the terms slight and gross. *N. Ch. Rolling Mills Co.* v. *Monka*, 4 B. 664.

1976. Where both parties are guilty of negligence, the plaintiff cannot recover unless that of the defendant is gross and that of the plaintiff slight in comparison. *Ch. City Ry.* v. *Lewis*, 5 B. 242; *Winchester* v. *Case*, 5 B. 486; *Glover* v. *Gray*, 9 B. 329.

1977. Rule of comparative negligence applies in the use of a tumbling rod. *W., St. L. & P. Ry.* v. *Thompson*, 10 B. 271.

1978. The law of contributory negligence does not authorize the jury to weigh the degrees of negligence and find for the party least in fault. *Wabash Ry.* v. *Jones*, 5 B. 607.

1979. A plaintiff guilty of negligence contributing in a slight degree to the injury, may recover of a defendant who has been guilty of gross negligence, if the negligence of the plaintiff is slight and that of the defendant gross in comparison; and both the terms gross and slight, or their equivalent, should be used in the instructions. *C., B. & Q. R. R.* v. *Avery*, 8 B. 133.

1980. If the plaintiff has exercised ordinary care, and the defendant was negligent, though not to the extent of being grossly so, the plaintiff may recover, although his care was not of that extreme degree denominated great care. *C., B. & Q. R. R.* v. *Dougherty*, 12 B. 181.

1981. A plaintiff who by want of ordinary care has contributed to the injury, cannot recover, no matter what may have been the degree

—17

of the defendant's negligence, provided it does not amount to a willful and intentional wrong. *C., B. & Q. R. R.* v. *Dougherty*, 12 B. 181; *Union Ry. & T.* v. *Kallaher*, 12 B. 400.

1982. The element of comparison is as indispensable to a proper statement of the rule of comparative negligence as are the degrees of the negligence of the respective parties; and an instruction which fails to institute a comparison, is erroneous. *C. & E. I. R. R.* v. *O'Connor*, 13 B. 62.

1983. A plaintiff guilty of negligence contributing to the injury may recover, if his negligence is slight and that of the defendant gross in comparison. But there must be no want of ordinary care on his part. *W., St. L. & P. Ry.* v. *Moran*, 13 B. 72.

1984. If the negligence of the plaintiff is slight and that of the defendant gross, and it so appears when compared with each other, the plaintiff may recover. *First Nat. Bank* v. *Eitemiller*, 14 B. 22.

1985. Before the plaintiff can recover it must appear that his own negligence was no greater than that defined by the law as slight, and that the defendant was guilty of gross negligence. *St. L., A. & T. H. R. R.* v. *Andres*, 16 B. 292.

1986. Before the rule of comparative negligence can have any application it must appear that the plaintiff exercised ordinary care and that the defendant was guilty of gross negligence. *Gardner* v. *C., R. I. & P. Ry.* 17 B. 262.

1987. An instruction which requires the jury to find whether the negligence of the plaintiff was slight and that of defendant gross, but does not require the jury to compare the negligence of the parties, and determine from such comparison whether the one is slight and the other gross, is erroneous. *C. & A. R. R.* v. *Dillon*, 17 B. 355.

1988. The rule of comparative negligence has no application, and cannot be invoked, except in cases where the party injured observed ordinary care with reference to the circumstances involved, for his safety. *Garfield Manf. Co.* v. *McLean*, 18 B. 447.

1989. If the plaintiff was alone in fault, or if both parties were equally in fault, the plaintiff cannot recover. *Aurora Branch R. R.* v. *Grimes*, 13 Ill. 585, 591.

1990. A person guilty of negligence in attempting to cross a railroad track cannot recover for an injury, unless the company has been guilty of negligence or misconduct still more gross and willful than his own. *C. & R. I. R. R.* v. *Still*, 19 Ill. 499.

1991. The plaintiff's negligence must be, as compared with that of the defendant, so much less culpable as to incline the balance in his favor, both being in some fault. *Peoria Bridge Assoc.* v. *Loomis*, 20 Ill. 235, 251.

1992. If the negligence of the plaintiff is only slight and that of the defendant is gross, a recovery may be had. *C., B. & Q. R. R.* v. *Dewey*, 26 Ill. 255, 258.

1993. The degrees of negligence of the parties may be measured, and if that of the plaintiff is comparatively slight and that of the defendant gross, a recovery may be had. *G. & C. U. R. R.* v. *Jacobs*, 20 Ill. 478; *C. B. & Q. R. R.* v. *Hazzard*, 26 Ill. 373, 387.

1994. A plaintiff whose negligence has contributed to the injury, may recover, if the defendant has been guilty of a higher degree of negligence amounting to willful injury. *St. L., A. & T. H. R. R.* v. *Todd*, 36 Ill. 409, 414.

1995. Although the plaintiff may be chargeable with some degree of negligence, yet if it is but slight as compared with that of the defendant, the plaintiff may recover, even where the slight degree of negligence to some extent contributed to the injury. *Coursen* v. *Ely*, 37 Ill. 338; *C. & A. R. R.* v. *Hogarth*, 38 Ill. 370; *C., B. & Q. R. R.* v. *Cauffman*, 38 Ill. 424; *C., B. & Q. R. R.* v. *Triplett*, 38 Ill. 482.

1996. If the owner of property burned by the emission of sparks from an engine is guilty of negligence in failing to take proper precautions to protect the same from fire, he cannot recover of the railway company for its destruction, unless the negligence of the latter is more gross than his own. *Great Western R. R.* v. *Haworth*, 39, Ill. 346.

1997. Railway company not liable for killing stock, if the owner has been guilty of negligence contributing to the injury, unless the company has been guilty of negligence more gross than that of the plaintiff. The jury in such a case may compare the degrees of negligence. *I. C. R. R.* v. *Middlesworth*, 43 Ill. 64.

1998. If the owner of stock killed by a train of cars, while crossing the track is guilty of as great negligence as the company, no recovery can be had for an injury. *O. & M. R. R.* v. *Eaves*, 42 Ill. 288.

1999. To recover, the plaintiff must show that the injury resulted from the negligence, of the defendant, and not from any fault on his part which materially contributed to it; or, if not wholly free from fault himself, that his negligence was slight in comparison with that of the defendant. *Ortmayer* v. *Johnson*, 45 Ill. 469; *C. & A. R. R.* v. *Gretzner*, 46 Ill. 74.

2000. When both parties are at fault the plaintiff may in some cases recover, as where his negligence is slight and that of the defendant is gross. *C. & A. R. R.* v. *Gretzner*, 46 Ill. 74.

2001. This rule holds good even where the plaintiff's slight negligence in some degree contributed to the injury. If the defendant has been guilty of a higher degree of negligence, slight negligence of the plaintiff will not absolve the defendant from the use of all reasonable efforts to avoid the injury. *Ib.*

2002. Negligence on the part of the owner of mules in penning them alongside of a railway fence of the right of way, over which they broke and got upon the track, where they were killed, will not defeat his right to recover for the injury, where the exercise of ordinary care by the company might have prevented the injury. *I. C. R. R.* v. *Middlesworth*, 46 Ill. 494.

2003. Unless the negligence of a railway company in suffering weeds and grass to accumulate on its right of way is greater than that of the adjoining land owner-in the same respect, the latter cannot recover for an injury by fire communicated from an engine. *O. & M. R. R.* v. *Shanefelt*, 47 Ill. 497; *I. C. R. R.* v. *Frazier*, 47 Ill. 505; *Ch. & N. W. Ry.* v. *Simonson*, 54 Ill. 504.

2004. In case of mutual negligence that of the defendant must be so much greater than that of the plaintiff as to clearly predominate, to authorize a recovery. *C., B. & Q. R. R.* v. *Payne*, 49 Ill. 499. (Overruled.)

2005. For an injury received in getting off a steamboat, the plaintiff, if guilty of negligence, cannot recover, unless that of the defendant was much greater. *Keokuk Packet Co.* v. *Henry*, 50 Ill. 264.

2006. Where the degree of negligence of the plaintiff is slight as compared with that of the defendant contributing to the injury, he may recover. *C. & A. R. R.* v. *Pondrom*, 51 Ill. 333.

2007. The negligence of a passenger in resting his arm on a car window with his elbow slightly projecting out, is slight as compared with that of the company in permitting its freight cars to stand so near the passenger track as to injure the arm of the passenger, and a recovery may be had. *Ib.*

2008. Liability for a personal injury does not depend upon the absence of all negligence on the part of the plaintiff or defendant, but upon the relative degree of care, or want of care as manifested by both parties. *C. & N. W. Ry.* v. *Sweeney*, 52 Ill. 325.

2009. In an action for personal injury caused by the negligence of the defendant, the plaintiff cannot recover, if he has been guilty of contributory negligence, unless it is far less in degree than that of the defendant. *C., B & Q. R. R.* v. *Dunn*, 52 Ill. 451.

2010. Where the negligence of the defendant in placing and leaving obstructions upon a sidewalk, is much greater than that of a twelve year old boy, who is injured thereby, the father of such boy may recover for expenses incurred and loss of services. *Kerr* v. *Forgue*, 54 Ill. 482.

2011. If the negligence of a plaintiff is slight in failing to keep his horses up and in his efforts to find them after their escape, and that of a railway company is gross in permitting a gate in its fence to stand open a long time, whereby the horses are injured on its track, the plaintiff may recover. *C. & N. W. Ry.* v. *Harris*, 54 Ill. 528.

2012. Where stock are killed upon a railway track and the injury might have been avoided by ordinary care on the part of the company, it will be liable, even though the stock were upon the track without the fault of the company. *C. & N. W. Ry.* v. *Barrie*, 55 Ill. 226.

2013. Although a person killed by a train of cars may have been guilty of some negligence contributing to the injury, yet if the company was guilty of a higher degree of negligence, with which, when compared, that of the deceased is slight, or greatly disproportionate, his personal representative may recover. *I. C. R. R.* v. *Baches*, 55 Ill. 379.

2014. Where the person killed and the servants of the company are both guilty of gross negligence contributing to the injury, no recovery can be had, as a general rule. If the negligence of each is equal, no recovery can be had. *Ib.*

2015. Where the negligence of the plaintiff contributing to the injury is greater than that of the defendant, the former cannot recover. *W. U. T. Co.* v. *Quinn*, 56 Ill. 319.

2016. If both parties are equally in fault the plaintiff cannot recover. *St. L., A. & T. H. R. R.* v. *Manly*, 58 Ill. 300, 306.

2017. Where there has been fault on both sides, the plaintiff may recover if his negligence is slight and that if the defendant is gross in comparison with that of the plaintiff. The fact that a party has been guilty of some negligence, does not excuse gross negligence, or authorize the other party to recklessly and wantonly destroy his property or commit a personal injury. *St. L., A. & T. H. R. R.* v. *Manly*, 58 Ill. 306.

2018. Where a child not quite five years old was struck by a passing train in a village, which was running at great speed, the child not being old enough to be chargeable with negligence, and its mother's negligence being slight, and the company being guilty of great negligence: *Held*, that the company was liable. *C. & A. R. R.* v. *Gregory*, 58 Ill. 226.

2019. Negligence resulting in injury is comparative, and it is not necessary that the plaintiff shall be free from all negligence, or that he shall exercise the highest possible degree of prudence and caution to entitle him to recover, if it appears that the defendant was guilty of a higher degree of negligence. *C., B. & Q. R. R.* v. *Payne*, 59 Ill. 534.

2020. An instruction that if the bell, &c., was not sounded as required by law, the plaintiff might recover for the killing of her husband, unless he was guilty of a greater degree of negligence, is too broad. The liability of the company should be limited to the injury caused by the neglect to give the statutory warning, and to the fact that the negligence of the deceased must have been slight as compared with that of the company. *C., B. & Q. R. R.* v. *Lee*, 60 Ill. 501.

2021. It is error to instruct that the plaintiff may recover if the negligence of the defendant was greater than his. Where there is mutual negligence, the plaintiff may recover if his is slight when compared with the defendant's; but there must be more than a bare preponderance against the defendant. *C., B. & Q. R. R.* v. *Dunn*, 61 Ill. 385.

2022. If the negligence of the parties producing the injury is equal, or nearly so, or that of the plaintiff is greater than that of the defendant, he cannot recover. *C. & A. R. R.* v. *Murray*, 62 Ill. 326; *C., B. & Q. R. R.* v. *Van Patten*, 64 Ill. 510, 517; *O. & M. R. R.* v. *Eaves*, 42 Ill. 288.

2023. Partial or slight negligence and inattention of the party injured, will not bar a recovery when palpable negligence of the employer is proven. *C. & A. R. R.* v. *Sullivan*, 63 Ill. 293.

2024. A plaintiff while walking along the track of a railway company in a village, was overtaken and struck by an engine, without any head-light and running at a high rate of speed and no bell was rung or whistle sounded: *Held*, that the negligence of the plaintiff was slight, when compared with the gross and criminal negligence of the company. *I. & St. L. R. R.* v. *Galbreath*, 63 Ill. 436.

2025. An instruction to find for the plaintiff if the "defendant was guilty of considerable negligence and plaintiff was guilty of but little negligence," is bad in the use of the words considerable and little in respect to the negligence of the parties. *I. C. R. R.* v. *Shultz*, 64 Ill. 172.

2026. An instruction that if the deceased failed to use ordinary care and prudence in going upon the railroad track, yet if the company was guilty of a greater degree of negligence, the plaintiff might recover, does not state the rule of comparative negligence with sufficient accuracy. It is not the law, that if the person injured is guilty of gross negligence, he may recover on proof of a higher degree of gross negligence on the part of the defendant. *C., B. & Q. R. R.* v. *Van Patten*, 64 Ill. 510.

2027. If the plaintiff alone is guilty of negligence, or the negligence of the parties is equal, or the plaintiff's negligence is gross, no action will lie in his favor, unless the injury is willfully inflicted. *C., B. & Q. R. R.* v. *Lee*, 68 Ill. 576.

2028. The failure to give the statutory signal before reaching a highway crossing, will authorize a less degree of care by a traveller attempting to cross the track, and any injury to him at such crossing must naturally be attributable in a greater degree to the negligence of the company. *C. & A. R. R.* v. *Elmore*, 67 Ill. 176.

2029. An instruction to find for the plaintiff, even if guilty of great negligence, provided the defendant was only guilty of some more negligence, does not state the law correctly. *I. C. R. R.* v. *Maffit*, 67 Ill. 431.

2030. An instruction that the jury may find for the plaintiff, unless his negligence was equal to, or greater than that of the defendant, is not the law, and is erroneous. *I. C. R. R.* v. *Benton*, 69 Ill. 174.

1031. In case of a collision through the negligence of the plaintiff, he cannot recover, even if the company also was in default, unless the company or its servants willfully caused the injury, or was guilty of such negligence or reckless conduct as that ¡the plaintiff's was slight when compared with it. *C. W. Div. Ry.* v. *Bert,* 69 Ill. 388.

2032. Although the plaintiff may have been guilty of some negligence, still if it is slight as compared with that of the defendant, he may recover; but he cannot unless the negligence of the defendant clearly and largely exceeds that of the plaintiff. *Ch. & N. W. Ry.* v. *Clark,* 70 Ill. 276.

2033. It is not sufficient to entitle the plaintiff to recover where he has been guilty of contributory negligence, that there is a mere preponderance in the degrees of negligence against the defendant. *Ib.*

2034. In an action for causing death by negligence, if the company was guilty of negligence, and the deceased used ordinary care, or was guilty of slight negligence in comparison with that of the company, which ¦was gross, a recovery may be had. *I. C. R. R.* v. *Cragin,* 71 Ill. 177.

2035. Where a person killed by a train of cars was guilty only of slight negligence as compared with that of the company, which was gross, a recovery may be had. *P., C. & St. L. Ry.* v. *Knutson,* 69 Ill. 103.

2036. This court has never held that a plaintiff may recover on account of the negligence of the defendant being greater than his. *C. & A. R. R.* v. *Mock,* 72 Ill. 141.

2037. Where the plaintiff's own negligence was the cause of the injury, or the negligence of the parties is equal, or nearly so, there can be no recovery. It is only where the negligence of the plaintiff is slight in comparison, and that of the defendant is gross, that a recovery is warranted, except where the injury is willfully inflicted. *I. C. R. R.* v. *Hall,* 72 Ill. 222.

2038. A plaintiff free from all negligence may recover for an injury resulting from negligence of the defendant, or a plaintiff who is even guilty of slight negligence, may recover of a defendant who has been grossly negligent, or whose conduct has been wanton or willful. *I. C. R. R.* v. *Hammer,* 72 Ill. 347.

2039. Where a father sues for an injury to his child, his conduct must be free from blame, or his negligence, at least, should be slight, and that of the defendant gross, to entitle him to recover. *Hund* v. *Geier,* 72 Ill. 393.

2040. Where there is evidence of contributory negligence on the part of the plaintiff, it is improper to give an instruction which assumes that a mere preponderance of negligence on the part of the defendant, will entitle the plaintiff to recover. *R., R. I. & St. L. R. R.* v. *Irish,* 72 Ill. 404.

2041. Error to instruct that the plaintiff can recover if the negligence of the defendant was of a higher degree than that of the plaintiff. *I. C. R. R.* v. *Goddard,* 72 Ill. 567.

2042. Where a servant is employed in a business, and at a place not at all dangerous, and the employer creates a peril at the place where the servant is at work, and the servant in the performance of his regular duty, has occasion to pass where the peril is, and is guilty of negligence in so doing, and is injured, his negligence is slight as compared with that of his employer, which is gross, and the servant may recover. *Fairbank* v. *Haentzsche,* 73 Ill. 236.

2043. Where a party killed was guilty of contributory negligence, his personal representative cannot recover, unless the negligence of the defendant contributing to cause the death was gross, in comparison with which, the negligence of the intestate was slight. *C., B. & Q. R. R.* v. *Van Patten,* 74 Ill. 91.

2044. If a railway company is guilty of gross negligence resulting in the death of a person, and the latter is guilty of only slight negligence, this will give a right of recovery. *T., W. & W. Ry.* v. *O'Connor*, 77 Ill. 391.

2045. An instruction for the defendant that the deceased must have been free from contributory negligence, to authorize a recovery, is too broad. If his negligence was slight and that of the company was gross, it will be liable. *Ib.*

2046. In a case of mutual negligence, it is error to instruct that the plaintiff may recover if the defendant is guilty of more negligence than that of the plaintiff, in causing the injury. *I., B. & W. R. R.* v. *Flanigan*, 77 Ill. 365.

2047. If a person killed was guilty of slight negligence, a recovery may be had, if the railway company was guilty of gross negligence causing the injury. *St. L., V. & T. H. R. R.* v. *Dunn*, 78 Ill. 197, 202.

2048. The omission of proper precautions on the part of a railway company, under some circumstances may relieve the plaintiff from the charge of negligence on his part. *Ib.*

2049. A party driving upon a railroad track without looking for an approaching train is guilty of such negligence as to bar an action for an injury, unless the company is guilty of gross negligence. *C. & N. W. Ry.* v. *Hatch*, 79 Ill. 137.

2050. Where the plaintiff's negligence is comparatively slight and that of the defendant gross, the plaintiff will not be deprived of his action; but even if the negligence of the defendant is gross, yet if the negligence of the plaintiff is not slight as compared with that of defendant, the plaintiff cannot recover. *Sterling Bridge Co.* v. *Pearl*, 80 Ill. 251.

2051. It is error to instruct that a plaintiff who has been guilty of negligence which contributed to the injury is entitled to recover, unless his negligence contributed to a considerable degree to such injury. *Ib.*

2052. Where an action is brought to recover for an injury resulting from negligence of another, which was not wanton or willful, it is an essential element to a recovery that the plaintiff or person injured must have exercised ordinary care to avert the injury; but where the injury has been willfully inflicted, an action lies, although the plaintiff. or party injured, may not have been free from negligence. *Litchfield Coal Co.* v. *Taylor*, 81 Ill. 590.

2053. A plaintiff may recover for the death of his intestate, although the latter was guilty of contributory negligence, provided it was slight, and that of the defendant gross, in comparison with each other; but if the negligence of the intestate was not slight and the defendant's gross when compared, no recovery can be had. *R., R. I. & St. L. R. R.* v. *Delaney*, 82 Ill. 198.

2054. Whether the negligence of a person killed by a railway train was slight as compared with that of defendant, is a question of fact for the jury. *Schmidt* v. *C. & N. W. Ry.*, 83 Ill. 405; *Chicago* v. *Kimball*, 18 B. 240; *Penn. Co.* v. *Frana*, 112 Ill. 398; *Wabash Ry.* v. *Elliott*, 98 Ill. 481.

2055. A mere preponderance of negligence on the part of the defendant is not sufficient to render him liable. Plaintiff's must be slight as compared with defendant's, which must be gross. *Schmidt* v. *C. & N. W. Ry.*, 83 Ill. 405; *R., R. I. & St. L. R. R.* v. *Irish*, 72 Ill. 404; *C. & N. W. Ry.* v. *Clark*, 70 Ill. 276.

2056. It is indispensible to a right of recovery that the injured party shall have exercised ordinary care for the security of his person or property, or that the injury be willfully or wantonly inflicted. *I. C. R. R.* v. *Hetherington,* 83 Ill. 510.

2057. In a case of contributory negligence it is error to instruct that the plaintiff may recover, though guilty of slight negligence, if the defendant's employes fell short in *any* degree of the exercise of that high degree of care as under the circumstances, it was rea-' sonable to have used to prevent the injury. *I. C. R. R.* v. *Hammer,* 85 Ill. 526.

2058 An instruction that even if the plaintiff was guilty of negligence, that fact does not destroy his right to recover, if the negligence of the defendant was so much greater than that of the plaintiff as to clearly preponderate and outweigh it, is clearly erroneous. *Joliet* v. *Seward,* 86 Ill. 402.

2059. Where the negligence of a railway company was gross, even if the person killed, while crossing the track was guilty of negligence in failing to listen and look for a train out of time, which negligence is slight, a recovery may be had. *C. B. & Q. R. R.* v. *Lee,* 87 Ill. 454.

2060. Error to instruct that a plaintiff may recover, unless his negligence contributing to the injury was equal to or greater than that of defendant. *I. & St. L. R. R.* v. *Evans,* 88 Ill. 63; *I. C. R. R.* v. *Benton,* 69 Ill. 174.

2061. In instructing the jury in a case of mutual negligence, both the terms slight and gross should be used. *E. St. L., P. & P. Co.* v. *Hightower,* 92 Ill. 139; *C. B. & Q. R. R.* v. *Avery,* 8 B. 133.

2062. Where there is negligence on the part of a child which is injured, or those having its care, contributing directly to the injury, there can be no recovery, unless such negligence is slight and that of the defendant is gross, in comparison in respect to that which produced the injury. It is not sufficient that the defendant may have been guilty of a greater degree of negligence in that respect. *T., W. & W. Ry.* v. *Grable,* 88 Ill. 441.

2063. Error to instruct that the plaintiff may recover though negligent, provided his negligence is slight in comparison with that of the defendant. *C., B. & Q. R. R.* v. *Harwood,* 90 Ill. 425.

2064. A person struck and injured by a train of cars within the limits of a city at a street crossing, may recover of the company, if at the time of the collision the train was running at an improper rate of speed with reference to the plaintiff's safety, even if he was guilty of slight negligence, provided the negligence of the company was gross, when compared with that of the plaintiff. *Wabash R. R.* v. *Henks,* 91 Ill. 406.

2065. An instruction that the jury cannot find for the plaintiff unless they "believe from the evidence that the injury complained of was caused by the negligence of the defendant and the plaintiff was without fault," is stronger than the law will justify, as ignoring the doctrine of comparative negligence. *O. & M. R. R.* v. *Porter,* 92 Ill. 437.

2066. Where the plaintiff is guilty of contributory negligence he cannot recover, unless it appears that his negligence was slight and that of the defendant gross in comparison with each other; and in instructing both these terms should be used. *E. St. L., P. & P. Co.* v. *Hightower,* 92 Ill. 139.

2067. In an action to recover for the burning of a building placed near the defendant's railroad, by the escape of sparks, an instruction placing the right of recovery alone on the defendant's negligence, and which entirely ignores the question of due care on the part of the plaintiff in trying to save the property, is erroneous. *C. & A. R. R.* v. *Pennell,* 94 Ill. 448.

2068. Improper to state in an instruction that where a person is injured for want of proper care on his part, no action will lie, unless the injury was willfully inflicted by the defendant, or that if it were reasonably possible for the plaintiff under all the circumstances, to have prevented the injury by the exercise of *proper caution*, and if such care would have averted the injury, in such case he was guilty of gross negligence and cannot recover, unless defendant willfully inflicted the injury. *Stratton* v. *C., C. H. Ry.*, 95 Ill. 25.

2069. In a case of mutual negligence, a mere preponderance in degree will not render the defendant liable. It is error to instruct, that although the plaintiff by his own negligence may have contributed to the injury, yet if the negligence of the defendant was of a higher degree, or so much greater than that of the plaintiff, that the negligence of the latter was slight in comparison, the plaintiff may recover. *C. & N. W. Ry.* v. *Dimick*, 96 Ill. 42.

2070. In an action for causing the death of a person, it should be left to the jury by instructions to say whether the negligence of the deceased in passing under a freight car, was slight and that of the agents of the company gross, in obstructing the passage to the depot, and in inviting the deceased to pass under the freight car. *C., B. & Q. R. R.* v. *Sykes*, 96 Ill. 162.

2071. A servant of a railway company, to recover for a personal injury growing out of the negligence of the company, must have used ordinary care on his part, considering his surroundings. *Wabash Ry.* v. *Elliott*, 98 Ill. 481.

2072. Where the question of contributory negligence on the part of the plaintiff's intestate is fairly raised, it is error to ignore entirely that question in the instructions. *N. Ch. Rolling Mill Co.* v. *Morrissey*, 111 Ill. 646; *W., St. L. & P. Ry.* v. *Shacklet*, 105 Ill. 364; *Peoria* v. *Simpson*, 110 Ill. 294.

2073. An instruction that if the plaintiff was guilty of some negligence, but that the defendant was guilty of gross negligence contributing to the injury and that the plaintiff's negligence was slight as compared with that of defendant, a recovery could be had, states the law correctly. *Chicago* v. *Stearns*, 105 Ill. 554.

2074. An instruction speaking of negligence and also gross negligence of the defendant and then referring to the slight negligence of the person injured as compared with the defendant, is too loose and inaccurate in not stating definitely which degree of negligence of the defendant the jury should compare with the injured party's negligence. *C., R. I. & P. R. R.* v. *Clark*, 108 Ill. 113.

2075. An instruction seeming to import that *any* negligence on the part of the plaintiff contributing to his injury, might defeat a recovery, is faulty. It should read that if the plaintiff so far contributed to the injury by his own negligence, or want of *ordinary* care and caution, he could not recover. *C., B. & Q. R. R.* v. *Avery*, 109 Ill. 314.

2076. Although a bell is not rung or whistle sounded, at a public crossing, still a party claiming to recover for an injury in consequence of such omission, must have used due care and caution. To recover he is required to exercise such care as might be expected of prudent men generally under like circumstances. *W., St. L. & P. Ry.* v. *Wallace*, 110 Ill. 114.

2077. The law does not require a servant working in a dangerous place to use the highest degree of care and caution, to entitle him to recover for an injury received from the negligence of other servants or agents of his employer. *L. S. & M. S. R. R.* v. *O'Connor*, 115 Ill. 254.

2078. If a fireman of a railway company while in the discharge of his duty, and using ordinary care for his safety, is injured and killed by the explosion of a boiler of the company, and the explosion was the result of the defendant's negligence, the fireman's personal representative may recover. *Calumet Iron and Steel Co.* v. *Martin*, 115 Ill. 358.

2079. Proof of the negligent acts or omissions of the defendant charged, which fails to show negligence in the plaintiff, makes a *prima facie* right of recovery, and it is then incumbent on the defendant to show such negligence on the part of the plaintiff as will defeat a recovery, or to give proof tending to show it, in order to warrant the court in giving an instruction relating to contributory negligence. *U. S. Rolling Stock Co.* v. *Wilder*, 116 Ill. 100.

2080. It is not necessary in an instruction to state the law of comparative negligence as applicable to an infant incapable of observing ordinary care. *Ch., St. L. & P. R. R.* v. *Welsh*, 118 Ill. 572.

2081. Although a plaintiff may have been guilty of some negligence contributing to the injury, he may nevertheless recover, if his negligence was slight when compared with that of the defendant which was gross. (This applies generally to cases involving injury to person or property, and to actions for death by negligence.) *C., R. I. & P. R. R.* v. *Dignan*, 56 Ill. 487; *C., B. & Q. R. R.* v. *Gregory*, 58 Ill. 272; *St. L., A. & T. H. R. R.* v. *Manly*, 58 Ill. 306; *C., B. & Q. R. R.* v. *Dunn*, 61 Ill. 385; *I. & St. L. R. R.* v. *Stables*, 62 Ill. 313; *C. & A. R. R.* v. *Murray*, 62 Ill. 326; *C., B. & Q. R. R.* v. *Van Patten*, 64 Ill. 517; *T., W. & W. Ry.* v. *Spencer*, 66 Ill. 528; *I. C. R. R.* v. *Hoffman*, 67 Ill. 287; *R., R. I. & St. L. R. R.* v. *Coultas*, 67 Ill. 398, 401; *I. C. R. R.* v. *Maffit*, 67 Ill. 431; *C., B. & Q. R R.* v. *Lee*, 68 Ill. 576; *I. C. R. R.* v. *Benton*, 69 Ill. 174; *P., C. & St. L. Ry.* v. *Knutson*, 69 Ill. 103; *Ch. W. D. Ry.* v. *Bert*, 69 Ill. 388; *Ch. City Ry.* v. *Lewis*, 5 Bradw. 242; *I. C. R. R.* v. *Cragin*, 71 Ill. 177; *T., W. & W. Ry.* v. *McGinnis*, 71 Ill. 346; *C. & A. R. R.* v. *Mock*, 72 Ill. 141; *I. C. R. R.* v. *Hall*, 72 Ill. 222; *R., R. I. & St. L. R. R.* v. *Hillmer*, 72 Ill. 235; *I. C. R. R.* v. *Hammer*, 72 Ill. 347; *G. T. M. & T. Co.* v. *Hawkins*, 72 Ill. 386; *Hund* v. *Geier*, 72 Ill. 393; *I. C. R. R.* v. *Goddard*, 72 Ill. 567; *Fairbank* v. *Haentzsche*, 73 Ill. 236; *C., B. & Q. R. R.* v. *Van Patten*, 74 Ill. 91; *T., W. & W. Ry.* v. *O'Connor*, 77 Ill. 391; *St. L., V. & T. H. R. R.* v. *Dunn*, 78 Ill. 197; *Kewanee* v. *Depew*, 80 Ill. 119; *Sterling Bridge Co.* v. *Pearl*, 80 Ill. 251; *R., R. I. & St. L. R. R.* v. *Delaney*, 82 Ill. 198; *Schmidt* v. *C. & N. W. Ry.*, 83 Ill. 405; *I. C. R. R.* v. *Hetherington*, 83 Ill. 510; *Quinn* v. *Donovan*, 85 Ill. 194; *I. C. R. R.* v. *Hammer*, 85 Ill. 526; *C., B. & Q. R. R.* v. *Lee*, 87 Ill. 454; *I. & St. L. R. R.* v. *Evans*, 88 Ill. 63; *T., W. & W. Ry.* v. *Grable*, 88 Ill. 441; *C., B. & Q. R. R.* v. *Harwood*, 90 Ill. 425; *Wabash Ry.* v. *Henks*, 91 Ill. 406; *E. St. L., P. & P. Co.* v. *Hightower*, 92 Ill. 139; *I. C. R. R.* v. *Patterson*, 93 Ill. 290; *Stratton* v. *C., C. H. Ry.*, 95 Ill. 25; *C. & N. W. Ry.* v. *Dimick*, 96 Ill. 42; *C., B. & Q. R. R.* v. *Sykes*, 96 Ill. 162; *C., B. & Q. R. R.* v. *Johnson*, 103 Ill. 512; *Chicago* v. *Stearns*, 105 Ill. 554; *W., St. L. & P. Ry.* v. *Wallace*, 110 Ill. 114; *C. & A. R. R.* v. *Johnson*, 116 Ill. 206; *L. S. & M. S. R. R.* v. *Berlink*, 2 B. 427; *C. & A. R. R.* v. *Langley*, 2 B. 505; *I. C. R. R.* v. *Brookshire*, 3 B. 225; *Winchester* v. *Case*, 5 B. 486; *C., B. & Q. R. R.* v. *Avery*, 8 B. 133; *Glover* v. *Gray*, 9 B. 329; *W., St. L. & P. Ry.* v. *Moran*, 13 B. 72; *St. L., A. & T. H. R. R.* v. *Andres*, 16 B. 292; *Ch., St. L. & P. R. R.* v. *Welsh*, 118 Ill. 572; *Ch. & E. Ill. R. R.* v. *O'Connor*, 119 Ill. 586; *Ch. & N. W. Ry.* v. *Goebel*, 119 Ill. 515; *Union Ry. & Transit Co.* v. *Shacklet*, 119 Ill. 232; *Ch., St. L. & P. R. R.* v. *Hutchinson*, 120 Ill. 587.

WILLFUL INJURIES.

2082. A party failing to exercise ordinary care may recover for an injury wantonly or willfully inflicted, or resulting from gross culpable and criminal negligence amounting to a wanton, reckless or willful wrong. *C. & N. W. Ry.* v. *Clark,* 2 B. 116; *C., B. & Q. R. R.* v. *Colwell,* 3 B. 545; *C., B. & Q. R. R.* v. *Dougherty,* 12 B. 181; *Union Ry. & Tr. Co.* v. *Kallaher,* 12 B. 400; *C. B. & Q. R. R.* v. *Triplett,* 38 Ill. 482; *St. L., A. & T. H. R. R.* v. *Todd,* 36 Ill. 414; *St. L., A. & T. H. R. R.* v. *Manly,* 58 Ill. 306; *C., B. & Q. R. R.* v. *Payne,* 59 Ill. 534; *C. & A. R. R.* v. *Wilson,* 63 Ill. 167; *I. & St. L. R. R.* v. *Galbreath,* 63 Ill. 436; *C., B. & Q. R. R.* v. *Lee,* 68 Ill. 576; *C. W. D. Ry.* v. *Bert,* 69 Ill. 388; *I. C. R. R.* v. *Godfrey,* 71 Ill. 500; *I. C. R. R.* v. *Hall,* 72 Ill. 222; *I. C. R. R.* v. *Hammer,* 72 Ill. 347; *Litchfield Coal Co.* v. *Taylor,* 81 Ill. 590; *I. C. R. R.* v. *Hetherington,* 83 Ill. 510; *C., B. & Q. R. R.* v. *Dickson,* 88 Ill. 431; *Austin* v. *C., R. I. & P. R. R.,* 91 Ill. 35; *Stratton* v. *C. C. H. Ry.* 95 Ill. 25.

2083. The fact that the owner of stock may be guilty of negligence in permitting the same to go at large, or to break through the fencing and get upon the railroad track, will not relieve the railway company from its duty to use ordinary care to avoid injuring them, or defeat an action for killing or injuring them, where this might have been prevented by the exercise of ordinary care. *Ch. & Miss. R. R.* v. *Patchin,* 16 Ill. 198; *Great Western R. R.* v. *Thompson,* 17 Ill. 131; *C.M.T. R. R.* v. *Rockafellow,* 17 Ill. 541; *J. & N.I.R.R.* v. *Jones,* 20 Ill. 221; *G. & C. U. R. R.* v. *Crawford,* 25 Ill. 529; *C.,B. & Q.R.R.* v. *Cauffman,* 28 Ill. 513; *I. & C. R. R.* v. *Phelps,* 29 Ill. 447; *I.C.R.R.* v. *Goodwin,* 30 Ill. 117; *G. W.R.R.* v. *Morthland,* 30 Ill. 451; *St.L., A. & T.H.R. R.* v. *Todd,* 36 Ill. 409; *C.,B.&Q. R.R.* v. *Cauffman,* 38 Ill. 424; *I.C. R.R.* v. *Whalen,* 42 Ill. 396; *I.C.R.R.* v. *Wren,* 43 Ill. 78; *I.C.R.R.* v. *Middlesworth,* 46 Ill. 494; *I.C.R.R.* v. *Baker,* 47 Ill. 295; *T.P.& W. Ry.* v. *Bray,* 57 Ill. 514; *T.P.& W.Ry.* v. *Ingraham,* 58 Ill. 120; *R.R. I.& St.L.R.R.* v. *Lewis,* 58 Ill. 49; *C., B. & Q.R.R.* v. *Van Patten,* 64 Ill. 517; *St. L.A.& T.H.R.R.* v. *Manly,* 58 Ill. 303; *C. & A. R. R.* v. *McMorrow,* 67 Ill. 218; *C. & A. R. R.* v. *Becker,* 76 Ill. 25; *P., P. & J. R.R.* v. *Champ,* 75 Ill. 577.

2084. KILLING STOCK—FRIGHTENING TEAM. § 6½. Any engineer, or person having charge of and running any railroad engine or locomotive, who shall willfully or maliciously kill, wound or disfigure any horse, cow, mule, hog, sheep or other useful animal, shall, upon conviction, be fined in the sum of not less than the value of the property so killed, wounded or disfigured, or confined in the county jail for a period of not less than ten days; and any such engineer or fireman, or other person, who shall wantonly or unnecessarily blow the engine whistle, so as to frighten any team, shall be liable to a fine of not less than $10 nor more than $50. [See Criminal Code, ch. 38, § 191, of R. S. 1887. R. S. 1887, p. 1014, § 69; S. & C., p. 1937, § 69; Cothran, p. 1154, § 55. See ante, 175.]

2085. If the alarm whistle is needlessly sounded in the rear of a team traveling in a narrow lane near the railroad track and thereby causes the team to run away and injures the plaintiff, the company will be liable for the injury. *C., B. & Q. R. R.* v. *Dickson,* 88 Ill. 431.

2086. As to liability for carelessly and recklessly sounding the whistle at an improper place, or where not required, resulting in injury. See also *T., W. & W. Ry.* v. *Harmon,* 47 Ill. 298; *P.,W.& B. R. R.* v. *Stinger,* 78 Pa. St. 219; *P. & R. R. R.* v. *Killips,* 88 Pa. St. 405; *Georgia R. R.* v. *Newsome,* 60 Ga. 492; *Penn. R. R.* v. *Barnett,* 59 Pa. St. 259; *C. & N. W. Ry.* v. *Clark,* 2 B. 116; *Hill* v. *P. & R. R. R.,*

55 Me. 438. See also Crim. Code; S. & C., § 243. *Hudson* v. *L. & N. R. R.*, 14 Bush. 303; *Billman* v. *I.*, *C. & L. R. R.*, 76 Ind. 176; *City of Joliet* v. *Seward*, 86 Ill. 402.

2086a. For an injury caused by the blowing of a whistle at a proper time and place, but so near a team as to cause it to run away with the plaintiff in his wagon, exemplary damages not proper. *C., B. & Q. R. R.* v. *Dunn*, 52 Ill. 451. See *W., St. L. & P. Ry.*, v. *Thompson*, 15 Bradw. 118; 10 *Id.* 271.

2087. STARTING TRAIN WITHOUT SIGNAL. § 7. If any engineer on any railroad shall start his train at any station, or within any city, incorporated town or village, without ringing the bell or sounding the whistle a reasonable time before starting, he shall forfeit a sum not less than $10 nor more than $100, to be recovered in an action of debt in the name of the people of the state of Illinois, and such corporation shall also forfeit a like sum, to be recovered in the same manner. [R. S. 1887, p. 1014, § 70; S. & C., p. 1937, § 70; Cothran, p. 1154, § 56.]

2088. Not the duty of yard master to give any signal before loosing the brakes of a freight car, although boys are near by. *C. & A. R. R.* v. *McLaughlin*, 47 Ill. 265.

2088a. Injury to servant unloading iron from car, by running train on switch without giving any signal, no recovery because a fellow servant. *C. & A. R. R.* v. *Keefe*, 47 Ill. 108. Not duty of yard master to give signal before loosing brakes on freight car. *C. & A. R. R.* v. *McLaughlin*, 47 Ill. 265. Train on a high trestle, starting with a jerk. *R., R. I. & St. L. R. R.* v. *Coultas*, 67 Ill. 398; *I. C. R. R.* v. *Green*, 81 Ill. 19. Setting a car in motion negligently. *Noble* v. *Cunningham*, 74 Ill. 51. Starting street car as passenger is getting off without notice. *Ch. W. Div. Ry.* v. *Mills*, 91 Ill. 39; *Ch. City. Ry.* v. *Mumford*, 97 Ill. 560; *Ch. W. Div. Ry.* v. *Mills*, 105 Ill. 63. Starting car suddenly without notice to servant attempting to get on same. *Ch. & W. Ind. R. R.* v. *Bingenheimer*, 116 Ill. 226; same case, 14 Bradw. 125.

2089. APPROACHES AT CROSSINGS. § 8. Hereafter, at all of the railroad crossings of highways and streets in this state, the several railroad corporations in this state shall construct and maintain said crossings and the approaches thereto, within their, respective rights of way, so that at all times they shall be safe as to persons and property. [L. 1869, p. 312, § 1, (re-written); R. S. 1887, p. 1014, § 71; S. & C., p. 1937, § 71; Cothran, p. 1154, § 57. See ante, 118, 145, 149.]

2090. The option vested in a railway company by charter to change highway crossings will not be interfered with by a court of equity where the company has not failed to exercise proper care, skill and precaution. *I. C. R. R.* v. *Bentley*, 64 Ill. 438.

2091. In the absence of statutory enactment, a railway company is under no obligation to leave every highway that it crosses in a safe condition. *People* v. *C. & A. R. R.*, 67 Ill. 118.

2092. Duty by charter to make suitable crossings and keep the same in repair is binding on the successor of the company. *Ib.*

2093. Change in crossing by consent of the public authorities does not relieve company of this duty. *Ib.*

2094. Duty to restore stream crossed to its former state. *C., R. I. & P. R. R.* v. *Moffit,* 75 Ill.524; *T., W. & W. Ry.* v. *Morrison,* 71 Ill.616.

2095. Company chargeable with notice of all perilous circumstances of a crossing made by it. *R., R. I. & St. L. R. R.* v. *Hillmer,* 72 Ill. 235; *I. & St. L. R. R.* v. *Stables,* 62 Ill. 313.

2096. City cannot extend a new street and require a railway company to construct a crossing over the same, without regard to benefits. *I. C. R: R.* v. *Bloomington,* 76 Ill. 447.

2097. Although the company may be in fault in not making a road crossing safe, it will not be liable for the death of one killed at such crossing, if he was guilty of gross negligence. *R., R. I. & St. L. R. R.* v. *Byam,* 80 Ill. 528.

2097a. Action against city for a personal injury from defective street crossing. *Peru* v. *French,* 55 Ill. 317; *Centralia* v. *Scott,* 59 Ill. 129. Where duty of company to make a bridge over highway. *I. & St. L. R. R.* v. *Stables,* 62 Ill. 313. As to liability of carriers and warehousemen for not keeping safe approaches to their cars and places of business. See *Buckingham* v. *Fisher,* 70 Ill. 121.

2098. NEGLECT TO MAKE, ETC., CROSSINGS—NOTICE. § 9. Whenever any railroad corporation shall neglect to construct and maintain any of its crossings and approaches, as provided in section 8 of this act, it shall be the duty of the proper public authorities, having the charge of such highways or streets, to notify, in writing, the nearest agent of said railroad corporation of the condition of said crossing or approaches, and direct the same to be constructed, altered or repaired in such manner as they shall deem necessary for the safety of persons and property. [R. S. 1887, p. 1015, § 72; S. & C., p. 1938, § 72; Cothran, p. 1154, § 58.]

2099. ROAD CROSSINGS—FAILURE TO CONSTRUCT OR REPAIR. § 10. If any railroad corporation of this State shall, after having been notified, as provided in section 9 of this act, neglect or refuse to construct, alter or repair such crossing or approaches within thirty days after such notice, then said public authorities shall forthwith cause such construction, alteration or repairs to be made. [R. S. 1887, p. 1015, § 73; S. & C., p. 1938, § 73; Cothran, p. 1154, § 59.]

2100. ROAD CROSSING—PENALTY FOR NEGLECT OF DUTY. § 11. Said railroad corporation shall be holden for all necessary expenses incurred in making such construction, alteration and repairs, and in addition thereto shall be liable to a fine of $100 for such neglect to comply with the requirements of this act, which fine shall be enforced by the said public authorities, in the name of the people of the state of Illinois, before any court of competent jurisdiction in the county. Such fine, when collected, to be paid into the treasury of the authorities enforcing the fine. [R. S. 1887, p. 1015, § 74; S. & C., p. 1938, § 74; Cothran, p. 1154, § 60.]

2101. STOPPAGE BEFORE REACHING RAILROAD CROSSING, ETC. § 12. All trains running on any railroad in this State, when approaching a crossing with another railroad upon the same level, or when approaching a swing or draw bridge, in use as such, shall be brought to a full stop before reaching the same, and within eight hundred (800) feet therefrom, and the engineer or other person in charge of. the engine attached to the train shall positively ascertain that the way is clear and that the train can safely resume its course before proceeding to pass the bridge or crossing. [As amended by act of June 19, 1885. In force July 1, 1885. R. S. 1887, p. 1015, § 75; S. & C., p. 1938, § 75; Cothran, p. 1155, § 61.]

2102. Engine driver having the right to the road at a crossing will be criminally culpable for exercising such right if he knows or has reason to expect a collision. *C. & A. R. R.* v. *R., R. I. & St. L. R. R.,* 72 Ill. 34.

2103. ACTION FOR PENALTY—LIMITATION. § 13. Every engineer or other person having charge of such engine, violating the provisions of the preceding section, shall be liable to a penalty of two hundred dollars for each offense, to be recovered in an action of debt in the name of the people of the state of Illinois, and the corporation on whose road such offense is committed, shall be liable to a penalty of not exceeding two hundred dollars, to be recovered in like manner, the amount so recovered to be paid into the treasury of the county in which the offense occurs, but no recovery shall be had in any case for any offense committed more than sixty days prior to the commencement of the action. The provisions of this and of the preceding section shall extend to and govern all cases of neglect or failure to stop the train as required by law before passing any bridge or railroad crossing, whether occurring before or after the said provisions shall take effect, and no act or part of an act inconsistent with such operation and effect being given to this law shall in any way apply hereto. [As amended by act June 19, 1885. In force July 1, 1885. L. 1885. R. S. 1887, p. 1015, § 76; S. & C., p. 1938, § 76; Cothran, p. 1155, § 62; 3 S. & C., p. 442.]

2104. EFFECT OF CHANGE OF THE LAW—*on action for penalty.* The change of §§ 12 and 13 of the railroad law in respect to the stopping of trains before crossing other railways, and as to the penalty for a failure to perform the duty, had the effect to extinguish all right of action in a suit brought under the old law, although it did not repeal §§ 2 and 4 of ch. 131 of the revised statutes. *Mix* v. *People,* 116 Ill. 502.

2105. The people not bound to unite two causes of action and defeat the jurisdiction of a justice of the peace. *I. & St. L. R. R.* v. *People,* 91 Ill. 452.

2106. A justice of the peace has jurisdiction of an action to recover the penalty provided for in this section of the statute. *Ib.*

2107. A suit in the name of the people of the state of Illinois for the use of C. D. against a railway company to recover the penalty, will not be dismissed because brought for the use of an individual. *Ib.*

2108. In an action to recover the penalty, the neglect of the servants of the company to obey orders, is no defence. *Ib.*

2109. A recovery in such a case is a bar to a future action to recover the same penalty. *Ib.*

2110. If the people recover judgment they are entitled to judgment for costs. *Ib.*

2111. OBSTRUCTION OF HIGHWAY—CARS ON TRACK. § 14. No railroad corporation shall obstruct any public highway by stopping any train upon, or by leaving any car or locomotive engine standing on its track, where the same intersects or crosses such public highways, except for the purpose of receiving or discharging passengers, or to receive the necessary fuel and water, and in no case to exceed ten minutes for each train, car or locomotive engine. [R. S. 1887, p. 1016, § 77; S. & C., p. 1939, § 77; Cothran, p. 1155, § 63.]

2112. Liability of railway company for obstructing a public road to plaintiff's inn. *I. C. R. R.* v *White*, 18 Ill. 164.

2113. Liability of railway company for breach of an ordinance to prevent the obstruction of a highway by leaving freight cars standing on the same. *Gr. Western R R.* v. *Decatur*, 33 Ill. 381; *Ill. C. R. R.* v. *Galena*, 40 Ill. 344; *T., P. & W. Ry.* v. *Town of Chenoa*, 43 Ill. 209.

2114. This section does not apply where the highway is obstructed by cars left standing on the same by strangers, without the knowledge or assent of the company. *Peoria, Decatur & Evansville R. R.* v. *Lyons*, 9 B. 350.

2115. If a railway company unnecessarily obstructs the streets of a town with its cars contrary to an ordinance, it will be liable to the penalty prescribed for so doing. *I. C. R. R.* v. *Galena*, 40 Ill. 344.

2116. Ordinance that "no person shall put or cause to be put in any street, sidewalk or other public place within the city, any dust, dirt, filth, shavings or other rubbish or obstructions of any kind," is broad enough to embrace the obstruction of a street by a railroad company with its cars. *Ib.*

2117. Where an ordinance prohibited a railway company from obstructing a public street, by permitting its cars to remain stationary therein for more than fifteen minutes, but referred to another ordinance, which as copied into the record, bore date subsequently to the first: *Held*, that the town failed to establish a right of recovery. *T., P. & W. Ry.* v. *Chenoa*, 43 Ill. 209.

2117a. The corporation and its engineer or conductor are placed under the same liability as respects the fine, and the liability of the former is not limited by the recovery against the other. *T., W. & W. Ry.* v. *People*, 81 Ill. 141.

2118. STONING, ETC., TRAIN. Any person who shall throw any stone or other hard substance at any railroad car, train or locomotive, shall be deemed guilty of a misdemeanor, and, on conviction thereof, shall be fined in any sum not more than $200, and shall stand committed to the county jail until such fine and costs shall be paid. [R. S. 1887, p. 1016, § 77; S. & C., p. 1939, § 77; Cothran, p. 1155, § 63.]

2119. PENALTY FOR OBSTRUCTING HIGHWAY. § 15. Every engineer or conductor violating the provisions of the preceding section shall, for each offense, forfeit the sum of not less than $10 nor more than $100, to be recovered in an action of debt, in the name of the people of the state of Illinois, for the use of any person who may sue for the same, and the corporation on whose road the offense is committed shall be liable for the like sum. [R. S. 1887, p. 1016, § 78; S. & C., p. 1939, § 78; Cothran, p. 1155, § 64.]

2120. The intention is to subject the engineer, conductor and the company indifferently to a fine of not less than $10, nor more than $100 for the obstruction, and not that the corporation shall be liable for the like sum for which the engineer or conductor shall have been convicted. *T. W. & W. Ry.* v. *People*, 81 Ill. 141.

2121. The state's attorney cannot maintain the suit in the name of the people for his use, for the penalty. *People* v. *Wabash, St. L. & P. R. R.*, 12 B. 263. ·

2122. MINORS TO KEEP OFF CARS. § 17. No person or minor shall climb, jump, step, stand upon, cling to, or in any way attach himself to any locomotive engine or car, either stationary or in motion, upon any part of the track of any railroad, unless in so doing he shall be acting in compliance with law, or by permission, under the lawful rules and regulations of the corporation then owning or managing such railroad. [R. S. 1887, p. 1016, § 79; S. & C., p. 1939, § 79; Cothran, p. 1155, § 65.]

2123. This section applies only to climbing, stepping, standing upon, clinging to or in any way attaching one's self to a locomotive engine or car, either stationary or in motion on the track. It does not affect the question of the negligence of a person who attempts to pass under a freight car on the invitation of the conductor. *C., B. & Q. R. R.* v. *Sykes*, 96 Ill. 162.

2124. A violation of this section will not absolve the company from the consequences of its act, unless directly contributing to the injury. Case of a deaf mute less than fourteen years old injured, *held*, not within the statute. *Lammert* v. *C. & A. R. R.*, 9 B. 388.

2125. RAILROAD AGENT, ETC., TO MAKE COMPLAINT. § 18. Whenever any officer, agent, or employe of any railroad corporation shall have any information that any person or minor has violated any of the provisions of the preceding section, and has thereby endangered himself, or caused reasonable alarm to others, said officer, agent, or employe shall, without unnecessary delay, make complaint of such offense against such person or minor before some justice of the peace. [R. S. 1887, p. 1016, §80; S. & C., p. 1939, §80; Cothran, p. 1155, §66.

2126. PENALTY. § 19. Any person or minor who shall violate any of the provisions of the seventeenth section of this act shall be punished by a fine not exceeding $25, to be recovered in an action of debt, in the name of the people of the state of Illinois, before a justice of the peace, or, upon conviction, by imprisonment in the county jail, or other place of confinement, for a period not exceeding twelve hours. [R. S. 1887, p. 1016, § 81; S. & C., p. 1939, § 81; Cothran, p. 1156, § 67.

2127. A minor under fourteen years of age, presumed ignorant of this law, and therefore proof of capacity to commit crime is necessary. *Lammert v. C. & A. R. R.*, 9 Bradw. 388.

2128. THREE PRECEDING SECTIONS POSTED. § 20. The several railroad corporations in this state shall, without unnecessary delay, cause printed copies of the three preceding sections of this act to be kept posted in conspicuous places at all their stations along their lines of railroad in this state. Every railroad corporation that shall neglect to post, and keep posted, such notices as required by this section, shall, for each offense, forfeit the sum of $50, to be recovered in an action of debt, in the name of the people of the state of Illinois. [R. S. 1887, p. 1016, § 82; S. & C., p. 1940, § 82; Cothran, p. 1156, § 68.]

2129. FREIGHT CARS BEHIND PASSENGER TRAINS—PROHIBITED. § 21. In no train shall freight, merchandise or lumber cars be run in the rear of passenger cars, and if such cars, or any of them, shall be so run, the officer or agent who so directed, or knowingly suffered such arrangement to be made, shall each be deemed guilty of a misdemeanor, and punished accordingly. [2d L. 1849, p. 26, § 37. (Penalty reduced.) R. S. 1887, p. 1016, § 83; S. & C., p. 1940, § 83; Cothran, p. 1156, § 69.]

2130. FURNISHING MEANS OF TRANSPORTATION—KEEPING DEPOTS OPEN AND LIGHTED AND WARMED. § 22. Every railroad corporation in the state shall furnish, start and run cars for the transportation of such passengers and property as shall, within a reasonable time previous thereto, be ready or be offered for transportation at the several stations on its railroads and at the junctions of other railroads, and at such stopping places as may be established for receiving and discharging way-passengers and freights; and shall take, receive, transport and discharge such passengers and property, at, from and to such stations, junctions and places, on and from all trains advertised to stop at the same for passengers and freight, respectively, upon the due payment, or tender of payment of tolls, freight or fare legally authorized therefor, if payment shall be demanded, and such railroad companies shall at all junctions with other railroads, and at all depots where said railroad companies stop their trains regularly to receive and discharge passengers in cities and villages, for at least one-half hour before the arrival of, and one-half hour

—18

after the arrival of any passenger train, cause their respective depots to be open for the reception of passengers; said depots to be kept well lighted and warmed for the space of time aforesaid. [In lieu of 2d Laws of 1849, p. 26, § 35; Laws of 1883, p. 125. R. S. 1887, p. 1017, § 84; S. & C., p. 1940, § 84; Cothran, p. 1156, § 70.

2130a. A railway company is bound to construct its road to and from the several points named in its charter, and when built to run its trains over its entire line in such manner as to afford reasonable facilities for the prompt and efficient transaction of such legitimate business as may be offered on any and every part of the road; and this obligation is binding on its successor. *People* v. *L. & N. R. R.*, 120 Ill. 48.

2130b. Contract to furnish passenger facilities at a particular point is personal and does not bind company's successor. *Ib.*

2131. ACCOMMODATIONS AT STATIONS. Duty of railway companies to furnish safe and convenient platforms and approaches to their passenger coaches. *C. & A. R. R.* v. *Wilson*, 63 Ill. 167. Duty to furnish safe places for alighting from cars. *C., R. I. & P. R. R.* v. *Dingman*, 1 Bradw. 162.

2132. Company setting a passenger down upon a platform used by it in connection with another company, liable for an injury to him while waiting for passage on the other road, caused by want of proper safeguards or lights. *W., St. L. & P. Ry.* v. *Wolff*, 13 Bradw. 437; *Seymour* v. *C., B. & Q. R. R.*, 3 Biss. 43.

2133. Stations and depots must be arranged with care, properly lighted when dark, and made safe and convenient. Ordinary care only required, except in favor of passengers. *T., W. & W. Ry.* v. *Grush*, 67 Ill. 262.

2134. Injury of one looking after freight stepping into a hole in platform, company held liable. *Ib.*

2135. Liability for injury caused by defective platform. *McDonald* v. *Ch. & N. W. R. R.*, 26 Iowa 124; *Dobiecki* v. *Sharp*, 88 N. Y. 203; *Louisville & Nashville R. R.* v. *Wolfe*, 80 Ky. 82: *Held,* liable for not keeping it properly lighted. *Patten* v. *Ch. & N. W. Ry.*, 32 Wis. 524.

2136. Company liable for a personal injury caused by the narrowness of the platform. *Ch. & A. R. R.* v. *Wilson*, 63 Ill. 167.

2137. Liability to servant for injury caused by awning of station house being too near the cars passing. *I. C. R. R.* v. *Welch*, 52 Ill. 183.

2138. As to duty in respect to platform see *C. & N. W. Ry.* v. *Scates*, 90 Ill. 586; *Liscomb* v. *New Jersey R. R. & Transp. Co.* 6 Lans. 75; *Gillis* v. *Penn. R. R.*, 59 Pa. St. 129.

2139. Duty to keep open a safe passage to and from passenger trains. *C. & N. W. Ry.* v. *Coss*, 73 Ill. 394; *Barrett* v. *Black*, 56 Me. 498; *Caswell* v. *B. & W. R. R.*, 98 Mass. 194.

2140. DUTY TO FURNISH CARS FOR TRANSPORTATION. Company not bound under all circumstances to furnish a sufficient number of cars, so that all applying may have seats. *C. & N. W. Ry.* v. *Carroll*, 5 Bradw. 201. Duty to furnish suitable sitting accommodation for its ordinary number of passengers. *Ib.*

2141. Must furnish reasonable and ordinary facilities for transportation, such as will meet the ordinary demands of the public. *G. & C. U. R. R.* v. *Rae*, 18 Ill. 488.; *O. & M. Ry.* v. *People*, 120 Ill. 200.

2142. DELAY IN TRANSPORTING. Liability for delay in transporting freight. *I. C. R. R.* v. *McClellan*, 54 Ill. 58; *T., W. & W. Ry.* v. *Lockhart*, 71 Ill. 627; *I. C. R. R.* v. *Waters*, 41 Ill. 73; Hutchinson on Carriers, § 292.

2143. EXCUSE FOR DELAY. *I. C. R. R.* v. *McClellan*, 54 Ill. 58; *I. & St. L. R. R.* v. *Juntgen*, 10 Bradw. 295; Hutchinson on Carriers, § 293.

2144. Cannot delay transportation of goods already delivered and waiting shipment, in order to receive and forward other goods. *Great Western R. R.* v. *Burns*, 60 Ill. 284; *Ch. & N. W. Ry.* v. *People*, 56 Ill. 365.

2145. PENALTY FOR BREACH OF PRECEDING SECTION. § 23. In case of the refusal of such corporation or railroad company, or its agents, to take, receive and transport any person or property, or to deliver the same within a reasonable time, at their regular or appointed time and place, or to keep their said depots open, lighted and warmed according to the provisions of the preceding section of this act, such corporation or railroad company shall pay to the party aggrieved, treble the amount of damages sustained thereby, with costs of suit; and in addition thereto, said corporation or railroad company shall forfeit a sum of not less than twenty-five dollars nor more than one thousand dollars for each offense, to be recovered in an action of debt, in the name of the people of the state of Illinois—the treble damages for the use of the party aggrieved, and the forfeiture for the use of the school fund of the county in which the offense is committed. [Laws of 2d 1849, p. 26, § 36, as amended by act of 1883, p. 125. In force July 1, 1883. R. S. 1887, p. 1017, § 85; S. & C., p. 1940, § 85; Cothran, p. 1156, § 71.]

2146. The treble damages is in the nature of a penalty and can only be recovered when specially declared for in the manner prescribed by the statute. *I. & St. L. R. R. & Coal Co.* v. *People*, 19 B. 141.

2147. STATUTE TO BE STRICTLY CONSTRUED. The language "as shall within a reasonable time previous thereto *be ready* or *be offered* for transportation," does not include coal in the earth to be dug and raised after the cars are furnished. *Ib.*

2147a. LIMITATION. The two-year limitation law is a good plea to an action seeking treble damages. So held where the treble damages are set up in an amendment to the declaration. *I. & St. L. R. R. & Coal Co.* v. *People*, 19 Bradw. 141.

2148. TEXAS CATTLE. § 23½. In any suit brought for a violation of "An act concerning the transportation of Texas or Cherokee cattle," approved April 16, 1869, the consignor of any live stock, the bringing of which into this state shall constitute the offense created by this act, if he be a citizen of this state, and if not the consignee, if he shall have knowledge of and consent to such consignment, of any such live stock, shall be made a joint defendant with any railroad or transportation company which may be sued for the offense aforesaid, and the said consignor, or consignee, shall suffer jointly any penalty passed upon any such railroad or transportation company for any violation of the act aforesaid. Any action brought for a violation of the act aforesaid, must be commenced within the eighteen months next succeeding the bringing of the cattle into this state, on account of which the action may be brought. Any railroad company who shall transport any Texas, Cherokee or diseased cattle in violation of the aforesaid act, without knowing them to be such, may recover from any consignor or consignee any sum of money it may be compelled by the judgment of any court to pay for the transportation of such cattle, and the record of the judgment against the said company shall, in any suit against any such consignor or consignee, be evidence of the amount of damages to be recovered, with interest from the time of payment: *Provided,* that nothing in this section shall be construed to affect any right existing or suit pending. [R. S. 1887, p. 1017, § 86; S. & C., p. 1941, § 86; Cothran, p. 1157, § 72.]

See also, R. S. 1887, p. 141, §§ 15–26; S. & C., p. 282, § 12–23. Overruled and superceded decisions on *Sangamon Distilling Co.* v. *Young*, 77 Ill. 197; *Smith* v. *Race*, 76 Ill. 490; *Frye* v. *Chicago, &c.*, 73 Ill. 399; *Ch. & A. R. R.* v. *Gassaway*, 71 Ill. 570; *Hatch* v. *Marsh*, 71 Ill. 370; *Davis* v. *Walker*, 60 Ill. 452; *Yeazel* v. *Alexander*, 58 Ill. 254; *Stevens* v. *Brown*, 58 Ill. 289; *Somerville* v. *Marks*, 58 Ill. 371.

2149. CONSTITUTIONALITY. The statute relating to Texas and Cherokee cattle, and making a party having them liable for diseases communicated by them, is unconstitutional. *Jarvis* v. *Riggin*, 94 Ill. 164.

2150. Act to prevent the importation of Texas cattle into this state is unconstitutional. *Salzenstein* v. *Mavis*, 91 Ill. 391, overruling *Yeazel* v. *Alexander*, 58 Ill. 254.

2151. The act is unconstitutional as being repugnant to § 8 art. 1, of the constitution of the United States. *Ch. & A. R. R.* v. *Erickson*, 91 Ill. 613; *Railroad Co.* v. *Husen*, 5 Otto (95 U. S.) 465.

2152. LIABILITY FOR SPEED IN EXCESS OF THAT LIMITED BY ORDINANCE. § 24. Whenever any railroad corporation shall by itself or agents, run any train, locomotive engine, or car, at a greater rate of speed in or through the incorporated limits of any city, town or village, than is permitted by any ordinance of such city, town or village, such corporation shall be liable to the person aggrieved for all damages done the person or property by such train, locomotive engine or car; and the same shall be presumed to have been done by the negligence of said corporation or their agents; and in addition to such penalties as may be provided by such city, town or village, the person aggrieved by the violation of any of the provisions of this section, shall have an action against such corporation so violating any of the provisions to recover a penalty of not less than one hundred dollars ($100) nor more than two hundred dollars ($200), to be recovered in any court of competent jurisdiction; said action to be an action of debt, in the name of the people of the state of Illinois, for the use of the person aggrieved; but the court or jury trying the case may reduce said penalty to any sum, not less, however, than fifty dollars ($50), where the offense committed by such violation may appear not to be malicious or willful: *Provided*, that no such ordinance shall limit the rate of speed, in case of passenger trains to less than ten miles per hour, nor in any other case to less than six miles per hour. [Laws of 1865, p. 103, §§ 1, 2 and 3, as amended by act of May 22, 1877. In force July 1, 1877. Laws of 1877, p. 165. R. S. 1887, p. 1017, § 87; S. & C., p. 1941, § 87; Cothran, p. 1157, § 73.]

2153. FRIGHTENING HORSE. Where a railway train is run within a city at a speed in excess of that allowed by ordinance, whereby the horses of a person about to cross the railway track are frightened, and his carriage is upset, and he is injured in person and property, the company operating such train will become liable to the party so aggrieved, for the penalty provided for in § 2 of the act relating to railroads, although the train may not have struck such person or his horses or carriage. *Ch. & E. Ill. R.R.* v. *People*, 120 Ill. 667.

2154. In an action to recover of a railway company the penalty given by § 62 of the railroad act, it is not necessary for the plaintiff to show that he was injured by actual collision with the train, running at a greater speed than allowed by law. It is sufficient for a recovery, to show that the train was run faster than was allowed by ordinance, and that in consequence thereof, he was aggrieved, by the frightening of his team. *Ib.*

2155. CONSTITUTIONALITY OF STATUTE. The act of 1865 making railway companies liable for all damages done to any individual, and for stock killed by any train or engine, in an incorporated city or town, where their trains are permitted to be run at a greater speed through such city or town, than is permitted by ordinance, is not unconstitutional. It is no objection that the statute gives the penalty to the injured party. *C., R. I. & P. R. R.* v. *Reidy*, 66 Ill. 43.

2156. SAME—*police regulation*. A charter to operate railroads by steam power does not confer unlimited discretion in the regulation of the speed of trains. The power to regulate the speed of trains, as a police power is inherent in the state and cannot be granted away. *T., P. & W. Ry.* v. *Deacon*, 63 Ill. 91; *C., B. & Q. R. R.*, v. *Haggerty*, 67 Ill. 113.

2157. REGULATION OF SPEED BY ORDINANCE—*sufficiency*. An ordinance "that it shall be unlawful for any railroad company, by themselves or their agents, to run at a greater rate of speed within the corporate limits of the town of C, than five miles per hour," and providing a penalty for its violation of not less than $10, nor more than $100, while informal, is valid. *T., P. & W. Ry.* v. *Deacon*, 63 Ill. 91.

2158. SAME—*presumption*. Greater speed than is allowed by ordinance raises a presumption of negligence as the cause of the injury. Where proof of contributory negligence is shown, this presumption is rebutted, and the plaintiff can recover only when his negligence is slight and that of defendant gross. *L. S. & M. S. R. R.* v. *Berlink*, 2 Bradw. 427.

2159. ORDINANCE, to raise the statutory *presumption* of negligence, must conform to the statute. If it limits the speed of a passenger train to less than ten miles per hour, it is not admissible as evidence. *C., B. & Q. R. R.* v. *Dougherty*, 12 Bradw. 181.

2160. Killing stock in a town does not raise a presumption that the train was running at a prohibited rate of speed; but proof that train was run in the town at a greater speed than that allowed by a valid ordinance, and that the injury occurred while the train was so running, makes a case of presumptive negligence. *C. & A. R. R.* v. *Engle*, 58 Ill. 381

2161. If stock is killed by a train while running through an incorporated town faster than is allowed by ordinance, the killing will be presumed to have been done through negligence of the company; but this presumption may be rebutted. *T., P. & W. Ry.* v. *Deacon*, 63 Ill. 91.

2162. An instruction that if the cow of the plaintiff was killed within the corporate limits by a train of the defendant, and that such train was, at the time of the killing, being run at a greater rate of speed than that prescribed by the ordinance of the town, then the defendant is presumed to have been guilty of negligence in killing the cow, *held*, not erroneous. *C., B. & Q. R. R.* v. *Haggerty*, 67 Ill. 113.

2163. LIABILITY. Where a person is killed by a train running much faster than is allowed by ordinance, and no bell is rung or whistle sounded, the company will be liable, unless the proof shows the injury was not the result of such negligence. *St. L., V. & T. H. R. R.* v. *Morgan*, 12 Bradw. 256.

2164. SAME — *contributory negligence.* Although a train was running through a town at a higher speed than allowed by ordinance, yet if the deceased was guilty of gross negligence contributing to his death, no recovery can be had. *W., St. L. & P. Ry.* v. *Weisbeck*, 14 Bradw. 525.

2165. ORDINANCE — *power to pass.* Power in municipal corporation to define and abate nuisances, &c.; authorizes an ordinance prohibiting the running of engines, &c., within its limits, exceeding six miles per hour. *C., B. & Q. R. R.* v. *Haggerty*, 67 Ill. 113.

2166. If an ordinance prohibits a speed of more than six miles an hour, the running of a train at fifteen miles an hour, resulting in the death of one wrongfully on the track, will make the injury willful or wanton, and render the company liable. *I. C. R. R.* v. *Hetherington*, 83 Ill. 510.

2167. It is gross negligence to run a train through a town at a speed prohibited by law, and if death of a person results therefrom the company will be liable. *C. & A. R. R.* v. *Becker*, 84 Ill. 483.

2168. Engineer running his train in a city at twenty miles an hour, the ordinance limiting it to eight, and seeing a switchman on or near the track, will have no right to assume that he will get out of the way, or keep off the track and avoid danger, as might be if the train was being run at only eight miles an hour. *L. S. & M. S. R. R.* v. *O'Connor*, 115 Ill. 254.

2169. SPEED AS NEGLIGENCE—*in absence of ordinance.* Running train at a high rate of speed, is not of itself reckless or wanton disregard of the public safety, or a willful attempt to injure, where there is no ordinance violated. *Garland* v. *C. & N. W. Ry.*, 8 Bradw. 571.

2170. The general law of the state imposes no restrictions on railway companies as to the rate of speed their trains may run. If not prohibited by ordinance, they may adopt such rate of speed as they may desire, provided it be reasonably safe for its passengers and the public. *C. & A. R. R.* v. *Robinson*, 9 Bradw. 89; *W., St. L. & P. Ry.* v.

Neikirk, 13 Bradw. 387; same case, 15 Bradw. 172; *C., R.I. & P.Ry.*
v. *Givens*, 18 Bradw. 404; *C., B. & Q. R. R.* v. *Lee*, 68 Ill. 576.

2171. A railway company may run its trains at any speed it
chooses, so that when taken in connection with the character of the
road, its grades, curves, &c., it appears not to increase the ordinary
risks of travel. So long as the increased speed adds nothing to the
risks and danger of the traveling public, the courts have no right to
interfere. *I., B. & W.Ry.* v. *Hall*, 106 Ill. 371.

2172. Speed in the transit and punctuality in the arrival and connections of trains, is required and is lawful. Speed may be regulated
by the companies to suit the times and places. *C. & M. R. R.* v.
Patchin, 16 Ill. 198.

2173. NEGLIGENCE IN SPEED—*depends on circumstances.* A rate
of speed that would be highly dangerous, or even reckless in a populous city with numerous street crossings, may not be hazardous in
leaving a town, after reaching its sparsely settled suburbs. *P., D. & E.
Ry.* v. *Miller*, 11 Bradw. 375.

2174. A railway company must conform the speed of its trains to
the safety of the public at all places in a city where persons have an
equal right to travel with it. *C., B. & Q. R. R.* v. *Dougherty*, 12 Brad.
181.

2175. Whether the rate of speed at the time a collision occurred,
was dangerous and negligent, is a question of fact. *W., St. L. & P.
Ry.* v. *Hicks*, 13 Bradw. 407.

2176. Where a traveler at a highway crossing cannot see the approach of a train, owing to a deep cut or other cause, the company
should run at a low rate of speed and give the statutory signals. *C.,
B. & Q. R. R.* v. *Triplett*, 38 Ill. 482.

2177. Whether a train which killed a cow was run at too great a
speed through a populous town, so as to amount to negligence, is a
question of fact. *T., P. & W. Ry.* v. *Foster*, 43 Ill. 415.

2178. It is great negligence in a railway company to run one of its
fastest trains with unabated speed through a town, where persons are
liable at all times to be upon its track; and if while so running, it
injures a child, it will be liable. *C. & A. R. R.* v. *Gregory*, 58 Ill. 226.

2179. Running a freight train at a rapid speed, down grade, at a
dangerous crossing, without giving the statutory warning, is gross
negligence. *C., B. & Q. R. R.* v. *Payne*, 59 Ill. 534.

2180. The speed of a train may be considered in connection with
its location and that of a highway, and other surrounding circumstances, on the question of negligence. *I. & St. L. R. R.* v. *Stables*, 62
Ill. 313.

2181. Cases in which the speed of the train figures on the question
of liability for negligence. *I. & St. L. R. R.* v. *Galbreath*, 63 Ill. 436;
C. & N. W.Ry. v. *Ryan*, 70 Ill. 211; *C. & A. R. R.* v. *Becker*, 76 Ill. 25;
I. & St. L. R. R. v. *Peyton*, 76 Ill. 340.

2182. Railway company is guilty of gross negligence in running a
train at the rate of ten miles an hour in a populous city, contrary to
an ordinance thereof, especially where there are many tracks, &c., to
cross. *P., C. & St. L. Ry.* v. *Knutson*, 69 Ill. 103.

2183. The jury may consider all the circumstances and from them
determine whether a train was run at an improper speed in reference
to the safety of one killed. *I. C. R. R.* v. *Cragin*, 71 Ill. 177.

2183a. DUE CARE OF PLAINTIFF—*not excused by great speed.*
The fact that a railway company violates the statute by running its
trains at a speed prohibited by ordinance, does not relieve the plaintiff

seeking to recover for a personal injury from the necessity of proving due care. *St. L., A. & T. H. R.R.* v. *Andres,* 16 Bradw. 292.

2184. Where animals at large contrary to law, are killed at a road crossing, proof that the train was running at an unreasonable rate of speed, is not sufficient to authorize a recovery. *T., W. & W. Ry.* v. *Barlow,* 71 Ill. 640.

2185. Company should so regulate the speed of its trains in crossing public highways and give warning, that all passing may be apprised of the danger, and a neglect to do so, makes it liable. *R., R. I. & St. L. R. R.* v. *Hillmer,* 72 Ill. 235.

2186. Running at prohibited rate of speed in a populous city, is gross negligence. *St. L., V. & T. H. R. R.* v. *Dunn,* 78 Ill. 197.

2187. It is the duty of a railway company to run its trains through a village at such a speed as to have them under control, even if there is no ordinance on the subject. *C. & A. R. R.* v. *Engle,* 84 Ill. 397.

2188. The speed of a train approaching a highway crossing should not be so great at the crossing as to render unavailing the warning of its bell or whistle. *C., B. & Q. R. R.* v. *Lee,* 87 Ill. 454.

2189. A railway company must regulate the speed of its trains in cities and public thoroughfares, with reference to the safety of the public, or be liable for the damages resulting from its negligence or willfulness in this respect. Running at greater speed than is allowed by law, is not only carelessness, but the act is willful. *Wabash Ry.* v. *Henks,* 91 Ill. 406.

2190. The law prohibiting the running of trains at a greater speed than ten miles an hour in cities, is not a license to run at such speed in all cases. In some places within a city, that would be a dangerous speed. The rate of speed must conform to the safety of the public at all places in a city where others have equal rights. *Ib.*

2191. Where a person was struck at a street crossing, proof that the train was run at an unusual speed; that no bell or whistle was sounded, and no light on the forward car that struck the plaintiff, and that plaintiff was using proper care, makes out a clear right of recovery. *L., E. & W. Ry.* v. *Zoffinger,* 107 Ill. 199.

2192. Whether a given rate of speed is dangerous or not, is to be determined by the surrounding circumstances, such as the condition of the track, fencing of the road, &c. *I., B. & W. Ry.* v. *Hall,* 106 Ill. 371.

2193. If an engineer drives his engine at a negligent or high rate of speed which materially contributes to an injury received by him, he cannot recover. *I. C. R. R.* v. *Patterson,* 69 Ill. 650; same case, 93 Ill. 290.

2194. EVIDENCE—*ordinance.* In order to recover on the ground that a train was run at a greater rate of speed than authorized by ordinance, the ordinance must be put in evidence. *C. & N. W. Ry.* v *Schumilowsky,* 8 Bradw. 613.

2195. Under an averment of negligence generally, proof of a violation of an ordinance is admissible. *C., R. I. & P. R. R.* v. *Reidy,* 66 Ill. 43.

2196. If the declaration contains no averment that there was a city ordinance regulating the speed of trains at the place of the accident, evidence that the speed of the train was greater than that prescribed by ordinance, is inadmissible. *I. C. R. R.* v. *Godfrey,* 71 Ill. 500.

2197. To recover for killing an animal in a town on the ground the train was run at a prohibited speed, the plaintiff must prove that the ordinance was in force by due publication, at the time of the injury. *C. & A. R. R.* v. *Engle*, 76 Ill. 317.

2198. An ordinance of a city limiting speed of trains to six miles an hour, is proper evidence to go to the jury, on the question of negligence. *T., W. & W. Ry.* v. *O'Connor*, 77 Ill. 391.

2199. Testimony showing how far a train of cars ran after striking a person, is competent evidence to show that the train was running at a greater speed that allowed by ordinance of the city, and was not under proper control. *Penn. Co.* v. *Conlan*, 101 Ill. 93.

2200. Declaration informally drawn, held sufficient to admit in evidence proof of an ordinance regulating the rate of speed of a railway. *L. S. & M. S. R. R.* v. *O'Connor*, 115 Ill. 254.

2201. INSTRUCTIONS, wholly ignoring the question whether plaintiff was injured in consequence of the negligent acts or omissions of duty are erroneous. *C., B. & Q. R. R.* v. *Dvorak*, 7 Brad. 555.

2202. Instructions based on negligence of company in running in disregard of ordinance and failure to give warnings, and the care of the deceased, are bad in excluding the rule of comparative negligence. *Schmidt* v. *C. & N. W. Ry.*, 83 Ill. 405.

2203. Failure of servants to obey orders no defense to company. *I. & St. L. R. R.* v. *People*, 91 Ill. 452.

2204. DUTY TO STOP AT ALL STATIONS. § 25. Every railroad corporation shall cause its passenger trains to stop upon its arrival at each station, advertised by such corporation as a place for receiving and discharging passengers, upon and from such trains, a sufficient length of time to receive and let off such passengers with safety: *Provided*, all regular passenger trains shall stop a sufficient length of time at the railroad station of county seats, to receive and let off passengers with safety. [As amended by Laws of 1879, p. 225. Amendment is original section with proviso added. R. S. 1887, p. 1018, § 88; S & C., p. 1943, § 88; Cothran, p. 1158, § 74.]

2205. STOPPING AT STATION CALLED FOR IN TICKET. Railway companies may have passenger trains that stop only at the principal stations. They may also run freight trains which only stop at certain stations for fuel and water, or at such other stations as the transportation of stock or freight may require. *C. & A. R. R.* v. *Randolph*, 53 Ill. 510.

2206. They may exclude all passengers from such freight trains, or only carry them to the places at which they are accustomed to stop. Taking up passenger's ticket does not amount to a contract to stop train at his station. *Ib.*

2207. If passenger guilty of want of ordinary care in leaping from train while being carried beyond his station, he cannot recover for a personal injury, although the conductor may have given his opinion that he might get off safely. *Ib.*

2208. SAME—*time for getting off*. Reasonable time must be allowed for passengers, whether old or young, to alight in safety. Liability for an injury caused by want of time to get off. *T., W. & W. Ry.* v. *Baddeley*, 54 Ill. 19.

2209. Where train stops at station a reasonable time for passengers to get off, and a passenger neglects to get off until train starts, when he attempts on his own motion to get off and is killed, and no negligence on the part of the company is shown, no recovery can be had. *I. C. R. R.* v. *Slatton,* 54 Ill. 133.

2210. Carrying a passenger beyond his station without giving him a reasonable opportunity of leaving the train, gives him a right of action for damages. *I. C. R. R.* v. *Able,* 59 Ill. 131; *O. & M. Ry.* v. *Stratton,* 78 Ill. 88, 94.

2211. CONTRIBUTORY NEGLIGENCE. If passenger while being carried beyond his station against his will, leaps from the train while in rapid motion, or attempts to get off under such circumstances as to make the act dangerous, he cannot recover for an injury received in such attempt. *Aliter,* if act is apparently safe and free from appearance of danger. *I. C. R. R.* v. *Able,* 59 Ill. 131.

2212. LIABILITY —*for carrying beyond station.* Where a freight train was in the habit of carrying passengers to a certain station, and before the company had made any different rule or regulation in this respect, the plaintiff purchased a ticket for such station, but was informed by the conductor that he would not stop at such station, and advised him to take passage with another extra train, to which he applied and was refused passage, and the plaintiff entered the first train, informing the conductor of the facts, and was by it carried to the next station beyond the one named in his ticket: *Held,* that the company was liable to the plaintiff in compensatory damages. *C., R. I. & P. R. R.* v. *Fisher,* 66 Ill. 152.

2213. LIABILITY FOR NOT STOPPING TO TAKE ON PASSENGER. If a railway company wrongfully fails to stop at a station to take on a passenger, he will be entitled to recover nominal and such actual damage as he may sustain by reason of the delay. *I., B. & W. Ry.* v. *Birney,* 71 Ill. 391.

2214. CARRYING BEYOND STATION. Company liable for damages for carrying passenger against his will beyond his station, by not affording him a chance to get off. *I. C. R. R.* v. *Chambers,* 71 Ill. 519.

2215. If passenger under the apprehension that he is being carried beyond his station, leaps from the train in motion under such circumstances as to make his act probably dangerous, he cannot recover. *Ib.*

2216. Passenger has no right to attempt to get off a train while in motion, and if he does so without the knowledge or direction of any employe of the company, he must bear the consequences. *O. & M. Ry.* v. *Stratton,* 78 Ill. 88.

2217. Where a passenger while asleep, is carried beyond his station and when the train arrives at a bridge where it stops to take on water, he gets up and without any advice from any one connected with the company, goes out of the car in a dark night, and finding no brakeman put out his foot to reach the platform, and there being none there, when the train gave a jerk, which pulled both feet off the car and he fell through the bridge and was injured: *Held,* that his negligence was such as to bar a recovery. *I. C. R. R.* v. *Green,* 81 Ill. 19.

2218. If a passenger is negligently carried past his station, this will not justify him in needlessly exposing himself to danger, and his injury in getting off train while in motion, or upon a high bridge, is too remote to be recovered in an action for the negligence of the company. *I. C. R. R.* v. *Green,* 81 Ill. 19.

2219. The fact that a passenger is in danger of being carried past his station, will not justify him in getting off while the train is in motion, or imprudently exposing himself to danger. If he does so and is injured, he cannot recover for the same. *I. C. R. R.* v. *Lutz,* 84 Ill. 598.

2220. If train fails to stop at passenger station, it furnishes no excuse for passenger to leap from the train some three miles beyond, while the train is running at the rate of fifteen miles an hour, and if he does so and is injured, he cannot recover. He should have remained on train, and sued for damages for being carried beyond his proper station. *Dougherty* v. *C., B. & Q. R. R.*, 86 Ill. 467.

2221. Passenger having ample time to get aboard train who waits until it starts, and is injured in attempting to get on, has no cause of action for his injury. *C. & N. W.* v. *Scates*, 90 Ill. 586.

2222. It may be true, that alighting from a train of cars while in motion is negligence, where the railway company is not in fault, and the train has considerable speed, but it is not necessarily true when it is a question of comparative negligence. *C. & A. R. R.* v. *Bonifield*, 104 Ill. 223.

2223. It is the duty of every railroad company to cause its passenger trains to stop at each station advertised as a place for receiving and discharging passengers, a sufficient time to receive and let off passengers with safety, and to provide a reasonably safe way of reaching and departing from their cars at all usual stations, and it is the duty of passengers to exercise ordinary care in attempting to take passage on railway cars. *W., St. L. & P. Ry.* v. *Rector*, 104 Ill. 296.

2224. The act of 1879 requiring all passenger trains to stop at county seats long enough to take and discharge passengers, is not a regulation of *inter*-state commerce. It is a proper exercise of police power and is a valid law. *C. & A. R. R.* v. *People*, 105 Ill. 657.

2225. Regular express train not stopping at all intermediate stations is a regular passenger train within the meaning of the statute, and must stop at county seats. *Ib.*

2226. LIABILITY FOR INJURY caused by starting car suddenly. *C., B. & Q. R. R.* v. *Hazzard*, 26 Ill. 373; *C. & W. I. Ry.* v. *Bingenheimer*, 116 Ill. 226.

2227. Liability of street railway company for suddenly starting cars while passengers getting on or off. *Ch. W. Div. Ry.* v. *Mills*, 91 Ill. 39; *Ch. C. Ry.* v. *Mumford*, 97 Ill. 560.

2228. STOPPING AT COUNTY SEAT. Where a railroad is built to a town, as required by its charter, and a depot is established at the end of its line within such town, which is a county seat, the company operating such road will have no discretion as to which of its passenger trains shall stop there and which shall not, as it would have, within certain reasonable limitations, if such town was not a county seat, but all its passenger trains must stop at such place. It is not sufficient that all its trains may stop at a new depot located at a junction with another road, a quarter of a mile beyond the corporate limits of such town. *People* v. *L. & N. R. R.*, 120 Ill. 48.

2229. BRAKEMAN ON PASSENGER CARS. § 26. No railroad corporation shall run or permit to be run upon its railroad any train of cars moved by steam power, for the transportation of passengers, unless there is placed upon the train one trusty and skillful brakeman for every two cars in the train, or unless the brakes are efficiently operated by power applied from the locomotive. [R. S. 1887, p. 1018, § 89; S. & C., p. 1943, § 89; Cothran, p. 1158, § 75.]

2230. If a brakeman is injured by reason of the nut on a brake being gone, of which he had no knowledge, he will not be guilty of contributory negligence. *Ch. & E. Ill. R. R.* v. *Hagar*, 11 Bradw. 498.

2231. If fellow servants of a brakeman neglect to discover and report defective brake, and he is killed through such defect, no recovery can be had. *C. & A. R. R.* v. *Bragonier*, 11 Brad. 516.

2232. Where a brakeman was killed by a defect in the brake, the nut which kept the wheel in its place on the upright shaft, having become loose, and in the effort to work the brake, the wheel came off and the deceased was thrown to the ground: *Held*, that it was the duty of the brakeman to see that the brake was in a fit condition for use, and the company was not to suffer for his neglect of duty. *I. C. R. R.* v. *Jewell*, 46 Ill. 99.

2233. To run a train of six or eight cars without a brakeman, is gross negligence. So a failure to apply brakes on signal of danger by engineer implies gross negligence. *T., W. & W. Ry.* v. *McGinnis*, 71 Ill. 346.

2234. BRAKEMAN ON FREIGHT CARS. § 27. No railroad corporation shall run or permit to be run upon its railroad any train of cars, for the transportation of merchandise or other freight, without a good and sufficient brake attached to the rear or hindmost car of the train, and a trusty and skillful brakeman stationed upon said car, unless the brakes are efficiently operated by power applied from the locomotive. [R. S. 1887, p. 1018, § 90; S. & C., p. 1943, § 90; Cothran, p. 1158, § 76.]

2235. DAMAGES—PENALTY. § 28. If any railroad corporation shall violate any of the provisions of the three preceding sections, it shall be liable to the person aggrieved for all damages done to person or property by reason thereof, with costs of suit; and in addition thereto, said corporation shall forfeit the sum of not less than $100 nor more than $500, for each offense, to be recovered in an action of debt, in the name of the people of the state of Illinois, for the use of any person aggrieved, before any court of competent jurisdiction. [R. S. 1887, p. 1018, § 91; S. & C., p. 1943, § 91; Cothran, p. 1158, § 77.]

2236. CHECKS OR RECEIPTS FOR BAGGAGE. § 29. Every railroad corporation, when requested, shall give checks or receipts to passengers for their ordinary baggage, when delivered for transportation on any passenger train, which baggage shall, in no case, exceed one hundred pounds in weight for each passenger, and shall deliver such baggage to any passenger upon the surrender of such checks or receipts. Any such corporation willfully refusing to comply with the requirements of this section, shall pay a fine of not less than $10 nor more than $100, which may be recovered before any court of competent jurisdiction, in an action of debt, in the name of the people of the state of Illinois, for the use of the person aggrieved: *Provided*, that no passenger shall be entitled to receive checks or receipts for any baggage unless he shall have paid or tendered the lawful rate of fare for his transportation to the proper agent for such corporation. [R. S 1887, p. 1018, § 92; S. & C., p. 1943, § 92; Cothran, p. 1158, § 78.]

2237. Check as evidence. In an action for lost baggage, a nickel plated check was given in evidence, and a witness, a baggage-master of the defendant, testified that nickel plated checks had never been used on through baggage to his knowledge, and was then asked whether his position was such that he would have known, if they had been so used: *Held*, that the question was proper. *L. S. & M. S. Ry.* v. *Lassen*, 12 Bradw. 659.

2238. Where a baggagemaster, having testified that a particular kind of check was alone available to carry baggage between two designated stations, was asked to state whether it was possible for the check held by the plaintiff to have been used in the usual course of business: *Held*, that the question was competent. *Ib.*

2239. Carrier by receiving passenger's baggage becomes immediately responsible for its safe delivery at its destination. *Woods* v. *Devin*, 13 Ill. 746.

2240. He is responsible for the baggage of a passenger the same as a common carrier of goods. He can only excuse himself for non-delivery by showing loss by the act of God or the public enemy. *Ib.*

2241. His responsibility commences when the baggage is delivered to him or his agent, and prepayment of fare is not necessary to charge him for its loss. His compensation is included in the passenger fare. *Ib.*

2242. A passenger's baggage may include a reasonable sum of money for traveling expenses and such articles of necessity and convenience as are usually carried by passengers for personal use, comfort, instruction, amusement or protection. It may include a pocket pistol and a pair of dueling pistols, not carried as merchandise. *Ib.*

2243. Common carrier not liable for the loss of money packed among other goods in a box in such way as to mislead and deceive the carrier. If to be held liable, he should be informed of the contents. *C. & A. R. R.* v. *Thompson*, 19 Ill. 578.

2244. Railway corporation will not be liable for lost baggage, unless it is shown to have been in its possession, or it has contracted in some way to transport it. *M., S. & N. Ind. R. R.* v. *Meyres*, 21 Ill. 627.

2245. In an action for lost baggage, it is proper to instruct that a recovery maybe had for such articles of necessity and convenience as passengers usually carry for their personal use, comfort, instruction, amusement and protection, having regard to the length and object of their journeys. *Parmelee* v. *Fischer*, 22 Ill. 212.

2246. The delivery of a baggage check, is *prima facie* evidence that the company has the baggage. A revolver included in personal baggage. *Davis* v. *M., S. & N. Ind. R. R.*, 22 Ill, 278.

2247. A reasonable amount of bank bills may be carried in a trunk, and their value recovered as lost baggage. *I. C. R. R.* v. *Copeland*, 24 Ill. 332.

2248. Declaration need not aver the plaintiff was a passenger to admit proof of that fact, which may be shown by the possession of the baggage check and ticket, or by the check alone, if such checks are not given until the passenger tickets are shown. *Ib.*

2249. Railway company selling through tickets over its own and other roads, is liable for the safety of passengers and baggage to the point of destination. *Ib.*

2250. The purchase of a railroad ticket includes the payment for the transportation of the person's baggage, not exceeding a specified weight. *C. & C., A. L. R. R.* v. *Marcus*, 38 Ill. 219.

2251. Baggage consists of such articles as are necessary for a person's comfort and convenience, with the necessary amount of money for expenses. *Ib.*

2252. The owner, who under the pretense of having baggage transported, places in the hands of the agent of a railway company, merchandise, jewelry and other valuables, is guilty of fraud, which releases the company from liability as common carrier for a loss. *Ib.; M. C. R. R.* v. *Carrow*, 73 Ill. 348.

2253. The price paid for a ticket includes the carrying of the passenger's baggage, and the recognition of the ticket by the railroad company, is an admission that the check given for the baggage is equally binding. *C. & R. I. R. R.* v. *Fahey*, 52 Ill. 81.

2254. Where a passenger's ticket entitles him to travel over different lines of road to his destination, and to which his baggage is checked, all of the lines recognizing the validity of the ticket, each company into whose possession the baggage may come, will be liable to the owner for its loss. *C. & R. I. R. R.* v. *Fahey*, 52 Ill. 81.

2255. A Chicago grocer, who went into the country in quest of butter, sought to recover of a railroad company the value of two revolvers, among other things, which he claimed were in the trunk as a part of his baggage, which was lost by the company: *Held*, with due regard to the habits and condition in life of the passenger, more than one revolver was not reasonably necessary for his personal use and protection. *C., R. I. & P. R. R.* v. *Collins*, 56 Ill. 212.

2256. Where plaintiff ships as personal baggage merchandise to be used in his trade, which in no sense is capable of being considered personal baggage, the company not having notice of the contents, will be released from their liability as a common carrier. *M. S. & N. Ind. R. R.* v. *Oehm*, 56 Ill. 293.

2257. Where the passenger leaves his baggage in charge of the carrier, the liability of the latter will not be changed to that of warehouseman, until the baggage is stored in a safe and secure warehouse. If placed in an insecure room and is there stolen, the company will be liable as a carrier and not as a warehouseman. *Bartholomew* v. *St. L., J. & C. R. R.*, 53 Ill. 227.

2258. Where baggage after reaching its destination, is not for any cause delivered to the passenger or his agent, it is the duty of the company to deposit it in their baggage room, when their liability is changed to that of warehousemen. *C., R. I. & P. R. R.* v. *Fairclough*, 52 Ill. 106.

2259. Railway company not liable for loss by fire of costly jewelry checked in a trunk as ordinary baggage, unless it was guilty of gross negligence. The fraud of passenger releases the carrier of his extraordinary liability. *M. C. R. R.* v. *Carrow*, 73 Ill. 348.

2260. Carrier of passengers is not bound to inquire as to the contents of a trunk delivered to it as ordinary baggage, such as travelers usually carry, even if the same is of considerable weight, but may rely upon the representations, arising by implication, that it contains nothing more than such baggage. *M. C. R. R.* v. *Carrow*, 73 Ill. 348.

2261. Owners of sleeping cars, are not carriers, and cannot be held liable as such for property lost by or stolen from lodgers while on their cars. *Pullman Palace Car Co.* v. *Smith*, 73 Ill. 360.

2262. A sacque and muff and silver napkin rings, cannot be said to constitute any part of a gentleman's traveling baggage, and no recovery can be had for their value in case of a loss. *C., R. I. & P. R. R.* v. *Boyce,* 73 Ill. 510.

2263. The carrier's strict liability for baggage continues until the owner has had a reasonable time and opportunity to come and take the same away. If not called for in a reasonable time, it may be stored in a secure warehouse, and the liability as carrier will cease and that of warehouseman attach. *Ib.*

2264. What is a reasonable time and opportunity to call for baggage is a mixed question of law and fact, depending greatly upon the peculiar facts of each case; but where the facts are undisputed, it is purely a question of law. *Ib.*

2265. Unless the carrier is at fault, the passenger cannot, for purposes of his own convenience, or by reason of any inevitable accident to himself, be permitted to extend the strict liability incident to a common carrier in respect to baggage after it has reached its destination. *Ib.*

2266. The delivery of a baggage check to a passenger is *prima facie* evidence that the carrier has received the baggage it represents. Such evidence may be overcome by proof to the contrary; but the burden of proof is upon the carrier to show a non-delivery. *C., R. I. & P. R. R.* v. *Clayton,* 78 Ill. 616.

2267. It is immaterial when baggage comes to the possession of the carrier, whether at the time the check is issued, or at a subsequent time. In either case its liability as an insurer becomes fixed in case of a loss. *C., R. I. & P. R. R.* v. *Clayton,* 78 Ill. 616.

2268. Where a railway company received a passenger's check for baggage which had not then arrived by another road, and gave its own check for the same, and it appeared that it surrendered the passenger's first check to the other railway company, it was held that this was sufficient, in the absence of proof to the contrary, to show that the baggage was received by the company so surrendering the first check. *Ib.*

2269. Where the carrier has taken the baggage to its destination, if not called for, it should be stored in a safe and secure warehouse. *St. L. & C. R. R.* v. *Hardway,* 17 Bradw. 321.

2270. A passenger's trunk was carried to its destination, and not being called for, was {placed over night by the carrier in the ladies' waiting room, which was broken into and the contents of the trunk stolen: *Held,* that the company was liable for the contents, except as to $10 for a silk quilt, that not being baggage. *Id.*

2271. Railway company is an insurer of baggage until its arrival and discharge at the place of destination, and until the owner has had reasonable time and opportunity to claim and take it away. If not called for in a reasonable time it may be stored in a safe warehouse, and the liability of carrier ceases and that of warehouseman attaches. *C. & A. R. R.* v. *Addizoat,* 17 Bradw. 632.

2272. The reasonable time to apply for baggage transported on the same train with the passenger, is directly after its arrival and transfer to the platform, making due allowance for the confusion occasioned by the arrival and departure of the train and for the delay caused by the crowd on the platform. *Ib.*

2273. The passenger should not prolong the strict liability of the carrier any longer than is necessary under the circumstances. If informed that his baggage has not arrived and he gives no directions, no notice of its arrival can be given and he should inquire again in a reasonable time. *Id.*

2274. BAGGAGE SMASHING. § 30. Any person employed

by a railroad corporation in this state, who shall willfully, carelessly or negligently break, injure or destroy any baggage, shall be liable for the amount of damage to the owner thereof, and may be arrested, and, on conviction before a justice of the peace, be fined in any sum not exceeding $200, and held in custody or confined in the county jail until such fine shall be paid: *Provided*, that the remedy hereby given against such employe shall not lessen the liability of such corporation. [R. S. 1887, p. 1018, § 93; S. & C., p. 1944, § 93; Cothran, p. 1159, § 79. See Crim. Code, Ch. 38, § 245.]

2275. EJECTING PASSENGER—AT WHAT PLACES AND FOR WHAT CAUSE. § 31. If any passenger on any railroad car or train shall refuse, upon reasonable demand, to pay his lawful fare—or shall, upon such car or train, use abusive, threatening, vulgar, obscene, or profane language thereon—or shall so conduct himself as to make his presence offensive or unsafe to passengers thereon, it shall be lawful for the conductor of the train to remove, or cause to be removed, such passenger from the train at any regular station; but if such conductor shall use, cause or permit to be used unreasonable force or violence, he shall be liable for all damages to the person injured thereby: *Provided*, that the recovery and satisfaction of damages, under the provisions of this section, shall not lessen the liability of, or the amount of the damages that such corporation may be liable to, for such acts. [R. S. 1887, p. 1018, § 94; S. & C., p. 1944, § 94; Cothran, p. 1159, § 80.]

2276. PLACE OF EXPULSION. The rule that a passenger cannot be expelled from the train except at a regular station, does not apply in all cases. Where a passenger has been once expelled at a regular station and on the starting of the train again leaped on the same, he will not be entitled to the same consideration, as if he had not once been expelled. *C., B. & Q. R. R. v. Boger,* 1 Bradw. 472.

2277. The rule requiring a ticket or full fare from passengers, is a reasonable one, and necessary to the proper transaction of business. *St. L. & C. R. R. v. Carroll,* 13 Bradw. 585.

2278. To avail of a reduced rate of fare to a place and return, the passenger must procure a special ticket. Attempting to ride at such reduced rate without ticket, passenger may be put off. *St. L. & C. R. R. v. Carroll,* 13 Bradw. 585.

2279. A husband bought a non-transferable 1,000 mile ticket and told the agent to issue it to "E. Bannerman," and the agent supposing it was intended for a man, inserted "Mr." before the name, and the ticket was presented by the husband to pay his wife's fare, he stating at the time to the conductor that it was bought for his wife "Elsa," and the conductor refused to receive the ticket, and upon refusal to pay fare, put her off at the next station using no unnecessary force: *Held*, that the wife could not recover for her expulsion. *C. & N. W. Ry. v. Bannerman,* 15 Bradw. 100.

2280. Implied contract that passenger shall be humanely treated, and not assaulted or maltreated by servants of railway company, and it will be liable for breach of such contract. *C., R. I. & P. Ry.* v. *Barrett,* 16 Bradw. 17.

2281. While the carrier may rescind the contract of carriage for certain misconduct of the passenger, the penalty for such misconduct must not be enforced unreasonably or oppressively. If unnecessary force is used inflicting unnecessary damage, the company will be liable. *Ib.*

2282. ASSAULT AND BATTERY. The use of mere words by a passenger to a conductor, will not justify an assault and battery of the passenger by the conductor, or relieve the company from liability for the same. *Coggins* v. *C. & A. R. R.,* 18 Bradw. 620.

2283. Passengers who neglect to purchase tickets at stations before embarking on cars, may be charged additional fare, if proper conveniences and facilities are furnished them for procuring tickets. *C., B. & Q. R. R.* v. *Parks,* 18 Ill. 460.

2284. If a passenger pays fare only from one station to another, without a ticket, he may be compelled to pay an extra charge at each station. *C., B. & Q. R. R.* v. *Parks,* 18 Ill. 460.

2285. EXPULSION. If a passenger refuses to pay the fare required by the tariff of the company, he may be ejected from the cars at any regular station, but not elsewhere. *C., B. & Q. R. R.* v. *Parks,* 18 Ill. 460.

2286. The company must furnish proper facilities for procuring tickets, if it intends to charge extra fare, when tickets are not procured. If a ticket is applied for and not furnished, that fact may be shown by the station agent, and his certificate of it should be evidence to the conductor of the fact. *St. L., A. & Ch. R. R.* v. *Dalby,* 19 Ill. 353.

2287. If a passenger refuses to pay his fare, the conductor may lawfully put him off the train at a proper place; but when he does not refuse to pay the fare he is legally bound to pay, his removal from the cars by the conductor, is unlawful, and trespass lies for an assault and battery. *Ib.*

2288. A passenger offered a ticket which was void by reason of having a hole punched through it, and refusing to pay fare, was ejected from the car, three or four miles from a station, without any aggravating circumstances: *Held,* (1) that attempting to use such a ticket without explaining how he obtained it was evidence of wrong on his part; (2) that the company had the right to eject him, but only at a station; (3) that his attempt to impose on the company must mitigate the damages, and (4) that $1,000 was excessive. *T. H., A. & St. L. R. R.* v. *Vanatta,* 21 Ill. 188.

2289. Where a passenger refuses to pay his fare, he may be ejected from the cars at any regular station but not elsewhere. If put off at any place other than a station, he will be entitled to at least nominal damages, but whether more, depends on the circumstances. *C. & A. R. R.* v. *Roberts,* 40 Ill. 503.

2290. A passenger wantonly refusing to pay fare was ejected by the conductor at a place two miles from any station, but in a manner free from indignity toward him, who was subjected to no other injury than being obliged to walk to the station: *Held,* that a verdict for $450 damages was excessive. *Ib.*

2291. EXPULSION—*want of ticket.* Where a railway company carries passengers on a freight train, and in such cases, requires tickets to be purchased before entering the train, and a passenger disregards the rule, he can only be expelled at a regular station. *I. C. R. R.* v. *Sutton,* 42 Ill. 438.

—19

2292. The willful neglect to purchase a ticket at the time and place required by the rules of the company, and a refusal to pay fare, are substantially the same offense against the rights of the company, and the penalty for one is no greater than for the other. *Id.*

2293. Where a passenger, before the departure of the train, was informed of the rule requiring tickets to be purchased before entering the train, and he then sought to buy them, but the office was closed, and then entered the train and offered to pay his fare to the conductor, which was refused, and he was expelled from the train at a place other than a regular station: *Held*, that the company was liable. *Ib.*

2294. OFFICE OPEN FOR TICKETS. Office should be kept open for the sale of tickets for a reasonable time before the departure of each train and up to the time fixed by its published rules for its departure, and not up to the time of actual departure. *St. L., A. & T. H. R. R. v. South*, 43 Ill. 176.

2295. Company bound to furnish a convenient and accessible place for the sale of tickets and afford a reasonable opportunity to purchase them, and parties who will not avail themselves of it, must pay the extra fare, or, on refusal, be ejected from the train. *Ib.*

2296. While the right of a railway company to discriminate in its fare, between those purchasing tickets and those who do not, is just and reasonable, still such right depends on the fact that a reasonable opportunity has been given to obtain tickets at the lowest rate. *St. L., A. & T. H. R. R. v. South*, 43 Ill. 176.

2297. EXPULSION FROM FREIGHT TRAIN. Railway company holding itself out as a carrier of passengers by a freight train, has no more right to expel a passenger therefrom without cause, than from a regular passenger train. *C. & A. R. R. v. Flagg*, 43 Ill. 364.

2298. They may as to certain classes of trains (as freight trains) require tickets to be purchased before entering the same. A passenger who knowingly disregards such a rule, is placed on the same footing with one who refuses to pay fare, and may be expelled at any regular station. *Ib.*

2299. If the passenger willfully neglects to purchase a ticket before entering the train, he cannot be expelled at a place other than a regular station. A water tank, even if a usual stopping place, is not a regular station. *Ib.*

2300. A failure to furnish reasonable facilities for procuring a ticket by keeping the office open a reasonable time prior to that fixed for the departure of the train, gives a person desiring to take passage, the right to enter the train and to be carried to his place of destination, by the payment of the regular fare to the conductor. Under such circumstances his expulsion would be unlawful. *Ib.*

2301. The refusal of a passenger to surrender his ticket to the conductor when demanded, does not constitute the same offense as the non-payment of fare, and the statutory prohibition against the expulsion of passengers for the latter offense, except at a regular station, does not apply to the former case. *I. C. R. R. v. Whittemore*, 43 Ill. 420.

2302. A railway company may expel a passenger at a place other than a regular station, for the violation of any reasonable rule, other than that of non-payment of fare. *I. C. R. R. v. Whittemore*, 43 Ill. 420.

2303. Where a passenger wantonly disregards any reasonable rule, the obligation to transport him ceases, and the company may expel him from the train, using no unnecessary force and not at a dangerous or inconvenient place. This is a common law right and has been restricted by statute only in cases of non-payment of fare. *I. C. R. R.* v. *Whittemore*, 43 Ill. 420.

2304. A rule requiring passengers to surrender their tickets when called for, is a reasonable one and may be enforced. *Ib.*

2305. It is unlawful to forcibly expel a passenger from a train between the usual stopping places for refusal to pay his fare, and trespass will lie for the injury, even though he agreed to get off if the train was stopped. *C. & N. W. Ry.* v. *Peacock*, 48 Ill. 253.

2306. In a suit to recover for being put off a freight train on which the plaintiff had taken passage without first procuring a ticket, it was objected that he had not proved such train was employed in carrying passengers: *Held*, it was sufficient that the evidence showed the defendants at the time of the occurrence were accustomed to carry passengers on freight trains; that notices were posted up around the window of the ticket office, that passengers on freight trains must first obtain tickets, and that there were persons on the train who had procured tickets. *I. C. R. R.* v. *Sutton*, 53 Ill. 397.

2307. Railway companies are liable for injuries caused to a person by reason of their servants putting him off, or compelling him to leave their train at any other than a regular station. *Ib.*

2308. A rule setting apart a car for the exclusive use of ladies, and gentlemen accompanied by ladies, is a reasonable rule and may be enforced. *C. & N. W. Ry.* v. *Williams*, 55 Ill. 185.

2309. In the absence of any other reasonable rule upon the subject the company cannot lawfully, from caprice, wantonness or prejudice, exclude a colored woman from the ladies' car on account of her color. *Ib.*

2310. If a passenger offers the conductor a worthless piece of paper, claiming it to be a pass, and on being informed it is not, refuses to pay fare or leave the train, the servants of the company will have the right to put him off at a regular station, and they may use the necessary force for that purpose. *C., R. I. & P. R. R.* v. *Herring*, 57 Ill. 59.

2311. It is not an unreasonable rule for a railroad company to require that persons desiring to ride on freight trains shall procure tickets sold expressly for such trains. *I. C. R. R.* v. *Nelson*, 59 Ill. 110.

2312. Where a person took passage on a freight train without first procuring the kind of ticket required by the rules of the company to entitle him to ride on that character of a train, it was held that the conductor had the right to require him to leave it at the usual place of getting on and off such trains, at a station. *Ib.*

2313. A railway company may require that passengers procure tickets before riding on freight trains, and conductors may expel from the cars at regular stations, such as neglect to comply with the regulation. *T., P. & W. Ry.* v. *Patterson*, 63 Ill. 304.

2314. If a passenger who neglects to procure a ticket to enable him to ride on freight train, is put off the train at a place other than a regular station, the company will be liable to him in compensatory damages. *Ib.*

2315. A railway company has the right to make a rule that no one shall be carried as a passenger on its freight trains. But when it is in the habit of carrying passengers on such a train, and has its regular hour for departure posted in its office at the station, it will not be justified in refusing to carry a passenger from such station, or in putting him off such train. *I. C. R. R.* v. *Johnson*, 67 Ill. 312.

2316. Company requiring tickets to ride on freight train, must keep ticket office open a reasonable time in advance of the hour of its departure. Failing in this, a person desiring passage, may enter the car to be carried to his destination on payment of the regular fare to the conductor. *Ib.* See *I. C. R. R.* v. *Cunningham*, 67 Ill. 316.

2317. If the holder of a ticket deports himself properly, the company have no right to refuse the ticket, or to admit him to the class of car his ticket designates, and when thus admitted, the company has no right, so long as he deports himself properly, to eject him from the train, before reaching his station. *Churchill* v. *C. & A. R. R.*, 67 Ill. 390.

2318. A railway company, not being obliged to give a lay over ticket, when it does so, it is upon the terms agreed upon by the parties, neither having the right to disregard them, when given and accepted. And when a passenger accepts a lay over ticket, marked good for thirty days only, he is bound by the terms imposed, and to make the same available, must use it within the time prescribed. *Ib.*

2319. The law does not require that the conductor in taking up a ticket shall give the holder a check, or punch the ticket and allow the passenger to hold it until all intermediate stations are passed. *C., B. & Q. R. R.* v. *Griffin*, 68 Ill. 499.

2320. If a passenger pays his fare to a certain station and the ticket agent by mistake gives him a ticket to an intermediate station, the demand of fare a second time by the conductor will be a breach of the implied contract on the part of the company to carry him to the proper station. By paying such demand his right of action will be as complete as if he resists the demand and suffers himself to be ejected. *Ib.*

2321. DISORDERLY CONDUCT. A passenger must observe proper decorum and observe all reasonable rules adopted by company. He is is not authorized to interpose resistance to every trivial imposition to which he may feel himself exposed, that must be overcome by counterforce, to preserve subordination. *Ib.*

2322. Where a passenger's ticket, by mistake, did not take him to the proper station, and 20 cents fare was demanded of him, which he refused to pay, and suffered himself to be forcibly ejected, and afterwards entered another car, and while the conductor was making change, used profane and obscene language in the presence of ladies, &c., for which he was again expelled, no unnecessary force being used : *Held*, that he was not entitled to recover anything. *Ib.*

2323. The use of grossly profane and obscene language by a passenger in a railway coach, where there are ladies, is such a breach of decorum, no matter if provoked to it, as will work a forfeiture of his right to be carried as a passenger, and the conductor will have the right to expel him from the cars, using no more force that is necessary. *C., B. & Q. R. R.* v. *Griffin*, 68 Ill. 499.

2324. Carriers of passengers may lawfully require those seeking to be carried to purchase tickets where convenient facilities to that end are afforded, to exhibit them to the persons designated by the carrier for that purpose, and surrender them after securing their seats. These are but reasonable rules. *Pullman Palace Car Co.* v. *Reed*, 75 Ill. 125.

2325. Case stated where passenger lost his berth ticket in a sleeping car, and was put out of the car. *Pullman Palace Car Co.* v. *Reed*, 75 Ill. 125.

2326. Expulsion of a man intruding himself into "ladies' private room" at a depot. *T., W. & W. Ry.* v. *Williams*, 77 Ill. 354.

2327. FAMILY TICKET. A railroad ticket, which on its face, purports to be for the exclusive use of a man and family, authorizes a son who is residing with him as a member of his family, to ride upon the road, although he may be over twenty years of age. *C. & N. W. Ry.* v. *Chisholm, Jr.*, 79 Ill. 584.

2328. If the purchaser of such ticket is at his purchase, informed that a son over twenty-one, would not under the regulations be permitted to ride on it, such regulation will form a part of the contract, and be binding on the purchaser or any one attempting to ride on it. *Ib.*

2329. To prove that a son of the holder of a family ticket had notice of a regulation that a son over age could not ride on the same, evidence that certain schedules were printed and furnished to the public by the company, is not admissable. It is proper, however, to show that such schedule was furnished to the purchaser of the ticket. *Ib.*

2330. DAMAGES. Where a person who is rightfully on a railway car and has paid his fare, is unlawfully expelled therefrom, he will be entitled to recover more than nominal damages, even though he sustains no pecuniary loss or actual injury to his person. *C. & N. W. Ry.* v. *Chisholm, Jr.*, 79 Ill. 584.

2331. Where a passenger conducts himself in an orderly and decent manner and offers to pay the fare fixed by the company, his expulsion by the conductor in a forcible manner, is unjustifiable, and the company will be liable civilly in an action for an assault and battery. *C., B. & Q. R. R.* v. *Bryan*, 90 Ill. 126.

2332. Where a passenger tenders the conductor a certain amount of fare to be carried to a certain station, which is less than the rate fixed by the company, saying he will pay no more, and the conductor retains a sum sufficient to take the passenger to an intermediate station, and returns the balance, the passenger will have the right on reaching such intermediate station, to pay the fare demanded from that point to the place of his destination, and upon his offering to pay the same, he cannot rightfully be put off the train. *Ib.*

2333. A railway company has no power to adopt rules and regulations prohibiting decently behaved persons from traveling on its road, who will pay their fare and conform to all reasonable requirements for the safety and comfort of passengers. *Ib.*

2334. When exemplary damages are recoverable for wrongful expulsion. *C., B. & Q. R. R.* v. *Bryan*, 90 Ill. 126.

2335. A carrier must not only protect his passengers against the violence and insults of strangers and co-passengers, but also against the violence and assaults of his own servants. If this protection is not afforded and the passenger is assaulted and beaten through the negligence of the carrier's servants, he will be responsible for the injury, especially for the assaults of his servants. *Ch. & E. R. R.* v. *Flexman*, 103 Ill. 546.

2336. Where a conductor of a railway company acting under instructions, refuses to accept a ticket issued by another company as agent of the former, and demands full fare, the passenger, if his ticket was issued by authority, may pay the fare again, and recover of the company to whom paid the amount so paid, as for a breach of the contract, or he may refuse to pay, and leave the train when so ordered and sue and recover of the company all damages sustained in consequence of his expulsion. But if he refuses to leave, he cannot recover for the force used to put him off, if no more is used than necessary. *Penn. R. R.* v. *Connell*, 112 Ill. 295.

2337. The responsibility of company for injury to passenger does not depend on his payment of fare. If he refuses to pay, the company may eject him. *O. & M. R. R.* v. *Muhling*, 30 Ill. 9. See also *C. & A. R. R.* v. *Randolph*, 53 Ill. 510;' *Arnold* v. *I. C. R. R.*, 83 Ill. 273; *C., B. & Q. R. R.* v. *McLallen*, 84 Ill. 109; *Shelton* v. *L. S. & M. S. Ry.*, 29 Ohio St. 214; *Crawford* v. *C., H. & D. R. R.*, 26 Ohio St. 580; *Townshend* v. *N. Y. C. & H. R. R.*, 56 N. Y. 296.

2338. BADGE. § 32. Every conductor, baggage-master, brakeman, or other servant of any railroad corporation in this state, employed on a passenger train, or about the passenger depots, shall wear upon his hat or cap a badge which shall indicate his office. No conductor without such badge shall demand, or be entitled to receive from any passenger, any fare, toll or ticket, or exercise any of the powers of his office; and neither shall any other of said officers or servants, without such badge, be authorized to meddle or interfere with any passenger, his baggage or property. [R. S. 1887, p. 1019, § 95; S. & C., p. 1944, § 95; Cothran, p. 1159, § 81.]

2339. COMMON LAW LIABILITY NOT TO BE LIMITED. § 33. That whenever any property is received by any railroad corporation to be transported from one place to another, within or without this state, it shall not be lawful for such corporation to limit its common law liability safely to deliver such property at the place to which the same is to be transported, by any stipulation or limitation expressed in the receipt given for the safe delivery of such property. [R. S. 1887, p. 1019, § 96; S. & C., p. 1945, § 96; Cothran, p. 1159, § 82. See ante, 162.]

2340. Railway companies as common carriers may restrict their common law liability by such contracts as may be specially agreed upon, except their liability for gross negligence or willful misfeasance. *I. C. R. R.* v. *Morrison*, 19 Ill. 136.

2341. WHEN AUTHORITY TO SIGN PRESUMED. Where a shipper of cattle made such a contract and delivered part of the cattle, it will be presumed that other persons, delivering the remainder acted as agents, and had authority to sign similar contracts. *Ib.*

2341a. RESTRICTION BY NOTICE. This liability cannot be restricted by notice, even when it is brought home to the knowledge of the owner. *Western Transp. Co.* v. *Newhall*, 24 Ill. 466.

2342. No distinction can be made in a notice in the newspapers or by hand bills, or one printed on the back of a receipt given. *Ib.*

2343. Notice printed on the back of carrier's receipt, forms no part of the contract, and need not be noticed in the declaration. The express assent of the shipper to such restriction must be proved, in order to give effect to it. *Ib.*

2344. PRESUMPTION AS TO SHIPPER'S ASSENT. As the carrier is bound to receive and carry all goods offered him subject to all the incidents of his employment, there can be no presumption that the owner intended to abandon any of his rights. *Ib.*

2345. BURDEN OF PROOF. The *onus* of proving the contract restricting the carrier's liability is upon him. *Ib.*

2345a. SPECIAL CARRIER. The rule is different with persons who are not common carriers and who are not bound to render the service required. They may make their own terms, and the owner of goods is presumed to assent to them by the delivery of the goods. *Ib.*

2346. WHAT MAY BE DONE BY NOTICE. The carrier may qualify his liability by a general notice to all who may employ him, of any reasonable requisition to be observed in regard to the manner of delivery and entry of parcels and various other matters, but he cannot avoid his liability as insurer by any such notice. *Ib.*

2347. RESTRICTION IN FREE PASS. The acceptance of a free pass with an indorsement printed thereon exempting the company from all liability for injury caused by negligence to the person or property of the holder, will protect the company for any injury not the result of gross or reckless negligence. *I. C. R. R.* v. *Read*, 37 Ill. 484.

2348. Railway companies have the right to restrict their common law liability as common carriers, by such contracts as may be agreed upon specially, they still remaining liable for gross negligence or willful misfeasance against which good morals and public policy forbid they should be permitted to stipulate. *Ib.*

2349. A railway company may restrict its liability for loss or injury to property placed in its charge for transportation, by special agreement, the carrier being still held responsible for gross negligence or willful misfeasance. *I. C. R. R.* v. *Smyser & Co.*, 38 Ill. 354.

2350. The rule is limited to cases where there is a special contract. It is not competent to limit the liability of the carrier by merely proving a usage on his part in giving bills of lading exempting him from certain classes of losses. *Ib.*

2351. Where goods are shipped under a verbal contract, and a day or two after their delivery, the subsequent making out and signing a freight bill with conditions and limitations, will not alter the carrier's liability under the verbal agreement, unless it was accepted as the contract of the parties, and this is a question of fact. *Baker* v. *M. S. & N. Ind. R. R.*, 42 Ill. 73.

2352. Where receipt is given for the goods containing a provision limiting the common law liability of the company, and the shipper accepts the same with a knowledge of its terms, and intending to assent to the restrictions contained in it, it becomes his contract as fully as if he had signed it. *Adams Express Co.* v. *Haynes*, 42 Ill. 89.

2353. The simple delivery of such a receipt to the shipper is not conclusive upon the latter. Whether he had knowledge of its terms and assented to its restrictions, is for the jury to determine as a question of fact upon evidence *aliunde*, and all the circumstances attending the giving of the receipt are admissible in evidence on that question. *Ib.*

2354. While a railway company may protect itself against certain risks assumed by common carriers, and belonging to their vocation, it is contrary to good morals and public policy that it should be allowed to stipulate against its own gross negligence or willful defaults. *I. C. R. R.* v. *Adams*, 42 Ill. 474.

2355. A contract for the shipment of hogs provided that the company should not be liable for loss "by delay of trains, or any damage said property might sustain, except such as might result from a collision of a train, or where cars were thrown from the track in course of transportation." During the trip, one car was thrown from the track by reason of a broken rail, while all the cars containing the hogs remained on the track: *Held*, that the company was liable for whatever hogs were lost or whatever shrinkage occurred by reason of the delay caused by the accident, but not for delay caused by cold weather. *I.C.R.R.* v. *Owens*, 53 Ill. 391.

2356. The carrier cannot claim exemption from liability for an injury to corn shipped caused by delay in transportation, under a clause in the bill of lading which relieves him from loss on perishable property. *I. C. R.R.* v. *McClellan*, 54 Ill. 58.

2357. While the carrier who first receives the goods to be carried over his and other lines may not by general notice, yet he may, by special contract with the shipper, limit his liability to such damage or loss as may occur on his own line of carriage. A carrier may by special contract relieve himself of his common law liability. *I. C. R. R.* v. *Frankenberg*, 54 Ill. 88.

2358. If a shipper takes a receipt from the carrier to whom the goods are delivered in the first instance, containing a clause that the carrier so receiving assumes no other liability for their safety or safe carriage, than may be incurred on its own road, with knowledge of its terms, and intending to assent to the restrictions contained in it, the carrier will be free from his common law liability for loss occurring beyond his own line. Whether he assented to it is a question of fact. *Ib.*

2359. A box of goods was delivered to a railway company marked to a point beyond its line of road. The bill of lading was called by the company its "through freight contract," and it contained this clause: "Which we promise to transport over the line of this railway, to the company's freight station at its terminus, and deliver to the consignee or owner, or to such company (if the same are to be forwarded beyond the limits of this railway) whose line may be considered a part of the route to the place of destination of said goods, it being distinctly understood that the responsibility of this company shall cease at the station where such goods are delivered to such person or carrier." Among the conditions printed in the bill of lading was this: "The responsibility of this company as a common carrier under this bill of lading to commence on the removal of the goods from the depot on the cars of the company and to terminate where unloaded from the cars at the place of delivery." This freight was never unloaded or delivered at this terminus, but proceeded to its destination in the cars in which it was received: *Held*, that this was a through freight contract, and the company liable beyond the terminus of their road. *T., P. & W. Ry* v. *Merriman*, 52 Ill. 123.

2360. RECEIPT AS A CONTRACT. The delivery of a receipt for the goods to the shipper is not conclusive upon him that the conditions therein set forth constitute a contract. Whether the shipper had knowledge of its terms and assented to its restrictions, is a question of fact for the jury to determine upon evidence *aliunde*, and all the circumstances attending the giving of the receipt are admissible in evidence on the question. *A. M. U. Express Co.* v. *Schier*, 55 Ill. 140.

2361. Railway company receiving goods to be shipped to a point beyond the terminus of its line, may by express agreement limit its liability to its own route and to its terminus. *C. & N. W. Ry.* v. *Montfort*, 60 Ill. 175.

2362. Where the shipper takes a receipt from the company restricting its liability to its own line of road, if he accepts it with a full knowledge of such conditions and intending to assent to them, it becomes his contract as fully as if he had signed it. *Ib.*

2363. Whether the shipper accepted a receipt with a knowledge of such restriction and with the intention to assent to it, is a question of fact for the jury. *C. & N. W. Ry.* v. *Montfort*, 60 Ill. 175.

2364. The insertion by the carrier in the shipping receipt that the company will not be liable for loss beyond a certain sum, being much less than the value of the goods shipped, will not release the carrier, unless it appears that the shipper knew of and assented to the limitation. *Adams Express Co.* v. *Stettaners*, 61 Ill. 184.

2365. Where the shipper has assented to the clause in the receipt that the carrier shall not be liable beyond a given sum, and as to that, only for gross negligence, the burden of proof of due care will rest on the carrier. It cannot by any contract excuse itself from reasonable care and diligence. *Ib.*

2366. Where no question is made as to the knowledge of the shipper of a provision in the bill of lading, that the carrier should not be liable for loss or damage to the property by fire or other casualty, while in transit, or at depots or landing, at the point of delivery, it will be inferred that the shipper received the bill with knowledge of its contents and agreed to its terms. *Anchor Line* v. *Knowles*, 66 Ill. 150.

2367. If the consignor of a package of money assents to a clause in the receipt stating that the carrier company undertook to forward the package "to the nearest point of destination reached by" such company, it becomes his contract as fully as if he had signed it, and he will be bound by its terms, and cannot hold the company liable for delays of another company taking the package. *United States Express Co.* v. *Haines*, 67 Ill. 137.

2368. If a shipper takes a receipt for his goods from a common carrier which contains conditions limiting the liability of the carrier, with a full understanding of such conditions, and intending to assent to them, it becomes his contract as fully as if he had signed it and he will be bound by the conditions. *Anchor Line* v. *Dater*, 68 Ill. 369.

2369. It does not follow because the carrier delivers to the shipper a receipt containing limitations of his liability, that the shipper assents thereto, as he has no alternative but to accept such a receipt as the carrier may give. Whether the shipper has assented to such conditions, is a question of fact for the jury. *Anchor Line* v. *Dater*, 68 Ill. 369.

2370. A limitation of the liability of an express company not to exceed $50, unless the value of the goods is truly stated, if brought to the knowledge of the consignor, is reasonable and consistent with public policy. *Openheimer & Co.* v. *United States Express Co.*, 69 Ill. 62.

2371. The established legal construction of conditions in the contract of carriers exempting them from liability is, not to treat them as providing against losses or injuries occasioned by actual negligence on their part. *Ib.*

2372. The fact that an express company has settled for other losses without insisting on the restrictions of its liability in the contract, will not preclude it from raising the question of its liability in a similar case subsequently arising. *Ib.*

2373. A distinction exists between the effect of those notices by a carrier by which it is sought to discharge him from duties which the law has annexed to his employment, and those designed simply to insure good faith and fair dealing on the part of his employer. In the former, notice without assent to the attempted restriction, is ineffectual, while in the latter, actual notice alone will be sufficient. *Ib.*

2374. Where a carrier delivers goods to a forwarder, who is its agent and the agent of the company to whom the same is delivered, and he gives a bill of lading limiting the duty of the latter to deliver the goods to another company, this will make the bill of lading a contract binding upon the first and second carriers, and the second will not be responsible for the delivery of the goods to the consignee by the last carrier. *C. & N. W. Ry.* v. *N. L. Packet Co.*, 70 Ill. 217.

2375. Common carriers may limit their common law liability. *Field* v. *C. & R. I. R. R.*, 71 Ill. 458.

2376. Where a railway company receives goods marked for a particular place, it is bound by the common law to deliver at that place; but it may restrict this liability by a contract fairly and understandingly made; and where so made, if in the form of a bill of lading, or otherwise, and the terms are understood and accepted by the shipper, it becomes the contract of the parties. *Field* v. *C. & R. I. R. R.*, 71 Ill. 458.

2377. The receipt or bill of lading of goods marked to New York, recited that the goods were to be transported over defendant's road to a certain station, and there delivered in good order to another company, whose line was a part of the route to the place of destination, and that the liability of the defendant should cease when the goods were so delivered at that station to the other company. The shipper accepted this receipt with the knowledge of its contents: *Held*, that it became a binding contract and that the liability of defendant ended with the delivery of the goods to the next carrier. *Field* v. *C. &. R. I. R. R.*, 71 Ill. 458.

2378. Whether the shipper has knowledge of and assents to a clause in a bill of lading or receipt for goods delivered to a common carrier, whereby the common law liability is limited, is a question of fact to be determined by the evidence in each case. *Field* v. *C. & R. I. R. R.*, 71 Ill. 458.

2379. Where goods are delivered to a carrier in Wisconsin, the contract to be performed there, the laws of that state will govern as to the construction of the contract, and determine the extent of the carrier's undertaking. *M. & St. P. Ry.* v. *Smith*, 74 Ill. 197.

2380. Where a carrier receives live stock for transportation, and a loss is sustained by the owner in consequence of their not being supplied with water, the burden of proof to show an exemption from liability rests upon the carrier. *T., W. & W. Ry.* v. *Hamilton*, 76 Ill. 393.

2381. It is only where the contract is for through transportation, that each connecting carrier will be entitled to the benefits and exemptions of the contract between the shipper and first carrier. *M. D. Transp. Co.* v. *Bolles*, 80 Ill. 473.

2382. Where a carrier receives goods for transportation marked to a place beyond the terminus of its line, without any special contract its liability as an insurer will continue until it delivers them to a connecting carrier. *M. Des. Transp. Co.* v. *Bolles*, 80 Ill. 473.

2384. Where goods are delivered to a carrier to be carried to a place and the charges for transporting to that place are paid in full, and there is no contract limiting its liability, such carrier will be responsible for the delivery of the goods at the place named, notwithstanding its line ends before reaching such place and the goods are delivered to another carrier in good order at the termination of its line. *Adams Express Co.* v. *Wilson*, 81 Ill. 339.

2385. The doctrine is settled in this state that railroad companies, may,¡ by contract, exempt themselves from liability on account of the negligence of their servants, other than that which is gross and willful. *Arnold* v. *I. C. R. R.*, 83 Ill. 273.

2386. The undertaking of a railway company to carry a passenger on a freight train, and the extra care and expense required in such case, form a sufficient consideration for a contract made with a passenger restricting and limiting its liability; but the same terms must be extended to and applied to all persons desiring to ride on such trains. *Ib.*

2387. The carrier may limit its obligation to carry safely over its own lines or only to points reached by its own carriages, and for safe storage and delivery to the next carrier in the route beyond, although the goods are marked to a point beyond its line. *Erie Ry.* v. *Wilcox*, 84 Ill. 239.

2388. A clause in the receipt given the owner for goods so restricting the carrier's obligation, if understandingly assented to by the shipper, will as effectually bind him as if he had signed it. *Ib.*

2389. Where the exemption is once established, the carrier will only be responsible on account of actual negligence or willful misconduct. The rule is the same if the goods are shipped to a point beyond the carrier's own line. *Ib.*

2390. LIMITING LIABILITY BY NOTICE. A carrier cannot discharge itself from duties which the law has annexed to the employment, by notice alone to the shipper. The shipper must assent to it, to make it effectual; but it is otherwise in respect to those duties designed simply to insure good faith and fair dealing. There a notice is sufficient. *Erie Ry.* v. *Wilcox*, 84 Ill. 239.

2391. The law on grounds of public policy, will not permit a common carrier of passengers or freight to contract against liability for its own actual negligence or that of its servants or employes. *Erie Ry.* v. *Wilcox*, 84 Ill. 239.

2392. If a passenger on a railway train while riding under a free ticket containing the usual restrictions, is injured by an accident he cannot hold the company liable, except for gross negligence, or a degree of negligence having the character of recklessness. *T., W. & W. Ry.* v. *Beggs*, 85 Ill. 80.

2393. A clause in a receipt or bill of lading exempting the carrier from a common law liability is not binding on the shipper unless it appears that he knew of and assented to the exemption, and this is a question of fact. *M. D. Transp. Co.* v. *Theilbar*, 86 Ill. 71.

2394. A common carrier is bound to accept and safely carry goods when properly tendered for shipment, unless destroyed by the act of God or the public enemy, and has no right to exempt itself from loss by fire, except by virtue of a special contract to that effect. It cannot limit its liability by its own act alone. *M. D. T. Co.* v. *Theilbar*, 86 Ill. 71.

2395. Where the bill of lading given shows the goods are to be forwarded to a particular place only, which is short of their destination, and the consignor has been a frequent shipper by the same line and was in the habit of receiving like bills of lading, it will be presumed he was familiar with its contents and assented to the same. *M. D. & Tr. Co.* v. *Moore*, 88 Ill. 136.

2396. The right of a carrier to limit its common law liability by contract, if made fairly and advisedly on behalf of the shipper, cannot be denied; but the mere fact that the bill of lading given contains a clause exempting the carrier from loss of the goods by fire, cannot be held conclusive of such contract. *M. D. Trans. Co.* v. *Leysor*, 89 Ill. 43.

2397. If a shipper with full knowledge of the terms and conditions of a bill of lading, assents to and accepts the same as the contract under which the goods are shipped, then the bill of lading will constitute a binding contract which will control the rights and liabilities of the parties. Whether the shipper knows the terms and conditions of a bill of lading and assents to the same, is a question of fact. *Ib.*

2398. Where the shipper has no knowledge that the bill of lading given contains a provision releasing the carrier from liability for loss by fire, and the goods are destroyed by fire before reaching their desti-

nation, and while in the custody of the carrier, the latter will be liable
to the owner for their value. *Ib.*

2399. A common carrier can only limit or restrict his liability by
agreement, and where the carrier gives a receipt for goods to be shipped,
containing a restriction of his liability, it must appear that the shipper was aware of such restrictions, otherwise there is no presumption
of his assent thereto. *M. D. Trans. Co.* v. *Jœsting*, 89 Ill. 152.

2400. To make such a restriction binding on the shipper, he must
expressly agree to it, or he must accept the receipt under such circumstances as clearly show his assent to the restriction. The receiving of
the receipt does not prove assent, but if the party reads the receipt
and makes no objection, his assent may be inferred. *Ib.*

2401. In the absence of evidence, it will be presumed that the shipper being the merchant who sold the goods, had only authority to ship
them with all the liabilities of the carrier attaching, without exceptions of any description. *Ib.*

2402. Where no receipt is given at the time a package is delivered
to an express company for transportation, the company cannot limit
its liability by a receipt afterwards given, where the proof negatives
all presumption of any knowledge on the part of the shipper that the
receipt contained a clause limiting the carrier's liability, or that the
carrier claimed any such limitation. *Am. Express Co.* v. *Spellman*,
90 Ill. 455.

2403. A shipper of goods is not bound by a clause in a carrier's
receipt or bill of lading given on the receipt of goods for transportation, limiting the common law liability of the carrier, unless the
shipper assents to the same. *Erie & Western Transp. Co.* v. *Dater*,
91 Ill. 195.

2404. The assent of a shipper to the conditions in a receipt or bill
of lading limiting the carrier's liability will not be inferred from the
mere fact of acceptance of the bill or receipt without objection. Nor
will it be conclusively inferred from the fact of the previous acceptance of a large number of similar bills not filled up by the shipper or
held in his possession to be filled up. *Ib.*

2405. The acceptance of a bill of lading containing a restriction of
the carrier's liability and the previous practice of giving and receiving similar bills of lading, are evidence tending to show that the
limitation of liability therein, was assented to by the shipper; but
neither one, nor both such facts will be conclusive evidence thereof.
Ib.

2406. The law of the state in which the contract is made for the
transportation of goods, must control as to its nature, interpretation
and effect. *M. C. R. R.* v. *Boyd*, 91 Ill. 268.

2407. An agent shipping goods a few days after the delivery of
the goods, and while they are in transit, cannot by taking a receipt
limiting the carrier's liability, bind the consignee. *M. C. R. R.* v.
Boyd, 91 Ill. 268.

2408. A common carrier can claim no exemption from liability
for the loss of goods entrusted to him, except such as is given by
express contract. *Boscowitz* v. *Adams Express Co.*, 93 Ill. 523.

2409. The contract must be assented to by the shipper with a view
to release the duties imposed by the law, and when this is once established the carrier, in case of loss, will only be responsible on account
of negligence or willful misconduct. He cannot contract against his
own actual negligence. *Boscowitz* v. *Adams Express Co.*, 93 Ill. 523.

2410. A clause in a receipt given to a shipper of goods limiting
and restricting the carrier's liability incident to its general employ-

ment, if understandingly assented to by the owner, will as effectually bind him as though he had signed it, but whether such restrictions have been assented to in a given case, is always a matter of evidence. *Ib.*

2411. The fact that the owner of goods, by himself or clerk, filled up a receipt taken for goods shipped, is evidence tending to show that the shipper had notice of the conditions and restrictions in the printed part thereof and assented to them, but it is not conclusive. It is still a question of fact. *Ib.*

2412. Where carrier gave a receipt for three separate distinct bales of furs, containing a printed clause that the company should not be liable for any loss or damage "of any box, package or thing," for over $50, unless the true value thereof was therein inserted: *Held*, that the limitation was not to be applied to the three bales, but as to each one separately. *Ib.*

2413. While it is true that a railroad carrier may by contract restrict its liability to its own line, there is no doubt that it may also extend its liability beyond its own line. *St. L. & I. M. R. R.* v. *Larned*, 103 Ill. 293.

2414. A railway company by accepting and acting under its charter, becomes a carrier of persons and property, and the law imposes all the duties and liabilities of a common carrier upon it, and such company cannot exonerate itself from such duty and responsibility by contract with others, nor in anywise escape or free itself from liability, unless released by the general assembly. *W., St. L. & P. Ry.* v. *Peyton*, 106 Ill. 534.

2415. A provision in a shipping contract voluntarily and understandingly entered into, that in consideration of reduced rates, no claim for damages should be made, unless made in writing verified by the affidavit of the shipper and delivered to the general freight agent of the carrier at his office within five days from the time the stock is removed from the cars, will be binding on the shipper, and is not void as being contrary to any law or public policy. *Black* v. *W., St. L. & P. Ry.*, 111 Ill. 351.

2416. A carrier may by special contract with the shipper, limit his liability to such damage or loss as may arise on his own line of carriage. *W., St. L. & P. Ry.* v. *Jaggerman*, 115 Ill. 407.

2417. Notwithstanding a provision in a bill of lading that the carrier should not be responsible for "damage to perishable property of any kind occasioned by delays from any cause," he may and will become liable for delay as the result of actual negligence. But proof of delay merely, is not sufficient to show negligence in transporting the goods. *Ib.*

2418. In an action by the shipper of apples under a bill of lading exempting the carrier from liability for damage to perishable property from delay, it is competent for the defendant to prove that prior to such shipment the plaintiff had filled up similar blank bills of lading for shipments which contained the same stipulation in regard to perishable property, as going to show the plaintiff's knowledge of and assent to such provision. *Ib.*

2419. The common law liability of a carrier may be limited by a special contract signed by both the contracting parties, except that public policy requires that the carrier should not be allowed to stipulate against the consequence of its own actual negligence or willful default. *C., B. & Q. R. R.* v. *Hale*, 2 Bradw. 150.

2420. Where the consignor merely takes a receipt containing a limitation of liability, it will not bind him, unless he knowingly assented to such restriction, but it is error to so instruct, where the consignee signs the contract. *Ib*.

2421. A condition in a contract for the transportation of goods forms no part of the contract, where it was not known or assented to by the shipper. *Adams Express Co.* v. *King*, 3 Bradw. 316.

2422. The statute prohibits a common carrier from limiting by contract, its liability to deliver the goods safely at their destination; but in this case, if the carrier was guilty of no negligence, it was not liable at common law, and the contract in that regard, is not within the statute and prohibited thereby. *I. & St. L. Ry.* v. *Jurey*, 8 B. 160.

2423. A carrier cannot by contract relieve himself from responsibility for his own negligence or that of his servants. Neither can he limit his common law liability to safely deliver property received for transportation. *W., St. L. & P. Ry.* v. *Black*, 11 Bradw. 465.

2424. A carrier is not restricted from providing in a shipping contract that in case any claim for damages is made, notice of the same shall be given within a prescribed time. Such a provision is reasonable and is not a limitation upon the common law duty of a carrier to safely deliver property received for shipment. *Ib*.

2425. Where goods are received by a common carrier, marked for transportation to a place beyond its line, and the bill of lading limited the common law liability of the carrier to safe carriage over its own line: *Held*, that the inhibition contained in Ch. 114, § 82, R. S. 1874, does not apply to a case where the carrier is under no obligation at common law to undertake to carry goods beyond its own line. *C. & N. W. Ry.* v. *Church*, 12 Bradw. 17.

2426. A custom that a carrier shall not be liable for injury to, or loss or destruction of live stock beyond the value of $100, is against public policy, and a custom which will excuse a carrier from acts of negligence, is invalid. *C., R. I. & P. R. R.* v. *Harmon*, 12 Bradw. 54.

2427. Notice to shipper of a rule that the carrier will not transport live stock unless the shipper signs a special contract limiting the carrier's liability to $100, does not create a contract, making the rule binding. The assent of the shipper must appear before he can be bound. *Ib*.

2428. A railway company has a right to exempt itself by contract from liability for loss or damage to goods delivered to it for carriage, except where the same is caused by negligence of its own servants. *I. C. R. R.* v. *Jonte*, 13 Bradw. 424.

2429. The contract in this case does not violate the statute which prohibits common carriers from limiting their common law liability to safely deliver, by a stipulation in the receipt, for the limitation to the liability of the carrier is not expressed in the receipt given by the carrier. Statute does not prohibit contracts limiting liability made independent of receipt. *I. C. R. R.* v. *Jonte*, 13 Bradw. 424.

2429a. Whether a shipper knew of the terms and conditions of a bill of lading and assented thereto, are questions of fact to be determined by the jury from the evidence. *L. S. & M. S. Ry.* v. *Davis*, 16 Bradw. 425.

2430. In attempting to prove the shipper's knowledge of or assent to the terms of the bill of lading, the carrier is not limited to evidence of any one particular fact, but may prove all the circumstances surrounding the transaction which have any legitimate tendency to establish the shipper's knowledge or assent. *Ib*.

2431. A common carrier cannot by contract stipulate for complete indemnity against his gross negligence: as to the right to stipulate for a partial exemption from his full liability authorities *pro* and *con* cited. *C., R. I. & P. R. R.* v. *Harmon*, 17 Bradw. 640.

2432. A special contract providing that no claim for loss by delay in transportation shall be payable or recoverable, but shall be absolutely barred, unless a written statement of it shall be made out and sent by the shipper to the general freight agent of the company within five days after such loss occurs, is valid and binding. *C. & A. R. R.* v. *Simms*, 18 Bradw. 68.

2433. A clause in a receipt restricting the carrier's liability to his own lines and for safe delivery to next carrier, is binding on the shipper the same as if signed by him, if knowingly assented to by him. *Fortier* v. *Penn. Co.*, 18 Bradw. 260.

2434. The limitation must be an affirmative one and the burden is on the company to show it. *Fortier* v. *Penn Co.*, 18 Bradw. 260.

2435. A common carrier can limit his ordinary liability only by a special contract, and the acceptance of a receipt or bill of lading with printed conditions, or notice limiting the carrier's liability by the owner or shipper of goods without dissent, will not establish such a contract. *Western Transit Co.* v. *Hosking*, 19 Bradw. 607.

2436. In such case, it is necessary to show that the shipper knew of and assented to the exemption, and such assent must be shown by other and additional evidence, and is not the subject of presumption from the terms of the receipt alone. *Ib.*

2437. The *lex loce* governs as to the validity and construction of the contract of shipment, but in the absence of proof, it will be presumed that the common law of another state is the same as in this state. *Fortier* v. *Penn. Co.*, 18 Bradw. 264; *Western Transit Co.* v. *Hosking*, 19 Bradw. 607.

2438. Carrier may only by special contract with the owner or shipper, limit his common law liability. *York Co.* v. *Ill. C. R. R.*, 3 Wall. 107; 70 U. S. 107; same case, 1 Biss. 377.

2439. May by special contract limit his liability in case of fire. *Van Schaack* v. *Northern Transp. Co.*, 3 Biss. 394.

2440. Statute prohibiting carriers from limiting their liability does not apply to limitation of amount of liability, where shipper fails to state the value. *Mather* v. *Am. Express Co.*, 9 Biss. 293.

2441. Statute does not prohibit carrier from contracting in bill of lading for benefit of insurance in case of loss. *Phœnix Ins. Co.* v. *Erie & Western Transp. Co.*, 10 Biss. 18.

2442. Railway company is liable as a common carrier for loss of a car of another carrier while such car is being hauled by railway company over its line. *P. & P. U. Ry.* v. *C., R. I. & P. Ry.*, 109 Ill. 135.

2443. To FURNISH AX, SAW, SLEDGE, ETC., FOR EACH CAR. § 34. That every railroad corporation shall furnish each car used for the transportation of passengers with one woodman's ax, one hand-saw, one sledge hammer and two leather buckets; said articles to be kept in good repair, ready for instant use, and in some convenient place in such car, easy of access in case of collision or other accident. [R. S., p. 1019, § 97; S. & C., p. 1948, § 97; Cothran, p. 1160, § 83.]

2444. FURNISH COUPLINGS—PENALTY. § 34½. It shall be the duty of all railroad corporations operating any railroad in this state, to provide such of their passenger cars as are used in trains with some suitable automatic coupling, or other coupling which will secure personal safety, within one year from the time this law goes into effect, and any company refusing or neglecting to provide such automatic coupling, or other couplings which will secure personal safety, for each passenger car so used in trains, shall be liable to a fine of not less than 25 nor more than $50. [R. S., p. 1019, § 98; S. & C., p. 1948, § 98; Cothran, p. 1160, § 84.]

2445. Company liable to brakeman for an injury to his hand caused by defective coupling on cars. *T., W. & W. Ry.* v. *Fredericks*, 71 Ill. 294.

2446. Company will not be liable for an injury to freight conductor received in consequence of the coupling of freight car suddenly getting out of repair, unless its attention has been called to it, or it could by great care have discovered the defect, and had opportunity to make the needed repairs. *I., B. & W. R. R.* v. *Flanigan*, 77 Ill. 365.

2447. Company will not be liable to an employe for personal injury received while coupling cars having double buffers, simply because a higher degree of care is required in using them than in those differently constructed. *Ib.*

2448. Where a brakeman in uncoupling a combination car to be left on a switch, which has a railing, instead of remaining on such car as it was his duty, gets upon a flat car next to it, and is injured in consequence of a jerk in starting, his own negligence will bar a recovery by him. *C. & A. R. R.* v. *Rush*, 84 Ill. 570.

2449. Injury to brakeman while coupling, from other defects. *C. & E. Ill. R. R.* v. *Rung*, 104 Ill. 641.

2450. FLAGMEN—SHELTER. § 35. In all cases where the public authorities having charge of any street over which there shall be a railroad crossing, shall notify any agent of the corporation owning, using or operating such railroad, that a flagman is necessary at such crossing, it shall be the duty of such railroad company, within sixty days thereafter, to place and retain a flagman at such crossing, who shall perform the duties usually required of flagmen; and such flagman is hereby empowered to stop any and all· persons from crossing a railroad track when, in his opinion, there is danger from approaching trains or locomotive engines; and any railroad company refusing or neglecting to place flagmen, as required by this section, shall be liable to a fine of $100 per day for every day they shall neglect or refuse to do so; and it is hereby made the duty of such public authorities having charge of such street, to enforce the payment of such fine, by suit, in the name of the town or municipal corporation wherein such crossing shall be situate, before any court of competent jurisdiction in the county, and the prosecuting attorney shall attend to the prosecution of all suits as directed by said public authorities. All the moneys collected under the provisions of this act shall be paid into the treasury of the town or municipal corporation in whose name such suits shall have been brought: *Provided,* that when any railroad company is required to keep a flagman at a crossing, it shall have the

right to erect and maintain in the highway or street crossed
a suitable house for the shelter of such flagman, the same to
be so located as to create the least obstruction to the use of
such street or highway, and afford the best view of the railroad track in each direction from such crossing. [L. 1869,
p. 314, § 8; R. S. 1887, p. 1019, § 99; S. & C., p. 1948, § 99;
Cothran, p. 1160, § 85. See ante, 145, 146.]

2451. Company liable to one injured by neglect to have a person
stationed on rear end of cars pushed through a city, or on ground to
give warning. *I. C. R. R.* v. *Ebert*, 74 Ill. 399.

2452. Company not liable for an injury to a switchman for not
providing rules whereby a watchman should have been kept on rear
end of train that produced the injury, a watch or look-out being kept
from the engine. *C. & N. W. Ry.* v. *Donahue*, 75 Ill. 106.

2453. Absence of watchman at a much frequented street crossing
whose duty it was to warn persons crossing the tracks of danger,
makes the company liable for an injury caused thereby. *St. L., V. &
T. H. R. R.* v. *Dunn*, 78 Ill. 197.

2454. Duty of railway companies to give warning at street crossing. *P. & P. U. Ry.* v. *Clayberg*, 107 Ill. 644; *C., R. I. & P. R. R.* v.
Eininger, 114 Ill. 79; *C. & A. R. R.* v. *Gretzner*, 46 Ill. 74; *C. & A. R. R.*
v. *McLaughlin*, 47 Ill. 265.

2455. PENALTIES. § 36. If any railroad corporation, or
any of its agents, servants or employes shall violate any of
the provisions of this act, such corporation, agent, servant or
employe shall, severally, unless otherwise herein provided,
be liable to a fine of not less than $10 nor more than $200, to
be recovered in an action of debt, in the name of the people
of the state of Illinois, for the use of any person aggrieved,
before any court of competent jurisdiction. [R. S., p. 1019,
§ 100; S. & C., p. 1948, § 100; Cothran, p. 1160, § 86.]

2456. Under prior statute giving a special remedy for failure to
ring a bell or whistle before reaching a public road crossing, and also
a general remedy for any failure to comply with act, it was *held* that
the action for the penalty might be prosecuted in either form of
action. *T., P. & W. Ry.* v. *Foster*, 43 Ill. 480.

2457. CORPORATION DEFINED. § 37. The word "corporation," as used in this act, shall be construed to include
all companies, lessees, contractors, persons, or association of
persons, owning operating or using any railroads in this
state. [R. S. 1887, p. 1019, § 101; S. & C., p. 1949, § 101;
Cothran, p. 1161, § 87.]

2458. CONTRACTORS. Corporation liable for the acts of its contractors exercising its corporate powers. *Lesher* v. *Wab. Nav. Co.*,
14 Ill. 85; *Hinde* v. *Wab. Nav. Co.*, 15 Ill. 72; *C., St. P. & F. D. L. R. R.*
v. *McCarthy*, 20 Ill. 385.

2459. Railway company cannot release itself from liability by leasing its road. *O. & M. R. R.* v. *Dunbar*, 20 Ill. 623.

2460. Liable for the torts and trespasses of its lessees, and for the
torts and acts of its contractors. *C. & R. I. R. R.* v. *Whipple,* 22 Ill.
105.

—20

2461. Where a railway company allows another company to use its unfenced road, and the latter kills stock upon the track each will be liable. *I. C. R. R.* v. *Kanouse*, 39 Ill. 272.

2462. Where two companies are using the same line of road, one company being the owner and the other using the road by its permission, the company owning the track is liable for damages done by reason of an unfenced track, by the trains of the other company, the same as if done by its own trains. and the other also will be liable. *T., P. & W. Ry.* v. *Rumbold*, 40 Ill. 143.

2463. As to liability of private owner for negligence of his contractor see *Schwartz* v. *Gilmore*, 45 Ill. 455; *Scammon* v. *Chicago*, 25 Ill. 424, 438; *Pfau* v. *Williamson*, 63 Ill. 16; *P. S. Loan and Trust Co.* v. *Doig*, 70 Ill. 52; *Hale* v. *Johnson*, 80 Ill. 185; *Kipperly* v. *Ramsden*, 83 Ill. 354.

2464. Lessor company liable for injuries to its passengers caused by the negligence of another company which it allows to use its road. *I. C. R. R.* v. *Barron*, 5 Wall. 90; 1 Biss. 412.

2465. The owner of posts taken and used by railway contractors in fencing the company's track, may maintain trover against the company for the value of the posts. *St. L., V. & T. H. R. R.* v. *Kaulbrumer*, 59 Ill. 152.

2466. Where the wrongful act is done by contractors or lessees of a chartered company in pursuance of the special powers and privileges conferred upon the company by its charter, and but for such charter they would have no right to prosecute the particular business, such contractors or lessees, as to third parties who may be injured by their acts will be regarded as the servants of the company acting under its direction, and the company will be held liable for any abuse of such of its privileges by its contractors or lessees. *West* v. *St. L., V. & T. H. R. R.*, 63 Ill. 545.

2467. Company not liable to a servant of contractors employed to build a freight house, who was poisoned by breathing a noxious exhalation from an ingredient in the paint. *Ib.*

2468. Where the contractors of a railway company are guilty of trespasses upon the land of another in constructing the road, the company will be liable for their acts; and if the injury is wanton or willful the company may be required to respond in exemplary damages. *R., R. I. & St. L. R. R.* v. *Wells*, 66 Ill. 321.

2469. Where lease is unauthorized by law, the lessees will only be regarded as the servants of the company owning the road, and the latter will not be released from any of its contracts and obligations. *O., O. & F. R. V. R. R.* v. *Black*, 79 Ill. 262.

2470. A railway company which fails to fence its track is liable for any damage resulting from such failure, whether caused by its own trains, or those of another company using its track. Either company is liable. *E. St. L. & C. Ry.* v. *Gerber*, 82 Ill. 632.

2471. A railway company holding the franchise and exclusive right to operate a railroad, must so use it as not to endanger passengers or property, whether the use be by themselves. or others they may permit to use the road. The company owning the road and franchise is liable for an injury to a passenger through the negligence of its lessees, or of another company using the road by its permission. *P. & R. I. R. R.* v. *Lane*, 83 Ill. 448.

2472. If a switch on a railroad is not properly locked, or otherwise secured, whether by the neglect of the employes of the company owning the same, or its lessees, or if the switch is not properly constructed and maintained, and injury is thereby occasioned to a passenger on a train operated by the lessees, the company owning the road and franchise will be liable. *Ib.*

2473. Company is liable for the trespass of hands employed by its contractors while engaged in the construction of its road, and where the fact appears that the trespass consists in entering upon the plaintiff's land and digging up the soil, and making embankments, it is not error to refuse evidence that the company had nothing to do in employing the hands doing the work, but that they were employed and paid by the contractors. *C. & St. L. R. R.* v. *Woosley*, 85 Ill. 370.

2474. The lessee of a railroad, who by contract, permits another company to use the road, is liable for the negligent acts of the latter company. *P., C. & St. L. Ry.* v. *Campbell*, 86 Ill. 443.

2475. By accepting and acting under its charter, the company becomes a common carrier and connot exonerate itself from its duty and liability by contract with others, or otherwise, unless released by the legislature. *W., St. L. & P. Ry.* v. *Peyton*, 106 Ill. 534.

2476. Where one company acquires the right to run its trains over a portion of another company's road by contract, in which it is agreed that the leased road shall be under the control and direction of the yard-master or other servant of the lessor, the yard-master of the lessor will at such place and for the time being, be the servant of the lessee, which will be liable for any injury caused by his negligence. *Ib.*

2477. Railway company is held to the same care for the safety of all persons while exercising its franchises, whether on its own road or that of another company. If it operates its trains over the road of another by contract or lease, it must see and know that the track is in a good and safe condition, not only for the safety of its passengers, but also for the safety of persons rightfully near to the track and liable to injury by its being used when in an unsafe condition. *Ib.*

2478. Where a railway company procures, by contract with another company, the right of running its trains into and out of a depot over the track of the latter, it thereby makes that portion of the track so used, its own, in so far that it will be responsible for all injuries resulting from negligence in keeping or permitting it to be in an unsafe condition. *Ib.*

2479. Where the trains of a railway corporation are made up by the employes of another company and on the track of the latter, and cars used to make up the same, belong to other companies, if the use of the cars and tracks and labor in making up such trains, is to enable such first named corporation to exercise its franchise, &c., such cars, tracks and servants, so far as the rights of its passengers who may receive an injury are concerned, must be regarded as the cars, tracks and servants of the company so using the same. *H. & St. J. R. R.* v. *Martin*, 111 Ill. 219. See, also, *Union Ry. & Transit Co.* v. *Kallaher*, 114 Ill. 325.

2480. Where a railway company operating its road in its own name, contracts with another company to make up its train in the depot of the latter, the former company is liable for an injury to a passenger occurring on its train while being made up by the servants of the latter, and it makes no difference that the servants were employed and paid by the latter road. *H. & St. J. R. R.* v. *Martin*, 11 Bradw. 386.

2481. Liable for defect in cars of foreign corporation which it uses. *C. & A. R. R.* v. *Bragonier*, 11 Bradw. 516.

2482. Railway corporations are liable for injuries by the wrongful acts of any lessee, contractor or other person, done in the exercise, by its permission, of any of its franchises; but this liability is limited to "wrongs done by them while in the performance of acts which they would have had no right to perform except under the charter of the company" sought to be made liable. *St. L., A. & T. H. R. R.* v: *Balsley,* 18 Bradw. 79.

2483. STREET RAILROADS. § 38. This act shall not apply to horse cars or street railroads. [§ 39, repeal, omitted. See "Statutes," ch. 131, § 5. R. S. 1887, p. 1019, § 102; S. & C., p. 1949, §§ 102 and 103; Cothran, p. 1161, § 88.]

An act relating to fires caused by locomotives. Approved and in force March 29, 1869.

2484. FIRES BY LOCOMOTIVES. § 1. *Be it enacted by the people of the state of Illinois, represented in the general assembly,* That in all actions against any person or incorporated company for the recovery of damages on account of any injury to any property, whether real or personal, occasioned by fire communicated by any locomotive engine while upon or passing along any railroad in this state, the fact that such fire was so communicated shall be taken as full *prima facie* evidence to charge with negligence the corporation, or person or persons who shall, at the time of such injury by fire, be in the use and occupation of such railroad, either as owners, lessees or mortgagees, and also those who shall at such time have the care and management of such engine; and it shall not, in any case, be considered as negligence on the part of the owner or occupant of the property injured, that he has used the same in the manner, or permitted the same to be used or remain in the condition it would have been used or remained had no railroad passed through or near the property so injured, except in cases of injury to personal property which shall be at the time upon the property occupied by such railroad. This act shall not apply to injuries already committed.

2484a. ACT TAKES EFFECT. § 2. This act shall take effect and be in force from and after its passage. [L. 1869, p. 312. R. S. 1887, p. 1020, §§ 103, 104; S. & C., p. 1949, §§ 104, 105; Cothran, p. 1161, §§ 89, 90. See ante, 1800–1806.]

2485. Negligence will be implied from the escape of fire from a locomotive, and the burden of proof lies on railway company to show that all the most approved mechanical appliances were used on the engine to prevent the escape of fire. *Buss* v. *C., B. & Q. R. R.,* 28 Ill. 9.

2486. In an action against a railroad company for an injury to property by fire escaping from one of its passing locomotives, the burden of proving that the engine was properly guarded to prevent the escape of sparks, is upon the company. It is bound to use all possible diligence to prevent such escape. *St. L., A. & T. H. R. R.* v. *Montgomery,* 39 Ill. 335.

2487. In an action against a railway company to recover for the burning of a warehouse and goods therein by the escape of sparks from a locomotive, the employment of an unnecessary amount of steam by which an undue quantity of sparks are emitted, constitutes negligence. *Great Western R. R.* v. *Haworth*, 39 Ill. 346.

2488. If it is true that sparks are emitted from a locomotive in proportion to the amount of steam applied, it will be negligence while passing near buildings, to apply to the engine an unnecessary amount of steam. *Ib.*

2489. CONTRIBUTORY. It is for the jury to determine from the evidence whether the injury resulted from an unnecessary exposure of the building by the owner, or by an undue amount of sparks emitted from the locomotive. *Id.*

2490. Where the owner of a building exposes it to such a degree of danger that it will most probably be destroyed, he cannot recover, unless the party causing the injury is shown to have been guilty of greater negligence; and such owner, when he permits the windows to remain open and unglazed, and other openings in the building to go unrepaired, so that fire emitted from a passing engine is liable to be blown into it, is guilty of negligence, and cannot recover for loss, unless greater negligence on the part of the company is shown. *Ib.*

2491. Railway companies should use all the appliances of science, and the highest degree of diligence to prevent the destruction of property contiguous to their lines by means of fire escaping from their passing trains. *St. L., A. & T. H. R.R.* v. *Gilham*, 39 Ill. 455.

2492. By failing to provide the most approved appliances for arresting sparks from their engines, by running poor engines, or those out of order, a railroad company becomes liable for all casualties occasioned thereby. *I. C. R. R.* v. *McClelland*, 42 Ill. 355.

2493. And an engine which throws sparks into a meadow 100 feet from the track of the road, is not provided with proper appliances for arresting its own sparks; and evidence of such fact is properly admitted to show the character of the engines in use on a road at a particular time. *Ib.*

2494. It is sufficient if the proof sustains substantially any one of the counts, and the plaintiff is not confined to the proof of the precise place where the fire originated. It is immaterial whether it commenced on the right of way of defendant or not. *Ib.*

2495. Where it appears that fire has escaped from a railroad locomotive, it will be presumed that the company were not employing the best known contrivances to retain the fire, and it will, to rebut the presumption, devolve upon the company to show that such machinery was thus employed and in repair. The design of the statute is that railway companies shall use all reasonable precautions to prevent the escape of fire, and they will be held to the discharge of that duty. *I. C. R. R.* v. *Mills*, 42 Ill. 407.

2496. It is not a conclusion of law that a railway company is guilty of negligence by permitting grass and weeds to remain on its right of way, and become dry and combustible, which ignite and communicate to adjoining lands. It is a question of fact to be determined by the jury in view of the extent to which grass and weeds have been allowed to accumulate in the particular locality, the season of the year and all other circumstances affecting the liability of fire to communicate. *Ib.*

2497. Same care to keep right of way clear of combustible matter as of an individual under same danger. *Ib.*

2498. The fact that fire has escaped from a locomotive by which plaintiff's property is burned, is not conclusive evidence of the company's liability. It is not an insurer against the escape of fire from its engines. *I. C. R. R.* v. *Mills*, 42 Ill. 407.

2499. It is not negligence *per se* for a railroad to suffer grass and weeds to accumulate on its right of way; the fact, however, is proper evidence for the jury, who may find negligence from it. *O. & M. R. R.* v. *Shanefelt*, 47 Ill. 497.

2500. If owner of cóntiguous lands suffer weeds and grass to accumalate thereon, so that a fire commencing on the right of way is communicated to his premises, his negligence will be held as contributing to his loss, and he cannot recover unless the negligence of the company is greater than his. *Ib.*

2501. A railway company is held to the same, but no higher duty to keep its right of way free from grass or weeds, than are the adjoining land-owners and proprietors, to keep the adjoining lands free from grass and weeds. Rule in *Bass* v. *C., B. & Q. R. R.,* 28 Ill. 9, was not concurred in by a majority of the court. *I. C. R. R.* v. *Frazier,* 47 Ill. 505.

2502. The question of comparative negligence on the part of a plaintiff and the railway company in respect to the accumulation of combustible material, are questions of fact and properly left to a jury. *I. C. R. R.* v. *Nunn,* 51 Ill. 78.

2503. Railway companies are required to provide and keep constantly in use, and in proper repair the most approved machinery to prevent the escape of fire from their engines, to the injury of property along their lines. If notwithstanding the use of such machinery, sparks escape and a fire is thereby communicated to buildings, the company will not be deemed guilty of negligence, unless the damage results from the neglect of some other duty. *T., P. & W. Ry.* v. *Pindar,* 53 Ill. 447.

2504. But even with the use of the best appliances to prevent the escape of fire, if through the overloading of the engine, the escape of fire and sparks is produced to a dangerous extent, the company will be deemed guilty of gross negligence. *Ib.*

2505. Where fire is communicated to a building through the negligence of a railway company, the owner cannot recover for the loss of such portion of the property as he could, easily and without danger, have saved from destruction. *T., P. & W. Ry.* v. *Pindar,* 53 Ill. 447.

2506. In this case it was claimed that a large sum of money was burned in a house to which fire had been communicated by the alleged negligence of a railway company. The money could have been secured with but slight effort and without danger to the owner: *Held,* that the company was not liable for the loss of the money by reason of the neglect of the owner to save the same. *Ib.*

2507. Whether or not the injury is not too remote is a question of fact for the jury under instructions. *Ib. Fent* v. *T., P. & W. Ry.,* 59 Ill. 349.

2508. Where fire is ignited on the right of way of a railroad company by reason of the accumulation of dry grass and weeds thereon, and communicated to the adjoining fields by the negligence of the owner in not keeping them free from combustible material, the owner cannot recover for the injury, unless the negligence of the company is greater than his own. *C. & N. W. Ry.* v. *Simonson,* 54 Ill. 504.

2509. It is erroneous in the instructions to base the plaintiff's right of recovery wholly on the question of the negligence of the company, ignoring the doctrine of contributory negligence on the part of the plaintiff. Where the adjoining land is wood land, that fact should be considered by the court in the instructions, as abating the degree of diligence required of the land-owner. *Ib.*

2510. Under the act of 1869 the mere proof of the fact that the fire was caused by sparks from the engine, constitutes *prima facie* evidence of negligence on the part of the company, and the burden of proof rests upon it to rebut this presumption. *C. & N. W. Ry.* v. *McCahill,* 56 Ill. 28.

2511. Proof of the fact that the engine threw out an unusual quantity of fire was held sufficient to overcome any direct evidence given that it was in good order, or if in good order, that it was skillfully managed by the engineer. *Ib.*

2512. If fire is communicated from a railway locomotive to the house of A. and from that to the house of B., it is not a conclusion of law that the fire from the locomotive is the remote and not the proximate cause of the injury to B., but that is a question of fact for the jury. *Fent* v. *T., P. & W. Ry.,* 59 Ill. 349.

2513. Where loss has been caused by an act, and it was under the circumstances a natural consequence which any reasonable person could have anticipated, then the act is the proximate cause. *Ib.*

2514. Experience having shown that railway companies by the use of certain mechanical contrivances can prevent the emission of fire sparks from locomotive engines, in such quantities at least, as not to be at all dangerous to property in the immediate vicinity, they must in every instance, be held to a strict performance of their duties in that regard. *C. & A. R. R.* v. *Quaintance,* 58 Ill. 389.

2515. If such companies use all proper and reasonable precaution to prevent the escape of fire from their engines by the application of the best and most approved mechanical appliances for that purpose, and keep the same constantly in good repair while in use, and carefully and skillfully managed by competent and prudent engineers, and nevertheless fire results, they will not be liable for the damage. *Ib.*

2516. The act of 1869 makes the fact of injury from the escape of fire from the engine full *prima facie* evidence of negligence on the part of the company, and throws the burden of proof on the company to show by affirmative evidence, that the engine at the time was equipped with the necessary and most effective appliances to prevent the escape of fire, and that the engine was in good repair, and was properly, carefully and skillfully handled by a competent engineer. *Ib.*

2517. It is not enough to rebut this *prima facie* case to show that the engine was originally constructed with the best and most approved inventions to prevent the escape of fire. The law requires a constant and vigilant watch to see that the engines are kept in proper repair, so as not to be dangerous, &c. *Ib.*

2518. It is negligence to use wood in a coal burning engine while running, for the reason that the meshes of the iron netting used to prevent the escape of fire sparks are made much larger when coal only is used for fuel, and the fine sparks from wood are much more dangerous, because they retain the fire for a much greater length of time. To use wood in such an engine in a dry time with a high wind prevailing is great carelessness and recklessness. *C. & A. R. R.* v. *Quaintance,* 58 Ill. 389.

2519. Evidence tending to prove the safe condition of the engine admissible. *Ib.*

2520. Where a railway company suffers a heavy growth of dry grass to remain on its right of way through plaintiff's premises, and fire is communicated from the locomotive of a freight train, while laboring to ascend a heavy grade, to the grass and weeds in the right of way, and from thence to the fence and grass of the plaintiff, which are destroyed, the company will be liable for the loss. *R., R. I. & St. L. R. R.* v. *Rogers,* 62 Ill. 346.

2521. Where fire is communicated from a locomotive engine of a railway company and thereby destroys the property of another, the presumption of negligence on the part of those having the care and management of the engine, created by statute, will not be sufficiently rebutted by proof, that the engine was at the time of the injury, provided with the best mechanical contrivances to prevent the escape of sparks, and that they were in good order. It should be further shown that the engine was properly managed. *C. & A. R. R.* v. *Clampit*, 63 Ill. 95.

2522. Permitting dry grass and weeds to accumulate on right of way whereby fire is communicated to plaintiff's premises, is negligence. *I. C. R. R.* v. *Frazier*, 64 Ill. 28.

2523. The law holds railway companies in the use of steam as a motive power, to a very high degree of care and skill in the use of the most effective appliances to prevent the emission of fire sparks and in the employment and retention of servants in charge of them, so as to prevent loss to property. *T., W. & W. Ry.* v. *Larmon*, 67 Ill. 68.

2524. Railway companies are not insurers against loss by fire from their engines. If they use the highest degree of care and skill to prevent such injury, any loss occurring must fall upon the owner. *T., W. & W. Ry.* v. *Larmon*, 67 Ill. 68.

2525. It is error to instruct that the destruction of property by the escape of fire from an engine is of itself, evidence of negligence on the part of the defendant. The statute makes such fact, only *prima facie* evidence of negligence. *Ib.*

2526. PLEADING—*declaration.* An averment in a declaration that it was the duty of the defendant to keep its right of way free from dry grass and weeds, and to so construct and operate its locomotives, as to prevent the escape of fire to the adjoining property, &c., is substantially an averment, that it was the duty of the company to provide its locomotives with the best appliances to prevent the escape of fire, and to so use them that it would not be liable to escape; and the performance of this duty is sufficiently negatived by an averment that the engine was so negligently used, that the fire did, by reason of such negligence, escape and produce the injury complained of. *T., W. & W. Ry.* v. *Corn*, 71 Ill. 493.

2527. Company not required to provide and use the best known appliances that mechanical skill and ingenuity have been able to devise and construct to prevent the escape of sparks from its locomotives, without reference to whether the company could by any degree of effort, know of such inventions or not, or whether they have been tested and proved to be the best. *Ib.*

2528. It is not bound to purchase the patent for every invention claimed to be an improvement on such machinery and test it; but when such an invention has been tested and approved as superior to that it is using, it is required to adopt and use the better machinery. *Ib.*

2529. Where property is destroyed by fire caused by sparks thrown from a passing engine, through the negligence of the servants of the company, and the destruction of the property is, under the circumstances of the case, a natural consequence, which any reasonable person could have anticipated, then the act of throwing the sparks which originated the fire, is a proximate cause, whether the property destroyed is the first or tenth, the latter being so situated that its destruction is a consequence reasonably to be anticipated from setting the first on fire, and the company will be liable. But if the destruction of the property is not the natural and proximate consequence of the escaping of the sparks, and consequent firing of the first building, then the company will not be liable. *T., W. & W. Ry.* v. *Muthersbaugh*, 71 Ill. 572.

2530. A warehouse standing near the railroad track was set on fire by sparks escaping from an engine of the company, there being at the time a strong wind blowing in the direction of the plaintiff's stable, which was situated 101 rods from the warehouse, and there was no combustible matter intervening. The high wind carried the brands from the warehouse to the stable which caused it to take fire and burn up: *Held,* that the burning of the stable was not the natural and proximate consequence of the burning of the warehouse, and that the company was not liable for the burning of the stable. *T., W. & W. Ry.* v. *Muthersbaugh*, 71 Ill. 572.

2531. Where a party erects his building at a reasonably safe distance from the railroad track, he cannot be held guilty of negligence because his building is so situated as to be liable to be set on fire by another subsequently erected in a dangerous proximity to the track. *T., W. & W. Ry.* v. *Maxfield*, 72 Ill. 95.

2532. VARIANCE. Where the declaration alleges that plaintiff's stacks were set on fire by sparks from the defendant's locomotive, evidence that they were destroyed by a fire which originated in another field, even though such fire was occasioned by sparks from the defendant's engine, will not sustain the averment and plaintiff cannot recover. *T., W. & W. Ry.* v. *Morgan*, 72 Ill. 155.

2533. It is not sufficient to overcome the statutory presumption of negligence from the escape of fire, to show that the engine was equipped with the proper appliances to prevent the escape of fire, and that the same was in good order, but it is also necessary to show that the engine was properly handled and managed by a competent and skillful engineer. *St. L., V. & T. H. R. R.* v. *Funk*, 85 Ill. 460.

2534. Under the statute a railway company in the use of a railroad as lessee or otherwise, is guilty of negligence if it fails to keep the right of way clear from all dead grass, weeds, &c., and for such neglect is made liable for injuries to others from the escape and transmission of fire from its engines. *P., C. & St. L. Ry.* v. *Campbell*, 86 Ill. 443.

2535. The communication of fire by any locomotive while on or passing over any railroad, affords full *prima facie* evidence to charge the corporation or persons in the use of such road as owner, lessee or mortgagee, under the statute with negligence in not keeping the right of way free from combustible matter, and in the use of the engines and for not having them in all respects in a good and safe condition. Proof of the communication of fire makes a case entitling the plaintiff to recover against any company using or occupying the road. *P., C. & St. L. Ry.* v. *Campbell*, 86 Ill. 443.

2536. If the servants of a railway company, to free its right of way from dry grass and combustible matter, put out a fire on the same on a day when the wind is high, and the fire escapes from them upon the lands of plaintiff without his negligence or fault, and injures his apple trees, the company will be liable. *O. & M. Ry.* v. *Porter*, 92 Ill. 437.

2537. The law requires a railroad company, in operating its trains, to use every possible precaution, by the use of all the best and most approved mechanical inventions to prevent loss from the escape of fire or sparks along the line of its road, and such company will be liable for a loss by fire caused by a neglect of such duty where the owner of the property is free from negligence. *C. & A. R. R.* v. *Pennell*, 94 Ill. 448.

2538. A party who erects a building on or near a railroad track knows the dangers incident to the use of steam as a motive power, and must be held to assume some of the hazards connected with its use on such thoroughfares. While he has the right to build near the track, yet if he does so, he is bound to a higher degree of care in providing proper means to protect his property, than if otherwise situated. He must also use all reasonable means to save his property in case of fire. *Ib.*

2539. Where a building near a railroad track is set on fire through the negligence of the railway company, the owner cannot recover for the loss of such property as he could easily and without danger have saved from destruction. *C. & A. R. R.* v. *Pennell*, 94 Ill. 448.

2540. The statute which declares that in actions for damages for injury to property "occasioned by fire communicated by any locomotive engine while passing along any railroad" shall be *prima facie* evidence "to charge with negligence," the owner or operator of the road at the time, was intended to charge upon the company using the locomotive all injuries which are shown to have resulted from fire from a passing train, unless the company can rebut such presumption by proof showing that the loss was not occasioned by its negligence. *C. & A. R. R.* v: *Pennell*, 110 Ill. 435.

2541. PROXIMATE CAUSE. Where a railway company through negligence, by the escape of fire from an engine, sets fire to a depot, from which a hotel in the vicinity is destroyed, to make the company liable to the owner of the hotel, it is not necessary that the burning of the hotel should be so certain to result from the burning of the depot that a reasonable person could have foreseen that the hotel *would* burn, or that it probably would. It is enough if it be a consequence so natural and direct that a reasonable person might and naturally would see that it was liable to result from the burning of the depot. *Ib.*

2542. In an action to recover the value of a stack of hay alleged to have been burned by fire communicated from a locomotive, an instruction that if the jury believe the hay was destroyed by fire communicated from one of defendant's engines, and that defendant's right of way was not free from dry grass and other combustible matter at the place where the fire started, &c., is erroneous, in assuming that the fire originated on defendant's right of way. *C. & A. R. R.* v. *Bloomfield*, 7 Bradw. 211.

2543. Error to instruct that defendant must show not only that the engine was supplied with the best and most approved appliances to prevent the escape of sparks at the time of the fire, but also that the engine was originally so constructed. Sufficient if it was properly constructed at time of fire. *C. & N. W. Ry.* v. *Boller*, 7 Bradw. 625.

2544. In an action for damages caused by fire escaping from an engine, an instruction for the plaintiff which fails to include the question whether the engine was supplied with proper appliances for arresting sparks, is erroneous, where there is testimony tending to prove that fact. *C. & A. R. R.* v. *Smith*, 10 Bradw. 359.

2545. Where the evidence shows that the engines causing the fire were equipped with the best and most approved appliances for preventing the escape of fire or sparks, and were properly and prudently managed, and no negligence on the part of the company is shown, no recovery can be had for the setting on fire an adjoining building. *C. & A. R. R.* v. *Smith*, 11 Bradw. 348.

2546. Where damage is caused by a fire communicated to property from sparks of an engine, the *prima facie* case made out under the statute (section 89) is rebutted by the company showing that at the time of the accident, the engine, smoke stack and spark arrester were all safe and in good order, and the engineer in charge of the loco-

motive was experienced and competent and properly performed his
duty. *I., B. & W. Ry.* v. *Craig*, 14 Bradw. 407.

2547. In an action for damages resulting from a fire set by defendant's locomotive in January, evidence that defendant cut and
burned the grass and weeds upon its right of way in September or
October previous, is not sufficient to show a compliance with the law.
Ind., B. & W. Ry. v. *Nicewander*, 21 App. Rep. 305.

2548. Whether corporation is guilty of negligence under this
statute is a question of fact. *Ib.*

PROTECTION OF PASSENGERS.

An act for the protection of passengers on railroads and steamboats. Approved
May 14, 1877. In force July 1, 1877. [Laws of 1877, p. 166. As amended by an act approved May 29, 1879. In force July 1, 1879. Laws of 1879, p. 223. R. S. 1887, p. 1020,
§§ 105-107; S. & C., p. 1950, §§ 106-108; Cothran, p. 1161, 1162, §§ 91-93.]

2549. § 1. *Be it enacted by the people of the state of
Illinois, represented in the general assembly*, That an act
entitled "An act for the protection of passengers on railroads," approved May 14, 1877, in force July 1, 1877, be
amended so as to read as follows: "An act for the protection
of passengers on railroads and steamboats."

2549a. CONDUCTORS INVESTED WITH POLICE POWERS. § 2.
That the conductors of all railroad trains, and captain or
master of any steamboat carrying passengers within the jurisdiction of this state, shall be invested with police powers while
on duty on their respective trains and boats.

2550. EJECTION OF PASSENGER FROM TRAIN. § 3. When
any passenger shall be guilty of disorderly conduct, or use
any obscene language, to the annoyance and vexation of passengers, or play any games of cards, or other games of chance
for money or other valuable thing, upon any railroad train or
steamboat, the conductor of such train, and captain or master
of such steamboat, is hereby authorized to stop his train or
steamboat, at any place where such offense has been committed and eject such passenger from the train or boat, using
only such force as may be necessary to accomplish such removal, and may command the assistance of the employes of
the railroad company or steamboat, or any of the passengers,
to assist in such removal; but before doing so he shall tender
to such passenger such proportion of the fare he has paid as
the distance he then is from the place to which he has paid
his fare, bears to the whole distance for which he has paid
his fare.

2551. WHEN PASSENGER MAY BE ARRESTED. § 4. When
any passenger shall be guilty of any crime or misdemeanor
upon any train, or steamboat, the conductor, captain or master, or employes of such train, or boat, may arrest such passenger and take him before any justice of the peace, in any
county through which such boat or train may pass, or in

which its trip may begin or terminate, and file an affidavit before such justice of the peace, charging him with such crime or misdemeanor.

STRIKES AND OBSTRUCTIONS OF RAILROADS.

An act to prohibit any person from obstructing the regular operation and conduct of the business of railroad companies or other corporations, firms or individuals. Approved June 2, 1877. In force July 1, 1877. [L. 1877, p. 167; R. S. 1887, p. 1020, §§ 108–111; S. & C., p. 1951, §§ 109–112; Cothran, p. 1162, §§ 94–97.]

2552. ENGINEER NOT TO ABANDON ENGINE. § 1. *Be it enacted by the people of the state of Illinois, represented in the general assembly,* If any locomotive engineer in furtherance of any combination or agreement, shall willfully and maliciously abandon his locomotive upon any railroad at any other point than the regular schedule destination of such locomotive, he shall be fined not less than twenty dollars, nor more than one hundred dollars, and confined in the county jail, not less than twenty days, nor more than ninety days.

2553. PERSONS OBSTRUCTING BUSINESS OF RAILROAD—FINE. § 2. If any person or persons shall willfully and maliciously, by any act or by means of intimidation, impede or obstruct, except by due process of law, the regular operation and conduct of the business of any railroad company or other corporation, firm or individual in this state, or of the regular running of any locomotive engine, freight or passenger train of any such company, or the labor and business of any such corporation, firm or individual he or they shall, on conviction thereof, be punished by a fine not less than twenty dollars, ($20.00) nor more than two hundred dollars ($200.00), and confined in the county jail not less than twenty nor more than ninety days.

2554. CONSPIRACY TO IMPEDE BUSINESS. § 3. If two or more persons shall willfully and maliciously combine or conspire together to obstruct or impede by any act, or by means of intimidation, the regular operation and conduct of the business of any railroad company or any other corporation, firm or individual in this state, or to impede, hinder or obstruct, except by due process of law, the regular running of any locomotive engine, freight or passenger train on any railroad, or the labor or business of any such corporation, firm, or individual, such persons shall, on conviction thereof, be punished by fine not less than twenty dollars ($20.00), nor more than two hundred dollars ($200.00), and confined in the county jail not less than twenty days, nor more than ninety days.

2555. CONSTRUCTION OF ACT. § 4. This act, shall not be construed to apply to cases of persons voluntarily quitting the employment of any railroad company or such other corporation, firm or individual, whether by concert of action or otherwise, e[x]cept as is provided in section one (1) of this act.

FRAUD IN RELATION TO TICKET.

An act to prevent frauds upon travelers and owner or owners of any railroad, steamboat or other conveyance for the transportation of passengers. Approved April 19, 1875. In force July 1, 1875. [Laws of 1875, p. 81; R. S. 1887, p. 1021, §§ 112–117; S. & C., p. 1951, §§ 113–118; Cothran, p. 1163, §§ 98–103.]

2556. OWNER TO FURNISH AGENT CERTIFICATE OF AUTHORITY TO SELL TICKETS. § 1. *Be it enacted by the people of the state of Illinois, represented in the general assembly,* That it shall be the duty of owner or owners of any railroad or steamboat for the transportation of passengers, to provide each agent, who may be authorized to sell tickets, or other certificates entitling the holder to travel upon any railroad or steamboat, with a certificate setting forth the authority of such agent to make such sales; which certificate shall be duly attested by the corporate seal of the owner of such railroad or steamboat.

2557. NOT LAWFUL FOR PERSON NOT HAVING SUCH AUTHORITY TO SELL TICKETS. § 2. That it shall not be lawful for any person not possessed of such authority, so evidenced, to sell, barter or transfer, for any consideration whatever, the whole or any part of any ticket or tickets, passes, or other evidences of the holder's title to travel on any railroad or steamboat, whether the same be situated, operated or owned within or without the limits of this state.

2558. PENALTY FOR VIOLATING ACT. § 3. That any person or persons violating the provisions of the second section of this act shall be deemed guilty of a misdemeanor, and shall be liable to be punished by a fine not exceeding five hundred dollars, and by imprisonment not exceeding one year, or either, or both, in the discretion of the court in which such person or persons shall be convicted.

2559. AGENT TO EXHIBIT CERTIFICATE ON REQUEST. § 4. That it shall be the duty of every agent who shall be authorized to sell tickets, or parts of tickets, or other evidences of the holder's title to travel, to exhibit to any person desiring to purchase a ticket, or to any officer of the law who may request him, the certificate of his authority thus to sell, and to keep said certificate posted in a conspicuous place in his office for the information of travelers.

2560. DUTY OF OWNER TO PROVIDE FOR REDEMPTION OF TICKETS. § 5. That it shall be the duty of the owner or owners of railroad or steamboat, by their agents or managers, to provide for the redemption of the whole, or any parts or coupons of any ticket or tickets, as they may have sold, as the purchaser, for any reason, has not used, and does not desire to use, at a rate which shall be equal to the difference between the price paid for the whole ticket and the cost of a ticket between the points for which the proportion of said ticket was actually used; and the sale by any person of the unused portion of any ticket otherwise than by the presentation of the same for redemption, as provided for in this section, shall be deemed to be a violation of the provisions of this act, and shall be punished as is hereinbefore provided: *Provided*, that this act shall not prohibit any person who has purchased a ticket from any agent authorized by this act, with the *bona fide* intention of traveling upon the same, from selling any part of the same to any other person.

2561. PENALTY FOR FAILURE TO REDEEM TICKETS. § 6. Any railroad or steamboat company that shall, by any of its ticket agents in this state, refuse to redeem any of its tickets or parts of tickets as prescribed in section five of this act, shall pay a fine of five hundred dollars for each offense, to the people of the state of Illinois, and it shall be unlawful for said company, subsequent to such refusal, to sell any ticket or tickets in this state until such fine is paid.

RECEIVING, CARRYING AND DELIVERING GRAIN.

An act regulating the receiving, transportation and delivery of grain by railroad corporations, and defining the duties of such corporations with respect thereto. Approved April 25, 1871. In force July 1, 1871. [L. 1871, p. 636; R. S. 1887, p. 1022, § 118; S. & C., p. 1952, § 119; Cothran, p. 1164, § 104.]

2562. RECEIVE AND CARRY GRAIN WITHOUT DISTINCTION. § 1. *Be it enacted by the people of the state of Illinois, represented in the general assembly,* That every railroad corporation, chartered by or organized under the laws of this state or doing business within the limits of the same, when desired by any person wishing to ship any grain over its road, shall receive and transport such grain in bulk, within a reasonable time, and load the same either upon its track, at its depot, or in any warehouse adjoining its track or side track, without distinction, discrimination or favor between one shipper and another, and without distinction or discrimination as to the manner in which such grain is offered to it for transportation, or as to the person, warehouse or place to whom or to which it may be consigned.

WEIGHING IN—RECEIPT. And at the time such grain is received by it for transportation, such corporation shall carefully and correctly weigh the same, and issue to the shipper thereof a receipt or bill of lading for such grain, in which shall be stated the true and correct weight.

WEIGHING OUT—SHRINKAGE. And such corporation shall weigh out and deliver to such shipper, his consignee or other person entitled to receive the same, at the place of delivery, the full amount of such grain, without any deduction for leakage, shrinkage or other loss in the quantity of the same.

DAMAGES. In default of such delivery, the corporation so failing to deliver the full amount of such grain shall pay to the person entitled thereto the full market value of any such grain not delivered at the time and place when and where the same should have been delivered.

EVIDENCE —SHORTAGE. If any such corporation shall, upon the receipt by it of any grain for transportation, neglect or refuse to weigh and receipt for the same, as aforesaid, the sworn statement of the shipper, or his agent having personal knowledge of the amount of grain so shipped, shall be taken as true, as to the amount so shipped; and in case of the neglect or refusal of any such corporation, upon the delivery by them of any grain, to weigh the same, as aforesaid, the sworn statement of the person to whom the same was delivered, or his agent having personal knowledge of the weight thereof, shall be taken as true, as to the amount delivered. And if, by such statements, it shall appear that such corporation has failed to deliver the amount so shown to be shipped, such corporation shall be liable for the shortage, and shall pay to the person entitled thereto the full market value of such shortage, at the time and place when and where the same should have been delivered.

2563. DISCRIMINATION—*as to person.* Where the company from a pressing cause, takes grain from wagons or boats, while grain remains for shipment in private warehouses, acting in good faith, and without partiality or oppression, it will not thereby incur liability. *G. & C. U. R. R.* v. *Rae,* 18 Ill. 488.

2564. If its servants, by reason of bribes or other improper motives, give preference to one person over another, the company may be held liable for damages thereby caused. *Ib.*

2565. DELAY. Must use proper diligence to transport freight offered, without delay, and unless it can excuse itself for the delay it will be liable in an action on the case. *Ib.*

2566. Must receive freights according to its usage and custom. If in the habit of running its cars upon side tracks to a private warehouse to receive freights, a readiness to deliver freights at such warehouse will impose on the company the duty to take the freight therefrom. *Ib.*

2567. TENDER OF CHARGES. A tender or readiness to pay the freight must be proved in an action to recover for non-transportation or delay in same. *Ib.*

2568. WAIVER. By omitting to demand prepayment of freight, the company will be bound to transmit freight according to its custom. Where not demanded slight evidence of willingness to pay will be sufficient, and readiness to pay may be presumed from the circumstances. *Ib.*

2569. DELAY. Where a box shipped at Adrian for Chicago on October 29th, arriving at Chicago on November 3d, the usual time for transportation being three days, and was not delivered by the freight agent until November 15th, this was *held* such an unreasonable delay as to entitle the owner to damage. *M. S. & N. I. R. R.* v. *Day,* 20 Ill. 375.

2570. Company bound to use every reasonable effort without delay to deliver at its destination in proper time, cattle loaded by the shipper, and for a failure to do so, will be liable for all proximate damages. *O. & M. R. R.* v. *Dunbar,* 20 Ill. 623.

2571. Where the goods are placed in a car of the company with its assent for shipment, it becomes liable for them the same as if delivered in its warehouse. *I. C. R. R.* v. *Smyser & Co.,* 38 Ill. 354.

2572. Where cattle are loaded in the cars by the owner, with the knowledge of the company, it should take them by the most regular cattle train, and failing to do so, will become liable for any damage thereby caused to the cattle. *I. C. R. R.* v. *Waters,* 41 Ill. 73.

2573. Measure of damage for delay in carrying cattle. *I. C. R. R.* v. *Waters,* 41 Ill. 73: or for not delivering in a reasonable time. *I. C. R. R.* v. *McClellan,* 54 Ill. 58; *I. C. R. R.* v. *Cobb,* 64 Ill. 128.

2574. Where the usual time for the transportation of corn was two and a half to three days, a delay of eleven days, and as to a part, of forty-five days, in its reaching its destination, is unreasonable, and renders the company liable for damages. *I. C. R. R.* v. *McClellan,* 54 Ill. 58.

2575. A railway company having received a large quantity of wool for transportation to Boston, carried it within fifty miles of the terminus of its road, where, owing to the obstruction of the road with which it connected, from snow, the wool was stored for two months, within which time the price declined in the Boston market: *Held,* that the company was liable, if for no other reason, because its agents knew that the road was so blocked with freight that the wool could not go through within a reasonable time, and failed to inform the shipper of this fact that he might have either sold at the point whence shipped, or have selected another route. *Great Western Ry.* v. *Burns,* 60 Ill. 284.

2576. A railway company having received goods for transportation without giving notice of facts that would cause delay, are required to carry the same through in a reasonable time, or respond in damages caused by the delay. *Ib.*

2577. A common carrier has no right to store a part of the freight received for transportation, and leave it there, while it receives new freight, and sends it through, and when it does so it must make compensation to the parties injured thereby. *Gr. West. Ry.* v. *Burns,* 60 Ill. 284.

2578. The fact that the latter shipments were of perishable property and live stock, will not furnish any excuse. Freight should not be received until it can be sent through without delaying other freight having the precedence. *Ib.*

2579. A delay of over thirty days in the transportation of grain, when the ordinary time required is only two or three days, is unreasonable, and not excusable on account of causes known at the time of accepting the same. *I. C. R. R.* v. *Cobb,* 64 Ill. 128.

2580. Where a limited military control was being exercised over a railroad during the war, and its shipments were immense, so that the side tracks of the road for a considerable distance were filled with loaded cars waiting to be unloaded by the military authorities: *Held,* that it was the right and duty of the company to refuse to accept perishable freights for shipment to its terminus, until its line was clear. *Ib.*

2581. The right conferred upon railroad corporations to carry passengers and property for a compensation, is coupled with a corresponding duty, that they shall receive and carry passengers and freights over their roads as they may be offered. *P. & R. I. Ry.* v. *C. V. M. Co.*, 68 Ill. 489.

2582. The duties railroad corporations owe to the public and which are the considerations upon which their privileges are conferred, cannot be avoided by neglect, refusal or by agreement with other persons or corporations. Therefore any contract to prevent the faithful discharge of any such duties will be against public policy and void. *Ib.*

2583. It is the duty of a common carrier to forward and deliver goods at the point it contracts to convey them to, within a reasonable time, and if it fails to do so, it will be liable, whether it knew that its connecting line could not without unreasonable delay, forward the goods or not. Crowded condition of connecting road, no excuse. *T., W. & W. Ry.* v. *Lockhart*, 71 Ill. 627.

2584. MEASURE OF DAMAGES. The owner of grain shipped for market is entitled to recover the difference between the market price at the point of destination, when it should have arrived, and the time it does arrive. If in consequence of the delay there ceases to be a market for the grain at such place, the owner may without unreasonable delay, ship the same to some other point, and sell and hold the company liable for the loss. *I. C. R. R.* v. *Cobb*, 72 Ill. 148.

2585. DELAY IN FURNISHING TRANSPORTATION. Where a person desirous of shipping a large quantity of corn over a railroad to Cairo, stored the same in a warehouse on the premises of the company to be transported as soon as cars could be procured, but the company never received or receipted for the same, and was unable to forward the same for want of cars and because the road was controlled by military authorities of the United States who refused to give permits to ship the same, and in consequence of which the grain was injured by exposure, &c: *Held*, that the company was not liable to the owner of the grain for the delay in furnishing transportation. *I. C. R. R.* v. *Hornberger*, 77 Ill 457.

2586. EXCUSE FOR DELAY. Company may show in defense that the delay was caused solely by the lawless, irresistible violence of men who were not in its employ. *P., Ft. W. & C. R. R.* v. *Hazen*, 84 Ill. 36.

2587. Liable for delay caused by a strike among its servants, &c. *P., Ft. W. & C. R. R.* v. *Hazen*, 84 Ill. 36.

2588. As to negligence in not forwarding by connecting line. *Erie Ry.* v. *Wilcox*, 84 Ill. 239.

2589. Where a railway company refused to furnish cars for the transportation of grain to Cairo during the war, on account of the large accumulation of cars on its track at that point waiting to be unloaded, and finally furnished cars on the promise of the shipper to unload the same, which was not done, either by him or the consignee, but refused, it was held in a suit against the company to recover damages for delay in transporting the grain, that the jury were justified in finding for the defendant. *Cobb* v. *I. C. R. R.*, 88 Ill. 394.

2590. Company is bound to receive and transport cattle when they are first offered for shipment, unless it has a reasonable excuse for its refusal, and when its refusal to take and ship cattle is without such excuse, it will be liable in damages to the owner for the deterioration in the value of the cattle. *C. & A. R. R.* v. *Erickson*, 91 Ill. 613.

2591. An unconstitutional law prohibiting the shipping or carrying of Texas or Cherokee cattle into or through the state, being void,

—21

will afford no excuse for a refusal or delay in receiving and shipping such cattle when offered. *C. & A. R. R.* v. *Erickson*, 91 Ill. 613.

2592. A railway company under military control and operated by the military in the transportation of troops, munitions of war, &c., so as not to be in the free exercise of its franchise, is not liable for refusing to receive freights for transportation, it not being safe to undertake their carriage. But if it accepts and undertakes to carry freights such military interference will not excuse its delay to transport. *Phelps* v. *I. C. R. R.*, 94 Ill. 548.

2593. Where a railroad is under the military control of the United States and operated by its officers, the company is not in the free use of its franchise, and its duty to the public to receive and transport freight is for the time suspended, and it is not liable for not receiving freight so long as it is not in the control of its road. *I. C. R. R.* v. *Phelps*, 4 Bradw. 238.

2594. For a delay occasioned by the refusal of the company's servants to do their duty, the company is responsible; but for a delay resulting solely from the lawless violence of men not in its employ, it is not responsible. *I. & St. L. R. R.* v. *Juntgen*, 10 Bradw. 295.

2595. For a failure of carrier to transport cattle to their destination within a reasonable time, an action lies. *W., St. L. & P. Ry.* v. *McCasland*, 11 Bradw. 491.

2596. Nothing but the act of God or the public enemy will excuse a carrier from the ultimate delivery of goods entrusted to its care; but it is not to the same extent liable for every delay in reaching the place of destination. *Ib.*

2597. Custom or usage in the shipment of certain classes of freight may fix the liability of a carrier for a refusal to transport that kind of freight in conformity to the custom; but custom and usage cannot be held to extend the terms of a penal statute. *I. & St. L. Coal Co.* v. *People*, 19 Bradw. 141.

2598. The fact that a railway is under military control in time of war, is a sufficient excuse for delay in making shipment of goods. *I. C. R. R.* v. *Ashmead*, 58 Ill. 487. As to discrimination see post, 2653, 2725.

2599. SCALES—WEIGHING—PENALTIES. § 2. At all stations or places from which the shipments of grain by the road of such corporation shall have amounted during the previous year to fifty thousand (50,000) bushels or more, such corporation shall, when required so to do by the persons who are the shippers of the major part of said fifty thousand bushels of grain, erect and keep in good condition for use, and use in weighing grain to be shipped over its road, true and correct scales, of proper structure and capacity for the weighing of grain by car load in their cars after the same shall have been loaded. Such corporation shall carefully and correctly weigh each car upon which grain shall be shipped from such place or station, both before and after the same is loaded, and ascertain and receipt for the true amount of grain so shipped. If any such corporation shall neglect or refuse to erect and keep in use such scales when required to do so as aforesaid, or shall neglect or refuse to weigh in the manner aforesaid any grain shipped in bulk from any station or place, the sworn statement of the shipper, or his agent, having personal knowledge of the amount of grain shipped, shall be taken as true as to the amount so shipped. In case any railroad corporation shall neglect or refuse to comply with any of the requirements of section first, second and fifth of this act, it shall, in addition to the penalties therein provided, forfeit and pay for every such offense and for each and every day such refusal or neglect is continued the sum of one hundred dollars ($100), to be recovered in an action of debt before any justice of the peace, in the name of the people of the state of Illinois, such penalty or forfeiture to be paid to the county in which the suit is brought, and shall also be required to pay all costs of prosecution, including such reasonable attorney's fees as may be assessed by the justice before whom the case may be tried. [As amended by act approved May 18, 1877. In force July 1, 1877. L. 1877, p. 168. (The act amending this section contains the following: § 2. All parts of said section in conflict with section one of this act are hereby repealed.) R. S. 1887, p. 1022, § 119; S. & C., p. 1953, § 120; Cothran, p. 1165, 105.]

2600. DELIVERY—PENALTY. § 3. Every railroad corporation which shall receive any grain in bulk for transportation to any place within the state, shall transport and deliver the same to any consignee, elevator, warehouse, or place to whom or to which it may be consigned or directed: *Provided*, such person, warehouse or place can be reached by any track owned, leased or used, or which can be used by such corporation; and every such corporation shall permit connections to be made and maintained with its track to and from any and all public warehouses where grain is or may be stored. Any such corporation neglecting or refusing to comply with the requirements of this section, shall be liable to all persons injured thereby for all damages which they may sustain on that account, whether such damages result from any depreciation in the value of such property by such neglect or refusal to deliver such grain as directed, or in loss to the proprietor or manager of any public warehouse to which it is directed to be delivered, and costs of suit, including such reasonable attorney's fees as shall be taxed by the court. And in case of any second or later refusal of such railroad corporation to comply with the requirements of this section, such corporation shall be by the court, in the action on which such failure or refusal shall be found, adjudged to pay, for the use of the people of this state, a sum of not less than $1,000, nor more than $5,000, for each and every such failure or refusal, and this may be a part of the judgment of the court in any second or later proceeding against such corporation. In case any railroad corporation shall be found guilty of having violated, failed, or omitted to observe and comply with the requirements of this section, or any part thereof, three or more times, it shall be lawful for any person interested to apply to a court of chancery, and obtain the appointment of a receiver to take charge of and manage such railroad corporation until all damages, penalties, costs and expenses adjudged against such corporation for any and every violation shall, together with interest, be fully satisfied. [R. S. 1887, p. 1023, § 120; S. & C., p. 1954, § 121; Cothran, p. 1166, § 106.]

PLACE OF DELIVERY.

2601. A railway company must receive grain according to its custom and usage. If that usage is to run its cars upon a side track to private warehouses, and there receive grain in the cars, a tender accordingly, or notice and readiness to so deliver, will impose an obligation on the company to take and carry the grain. Having adopted this mode it cannot capriciously require that the grain be delivered in a different manner, or at a different place. *G. & Ch. U. R. R.* v. *Rae*, 18 Ill. 488, 490.

2602. Under § 22 of act of 1867 entitled "warehousemen," railroad companies were positively inhibited from making delivery of any grain which they had received for transportation, into any warehouse, other than that to which it is consigned, without the consent of the owner or consignee thereof. *Vincent* v. *C. & A. R. R.*, 49 Ill. 33.

2603. Where a shipment of grain is made to a party having his warehouse on the line of the road by which the grain is transported, and such consignee is ready to receive it, it is the duty of the carrier to make a personal delivery to him, at the warehouse to which it is consigned. *Ib.*

2604. Where the owner of adjacent property, had with the consent of the company, for a valid consideration, been permitted to lay down a side track, connecting with the track of the company for the purpose of transporting to such property articles of freight, and such owner has erected thereon a warehouse, which is in readiness for the receipt of such freight, such side track is to be considered as a part of the line of the company for the purpose of delivery under this statute. *Ib.*

2605. In order to compel a railway company to deliver grain shipped on its road in bulk, at a particular elevator to which it may be consigned, such elevator must be connected by some track with the railroad line of the company, and be in fact, a portion thereof, or such as would be regarded as a portion thereof, for the purposes of such delivery under the act of 1867. *People ex rel.* v. *C. & A. R. R.*, 55 Ill. 95.

2606. Railway companies cannot disregard the custom which has obtained of conveying grain in bulk over the lines of their own roads, and delivering it at any elevator thereon to which it may be consigned. *Ib.*

2607. If consigned to an elevator or warehouse not on their road and beyond its *terminus*, or if there be no elevator on the road, then they may rightfully refuse to receive it in bulk. *People ex rel.* v. *C. & A. R. R.*, 55 Ill. 95.

2608. In a proceeding by mandamus to compel a railroad company to deliver at the elevator or grain warehouse of the relator, in the city of Chicago, whatever grain in bulk might be consigned to it, upon the line of their road, it appeared that the company entered the city from different points upon separate tracks or lines of road, being called divisions. The elevator was situate upon a track used by the company in connection with the business of one of these divisions exclusively, but could be reached from the other divisions, though by a very indirect route, and subjecting the company to great loss of time and pecuniary damage, in the delay that would be caused to their regular trains and business on the latter division: *Held*, that the roads constituting these divisions, though belonging to the same corporation and having a common name, were for the purposes of transportation, substantially different roads, constructed under different charters, and the track upon which the elevator was situated, having been laid for the convenience especially of one of those divisions, and only approachable from the other under the difficulties mentioned, it could not be regarded that the elevator was upon the line of the latter division in any such sense as to make it obligatory upon the company to deliver thereat freight coming over that division. *C. & N. W. Ry.* v. *People*, 56 Ill. 365.

2609. But the track upon which the elevator was situated was owned and used by the respondent company and another company in common, and was a direct continuation of the line of one of the respondent company's divisions, and of easy and convenient access from that division, and was used by the respondent, not only to deliver grain to other elevators thereon, some of which were of more difficult access than that of the relator, but also to deliver lumber and other freight coming over such division, thus making it not only legally, but actually, by positive occupation, a part of their road. So it was *held*, that in reference to grain coming over that division, the track upon which the relator's elevator was situated, was to be regarded as a part of the respondent's line of road, and it was their duty to deliver such grain to the elevator, if consigned to it. *Ib.*

2610. Where grain in bulk is consigned to a particular elevator on the line of a railroad, it is no sufficient excuse for the company to refuse to so deliver, that it cannot do so without a large additional expense caused by the loss of the use of motive power, labor of servants, and loss of use of cars, while the same is being delivered and unloaded at such elevator, and brought back, for it is precisely that expense for which the company is paid its freight. *C. & N. W. Ry.* v. *People*, 56 Ill. 365.

2611. By the rules of the common law, railway companies cannot be compelled to permit individuals to connect side tracks of their own with the tracks of the companies, in order to enable the latter to carry grain to warehouses or elevators which have been erected off their lines of roads. *People* v. *C. & N. W. Ry.*, 57 Ill. 436.

2612. And where it is sought to compel a railroad company to permit such connection upon the ground of an alleged custom among the companies whose lines concentrate at the place indicated, the custom must be clearly made to appear, and to have existed so long as to have the force of law. *Ib.*

2613. A contract by the owner of an elevator to connect the same with a railroad, personal in its nature, confers no rights upon a lessee of such owner. *People* v. *C. & N. W. Ry.*, 57 Ill. 436.

2614. To make a railway company liable under this section for not delivering grain to the consignee or place of consignment, the freight must be in bulk, and must be consigned to the warehouse or place in question at the time of shipment. A demand at the place of destination is not of itself sufficient. *C. & N. W. Ry.* v. *Stanbro*, 87 Ill. 195.

2615. DAMAGES. In a simple action on the case, without reference to the statute, against a railway company for not delivering grain shipped in bulk to a particular warehouse, the true measure of damages is the necessary cost of moving the cars to the place required. If the suit is under the statute, the depreciation in the price of the grain may be considered. *C. & N. W. Ry.* v. *Stanbro*, 87 Ill. 195.

2616. Statute being penal will not be extended by construction. *C. & N. W. Ry.* v. *Stanbro*, 87 Ill. 195.

2617. A railroad corporation will only be compelled to deliver grain in the particular warehouse or elevator to which it is consigned, when such warehouse or elevator is upon the line of its road. But the line of the road is not necessarily confined to such tracks, side tracks and switches by it owned or leased. *Hoyt* v. *C., B. & Q. R. R.*, 93 Ill. 601.

2618. If a railway corporation has already purchased or secured by contract or otherwise, the legal right to use the track of another road necessary to reach a particular warehouse or elevator, then such warehouse or elevator may be considered as being upon the line. But when it has to run over the track of another company for the use of which it has no license or contract, to reach such warehouse or elevator, it cannot be compelled to run over such track in order to deliver grain. *Ib.*

2619. The mandate of the constitution in respect to the delivery of grain shipped in bulk, at the warehouse or elevator to which it is consigned, must be understood to be confined to a delivery by the common carrier at the warehouse or elevator where consigned when such delivery can be made by availing itself of tracks it has the legal right to employ or use. *Hoyt* v. *C., B. & Q. R. R.*, 93 Ill. 601.

2620. Right to enjoin removal of a connecting side track. *Hoyt* v. *C., B. & Q. R. R.*, 93 Ill. 601.

2621. Where a company takes grain consigned to Chicago, its duty is to deliver it in Chicago at any warehouse upon its lines or side tracks, to which it has been consigned. *Vincent* v. *C. & A. R. R.* 49 Ill. 33.

2622. RIGHT TO CHANGE CONSIGNMENT. §4. All consignments of grain to any elevator or public warehouse shall be held to be temporary, and subject to change by the consignee or consignor at any time previous to the actual unloading of such property from the cars in which it is transported. Notice of any change in consignment may be served by the consignee on any agent of the railroad corporation having the property in possession who may be in charge of the business of such corporation at the point where such property is to be delivered; and if, after such notice, and while the same remains uncanceled, such property is delivered in any way different from such altered or changed consignment, such railroad corporation shall, at the election of the consignee or person entitled to control such property, be deemed to have illegally appropriated such property to its own use, and shall be liable to pay the owner or consignee of such property double the value of the property so appropriated; and no extra charge shall be permitted by the corporation having the custody of such property, in consequence of such change of consignment. [R. S. 1887, p 1023, § 121; S. & C., p 1955, § 121; Cothran, p. 1166, § 107.]

2623. RECEIVING ON TRACK—RIGHTS OF OWNERS SAVED.
§ 5. Any consignee or person entitled to receive the delivery of grain transported in bulk by any railroad, shall have twenty-four hours, free of expense, after actual notice of arrival by the corporation to the consignee, in which to remove the same from the cars of such railroad corporation, if he shall desire to receive it from the cars on the track; which twenty-four hours shall be held to embrace such time as the car containing such property is placed and kept by such corporation in a convenient and proper place for unloading. And it shall not be held to have been placed in a proper place for unloading, unless it can be reached by the consignee, or person entitled to receive it, with teams or other suitable means for removing the property from the car, and reasonably convenient to the depot of such railroad corporation at which it is accustomed to receive and unload merchandise consigned to that station or place. Nothing herein contained, however, shall be held to authorize the changing of any consignment of grain, except as to the place at which it is to be delivered or unloaded, nor shall such change of consignment, in any degree, affect the ownership or control of property in any other way. [R. S. 1887, p. 1023, § 122; S. & C., p. 1955, § 123; Cothran, p. 1167, § 108.]

2624. RECEIPT AND DELIVERY AT CROSSINGS, ETC. § 6. Every railroad corporation organized or doing business under the laws of this state, or authority thereof, shall receive and deliver all grain consigned to its care for transportion at the crossings and junctions of all other railroads, canals, and navigable rivers. Any violation of this section shall render any such railroad corporation subject to the same penalty as contained in section 3 of this act. [§ 7, repeal, omitted. See "Statutes," ch. 131, § 5. R. S. 1887, p. 1023, § 123; S. & C., p. 1955, § 124; Cothran, p. 1167, § 109.]

RAILROAD AND WAREHOUSE COMMISSIONERS.

An act to establish a board of railroad and warehouse commissioners, and prescribe their powers and duties. Approved April 13, 1871. In force July 1, 1871. [L. 1871-2, p. 618; R. S. 1887, p. 1035, §§ 167-185; S. & C., p. 1956, §§ 126-144; Cothran, p. 1184, §§ 153-171.]

2625. APPOINTMENT—TERM. § 1. *Be it enacted by the people of the state of Illinois, represented in the general assembly,* That a commission which shall be styled "Railroad and Warehouse Commission," shall be appointed as follows: Within twenty days after this act shall take effect, the governor shall appoint three persons as such commissioners, who shall hold their office until the next meeting of the general assembly, and until their successors are appointed and qualified. At the next meeting of the general assembly, and every two years thereafter, the governor, by and with the advice and consent of the senate, shall appoint three persons as such commissioners, who shall hold their offices for the term of two years from the first day of January in the year of their appointment, and until their successors are appointed and qualified.

2626. QUALIFICATIONS. § 2. No person shall be appointed as such commissioner who is at the time of his appointment in any way connected with any railroad company or warehouse, or who is directly or indirectly interested in any stock, bond, or other property of, or is in the employment of any railroad company or warehouseman; and no person appointed as such commissioner shall, during the term of his office, become interested in any stock, bond or other property of any railroad company or warehouse, or in any manner be employed by or connected with any railroad company or warehouse. The governor shall have power to remove any such commissioner at any time, in his discretion.

2627. OATH—BOND. § 3. Before entering upon the duties of his office, each of the said commissioners shall make and subscribe, and file with the secretary of state, an affidavit, in the following form:

I do solemnly swear (or affirm, as the case may be,) that I will support the constitution of the United States, and the constitution of the state of Illinois, and that I will faithfully discharge the duties of the office of commissioner of railroads and warehouses, according to the best of my ability.

And shall enter into bonds, with security to be approved by the governor, in the sum of $20,000, conditioned for the faithful performance of his duty as such commissioner.

2628. COMPENSATION — SECRETARY — OFFICE — EXPENSES. § 4. Each of said commissioners shall receive for his services a sum not exceeding $3,500 per annum, payable quarterly. They shall be furnished with an office, office furniture and stationery, at the expense of the state, and shall have power to appoint a secretary to perform such duties as they shall assign to him. Said secretary shall receive for his services a sum not exceeding $1,500 per annum. The office of the said commissioners shall be kept at Springfield, and all sums authorized to be paid by this act shall be paid out of the state treasury and only on the order of the governor: *Provided*, that the total sum to be expended by said commissioners for office rent and furniture and stationery shall, in no case, exceed the total sum of $800 per annum.

2629. RIGHT TO PASS ON TRAINS, ETC. § 5. The said commissioners shall have the right of passing, in the performance of their duties concerning railroads, on all railroads and railroad trains in this state.

2630. REPORT OF RAILROADS. § 6. Every railroad company incorporated or doing business in this state, or which shall hereafter become incorporated, or do business under any general or special law of this state, shall, on or before the first day of September, in the year of our Lord 1871, and on or before the same day in each year thereafter, make and transmit to the commissioners appointed by virtue of this act, at their office in Springfield, a full and true statement, under oath of the proper officers of said corporation, of the affairs of their said corporation, as the same existed on the first day of the preceding July, specifying—

First—The amount of capital stock subscribed, and by whom.

Second—The names of the owners of its stock, and the amounts owned by them respectively, and the residence of each stockholder as far as known.

Third—The amount of stock paid in, and by whom.

Fourth—The amount of its assets and liabilities.

Fifth—The names and place of residence of its officers.

Sixth—The amount of cash paid to the company on account of the original capital stock.

Seventh—The amount of funded debt.

Eighth—The amount of floating debt.

Ninth—The estimated value of the roadbed, including iron and bridges.

Tenth—The estimated value of rolling stock.

Eleventh—The estimated value of stations, buildings and fixtures.

Twelfth—The estimated value of other property.

Thirteenth—The length of single main track.

Fourteenth—The length of double main track.

Fifteenth—The length of branches, stating whether they have single or double track.

Sixteenth—The aggregate length of siding and other tracks not above enumerated.

Seventeenth—The number of miles run by passenger trains during the year preceding the making of the report.

Eighteenth—The number of miles run by freight trains during the same period.

Nineteenth—The number of tons of through freight carried during the same time.

Twentieth—The number of tons of local freight carried during the same time.

Twenty-first—Its monthly earnings for the transportation of passengers during the same time.

Twenty-second—Its monthly earnings for the transportation of freight during the same time.

Twenty-third—Its monthly earnings from all other sources respectively.

Twenty-fourth—The amount of expense incurred in the

running and management of passenger trains during the same
time.

Twenty-fifth—The amount of expense incurred in the running and management of freight trains during the same time;
also, the amount of expense incurred in the running and
management of mixed trains during the same time.

Twenty-sixth—All other expenses incurred in the running
and management of the road during the same time, including the salaries of officers, which shall be reported separately.

Twenty-seventh—The amount expended for repairs of road
and maintenance of way, including repairs and renewal of
bridges and renewal of iron.

Twenty-eighth—The amount expended for improvement,
and whether the same are estimated as a part of the expenses
of operating or repairing the road, and, if either, which.

Twenty-ninth—The amount expended for motive power and
cars.

Thirtieth—The amount expended for station houses, buildings and fixtures.

Thirty-first—All other expenses for the maintenance of
way.

Thirty-second—All other expenditures, either for management of the road, maintenance of way, motive power and
cars, or for other purposes.

Thirty-third—The rate of fare for passengers for each
month during the same time, through and way passengers
separately.

Thirty-fourth—The tariff of freights, showing each change
of tariff during the same time.

Thirty-fifth—A copy of each published rate of fare for
passengers and tariff of freight, in force or issued for the
government of its agents during the same time.

Thirty-sixth—Whether the rate of fare and tariff of freight
in such published lists are the same as those actually received by the company during the same time; if not, what
were received.

Thirty-seventh—What express companies run on its roads
and on what terms and on what conditions; the kind of business done by them, and whether they take their freights at
the depots or at the office of such express companies.

Thirty-eighth—What freight and transportation companies
run on its road, and on what terms.

Thirty-ninth—Whether such freight and transportation
companies use the cars of the railroad or the cars furnished
by themselves.

Fortieth—Whether the freight or cars of such companies
are given any preference in speed or order of transporation,
and if so, in what particular.

Forty-first—What running arrangements it has with other railroad companies, setting forth the contracts for the same.

2631. ADDITIONAL INQUIRIES. § 7. The said commissioners may make and propound to such railroad companies any additional interrogatories, which shall be answered by such companies in the same manner as those specified in the foregoing section.

2632. APPLIES TO OFFICERS OF ROAD. § 8. Sections 6 and 7 of this act shall apply to the president, directors and officers of every railroad company now existing or which shall be incorporated or organized in this state, and to every lessee, manager and operator of any railroad within this state.

2633. STATEMENT BY WAREHOUSEMAN. § 9. It shall be the duty of every owner, lessee and manager of every public warehouse in this state to furnish in writing under oath, at such times as such railroad and warehouse commissioners shall require and prescribe, a statement concerning the condition and management of his business as such warehouseman.

2634. REPORT BY COMMISSIONERS—EXAMINATION. § 10. Such commissioners shall, on or before the first day of December, in each year, and oftener if required by the governor to do so, make a report to the governor of their doings for the preceding year, containing such facts, statements and explanations as will disclose the actual workings of the system of railroad transportation and warehouse business in their bearings upon the business and prosperity of the people of this state, and such suggestions in relation thereto as to them may seem appropriate, and particularly, first, whether in their judgment the railroads can be classified in regard to the rate of fare and freight to be charged upon them, and if so, in what manner; second, whether a classification of freight can also be made, and if so, in what manner. They shall also, at such times as the governor shall direct, examine any particular subject connected with the condition and management of such railroads and warehouses, and report to him in writing their opinion thereon with their reasons therefor.

2635. EXAMINATIONS OF RAILROADS AND WAREHOUSES—SUITS. § 11. Said commissioners shall examine into the condition and management, and all other matters concerning the business of railroads and warehouses in this state, so far as the same pertain to the relation of such roads and warehouses to the public, and to the accommodation and security of persons doing business therewith; and whether such railroad companies and warehouses, their officers, directors, managers, lessees, agents and employes, comply with the laws of this state now in force, or which shall hereafter be in force concerning them. And whenever it shall come to their knowledge, either upon complaint or otherwise, or they shall have reason to believe that any such law or laws have been or are being violated, they shall prosecute or cause to be prosecuted all corporations or persons guilty of such violation. In order to enable said commissioners efficiently to perform their duties under this act, it is hereby made their duty to cause one of their number, at least once in six months, to visit each county in the state, in which is or shall be located a railroad station, and personally inquire into the management of such railroad and warehouse business.

2636. WHEN BOARD TO INVESTIGATE CAUSE OF ACCIDENT ON RAILROAD—BRIDGE, ETC., OUT OF REPAIR—MANDAMUS—PROCEEDINGS BY ATTORNEY GENERAL. § 11½. It shall be the duty of said board of commissioners to investigate the cause of any accident on any railroad resulting in the loss of life or injury to person or persons, which in their judgment shall require investigation, and the result of such investigation shall be reported upon in a special report to the governor as soon after said accident as may be practicable, and also in the annual report of said commissioners. And it is hereby made the duty of the general superintendent or manager of each railroad in this state, to inform said board of any such accident immediately after its occurrence. Whenever it shall come to the knowledge of said board, by complaint or otherwise, that any railroad bridge or trestle, or any portion of the track of any railroad in this state is out of repair, or is in an unsafe condition, it shall be the duty of such board to investigate, or cause an investigation to be made, of the condition of such railroad bridge, trestle or track and may employ such person or persons who may be civil engineer or engineers, as they shall deem necessary for the purpose of making such investigation, and whenever in the judgment of said board, after such investigation, it shall become necessary to rebuild such bridge, track or trestle, or repair the same, the said board shall give notice and information in writing to the corporation of the improvements and changes which they deem to be proper. And shall recommend to the corporation or person or persons owning or operating such railroad that it, or he, or they, make such repairs, changes or improvements, or rebuild such bridge or bridges on such railroad as the board shall deem necessary, to the safety of persons being transported thereon. And said board shall give such corporation or person or persons owning or operating said railroad an opportunity for a full and fair hearing on the subject of such investigation and recommendation.

And said board shall, after having given said corporation or person or persons operating such railroad an opportunity for a full hearing thereon, if such corporation or person shall not satisfy said board that no action is required to be taken by it or them, fix a time within which such changes or repairs shall be made, or such bridges, tracks or culverts shall be rebuilt, which time the board may extend. It shall be the duty of the corporation, person or persons owning or operating said railroad to comply with such recommendations of said board, as are just and reasonable. And the supreme court or the circuit court in any circuit, in which such railroad may be in part situated, shall have power in all cases of such recommendations by said board, to compel compliance therewith by mandamus. If any such corporation or person or persons owning or operating any such railroad, shall, after such hearing, neglect or refuse to comply with the recommendation or recommendations of said board as to making any repairs, changes or improvements, on any bridge, track or trestle, or to rebuild any bridge within the time which shall be fixed by said board therefor, said board shall report such neglect or refusal, together with the facts in such case as said board shall find the facts to be, touching the necessity for such repairs, changes or rebuilding to the attorney general of the state of Illinois, who shall thereupon take such action as may be necessary to secure compliance with such recommendations of said board. In all actions or proceedings brought by the attorney general to compel compliance with the recommendations of the board, the findings of the board shall be *prima facie* evidence of the facts therein stated, and the recommendations of the board shall be deemed *prima facie*, just and reasonable. Nothing herein contained shall impair the legal liability of any railroad company for the consequence of its acts. And all existing remedies therefor are hereby saved to the people and to individuals. [Added by act approved June 16; 1887. In force July 1, 1887. L. 1887, p. 255.]

2637. CANCELLATION OF WAREHOUSE LICENSES. § 12. Said commissioners are hereby authorized to hear and determine all applications for the cancellation of warehouse licenses in this state which may be issued in pursuance of any laws of this state, and for that purpose to make and adopt such rules and regulations concerning such hearing and determination as may, from time to time, by them be deemed proper. And if, upon such hearing, it shall appear that any public warehouseman has been guilty of violating any law of this state concerning the business of public warehousemen, said commissioners may cancel and revoke the license of said public warehouseman, and immediately notify the officer who

issued such license of such revocation and cancellation; and
no person whose license as a public warehouseman shall be
cancelled or revoked, shall be entitled to another license or
to carry on the business in this state of such public warehouseman, until the expiration of six months from the date
of such revocation and cancellation, and until he shall have
again been licensed: *Provided*, that this section shall not
be so construed as to prevent any such warehouseman from
delivering any grain on hand at the time of such revocation
or cancellation of his said license. And all licenses issued
in violation of the provisions of this section shall be deemed
null and void.

2638. POWER TO EXAMINE BOOKS, ETC. '§ 13. The property, books, records, accounts, papers and proceedings of all
such railroad companies, and all public warehousemen, shall
at all times, during business hours, be subject to the examination and inspection of such commissioners, and they shall
have power to examine, under oath or affirmation, any and
all directors, officers, managers, agents and employes of any
such railroad corporation, and any and all owners, managers,
lessees, agents and employes of such public warehouses and
other persons, concerning any matter relating to the condition and mangement of such business.

2639. MAY EXAMINE WITNESSES, ETC. § 14. In making
any examination as contemplated in this act, or for the purpose of obtaining information, pursuant to this act, said
commissioners shall have the power to issue subpœnas for
the attendance of witnesses, and may administer oaths. In
case any person shall willfully fail or refuse to obey such
subpœna, it shall be the duty of the circuit court of any
county, upon application of said commissioners, to issue an
attachment for such witness, and compel such witness to
attend before the commissioners, and give his testimony
upon such matters as shall be lawfully required by such
commissioners; and the said court shall have power to punish for contempt, as in other cases of refusal to obey the
process and order of such court.

2640. PENALTY AGAINST WITNESSES. § 15. Any person
who shall willfully neglect or refuse to obey the process of
subpœna issued by said commissioners, and appear and testify as therein required, shall be deemed guilty of a misdemeanor, and shall be liable to an indictment in any court of
competent jurisdiction, and on conviction thereof shall be
punished for each offense, by a fine of not less than 25 nor
more than $500, or by imprisonment of not more thirty days,
or both, in the discretion of the court before which such conviction shall be had.

2641. PENALTY AGAINST RAILROAD COMPANIES, WAREHOUSEMEN, ETC. § 16. Every railroad company, and every officer, agent or employe of any railroad company, and every owner, lessee, manager or employe of any warehouse, who shall willfully neglect to make and furnish any report required in this act, at the time herein required, or who shall willfully and unlawfully hinder, delay, or obstruct said commissioners in the discharge of the duties hereby imposed upon them, shall forfeit and pay a sum of not less than 100 nor more than $5,000 for each offense, to be recovered in an action of debt in the name and for the use of the people of the state of Illinois; and every railroad company, and every officer, agent or employe of any such railroad company, and every owner, lessee, manager, or agent or employe of any public warehouse, shall be liable to a like penalty for every period of ten days it or he shall willfully neglect or refuse to make such report.

2642. ATTORNEY GENERAL AND STATE'S ATTORNEY TO PROSECUTE SUITS. § 17. It shall be the duty of the attorney general, and the state's attorney in every circuit or county, on the request of said commissioners, to institute and prosecute any and all suits and proceedings which they, or either of them, shall be directed by said commissioners to institute and prosecute for a violation of this act, or any law of this state concerning railroad companies or warehouses, or the officers, employes, owners, operators or agents of any such companies or warehouses.

2643. IN NAME OF PEOPLE—PAY—QUI TAM ACTIONS. § 18. All such prosecution shall be in the name of the people of the state of Illinois, and all moneys arising therefrom shall be paid into the state treasury by the sheriff or other officer collecting the same; and the state's attorney shall be entitled to receive for his compensation, from the state treasury, on bills to be approved by the governor, a sum not exceeding ten per cent. of the amount received and paid into the state treasury as aforesaid: *Provided,* this act shall not be construed so as to prevent any person from prosecuting any *qui tam* action as authorized by law, and of receiving such part of the amount recovered in such action as is or may be provided under any law of this state.

2644. RIGHTS OF INDIVIDUALS SAVED. § 19. This act shall not be so construed as to waive or affect the right of any person, injured by the violation of any law in regard to railroad companies or warehouses, from prosecuting for his private damages in any manner allowed by law.

EXTORTION AND UNJUST DISCRIMINATION.

An act to prevent extortion and unjust discrimination in the rates charged for the transportation of passengers and freights on railroads in this state and to punish the same, and prescribe a mode of procedure and rules of evidence in relation thereto, and to repeal an act entitled "An act to prevent unjust discriminations and extortions in the rates to be charged by the different railroads in this state for the transportation of freights on said roads," approved April 7, A. D. 1871. Approved May 2, 1873. In force July 1, 1873. [R. S. 1887, p. 1024, §§ 124-133; S. & C., p. 1961, §§ 145-155; Cothran, p. 1167, §§ 110,-119. See ante, 94-97.]

2645. EXTORTION. § 1. *Be it enacted by the people of the state of Illinois, represented in the general assembly:* If any railroad corporation, organized or doing business in this state under any act of incorporation, or general law of this state, now in force or which may hereafter be enacted, or any railroad corporation organized or which may hereafter be organized under the laws of any other state, and doing business in this state, shall charge, collect, demand or receive more than a fair and reasonable rate of toll or compensation, for the transportation of passengers or freight, of any description, or for the use and transportation of any railroad car upon its track, or any of the branches thereof or upon any railroad within this state which it has the right, license or permission to use, operate or control, the same shall be deemed guilty of extortion, and upon conviction thereof shall be dealt with as hereinafter provided. See Const., art. 11, § 15; ante, 94–98.

2646. CONSTITUTIONAL POWER OF STATE TO REGULATE CHARGES— FEDERAL CONSTITUTION. The act against extortion and unjust discrimination is not in violation of the federal constitution. *Railroad Co.* v. *Fuller,* 17 Wall. 560; *C., B. & Q. R. R.* v. *Iowa,* 4 Otto, 155; *Peik* v. *Ch. & N. W. Ry.,* 4 Otto, 164; *Munn* v. *Illinois,* 4 Otto, 114; *Ch., &c., R. R.* v. *Ackley,* 4 Otto, 179; *W., St. L. & P. Ry.* v. *Blake,* 4 Otto, 180; *Stone* v. *Wisconsin,* 4 Otto, 181.

2647. POWER of state to regulate and fix charges for freight and passage. *B. & O. R. R.* v. *Maryland,* 21 Wall. 456; *Peik* v. *Ch. & N. W. Ry.,* 94 U. S. 164; 6 Biss. 177; *Ruggles* v. *People,* 91 Ill. 256; 108 U. S. 256; *C. & A. R. R.* v. *People,* 67 Ill. 11; *Parker* v. *Metropolitan R. R.,* 109 Mass. 506; *Ackley* v. *C., M. & St. P. Ry.,* 36 Wis. 252; 4 Otto, 179.

2648. RIGHTS IN CHARTER. A provision in a railway charter that the company may fix its rates of tolls, &c., does not prevent the regulation of rates by the state. *Ruggles* v. *Illinois,* 108 U. S. 526; *I. C. R. R.* v. *People,* 108 U. S. 541.

2649. The act of 1871 to establish a reasonable maximum rate of charges is not unconstitutional, but is a valid law. *Ruggles* v. *People,* 91 Ill. 256.

2650. The act of May 2, 1873, to prevent extortion and unjust discrimination in railroads, is a constitutional enactment, and is not violative of the contract between the state and the railway corporations growing out of the grant and acceptance of their charters, giving them power to establish such rates of toll as they might from time to time determine in their by-laws. *I. C. R. R.* v *People,* 95 Ill. 313.

2651. COMMON LAW POWER—*not affected by charter.* The rule forbidding unreasonable charges by common carriers exists at the common law, and charters giving railroad companies the right to establish rates of freight, &c., in general terms, are subject to the implied condition that those rates shall not be unreasonable. *C. & A. R. R.* v. *People,* 67 Ill. 18.

2652. CHARTERS SUBJECT TO IMPLIED LIMITATION. The "right to fix rates of tariff," &c., granted by a charter, must be construed as being with an implied limitation or restriction that in fixing rates of tariff, the company shall make them reasonable and not extortionate, and that in making discriminations, they shall be reasonably just. *St. L., A. & T. H. R. R.* v. *Hill,* 14 Bradw. 579.

2653. UNJUST DISCRIMINATION. § 2. If any such railroad corporation aforesaid shall make any unjust discrimination in its rates or charges of toll, or compensation, for the transportation of passengers or freight of any description, or for the use and transportation of any railroad car upon its said road, or upon any of the branches thereof, or upon any railroads connected therewith, which it has the right, license or permission to operate, control or use, within this state, the same shall be deemed guilty of having violated the provisions of this act, and upon conviction thereof shall be dealt with as hereinafter provided.

CONSTITUTIONALITY.

2654. Legislation to prevent unjust discrimination in freights is in no respect a violation of the charters of railway companies, and is constitutional. *C. & A. R. R.* v. *People,* 67 Ill. 11.

2655. UNJUST DISCRIMINATION. § 15, art. 11 of the constitution authorizing the passage of laws to correct abuses and to ¦prevent unjust discrimination and extortion, by implication, restrains the power of the legislature to the prohibition of such discrimination only as is unjust. *Ib.*

2656. An act of the legislature prohibiting any and all discrimination, whether just or unjust, and which does not permit the companies to show that the same is not unjust, but infers guilt conclusively from the mere fact of a difference in rates, is unconstitutional and void. *Ib.*

2657. An act prohibiting unjust discrimination in charges, making the charging of a greater compensation for a less distance, or for the same distance, *prima facie* evidence of unjust discrimination, and giving a trial by jury as to the facts, is within the constitutional power of the legislature. *Ib.*

2658. This section is not obnoxious to any constitutional objection, and a railway company will be liable to the penalty imposed for its violation in discriminating in the rates of charges as to different contracts for through transportation of freight from points in this state to a point in another state for different distances, charging a greater sum for the less distance of the entire carriage than for the greater distance. *Wabash, St. L. & P. Ry.* v. *People,* 105 Ill. 236; *People* v. *W., St. L. & P. Ry.,* 104 Ill. 476.

2659. There is nothing in the section or act which confines the unjust discrimination for charges for the transportation of property within the limits of the state. The language "within this state," in the last part of the section, has reference to the roads which a railway company may operate in this state. *Ib.* 104 Ill. 476.

—22

2660. INTER-STATE COMMERCE. A state law to prevent the unjust discrimination in rates for the transportation of passengers or freight from a point within to a point without the state, though it may incidentally affect commerce between states cannot be said to be a law regulating commerce between states within the meaning of the federal constitution, especially when it does not purport to exercise control over any railroad corporation except those that run or operate in the state. *People* v. *W., St. L. & P. Ry.*, 104 Ill. 476; *W., St. L. & P. Ry.* v. *People*, 105 Ill. 236.

2661. Statute must be construed to include a transportation of goods under one contract and by one voyage from the interior of the state of Illinois to New York. *Wabash, St. L. & P. Ry.* v. *Illinois*, 118 U. S. 557.

2662. INTER-STATE COMMERCE. Such a transportation is "commerce among the states," even as to that part of the voyage which lies within the state of Illinois. There may be a transportation of goods which is begun and ended within its limits and disconnected with any carriage outside of the state, which is *not* commerce among the states. *Ib.*

2663. The latter is subject to regulation by the state and this statute is void as applied to it; but the former is national in its character, and its regulation is confided to congress exclusively by that clause of the federal constitution which empowers it to regulate commerce among the states. *Ib.*

2664. This statute is void as to that part of the transmission of the freight which may be within the state where the contract is for its carriage beyond the limits of the state. Reversing the judgment of the state court. *Ib.*

2664a. Regulation of inter-state commerce. See *State* v. *Ch. & N. W. Ry.*, 70 Iowa 162; *Carton & Co.* v. *I. C. R. R.*, 59 Iowa 148; *Hart* v. *Ch. & N. W. Ry.*, 69 Iowa 485.

2665. POWER OF STATE OVER. The legislature has the undoubted power under the constitution, to prohibit unjust discrimination in charges for the transportation of persons and freights by railroads, whether between individuals, or localities or communities. *I. C. R. R.* v. *People*, 121 Ill. 304.

CONSTRUCTION.

2666. EFFECT ON EXISTING CONTRACTS. This act was not designed to reach a case where a contract existed prior to its passage, to carry on certain terms. It does not interfere with or abrogate pre-existing contracts. *C. & A. R. R.* v. *C. V. & W. Coal Co.*, 79 Ill. 121.

2667. This section is not limited to railroads organized under the laws of this state, but includes all railroad companies which operate railroads in this state. *People* v. *W., St. L. & P. Ry.*, 104 Ill. 476.

2668. AT COMMON LAW. Charges for freights and passengers must be uniform without favor or prejudice of the several classes established by the company. Extra charges may be made in case of neglect to procure tickets when an opportunity to do so is afforded. *C., B. & Q. R. R.* v. *Parks*, 18 Ill. 460.

2669. At common law all discriminations by common carriers are not forbidden, but only those which are unreasonable and unjust. They are required to serve all who properly apply for transportation in the order of their application. The authorities differ as to whether the common law rule requires an equality of charge. The weight of American authority requires that the charges shall be equal to all for the same service under like circumstances. *St. L., A. & T. H. R. R.* v. *Hill*, 14 Bradw. 579.

2670. The common law does not require, where there is a difference of distances in the services performed, that the charge shall be proportioned equally to the respective distances; or that where the greater charge is made for the greater distance, it shall bear any given proportion to that made for the shorter distance. *Ib.*

2671. NO SHIPMENT. The statute against unjust discrimination does not apply to a case where no shipment is made, but simply a demand for illegal freight on the one side and a refusal on the other to ship. *Kankakee Coal Co.* v. *I. C. R. R.*, 17 Bradw. 614.

2672. GIVING PREFERENCES. Action lies for damages caused by giving improper preferences to other shippers of grain. *G. & C. U. R. R.* v. *Rae*, 18 Ill. 488.

2673. DISCRIMINATING CHARGES. A railway company, although permitted to establish its rates of transportation, must do so without injurious discrimination as to individuals. *Vincent* v. *C. & A. R. R.*, 49 Ill. 33.

2674. And where it has fixed its rates for the transportation of grain from any given station on the line of its road to Chicago, it will not be permitted, on the grain being taken there, to charge one rate for delivery at the warehouse of one person, and a different rate for delivery at that of another, both warehouses being upon its line or side tracks. *Ib.*

2675. Among the duties of railway companies is the obligation to receive and carry for all persons alike, without injurious discrimination as to terms, and to deliver them in safety to the consignee. *C. & N. W. Ry.* v. *People*, 56 Ill. 365.

2676. A railway company can establish no custom inconsistent with the spirit and object of its charter. It can make such rules and contracts as it pleases, not inconsistent with its duties as a common carrier, and any general language used in its charter in respect to its powers, must be considered with that limitation. *Ib.*

2677. A railway company can make no such injurious or arbitrary discriminations between individuals in dealing with the public, by any custom or usage of its own, in not receiving shipments of grain in bulk, except on the condition that it may choose the consignee. It may not unjustly and arbitrarily discriminate in favor of any particular warehouse or consignee. *Ch. & N. W. Ry.* v. *People*, 56 Ill 365.

2678. WHAT IS UNJUST DISCRIMINATION. The establishment permanently of less rates of freight at points of competition with other roads than is fixed at other places for the same distance, is an unjust discrimination between places, even though the higher rates are reasonably low. Railway companies cannot use their power to benefit particular individuals or to build up particular localities by arbitrary discriminations in their favor to the injury of other persons or rival places. *C. & A. R. R.* v. *People*, 67 Ill. 11.

2679. The offense provided in this section consists in an unjust discrimination in the rates charged: *first*, for the transportation of passengers or freight of any description; *second*, for the use and transportation of any railroad car upon the road; *third*, for the use of any railroad car upon any of the branches of the road; *fourth*, upon any railroads connected with the road or its branches which it is authorized to use in this state. *People* v. *W., St. L. & P. Ry.*, 104 Ill. 476.

2680. A contract between a railroad company and a shipper that the latter shall pay the regular and established rates of freight, the same as all other shippers, and that the company shall pay back to him, by way of rebate, a certain portion of the freight so charged and paid, whereby such shipper will pay a less rate for transportation than that paid by others and the public generally for like services under similar circumstances and for like distances, is void, as being against public policy at the common law and in violation of the statute against unjust discriminations. *I., D. & S. R. R.* v. *Ervin*, 118 Ill. 250.

2682. The case of *T., W. & W. Ry.* v. *Elliott*, 76 Ill. 67, holding that a contract for a rebate of freight paid to a railway company was not in violation of the statute against unjust discriminations, was made under a different statute. That case was under the act of 1871 which provided only against unjust discriminations between places and not between individuals, as does the act of 1873. That case and *Erie & Pacif. Disp.* v. *Cecil*, 112 Ill. 180–185, are not authority. *Ib.*

2683. The seven specified acts of discrimination, in § 3 of the statute, defined the offense of unjust discrimination, "such" as is to "be deemed and taken" as "the unlawful discriminations prohibited by the provisions of the act," and the clause: "This section shall not be construed so as to exclude other evidence tending to show any unjust discrimination in freight and passage rates," the court regards as treating of the matter only in its evidential aspect, and not as intended to expand the definition of the offense so as to include any discrimination in freight or passenger rates that the court or jury may *deem* unjust. It is the intent of the clause not to confine the plaintiffs, to the *simple fact* which makes the *prima facie* case, but allow them to introduce "other evidence tending to show" the discrimination involved in the *prima facie* case was unjust. *St. L., A. & T. H. R. R.* v. *Hill*, 14 Bradw. 579.

2684. Where the distance from A. to B. was 14 miles and from C. to B. was 28 miles, and the schedule of reasonable maximum freight rates established by the commissioners for appellant's railroad was $14.22 per car load from A. to B. and $17.58 from C. to B., and the company charged $17.40 per car load for the latter distance and $5 for the former: *Held*, that this was no unjust discrimination within the meaning of the act of 1873. *Ib.*

2685. Where a railway company charged a shipper 2 cents more per 100 pounds per car load for carrying certain wheat from C. to B. than it charged others for wheat shipped from C. and consigned to and delivered in the B. elevator in B.: *Held*, that the case fell directly within the statute. The right of the company to compel the shipper under the penalty of a higher rate of toll, to ship his freights to a particular consignee, cannot be admitted. *Ib.*

2686. The law imposes a duty on a common carrier to make no unjust, injurious or arbitrary discriminations between individuals in its dealings with the public. The right to the transportation services of the common carrier, is a common right, belonging to every one alike. *Ib.*

2687. SAME—*distinctive purposes of sections 1 and 6 of the statute.* § 1 of the act against unjust discriminations by railway corporations, is directed against discriminations between localities through unequal charges for the same transportation, in the same direction, over equal parts of their roads; and it is violated when all are compelled to pay for transportation for the shorter distance a rate equal to or greater than that charged for the same transportation in the same direction for the longer distance, as well as when one or a few individuals are compelled to do so. *Ill. C. R. R.* v. *People*, 121 Ill. 304.

2688. § 6 of the act simply gives a right of action against a railway company to any person or corporation which has paid the company extortionate charges, or charges for receiving or handling freight in violation of the provisions of the act, and which has therefore been unjustly discriminated against by such railway company in its charges, for three times the amount of the damages sustained by the party aggrieved. That section has nothing to do with suits by the state, and its purpose is to afford a personal indemnity in cases of personal injury occasioned by the unjust discrimination. *Ib.*

2689. SAME—*discriminations as between localities, not involving the element of competition in trade.* On a prosecution by the people, against a railway company, to recover the penalty imposed by § 4 of the act of 1873, for an unjust discrimination in charges between localities, it is not incumbent on the people to prove a personal discrimination and a personal injury as between individuals, but it is sufficient merely to prove a discrimination as between localities, omitting specific evidence of its effect upon different individuals. *Ib.*

2690. So the fact that there may be no competition in a particular trade between two points upon a railroad, does not show that a discrimination in charges for transportation, as between such points, is not unjust. *Ib.*

2691. In this case it appeared that a railway company charged ten cents per hundred pounds for carrying green coffee in the sack from Chicago to Mattoon, a distance of one hundred and twenty-two miles, and on the same day charged another person sixteen cents per hundred for carrying coffee in the sack from Chicago to Kankakee, a distance of only fifty-six miles, the transportation in both instances being in the same direction and over the same road. In a suit by the state to recover the penalty for unjust discrimination, the defendant showed that there was no competition between Kankakee and Mattoon in the grocery trade, and claimed that the discrimination between these points was not unjust, and therefore allowable: *Held*, that the fact so shown constituted no defense to the action. *Ib.*

2692. SAME—*discrimination at competitive points.* The fact that at a given point there is competition among railroads for the transportation of freights, and some of them are charging reduced or "cut" rates, will not justify another railway company in discriminating in favor of such point as against other points on the line of its road. *Ib.*

2693. A reduced or "cut" rate by a railway company to meet a "cut" rate of a rival road, which reduced or "cut" rate discriminates against a non-competitive point, is not, of itself, within the meaning of the statute, a just discrimination. *Ib.*

2694. The rivalry at competing points, which the statute declares shall not justify a discrimination in charges, has not reference solely to such as competes for a common business to a common market. It may apply to the same or different markets, and to roads having different *termini. Ib.*

2695. SAME—*instances where there may be a just discrimination.* There may be some greater expense to a railway company in carrying goods a given short distance than for a longer distance per mile, owing to the stopping and starting of trains, loading and unloading, the wear of machinery, &c.; and when it has full loads for its cars in each direction, it may carry more cheaply than when it is obliged to run its cars empty, or only partially loaded, in one direction, or only partially loaded in both directions. For these and other reasons affecting the cost of carriage, a company may often afford to carry the longer distance to or from competitive points more cheaply, *pro rata*, than for the shorter distance. Discriminations made in good faith because of such circumstances, are just, and not within the statute. *Ib.*

2696. RIGHT OF ACTION. Before the railroad commissioners had assigned a railway company to any class as required by the act of 1871, a passenger sued the company for charging him fare exceeding three cents a mile, and there was no proof that the charge was unreasonable or as to what class the road did belong: *Held*, that the plaintiff could not recover. *Moore* v. *I. C. R. R.*, 68 Ill. 385.

2697. Same—before rates fixed. Until the railroad commissioners have fixed the maximum rates of charges, a railway company cannot incur any liability under the act of 1873 for unreasonable charges. After such rates are made, the taking of the rates named, or less rates will not subject the company to the penalty, even though the proof shows them to be more than fair and reasonable rates. *C., B. & Q. R. R.* v. *People*, 77 Ill. 443.

2698. To maintain the action it must be shown that the company has not only made a discrimination in its rates of tolls, but also that such discrimination is unjust. *St. L., A. & T. H. R. R.* v. *Hill*, 11 Bradw. 248.

2699. The schedule of rates required by the statute to be fixed by the railroad and warehouse commissioners, is only *prima facie* evidence that such rates are reasonable, and notwithstanding this, the company may traverse the allegation of extortion, and show that the rates charged are reasonable and not extortionate. *Ib.*

2700. The classification of freights made by the commissioners is a part of the schedule of maximum rates of charges, and is required to be published the same as the schedule. *St. L. & C. R. R.* v. *Blackwood*, 14 Bradw. 503.

2701. Carriers can change their rates of freight so as to operate on future contracts, but they cannot increase them so as to affect existing contracts. *T., W. & W. Ry.* v. *Roberts*, 71 Ill. 540; *North. Transf. Co.* v. *Sellick*, 52 Ill. 249.

2702. Pleading. A declaration to recover the penalty for extortion which fails to aver that a schedule of rates had been established, and that defendant had charged in excess thereof, is fatally defective. *C., B. & Q. R. R.* v. *People*, 77 Ill. 443.

2703. Declaration. In an action to recover the penalty for unjust discrimination, the declaration must show that the respective freights mentioned were of like quantity, of the same class, and that in respect to such freight, there was a higher charge for a less than for a greater distance. The description of the respective freights merely as one car load of ponies, and one car load of horses, does not sufficiently show them to be "like quantities of freight of the same class." *C., B. & Q. R. R.* v. *People*, 77 Ill. 443.

2704. This statute being a penal one, the declaration should clearly show that the precise statutory offense has been committed. *Kankakee Coal Co.* v. *I. C. R. R.*, 17 Bradw. 614.

2705. Limitation. Actions seeking the treble damages penalty for extortion, or unjust discrimination must be brought within two years next after the cause of action arises. *St. L., A. & T H. R. R.* v. *Hill*, 11 Bradw. 248.

2706. Evidence—of extortion or unjust discrimination. § 3. If any such railroad corporation shall charge, collect or receive, for the transportation of any passenger, or freight of any description, upon its railroad, for any distance, within this state, the same, or a greater amount of toll or compensation than is at the same time charged, collected or received for the transportation, in the same direction, of any passenger, or like quantity of freight of the same class, over a greater distance of the same railroad; or if it shall charge,

collect or receive, at any point upon its railroad, a higher
rate of toll or compensation for receiving, handling or delivering freight of the same class and quantity, than it shall,
at the same time, charge, collect or receive at any other
point upon the same railroad; or if it shall charge, collect
or receive for the transportation of any passenger, or freight
of any description, over its railroad, a greater amount as toll
or compensation than shall, at the same time, be charged,
collected or received by it for the transportation of any passenger, or like quantity of frieght of the same class, being
transported in the same direction, over any portion of the
same railroad, of equal distance; or if it shall charge, collect or recive from any person or persons, a higher or greater
amount of toll or compensation than it shall, at the same
time, charge, collect, or receive from any other person or
persons for receiving, handling or delivering freight of the
same class and like quantity, at the same point upon its
railroad; or if it shall charge, collect or receive from any person or persons, for the transportation of any frieght upon its
railroad, a higher or greater rate of toll or compensation
than it shall, at the same time, charge, collect or receive
from any other person or persons, for the transportation of the
like quantity of freight of the same class, being transported
from the same point, in the same direction, over equal distances of the same railroad; or if it shall charge, collect or
receive from any person or persons, for the use and transportation of any railroad car or cars upon its railroad, for
any distance, the same or a greater amount of toll or compensation than it at the same time charged, collected or received from any other person or persons, for the use and
transportation of any railroad car of the same class or number, for a like purpose, being transported in the same direction, over a greater distance of the same railroad; or if it
shall charge, collect or receive from any person or persons,
for the use and transportation of any railroad car or cars
upon its railroad, a higher or greater rate of toll or compensation than it shall, at the same time, charge, collect or receive from any other person or persons, for the use and
transportation of any railroad car or cars of the same class
or number, for a like purpose, being transported from the
same point, in the same direction, over an equal distance of
the same railroad; all such discriminating rates, charges,
collections or receipts, whether made directly, or by means
of any rebate, drawback, or other shift or evasion, shall be
deemed and taken, against such railroad corporation, as
prima facie evidence of the unjust discriminations prohibited by the provisions of this act; and it shall not be
deemed a sufficient excuse or justification of such discrimi-

nations on the part of such railroad corporation, that the railway station or point at which it shall charge, collect or receive the same cr less rates of toll or compensation, for the transportation of such passenger or frieght, or for the use and transportation of such railroad car the greater distance, than for the shorter distance, is a railway station or point at which their exists competition with any other railroad or means of transportation. This section shall not be construed so as to exclude other evidence tending to show any unjust discrimination in freight and passenger rates. The provisions of this section shall extend and apply to any railroad, the branches thereof, and any road or roads which any railroad corporation has the right, license or permission to use, operate or control, wholly or in part, within this state: *Provided, however,* that nothing herein contained shall be so construed as to prevent railroad corporations from issuing commutation, excursion or thousand-mile tickets, as the same are now issued by such corporations.

2707. The words used in the first clause of this section, "within this state," are not intended to limit the law to transportation within the state, but to provide and declare that certain things shall be *prima facie* evidence to sustain a charge of unjust discrimination. *People* v. *W., St. L. & P. Ry.*, 104 Ill. 476.

2708. The words "this section shall not be so construed as to exclude other evidence tending to show any unjust discrimination," &c., are not intended to enlarge the definition of the offense of unjust discrimination. It does not give the court plenary power to determine what is a reasonable charge, &c. The intent of the clause is to show that the plaintiff is not to be confined to the simple facts which make a *prima facie* case. *St. L., A. & T. H. R. R.* v. *Hill*, 14 Bradw. 579.

2709. A law similar to this suggested as not being subject to any constitutional objection. *C. & A. R. R.* v. *People*, 67 Ill. 11; *C., B. & Q. R. R.* v. *People*, 77 Ill. 448.

2710. PENALTIES. § 4. Any such railroad corporation guilty of extortion, or of making any unjust discrimination as to passenger or freight rates, or the rates for the use and transportation of railroad cars, or in receiving, handling or delivering freights, shall upon conviction thereof, be fined in any sum not less than one thousand dollars ($1,000), nor more than five thousand dollars ($5,000), for the first offense; and for the second offense not less than five thousand dollars ($5,000), nor more than ten thousand dollars ($10,000), and for the third offense not less than ten thousend dollars ($10,000), nor more than twenty thousand dollars ($20,000); and for every subsequent offense and conviction thereof, shall be liable to a fine of twenty-five thousand dollars ($25,000): *Provided,* that in all cases under this act either party shall have the right of trial by jury. [See "Quo Warranto," ch. 112, §§ 1–6.]

2711. APPEAL—*to what court.* In an action of debt by the state's attorney for extortion or unjust discrimination no appeal lies from an order dismissing the suit, to the supreme court. It should be taken to the appellate court. *People* v. *St. Louis & Cairo R. R.*, 106 Ill. 412.

2712. PROCEEDINGS TO RECOVER FINES. § 5. The fines hereinbefore provided for may be recovered in an action of debt, in the name of the people of the state of Illinois, and there may be several counts joined in the same declaration as to extortion and unjust discrimination, and as to passenger and freight rates, and rates for the use and transportation of railroad cars, and for receiving, handling or delivering freights. If, upon the trial of any cause instituted under this act, the jury shall find for the people, they shall assess and return with their verdict the amount of the fine to be imposed upon the defendant, at any sum not less than one thousand dollars ($1,000) nor more than five thousand dollars ($5,000), and the court shall render judgment accordingly; and if the jury shall find for the people, and that the defendant has been once before convicted of a violation of the provisions of this act, they shall return such finding with their verdict, and shall assess and return with their verdict the amount of the fine to be imposed upon the defendant, at any sum not less than five thousand dollars ($5,000) nor more than ten thousand dollars ($10,000), and the court shall render judgment accordingly; and if the jury shall find for the people, and that the defendant has been twice before convicted of a violation of the provisions of this act, with respect to extortion or unjust discrimination, they shall return such finding with their verdict, and shall assess and return with their verdict the amount of the fine to be imposed upon the defendant, at any sum not less than ten thousand dollars ($10,000) nor more than twenty thousand dollars ($20,000); and in like manner, for every subsequent offense and conviction, such defendant shall be liable to a fine of twenty-five thousand dollars ($25,000): *Provided*, that in all cases under the provisions of this act, a preponderance of evidence in favor of the people shall be sufficient to authorize a verdict and judgment for the people.

2713. ACTION. Before an action lies for the penalty, a schedule of rates must be established as provided in § 8 of the act classifying the freights, and the same published for the requisite period. Form of declaration. *C., B. & Q. R. R.* v. *People*, 77 Ill. 443; *St. L. & C. R. R.* v. *Blackwood*, 14 Bradw. 503.

2714. DAMAGES—TREBLE AND ATTORNEY'S FEE. § 6. If any such railroad corporation shall, in violation of any of the provisions of this act, ask, demand, charge or receive of any person or corporation any extortionate charge or charges for the transportation of any passengers, goods, merchandise or property, or for receiving, handling or delivering freights, or shall make any unjust discrimination against any person or corporation in its charges therefor, the person or corporation so offended against may, for each offense, recover of such railroad corporation, in any form of action, three times the amount of the damages sustained by the party aggrieved, together with cost of suit and a reasonable attorney's fee, to be fixed by the court where the same is heard, on appeal or otherwise, and taxed as a part of the costs of the case.

2715. The action given being penal in its nature, is barred in two years by the limitation law. *St. L., A. & T. H. R. R.* v. *Hill,* 11 Bradw. 248.

2716. The declaration must show a discrimination which was unjust to the plaintiff. Corporation may traverse charge of extortion. *Ib.*

2717. Action being highly penal the statute will be strictly construed and the case clearly brought within its provisions. *Ib.*

2718. Counts concluding "contrary to the form of the statute," shows the action is founded on the statute. *Ib.*

2719. DUTIES OF RAILROAD AND WAREHOUSE COMMISSIONERS. § 7. It shall be the duty of the railroad and warehouse commissioners to personally investigate and ascertain whether the provisions of this act are violated by any railroad corporation in this state, and to visit the various stations upon the line of each railroad for that purpose, as often as practicable; and whenever the facts, in any manner ascertained by said commissioners, shall in their judgment warrant such prosecution, it shall be the duty of said commissioners to immediately cause suits to be commenced and prosecuted against any railroad corporation which may violate the provisions of this act. Such suits and prosecutions may be instituted in any county in this state through or into which the line of the railroad corporation sued for violating this act may extend. And such railroad and warehouse commissioners are hereby authorized, when the facts of the case presented to them shall, in their judgment, warrant the commencement of such action, to employ counsel to assist the attorney general in conducting such suit on behalf of the state. No such suits commenced by said commissioners shall be dismissed, except said railroad and warehouse commissioners and the attorney general shall consent thereto.

2720. SCHEDULE OF RATES TO BE MADE—EVIDENCE § 8. The railroad and warehouse commissioners are hereby directed to make, for each of the railroad corporations doing business in this state, as soon as practicable, a schedule of reasonable maximum rates of charges for the transportation of passengers and freights, and cars of each of said railroads; and said schedule shall in all suits brought against such railroad corporations wherein is, in any way involved the charges of any such railroad corporation for the transportation of any

passenger or freight, or cars, or unjust discrimination in
relation thereto, be deemed and taken in all courts of this
state as *prima facie* evidence that the rates therein fixed, are
reasonable maximum rates of charges for the transportation
of passengers and freights, and cars upon the railroads for
which said schedules may have been respectively prepared.
Said commissioners shall, from time to time, as often as circumstances may require, change and revise said schedules.
When any schedule shall have been made or revised, as aforesaid, it shall be the duty of said commissioners to have the
same printed by the state printer under the contract governing the state printing, and said commissioners shall furnish
two copies of such printed schedule to the president, general
superintendent or receiver of each railroad company or corporation doing business in this state. All such schedules
heretofore or hereafter made shall be received and held in
all such suits as *prima facie* the schedules of said commissioners, without further proof than the production of the
schedule desired to be used as evidence, with a certificate of
the railroad and warehouse commissioners that the same is a
true copy of a schedule prepared by them for the railroad
company or corporation therein named. [As amended by
act approved June 30, 1885. In force July 1, 1885. L. 1885,
p. 232.

2721. Until the rates of charges are fixed by the commissioners no
liability can be incurred under the statute for extortionate charges.
C., B. & Q. R. R. v. *People*, 77 Ill. 443.

2722. Classification of the freights is a part of the schedule, and
should be published as such to give validity to the schedule of maximum charges. *St. L., &c., R. R.* v. *Blackwood*, 14 Bradw. 503.

2723. Special charters giving the right to fix charges held not to
prevent the state from fixing rates by general laws. *Ruggles* v. *Illinois*, 108 U. S. 526; *I. C. R. R.* v. *Illinois, Id.* 541.

2724. EVIDENCE—FINES—PRACTICE. § 10. In all cases
under the provisions of this act, the rules of evidence shall
be the same as in other civil actions, except as hereinbefore
otherwise provided. All fines recovered under the provisions
of this act shall be paid into the county treasury of the county
in which the suit is tried, by the person collecting the same,
in the manner now provided by law, to be used for county
purposes. The remedies hereby given shall be regarded as
cumulative to the remedies now given by law against railroad corporations, and this act shall not be construed as
repealing any statute giving such remedies. Suits commenced under the provisions of this act shall have precedence over all other business, except criminal business.

2725. "RAILROAD CORPORATION" DEFINED. § 11. The
term "railroad corporation," contained in this act, shall be

deemed and taken to mean all corporations, companies or individuals now owning or operating, or which may hereafter own or operate any railroad, in whole or in part, in this state; and the provisions of this act shall apply to all persons, firms and companies, and to all associations of persons, whether incorporated or otherwise, that shall do business as common carriers upon any of the lines of railways in this state (street railways excepted) the same as to railroad corporations hereinbefore mentioned. [§ 12, repeal, omitted. See "Statutes," ch. 131, § 5. See ante, 2457-2482.]

2725a. As to the ruling of the courts in some of the other states in respect to extortion and unjust discriminations see *State* v. *Concord R. R.*, 59 N. H. 85; *Commonwealth* v. *Housatonic R. R.*, 143 Mass. 264; *State* v. *C. & N. W. Ry.*, 70 Iowa, 162; *Scofield* v. *Ry. Co.*, 43 Ohio St. 571; *W. & St. P. R. R.* v. *Blake*, 94 U. S. 180; *Commonwealth* v. *Eastern R. R.*, 103 Mass. 258; *Sanford* v. *R. R.*, 24 Pa. St. 378; *Messenger* v. *Penn. R. R.*, 36 N. J. Law, 407, 412; *McDuffee* v. *R. R. Co.*, 52 N. H. 447; *Shipper* v. *R. R. Co.*, 47 Pa. St. 338; *Audenried* v. *P. R. R. Co.*, 68 Pa. St. 370; *New England Express Co.* v. *M. C. R.*, 57 Me. 188; *G. W. Ry.* v. *Sutton*, 4 Eng. & Ir. App. 226; *State* v. *D., L. & W. R. R.*, 48 N. J. (Law) 55; *Woodhouse* v. *R. G. Ry.*, 67 Tex. 416.

RAILROAD CROSSINGS.

An act in regard to the dangers incident to railroad crossings on the same level. Approved June 3, 1887. In force July 1, 1887. [Laws of 1887, p. 252. R. S. 1887, p 1015, §§ 76a, 76b; 3 S. & C., 1887, p. 449.

2726. TWO OR MORE RAILROADS CROSSING EACH OTHER ON SAME LEVEL—REQUIREMENTS. § 1. *Be it enacted by the people of the state of Illinois, represented in the general assembly,* That when, and in case two or more railroads crossing each other at a common grade, shall by a system of interlocking and automatic signals, or by other works, fixtures and machinery to be erected by them, or either of them, render it safe for engines and trains to pass over such crossing without stopping, and such system of interlocking and signals, works or fixtures, shall first be approved by the railroad and warehouse commissioners, or any two of them, and a plan of such interlocking and signals, works or fixtures, for such crossing designating the plan of crossing, shall have been filed with such railroad and warehouse commissioners, then, and in that case, it is hereby lawful for the engines and trains of any such railroad or railroads to pass over said crossing without stopping, any law, or the provisions of any law now in force, to the contrary notwithstanding; and all such other provisions of laws, contrary hereto, are hereby declared not to be applicable in such case: *Provided*, that the said railroad and warehouse commissioners shall have power in case such interlocking system, in their judgment, shall by experience prove to be unsafe or impracticable, to order the same to be discontinued.

2727. CIVIL ENGINEER TO EXAMINE SYSTEM, ETC—COMPENSATION. § 2. The said railroad and warehouse commissioners may appoint a competent civil engineer to examine such proposed system and plans, and report the result of such examination for the information of such railroad and warehouse commissioners; and said railroad and warehouse commissioners are hereby authorized to allow and reward five dollars per day as a compensation for the services of such civil engineer, or such reasonable sum as such commissioners shall deem fit, and to allow and reward such other and further sums, as they shall deem fit to pay all other fees, costs and expenses to arise under said application, to be paid by the railroad company or companies in interest, to be taxed and paid or collected as in other cases. And the said railroad and warehouse commissioners are also empowered, on application for their approval of any such system of interlocking and signals, works or fixtures, to require of the applicant security for such fees, costs and expenses, or the deposit, in lieu thereof, of a sufficient amount in money for that purpose to be fixed by them.

WEIGHING GRAIN IN BULK BY RAILROAD COMPANY.

An act relating to the receipt, shipment, transportation and weighing of grain in bulk by railroad companies. Approved June 15, 1887. In force July 1, 1887. [L. 1887, p. 253; R. S. 1887, p. 1040; 3 S. & C., p. 448.]

2728. ROAD RECEIVING FOR TRANSPORTATION SHALL FURNISH SUITABLE APPLIANCES FOR WEIGHING, ETC. § 1. *Be it enacted by the people of the state of Illinois, represented in the general assembly,* That in all counties of the third class, and in all cities having not less than 50,000 inhabitants, where bulk grain, millstuffs or seeds are delivered by any railroad transporting the same from initial points to another road for transportation to other points, such road or roads receiving the same for transportation to said points, or other connections leading thereto, shall provide suitable appliances for unloading, weighing and transferring such property from one car to another without mixing or in any way changing the identity of the property so transferred, and such property shall be accurately weighed in suitably covered hopper scales, which will determine the actual net weight of the entire contents of any carload of grain, millstuffs or seeds at a single draft, without gross or tare, and which weights shall always be given in the receipts or bills of lading and used as the basis of any freight contracts affecting such shipments between such railroad companies and the owners, agents or shippers of such grain, millstuffs or seeds so transported and transferred.

2729. WHERE ORIGINAL CAR RUNS THROUGH WITHOUT

TRANSFER. § 2. The practice of loading grain, millstuffs or seeds into foreign or connecting line cars at the initial point from which the grain, millstuffs or seeds are originally shipped, or the running of the original car through without transfer, shall not relieve the railroad making the contract to transport the same to its destination or connection leading thereto, from weighing and transferring such property in the manner aforesaid, unless the shipper, owner or agent of such grain, millstuffs or seeds shall otherwise order or direct.

2730. LIABILITY OF RAILROAD COMPANY FOR NEGLECT OR FAILURE — PROCEEDINGS. § 3. Any railroad company neglecting or refusing to comply promptly with any and all of the requirements of either sections one or two of this act, shall be liable in damages to the party interested, to be recovered by the party damaged in an action of assumpsit, and such party may proceed by mandamus against any railroad company so refusing or neglecting to comply with the requirements of this act; and if the shipper, owner or agent of any such grain, millstuffs or seeds shall fail or neglect to proceed by mandamus, it shall then be the duty of the railroad and warehouse commissioners of this state, upon complaint of the party or parties interested, to proceed against the railroad failing or refusing to comply with the provisions of this act; and all the powers heretofore conferred by law upon the board of railroad and warehouse commissioners of this state, shall be applicable in the conduct of any legal proceeding commenced by such commissioners under this act.

2731. PENALTY, HOW RECOVERED. § 4. Any railroad company so refusing or neglecting as aforesaid, shall be liable to a penalty of not less than $100 nor more than $500 for each neglect or refusal as aforesaid, to be recovered in an action of assumpsit in the name of the people of the state of Illinois for the use of the county in which such act or acts of neglect or refusal shall occur, and it shall be the duty of the railroad and warehouse commissioners to cause prosecutions for such penalties to be instituted and prosecuted.

WAREHOUSES.

An act to regulate public warehouses, and the warehousing and inspection of grain, and to give effect to article thirteen of the constitution of this state. Approved April 25, 1871. In force July 1, 1871. [L. 1871-2, p. 762; R. S. 1887, p. 1027; S. & C., p. 1966; Cothran p. 1172.]

2732. This act does not violate either the state or federal constitution. *Munn* v. *People*, 69 Ill. 80; *People* v. *Harper*, 91 Ill. 357; *Munn* v. *Illinois*, 94 U. S. 113.

2733. There is no constitutional provision which prohibits the legislature from committing the inspection of grain to a board created for that purpose. *People* v. *Harper*, 91 Ill. 357.

2734. CLASSIFIED. § 1. *Be it enacted by the poeple of the State of Illinois, represented in the general assembly,* That public warehouses, as defined in article 13 of the constitution of this state, shall be divided into three classes, to be designated as classes A., B. and C., respectively.

2735. CLASSES DEFINED. § 2. Public warehouses of class A. shall embrace all warehouses, elevators and granaries in which grain is stored in bulk, and in which the grain of different owners is mixed together, or in which grain is stored in such a manner that the identity of different lots or parcels cannot be accurately preserved, such warehouses, elevators or granaries being located in cities having not less than 100,000 inhabitants. Public warehouses of class B shall embrace all other warehouses, elevators or granaries in which grain is stored in bulk, and in which the grain of different owners is mixed together. Public warehouses of class C shall embrace all other warehouses or places where property of any kind is stored for a consideration.

2736. Statute relating to inspection of grain in Chicago, although in a certain sense local and special, is not within the constitutional prohibition against such legislation. *People* v. *Harper*, 91 Ill. 357.

2737. LICENSE. § 3. The proprietor, lessee or manager of any public warehouse of class A shall be required, before transacting any business in such warehouse, to procure from the circuit court of the county in which such warehouse is situated, a license, permitting such proprietor, lessee or manager to transact business as a public warehouseman under the laws of this state, which license shall be issued by the clerk of said court upon a written application, which shall set forth the location and name of such warehouse, and the individual name of each person interested as owner or principal in the management of the same; or, if the warehouse be owned or managed by a corporation, the names of the president, secretary and treasurer of such corporation shall be stated; and the said license shall give authority to carry on and conduct the business of a public warehouse of class A in accordance with the laws of this state, and shall be revocable by the said court upon a summary proceeding before the court, upon complaint of any person in writing, setting forth the particular violation of law, and upon satisfactory proof, to be taken in such manner as may be directed by the court.

2738. BOND. § 4. The person receiving a license as herein provided, shall file with the clerk of the court granting the same, a bond to the people of the state of Illinois, with good and sufficient surety, to be approved by said court, in the penal sum of $10,000, conditioned for the faithful performance of his duty as public warehouseman of class A, and his full and unreserved compliance with all laws of this state in relation thereto.

2739. PENALTY FOR DOING BUSINESS WITHOUT LICENSE. § 5. Any person who shall transact the business of a public warehouse of class A without first procuring a license as herein provided, or who shall continue to transact any such business after such license has been revoked (save only that he may be permitted to deliver property previously stored in such warehouse), shall, on conviction, be fined in a sum not less than $100 nor more than $500 for each and every day such business is so carried on; and the court may refuse to renew any license, or grant a new one, to any of the persons whose license has been revoked, within one year from the time the same was revoked.

2740. NOT TO DISCRIMINATE—NOT TO MIX GRADE—RECEIPTS. § 6. It shall be the duty of every warehouseman of class A to receive for storage any grain that may be tendered to him, in the usual manner in which warehouses are accustomed to receive the same in the ordinary and usual course of business, not making any discrimination between persons desiring to avail themselves of warehouse facilities— such grain, in all cases, to be inspected and graded by a duly authorized inspector, and to be stored with grain of a similar grade, received at the same time, as near as may be. In no case shall grain of different grades be mixed together while in store; but, if the owner or consignee so requests, and the warehouseman consent thereto, his grain of the same grade may be kept in a bin by itself, apart from that of the owners; which bin shall, thereupon, be marked and known as a "separate bin." If a warehouse receipt be issued for grain so kept separate, it shall state, on its face, that it is in a separate bin, and shall state the number of such bin; and no grain shall be delivered from such warehouses unless it be inspected on the delivery thereof by a duly authorized inspector of grain. Nothing in this section shall be so construed as to require the receipt of grain into any warehouse in which there is not sufficient room to accommodate or store it properly, or in cases where such warehouse is necessarily closed.

2741. MANNER OF ISSUING RECEIPTS. § 7. Upon application of the owner or consignee of grain stored in a public warehouse of class A, the same being accompanied with evidence that all transportation or other charges which may be a lien upon such grain, including charges for inspection, have been paid, the warehouseman shall issue to the person entitled thereto, a warehouse receipt therefor, subject to the order of the owner or consignee, which receipt shall bear date corresponding with the receipt of grain into store, and shall state upon its face the quantity and inspected grade of the grain, and that the grain mentioned in it has been received into store, to be stored with grain of the same grade by inspection, received at about the date of the receipt, and that it is deliverable upon the return of the receipt, properly indorsed by the person to whose order it was issued, and the payment of proper charges for storage. All warehouse receipts for grain, issued from the same warehouse, shall be consecutively numbered; and no two receipts, bearing the same number, shall be issued from the same warehouse during any one year, except in the case of a lost or destroyed receipt, in which case the new receipt shall bear the same date and number as the original, and shall be plainly marked on its face "duplicate." If the grain was received from railroad cars, the number of each car shall be stated upon the receipt, with the amount it contained; if from canal boat or other vessel, the name of such craft; if from teams or by other means, the manner of its receipt shall be stated on its face.

2742. CANCELING RECEIPTS. § 8. Upon the delivery of grain from store, upon any receipt, such receipt shall be plainly marked across its face with the word "canceled," with the name of the person canceling the same, and shall thereafter be void, and shall not again be put in circulation, nor shall grain be delivered twice upon the same receipt.

2743. FURTHER OF ISSUING AND CANCELING RECEIPTS. § 9. No warehouse receipt shall be issued, except upon the actual delivery of grain into store, in the warehouse from which it purports to be issued, and which is to be represented by the receipt; nor shall any receipt be issued for a greater quantity of grain than was contained in the lot or parcel stated to have been received; nor shall more than one receipt be issued for the same lot of grain, except in cases where receipts for a part of a lot are desired, and then the aggregate receipts for a particular lot shall cover that lot and no more. In cases where a part of the grain represented by the receipt is delivered out of store and the remainder is left, a new receipt may be issued for such remainder; but such new receipt shall bear the same date as the original, and shall state on its face that it is balance of receipt of the original number; and the receipt upon which a part has been delivered shall be canceled in the same manner as if it had all been delivered. In case it be desirable to divide one receipt into two or more, or in case it be desirable to consolidate two or more receipts into one, and the warehouseman consent thereto, the original receipt shall be canceled the same as if

—23

the grain had been delivered from store; and the new receipts shall express on their face that they are parts of other receipts, or a consolidation of other receipts, as the case may be; and the numbers of the original receipts shall also appear upon the new ones issued, as explanatory of the change, but no consolidation of receipts of dates differing more than ten days shall be permitted, and all new receipts issued for old ones canceled, as herein provided, shall bear the same dates as those originally issued, as near as may be.

2744. Not to limit liability. § 10. No warehouseman in this state shall insert in any receipt issued by him, any language in anywise limiting or modifying his liabilities or responsibility, as imposed by the laws of this state.

2745. As to the liability of a warehouseman, see *Myers* v. *Walker,* 31 Ill. 353; *St. L., A. & T. H. R. R.* v. *Montgomery,* 39 Ill. 335; *C. & A. R. R.* v. *Scott,* 42 Ill. 132; *Buckingham* v. *Fisher,* 70 Ill. 121; *Broadwell* v. *Howard,* 77 Ill. 305; *Bailey* v. *Bensley,* 87 Ill. 556; *German Nat. Bank* v. *Meadowcroft,* 95 Ill. 124.

2746. Sampler's ticket held not a warehouse receipt. *P. & P. U. R. R.* v. *Buckley,* 114 Ill. 337.

2747. Delivery of property. § 11. On the return of any warehouse receipt issued by him, properly indorsed, and the tender of all proper charges upon the property represented by it, such property shall be immediately deliverable to the holder of such receipt, and it shall not be subject to any further charges for storage, after demand for such delivery shall have been made. Unless the property represented by such receipt shall be delivered within two business hours after such demand shall have been made, the warehouseman in default shall be liable to the owner of such receipt for damages for such default, in the sum of one cent per bushel, and in addition thereto, one cent per bushel for each and every day of such neglect or refusal to deliver: *Provided,* no warehouseman shall be held to be in default in delivering if the property is delivered in the order demanded, and as rapidly as due diligence, care and prudence will justify.

2748. Warehouseman has a lien for storage. *Low* v. *Martin,* 18 Ill. 286.

2748a. Issue of receipts for grain not in store, does not deprive him of his lien for that actually stored. *Ib.* If he permits the grain to be removed before charges are paid, he does not lose his right to recover of the holder of the receipt. Purchaser takes subject to lien. *Cole* v. *Tyng.* 24 Ill. 99.

2749. Purchaser of receipt with notice that it is chargeable for storage becomes liable for charges. *Ib.*

2750. Such lien is lost by a delivery of the goods, and will not revive in case the goods accidentally be returned to warehouseman's possession. *Hale* v. *Barrett,* 26 Ill. 195.

2751. Where goods of different owners are shipped together the consignee will have no lien on the goods of one for the charges due on those of the other. *Ib.*

2752. Lien lost by agreement. *Board of Trade* v. *Buckingham,* 65 Ill. 72.

2753. Remedy against warehouseman for non-delivery. *Leonard* v. *Dunton,* 51 Ill. 482.

2754. Warehouse receipts as evidence of ownership. *Cool* v. *Phillips,* 66 Ill. 216; *Broadwell* v. *Howard,* 77 Ill. 305.

2755. POSTING GRAIN IN STORE—STATEMENT TO REGISTRAR—DAILY PUBLICATIONS—CANCELED RECEIPTS. § 12. The warehousemen of every public warehouse of class A shall, on or before Tuesday morning of each week, cause to be made out, and shall keep posted up in the business office of his warehouse, in a conspicuous place, a statement of the amount of each kind and grade of grain in store in his warehouse at the close of business on the previous Saturday; and shall, also, on each Tuesday morning, render a similar statement, made under oath before some officer authorized by law to administer oaths, by one of the principal owners or operators thereof, or by the bookkeeper thereof, having personal knowledge of the facts, to the warehouse registrar appointed as hereinafter provided. They shall also be required to furnish daily, to the same registrar, a correct statement of the amount of each kind and grade of grain received in store in such warehouse on the previous day; also the amount of each kind and grade of grain delivered or shipped by such warehouseman during the previous day, and what warehouse receipts have been canceled, upon which the grain has been delivered on such day, giving the number of each receipt, and amount, kind and grade of grain received and shipped upon each; also, how much grain, if any, was so delivered or shipped, and the kind and grade of it, for which warehouse receipts had not been issued, and when and how such unreceipted grain was received by them; the aggregate of such reported cancellations and delivery of unreceipted grain, corresponding in amount, kind and grade with the amount so reported, delivered or shipped. They shall also, at the same time, report what receipts, if any, have been canceled and new ones issued in their stead, as herein provided for. And the warehouseman making such statements, shall, in addition, furnish the said registrar any further information, regarding receipts issued or canceled, that may be necessary to enable him to keep a full and correct record of all receipts issued and canceled, and of grain received and delivered. [§ 13, repealed.]

2756. CHIEF INSPECTOR. § 14. 1. It shall be the duty of the governor to appoint by, and with the advice and consent of the senate, a suitable person, who shall not be a member of the board of trade, and who shall not be interested

either directly or indirectly, in any warehouse in this state,
a chief inspector of grain, who shall hold his office for the
term of two years, unless sooner removed as hereinafter provided, for in every city or county in which is located a warehouse of class A, or class B: *Provided*, that no such grain
inspector for cities or counties in which are located warehouses of class B, shall be appointed, except upon the application and petition of two or more warehousemen doing a
separate and distinct business, residing and doing business in
such city or county, and when there shall be a legally organized board of trade in such cities or counties; such application and petition shall be officially endorsed by such board of
trade, before such application and petition shall be granted.

2. HIS DUTIES. It shall be the duty of such chief inspector of grain to have a general supervision of the inspection
of grain as required by this act or laws of this state, under
the advice and immediate direction of the board of commissioners of railroads and warehouses.

3. ASSISTANT INSPECTORS. The said chief inspector shall
be authorized to nominate to the commissioners of railroads
and warehouses, such suitable persons, in sufficient number,
as may be deemed qualified for assistant inspectors, who shall
not be members of the board of trade, nor interested in any
warehouse, and also such other employes as may be necessary
to properly conduct the business of his office; and the said
commissioners are authorized to make such appointments.

4. CHIEF INSPECTOR'S OATH AND BOND. The chief inspector shall upon entering upon the duties of his office, be required to take an oath, as in cases of other officers, and he
shall execute a bond to the people of the state of Illinois, in
the penal sum of fifty thousand dollars when appointed for
any city in which is located a warehouse of class A, and ten
thousand dollars, when appointed for any other city or county
with sureties to be approved by the board of commissioners
of railroads and warehouses, with a condition therein that he
will faithfully and strictly discharge the duties of his said
office of inspector according to law, and the rules and regulations prescribing his duties; and that he will pay all damages
to any person or persons who may be injured by reason of
his neglect, refusal or failure to comply with law, and the
rules and regulations aforesaid.

5. ASSISTANT INSPECTOR'S OATH AND BOND. And each
assistant inspector shall take a like oath; execute a bond in
the penal sum of five thousand dollars, with like conditions,
and to be approved in like [manner as is provided in case
of the chief inspector, which said several bonds shall be filed
in the office of said commissioner; and suit may be brought

upon said bond or bonds in any court having jurisdiction thereof, in the county where the plaintiff or defendant resides, for the use of the person or persons injured.

6. RULES OF INSPECTION—CHARGES. The chief inspector of grain, and all assistant inspectors of grain, and other employes in connection therewith, shall be governed in their respective duties by such rules and regulations as may be prescribed by the board of commissioners of railroads and warehouses; and the said board of commissioners shall have full power to make all proper rules and regulations for the inspection of grain; and shall, also, have power to fix the rate of charges for the inspection of grain, and the manner in which the same shall be collected; which charges shall be regulated in such a manner as will in the judgment of the commissioners, produce sufficient revenue to meet the necessary expenses of the service of inspection, and no more.

7. PAY OF INSPECTOR AND ASSISTANTS, ETC. It shall be the duty of the said board of commissioners to fix the amount of compensation to be paid to the chief inspector, assistant inspectors, and all other persons employed in the inspection service, and prescribe the time and manner of their payment.

8. APPOINTMENT OF REGISTRAR AND ASSISTANTS. The said board of commissioners of railroads and warehouses are hereby authorized to appoint a suitable person as warehouse registrar, and such assistants as may be deemed necessary to perform the duties imposed upon such registrar by the provisions of this act.

9. GENERAL SUPERVISION—PAY, ETC. The said board of commissioners shall have and exercise a general supervision and control of such appointees; shall prescribe their respective duties; shall fix the amount of their compensation, and the time and manner of its payment.

10. REMOVAL FROM OFFICE. Upon the complaint, in writing, of any person, to the said board of commissioners, supported by reasonable and satisfactory proof, that any person appointed or employed under the provisions of this section has violated any of the rules prescribed for his government, has been guilty of any improper official act, or has been found insufficient or incompetent for the duties of his position, such person shall be immediately removed from his office or employment by the same authority that appointed him; and his place shall be filled, if necessary, by a new appointment; or, in case it shall be deemed necessary to reduce the number of persons so appointed or employed, their term of service shall cease under the orders of the same authority by which they were appointed or employed.

11. EXPENSES, HOW PAID. All necessary expenses incident to the inspection of grain, and to the office of registrar economically administered, including the rent of suitable offices, shall be deemed expenses of the inspection service, and shall be included in the estimate of expenses of such inspection service, and shall be paid from the funds collected for the same. [As amended by act approved and in force May 28, 1879. L. 1879, p. 226.]

2757. It is competent for the legislature to delegate to railroad and warehouse commissioners the power to control the subject of the inspection of grain. *People* v. *Harper*, 91 Ill. 357.

2758. The expenses of the inspection of grain may be required to be paid by those presumably benefited by it. *Ib.*

2759. Although the board of commissioners are only authorized to fix the fees for inspection at such rates as may be necessary to meet the expenses, yet if more is collected than necessary, the chief inspector cannot retain the same. *Ib.*

2760. INSPECTOR, APPOINTMENT OF. There can be no legally appointed inspectors of grain except they are appointed by the governor in the manner pointed out in the amendatory act of 1879, § 14.

2761. Sureties on chief inspector's bond are not responsible for moneys collected by him for inspection, where the duty of collecting and taking care of the same, is not imposed on him before the execution of his bond. *People* v. *Tompkins*, 74 Ill. 482.

2762. The people of the state of Illinois are proper parties plaintiff in an action upon such bond, although the sum when recovered must be paid into the inspection fund. *People* v. *Harper*, 91 Ill. 357.

2763. The allegation of the expiration of the principal's term of office and the appointment of his successor, is sufficient in a declaration upon inspector's bond, without showing the qualification of the successor. *Ib.*

2764. RATES OF STORAGE. § 15. Every warehouseman of public warehouses of class A shall be required, during the the first week in January of each year, to publish in one or more of the newspapers (daily, if there be such,) published in the city in which such warehouse is situated, a table or schedule of rates for the storage of grain in his warehouse during the ensuing year, which rates shall not be increased (except as provided for in section (16) of this act) during the year; and such published rates, or any published reduction of them, shall apply to all grain received into such warehouse from any person or source, and no discrimination shall be made directly or indirectly, for or against any charges made by such warehouseman for the storage of grain. The maximum charge for storage and handling of grain, including the cost of receiving and delivering, shall be, for the first ten days or part thereof, one and one-quarter (1¼) cents per bushel, and for each ten days, or part thereof after the first ten days one-half of one cent per bushel: *Provided, however,* that grain damp, or liable to early damage, as indicated by its inspection when received, may be subject to two cents per bushel storage, for the first ten days, and for each additional five days, or part thereof not exceeding one-half of one cent per bushel: *Provided, further*, that where grain has been received in any such warehouse prior to the first day of March, 1877, under any express or implied contract to pay and receive rates of storage different from those prescribed by law, or where it has been received under any custom or usage prior to said day to pay or receive rates of storage different from the rates fixed by law, it shall be lawful for any owner or manager of such warehouse to receive and collect such agreed or customary rates. [As amended by act approved May 21, 1877. In force July 1, 1877. L. 1877, p. 169.]

2765. As to the constitutional power of the legislature to fix and regulate charges. *Munn* v. *People*, 69 Ill. 80; *Munn* v. *Illinois*, 94 U. S. 113; *People* v. *Harper*, 91 Ill. 357.

2766. LOSS BY FIRE HEATING—ORDER OF DELIVERY—GRAIN OUT OF CONDITION. § 16. No public warehouseman shall be held responsible for any loss or damage to property by fire, while in his custody, provided reasonable care and vigilance be exercised to protect and preserve the same; nor shall he be held liable for damage to grain by heating, if it can be shown that he has exercised proper care in handling and storing the same, and that such heating or damage was the result of causes beyond his control; and, in order that no injustice may result to the holder of grain in any public warehouse of classes A or B, it shall be deemed the duty of such warehouseman to dispose of, by delivery or shipping, in the ordinary and legal manner of so delivering, that grain of any particular grade which was first received by them, or which has been for the longest time in store in his warehouse; and, unless public notice has been given that some portion of the grain in his warehouse is is out of condition, or becoming so, such warehouseman shall deliver grain of quality equal to that received by him, on all receipts as presented. In case, however, any warehouseman of classes A or B shall discover that any portion of the grain in his warehouse is out of condition, or becoming so, and it is not in his power to preserve the same, he shall immediately give public notice, by advertisement in a daily newspaper in the city in which such warehouse is situated, and by posting a notice in the most public place (for such a purpose) in such city, of its actual condition, as near as he can ascertain it; shall state in such notice the kind and grade of the grain, and the bins in which it is stored; and shall also state in such notice the receipts outstanding upon which such grain will be delivered, giving the numbers, amounts and dates of each—which receipts shall be those of

the oldest dates then in circulation or uncanceled, the grain
represented by which has not previously been declared or
receipted for as out of condition, or if the grain longest in
store has not been receipted for, he shall so state, and shall
give the name of the party for whom such grain was stored,
the date it was received, and the amount of it; and the
enumeration of receipts and identification of grain so dis-
credited shall embrace, as near as may be, as great a quantity of grain as is contained in such bins; and such grain
shall be delivered upon the return and cancellation of the
receipts, and the unreceipted grain upon the request of the
owner or person in charge thereof.   Nothing herein contained shall he held to relieve the said warehouseman from
exercising proper care and vigilance in preserving such
grain after such publication of its condition; but such grain
shall be kept separate and apart from all direct contact
with other grain, and shall not be mixed with other grain
while in store in such warehouse. Any warehouseman guilty
of any act or neglect, the effect of which is to depreciate property stored in the warehouse under his control, shall he held responsible as at common law, or upon the bond of such warehouseman, and in addition thereto, the license of such
warehouseman, if his warehouse be of class A, shall be revoked.   Nothing in this section shall be so construed us to
permit any warehouseman to deliver any grain stored in a
special bin, or by itself, as provided in this act, to any but
the owner of the lot, whether the same be represented by a
warehouse receipt or otherwise.   In case the grain declared
out of condition, as herein provided for, shall [not] be
removed from store by the owner thereof within two months
from the date of the notice of its being out of condition, it
shall be lawful for the warehouseman where the grain is
stored to sell the same at public auction, for account of said
owner, by giving ten days' public notice, by advertisement in
a newspaper (daily, if there he such), published in the city
or town where such warehouse is located.

2767. TAMPERING WITH GRAIN STORED—PRIVATE BINS—
DRYING, CLEANING, MOVING. § 17. It shall not be lawful for
any public warehouseman to mix any grain of different grades
together, or to select different qualities of the same grade for
the purpose of storing or delivering the same, nor shall he
attempt to deliver grain of one grade for another, or in any
way tamper with grain while in his possession or custody,
with a view of securing any profit to himself or any other
person; and in no case, even of grain stored in a separate bin,
shall he be permitted to mix grain of different grades together
while in store. He may, however, on request of the owner
of any grain stored in a private bin, be permitted to dry,

clean, or otherwise improve the condition or value of any such lot of grain; but in such case it shall only be delivered as such separate lot, or as the grade it was originally when received by him, without reference to the grade it may be as improved by such process of drying or cleaning. Nothing in this section, however, shall prevent any warehouseman from moving grain while within his warehouse for its preservation or safe keeping.

2768. EXAMINATION OF GRAIN AND SCALES—INCORRECT SCALES. § 18. All persons owning property, or who may be interested in the same, in any public warehouse, and all duly authorized inspectors of such property, shall at all times, during ordinary business hours, be at full liberty to examine any and all property stored in any public warehouse in this state, and all proper facilities shall be extended to such person by the warehouseman, his agents and servants, for an examination; and all parts of public warehouses shall be free for the inspection and examination of any person interested in property stored therein, or of any authorized inspector of such property. And all scales used for the weighing of property in public warehouses shall be subject to examination and test by any duly authorized inspector or sealer of weights and measures, at any time when required by any person or persons, agent or agents, whose property has been or is to be weighed on such scales—the expense of such test by an inspector or sealer to be paid by the warehouse proprietor if the scales are found incorrect, but not otherwise. Any warehouseman who may be guilty of continuing to use scales found to be in an imperfect or incorrect condition by such examination and test, until the same shall have been pronounced correct and properly sealed, shall be liable to be proceeded against as hereinafter provided.

2769. GRAIN MUST BE INSPECTED. § 19. In all places where there are legally appointed inspectors of grain, no proprietor or manager of a public warehouse of class B shall be permitted to receive any grain and mix the same with the grain of other owners, in the storage thereof, until the same shall have been inspected and graded by said inspector.

2770. This section applies only to places where there are inspectors of grain appointed under this act. Inspectors appointed by the board of trade are not legally appointed under this section. *E. St. L. Board of Trade* v. *People*, 105 Ill. 382.

2771. ASSUMING TO ACT AS INSPECTOR. § 20. Any person who shall assume to act as an inspector of grain, who has not first been so appointed and sworn, shall be held to be an imposter, and shall be punished by a fine of not less than $50 nor more than $100 for each and every attempt to so inspect grain, to be recovered before a justice of the peace.

2772. The offense created can only occur in a place where there are legally appointed inspectors under this law. In the absence of such appointment, any person may lawfully act as inspector. *Dutcher* v. *People*, 11 Bradw. 312; *E. St. L. Board of Trade* v. *People*, 105 Ill. 382.

2773. MISCONDUCT OF INSPECTOR—INFLUENCING. Any duly authorized inspector of grain who shall be guilty of neglect of duty, or who shall knowingly or carelessly inspect or grade any grain improperly, or who shall accept any money or other consideration, directly or indirectly, for any neglect of duty, or the improper performance of any duty as such inspector of grain; and any person who shall improperly influence any inspector of grain in the performance of his duties as such inspector, shall be deemed guilty of a misdemeanor, and, on conviction, shall be fined in a sum not less than 100 nor more than $1,000, in the discretion of the court, or shall be imprisoned in the county jail not less than three nor more than twelve months, or both, in the discretion of the court.

2774. OWNER, ETC., DISSATISFIED WITH INSPECTION—HIS RIGHTS. § 21. In case any owner or consignee of grain shall be dissatisfied with the inspection of any lot of grain, or shall, from any cause, desire to receive his property without its passing into store, he shall be at liberty to have the same withheld from going into any public warehouse (whether the property may have previously been consigned to such warehouse or not), by giving notice to the person or corporation in whose possession it may be at the time of giving such notice; and such grain shall be withheld from going into store, and be delivered to him, subject only to such proper charges as may be a lien upon it prior to such notice. The grain, if in railroad cars, to be removed therefrom by such owner or consignee within twenty-four hours after such notice has been given to the railroad company having it in possession: *Provided,* such railroad company place the same in a proper and convenient place for unloading; and any person or corporation refusing to allow such owner or consignee to so receive his grain shall be deemed guilty of conversion, and shall be liable to pay such owner or consignee double the value of the property so converted. Notice that such grain is not to be delivered into store may also be given to the proprietor or manager of any warehouse into which it would otherwise have been delivered, and if, after such notice, it be taken into store in such warehouse, the proprietor or manager of such warehouse shall be liable to the owner of such grain for double its market value.

2775. COMBINATION. § 22. It shall be unlawful for any proprietor, lessee or manager of any public warehouse, to enter into any contract, agreement, understanding, or combination, with any railroad company or other corporation, or with any individual or individuals, by which the property of any person is to be delivered to any public warehouse for storage or for any other purpose, contrary to the direction of the owner, his agent, or consignee. Any violation of this section shall subject the offender to be proceeded against as provided in section 23 of this act.

2776. SUITS. § 23. If any warehouseman of class A shall be guilty of a violation of any of the provisions of this act, it shall be lawful for any person injured by such violation to bring suit in any court of competent jurisdiction, upon the bond of such warehouseman, in the name of the people of the state of Illinois, to the use of such person. In all criminal prosecutions against a warehouseman, for the violation of any of the provisions of this act, it shall be the duty of the prosecuting attorney of the county in which such prosecution is brought, to prosecute the same to a final issue, in the name of and on behalf of the people of the state of Illinois.

2777. WAREHOUSE RECEIPT NEGOTIABLE. § 24. Warehouse receipts for property stored in any class of public warehouses, as herein described, shall be transferable by the indorsement of the party to whose order such receipt may be issued, and such indorsement shall be deemed a valid transfer of the property represented by such receipt, and may be made either in blank or to the order of another. All warehouse receipts for property stored in public warehouses of class C shall distinctly state on their face the brand or distinguishing marks upon such property.

2778. Delivery of receipt has the same effect in transferring the title to the grain as the delivery of the property itself, and no more nor less. *Burton* v. *Curyea*, 40 Ill. 320.

2779. Tender of warehouse receipt by the vendor of grain, is a sufficient tender of the grain, unless vendee insists on seeing the grain. *McPherson* v. *Gale*, 40 Ill. 368.

2780. By the act incorporating the Chicago Dock company, a warehouse receipt issued by that company, is made negotiable, and as such it absolutely vests in the holder the title to the property specified in it. *Ch. Dock Co.* v. *Foster*, 48 Ill. 507.

2781. Receipt being the contract of the parties cannot be varied by parol evidence. *Leonard* v. *Dunton*, 51 Ill. 482.

2782. Remedy against warehouseman refusing to deliver grain. *Leonard* v. *Dunton*, 51 Ill. 482; *Bailey* v. *Bensley*, 87 Ill. 556; *German Nat. Bank* v. *Meadowcroft*, 95 Ill. 124; *Canadian Bank* v. *McCrea*, 106 Ill. 281.

2783. Measure of damages for failure to deliver grain. *Leonard* v. *Dunton*, 51 Ill. 482.

2784. The addition of these words at end of the receipt: "Subject to their order, for all advances of money on the same," will not convert it into a mere pledge and render the grain liable to an execution against the party giving it, issued after the date of the receipt. *Cool* v. *Carmichael*, 66 Ill. 216.

2785. Where warehouseman purchases grain stored with him of the owner for another, and takes up his receipt and gives another to the person for whom he bought and whose money he used: *Held*, that the grain was not liable to be taken on execution against the warehouseman. *Broadwell* v. *Howard*, 77 Ill. 305.

2786. Commission merchant in Chicago may, by a custom of trade obtaining there, dispose of warehouse receipts for grain consigned to him, provided he keeps on hand other like receipts for the same quantity and quality of grain. The receipts do not represent the consignor's property. They are merely evidences of a debt to the consignee. *Bailey* v. *Bensley*, 87 Ill. 556.

2787. Person succeeding to the possession of warehouse and the grain stored therein, becomes liable to the holders of warehouse receipts, and subject to the same remedies as the former proprietor. *German Nat. Bank* v. *Meadowcroft*, 95 Ill. 124.

2788. Statute relating to negotiable instruments does not embrace warehouse receipts. The title vested in the assignee is not the same as that passed to the assignee of a note. *Canadian Bank* v. *McCrea*, 106 Ill. 281.

2789. By this section the indorsement of a warehouse receipt is made evidence of a transfer of the grain it represents, the same as the actual delivery of the property itself. It passes the assignee's' actual title and no more. *Ib.*

2790. The indorsement and delivery of a warehouse receipt for flour not only transfers the title to the flour to the assignee, but also gives him a right of action for any breach of duty by the warehouseman at *any time* during the bailment. *Sargent* v. *Central Warehouse Co.*, 15 Bradw. 553.

2791. An indorsement in blank of a warehouse receipt by the seller authorizes the purchaser to write over the indorsement only a contract of mere assignment of the legal title, unlike the case of a negotiable note. *Mida* v. *Geissmann*, 17 Bradw. 207.

2792. As to liability of assignor to assignee, where the warehouseman fails. *Mida* v. *Geissmann*, 17 Bradw. 207. See also *Hide & Leather Nat. Bank* v. *West*, 20 Bradw. 61.

2793. FALSE RECEIPTS — FRAUDULENT REMOVAL. § 25. Any warehouseman of any public warehouse who shall be guilty of issuing any warehouse receipt for any property not actually in store at the time of issuing such receipt, or who shall be guilty of issuing any warehouse receipt in any respect fraudulent in its character, either as to its date or the quantity, quality, or inspected grade of such property, or who shall remove any property from store (except to preserve it from fire or other sudden danger), without the return and cancellation of any and all outstanding receipts that may have been issued to represent such property, shall, when convicted thereof, be deemed guilty of a crime, and shall suffer, in addition to any other penalties prescribed by this act, imprisonment in the penitentiary for not less than one, and not more than ten years. [Restricted as to receipts issued before Oct. 8, 1871. L. 1871-2, p. 774.]

2794. COMMON LAW REMEDY SAVED. § 26. Nothing in

this act shall deprive any person of any common law remedy now existing.

2794a. PRINTED COPY OF ACT POSTED. § 27. All proprietors or managers of public warehouses shall keep posted up at all times, in a conspicuous place in their business offices, and in each of their warehouses, a printed copy of this act.

2795. REPEAL. § 28. All acts or parts of acts inconsistent with this act are hereby repealed.

An act to amend an act entitled "An act to regulate public warehouses and the warehousing and inspection of grain, and to give effect to article thirteen (13) of the constitution of the state," approved April 25, 1871, in force July 1, 1871, and to establish a committee of appeal, and prescribe their duties. Approved April 15, 1873. In force July 1, 1873. [Laws of 1873, p. 189; R. S. 1887, p. 1034; S. & C. p. 1975; Cothran, p. 1182.]

2796. COMMISSIONERS TO ESTABLISH GRADES. § 1. *Be it enacted by the people of the state of Illinois, represented in the general assembly,* That the board of railroad and warehouse commissioners shall establish a proper number and standard of grades for the inspection of grain, and may alter or change the same from time to time: *Provided,* no modification or change of grades shall be made, or any new ones established, without public notice being given of such contemplated change, for at least twenty days prior thereto, by publication in three daily newspapers printed in each city containing warehouses of class A: *And, provided further,* that no mixture of old and new grades, even though designated by the same name or distinction, shall be permitted while in store.

2797. COMMITTEE OF APPEALS. § 2. Within twenty days after this act takes effect, the board of railroad and warehouse commissioners shall appoint three discreet and competent persons to act as a committee of appeals, in every city wherein is located a warehouse of class A, who shall hold their office for one year and until their successors are appointed. And every year thereafter a like committee of appeals shall be appointed by said commissioners, who shall hold their office for one year and until their successors are appointed: *Provided,* said commissioners shall have power, in their discretion, to remove from office any member of said committee at any time, and fill vacancies thus created by the appointment of other discreet persons.

2798. APPEALS — NOTICES. § 3. In all matters involving doubt on the part of the chief inspector, or any assistant inspector, as to the proper inspection of any lot of grain, or in case any owner, consignee or shipper of grain, or any warehouse manager, shall be dissatisfied with the decision of the chief inspector or any assistant inspector, an appeal may be made to said committee of appeal, and the decision of a majority of said committee shall be final. Said board of commissioners are authorized to make all necessary rules governing the manner of appeals as herein provided. And all complaints in regard to the inspection of grain, and all notices requiring the services of the committee of appeal, may be served on said committee, or may be filed with the warehouse registrar of said city, who shall immediately notify said committee of the fact, and who shall furnish said committee with such clerical assistance as may be necessary for the proper discharge of their duties. It shall be the duty of said committee, on receiving such notice, to immediately act on and render a decision in each case.

2799. COMMITTEE ON APPEALS—OATH—BOND—WHO MAY SERVE ON. § 4. The said committee of appeals shall, before entering upon the duties of their office, take an oath, as in case of other inspectors of grain, and shall execute a bond in the penal sum of five thousand dollars; with like conditions as is provided in the case of other inspectors of grain, which said bonds shall be subject to the approval of the board of railroad and warehouse commissioners. *It is further provided*, that the salaries of said committee of appeals shall be fixed by the board of railroad and warehouse commissioners, and be paid from the inspection fund, or by the party taking the appeal, under such rules as the commission shall prescribe; and all necessary expenses incurred in carrying out the provisions of this act, except as herein otherwise provided, shall be paid out of the funds collected for the inspection service upon the order of the commissioners: *Provided*, that no person shall be appointed to serve on the committee of appeals who is a purchaser of, or a receiver of grain, or other articles to be passed upon by said committee. [As amended by act approved June 26, 1885. In force July 1, 1885. L. 1885, p. 253.]

2800. "REGISTERED FOR COLLECTION"—INSPECTION FEES. § 5. No grain shall be delivered from store from any warehouse of class A, for which or representing which warehouse receipts shall have been issued, except upon the return of such receipts stamped or otherwise plainly marked by the warehouse register with the words "registered for collection" and the date thereof; and said board of commissioners shall have power to fix the rates of charges for the inspection of grain, both into and out of warehouse; which charges shall be a lien upon all grain so inspected, and may be collected of the owners, receivers or shippers of such grain, in such manner as the said commissioners may prescribe.

2801. REPEAL. § 6. Section 13 of the act to which this is an amendment, is hereby repealed: *Provided*, the provisions contained in said section shall remain in force until the grades for the inspection of grain shall have been established by the commissioners, as provided in section 1 of this act. [Grades fixed by commissioners, July 1, 1873.]

STATE WEIGH-MASTERS.

An act to provide for the appointment of state weigh-masters. Approved June 23, 1883. In force July 1, 1883. [Laws of 1883, p. 172; R. S. 1887, p 1039; S. & C., p. 1976.]

2802. WEIGH-MASTER—APPOINTMENT OF. § 1. *Be it enacted by the people of the state of Illinois, represented in the general assembly,* That there shall be appointed by the railroad and warehouse commissioners in all cities where there is state inspection of grain, a state weigh-master and such assistance as shall be necessary.

2803. DUTIES OF. § 2. Said state weigh-master and assistants shall, at the places aforesaid supervise and have exclusive control of the weighing of grain and other property which may be subject to inspection, and the inspection of scales and the action and certificate of such weigh-master and assistants in the discharge of their aforesaid duties shall be conclusive upon all parties in interest.

2804. FIX FEES. § 3. The board of railroad and warehouse commissioners shall fix the fees to be paid for the weighing of grain or other property, which fees shall be paid equally by all parties interested in the purchase and sale of the property weighed, or scales inspected and tested.

2809. WEIGH-MASTER—QUALIFICATIONS—BOND—COMPENSATION. § 4. Said state weigh-master and assistants shall not be a member of any board of trade or association of like character; they shall give bonds in the sum of five thousand dollars ($5,000), conditioned for the faithful discharge of their duties, and shall receive such compensation as the board of railroad and warehouse commissioners shall determine.

2810. MAY ADOPT RULES. § 5. The railroad and warehouse commissioners shall adopt such rules and regulations for the weighing of grain and other property as they shall deem proper.

2811. NEGLECT OF DUTY—PENALTY. § 6. In case any person, warehouseman or railroad corporation, or any of their agents or employes, shall refuse or prevent the aforesaid state weigh-master or either of his assistants from having access to their scales, in the regular performance of their duties in supervising the weighing of any grain or other property in accordance with the tenor and meaning of this act they shall forfeit the sum of one hundred dollars ($100) for each offense, to be recovered in an action of debt, before any justice of the peace, in the name of the people of the state of Illinois; such penalty or forfeiture to be paid to the county in which the suit is brought, and shall also be required to pay all costs of prosecution.

STOCKHOLDERS—INDIVIDUAL LIABILITY.

(a.) FOR UNPAID SUBSCRIPTIONS.

2812. POWER OF SHAREHOLDER—*to make stock not liable to assessment.* It is not in the power of the shareholders by agreement with the corporation to make the shares of stock issued to them non-assessable, so as to excuse payment for such stock at its par value as against creditors. *Union Mut. L. Ins. Co.* v. *Frear Stone Manf. Co.,* 97 Ill. 537.

2813. REMEDY GIVEN—*no application to stockholders under special charters.* § 25 of the act of 1872, providing for proceedings in equity against the corporation and stockholders, whereby to make the latter pay their share of the debts to the extent of their unpaid stock, after exhausting the assets of the corporation, applies only to corporations organized under that law, and does not embrace bodies created by special charters. *Woodcock* v. *Turpin,* 96 Ill. 135.

2814. NATURE OF LIABILITY—*several—limitation.* The stockholder's liability is limited by the amount of his subscription unpaid at the time of the service of the garnishee process, and it is several—not joint. *Pease* v. *Underwriters' Union,* 1 Bradw. 287.

2815. SAME—*when secondary.* Under § 25 of the general incorporation act of 1872, where the proceeding must be by suit in equity, the stockholder's liability to pay anything on his unpaid stock is deferred until the assets of the corporation are exhausted. *Robertson* v. *Noeninger,* 20 Bradw. 227.

2816. RECOVERY LIMITED TO DEBT OF CORPORATION. A stockholder is not bound to pay more of his subscription, or notes given therefor than is necessary to pay outstanding debts, where the corporation is insolvent, and in the hands of a receiver. *Lamar Ins. Co.* v. *Moore,* 84 Ill. 575.

2817. OF THE DECREE—*award of execution before apportionment.* A creditor having exhausted his remedies against an insolvent corporation is, in equity, entitled to be subrogated to its rights against its debtors for stock; and in such case, a decree finding the sum due from a subscriber to the company and awarding an execution therefor, is not erroneous. It is not necessary to apportion the *pro rata* share of each stockholder necessary to discharge the debt of the company to the creditor. *Hickling* v. *Wilson,* 104 Ill. 54.

2818. WHO MAY ENFORCE LIABILITY—*receiver of corporation.* If a stockholder in an insurance company is a party to a decree appointing a receiver of the company, it will be conclusive on him, and the receiver may maintain a suit against him in his own name. *Rowell* v. *Chandler,* 83 Ill. 288. But to recover, the receiver must show an appointment by a decree which is conclusive on the defendant stockholder by his being a party to the suit against the corporation. *Chandler* v. *Brown,* 77 Ill. 333.

2819. SAME—*creditor with notice of defense.* Where the party seeking to enforce a stockholder's individual liability, was not a creditor of the corporation at the time when the latter paid for his stock in land, he will be considered as having given credit to the corporation in the condition it then was, and if an examination of the books would have shown that the stock was fully paid, he cannot recover. *Peck* v. *Coalfield Coal Co.,* 11 Bradw. 88.

2820. PAYMENT FOR STOCK—*in property binding.* Where a subscriber has paid for his stock in a corporation, he will not be liable under the statute for the debts of the company; and it is immaterial whether such stock is paid for in money or by the transfer of lands in good faith to the corporation. The directors cannot create a liability by declaring that the stock is not paid for. *Ib.*

2821. REMEDY TO ENFORCE—*garnishment.* It seems that a stockholder who owes the company for unpaid stock upon which a call has been made and notice given, is liable to be garnisheed on a judgment against the company. *Meints* v. *E. St. L. Co-operative Rail Mill Co.,* 89 Ill. 48.

2822. At the time of taking out summons against the corporation, the creditor may take process against any stockholder whose subscription is wholly or in part unpaid; and by service on the latter prevent further payment to the corporation for the stock, and hold the same in abeyance to abide the result of the trial in the original case; and where a recovery is had, the garnishee may be compelled to respond to such judgment creditor instead of paying his indebtedness to the corporation. *Pease* v. *Underwriters' Union,* 1 Bradw. 287.

2823. If the cause is commenced and conducted according to the statute, the whole proceeding will constitute but one case, and upon the trial of the issues formed upon the answers of the garnishee, the court will take judicial notice of the judgment against the corporation. But where the garnishment is by an independent proceeding there must be proof of the creditor's judgment, and where the garnishees deny being stockholders, the burden of proof is on the plaintiff to show their liability as such. *Ib.*

2824. Under § 8 of the chapter on corporations, by the proceeding in garnishment, stockholders may be compelled to pay to the garnisheeing creditor any balance unpaid upon stock owned by them respectively, whether such stock is called in or not. *Robertson* v. *Noeninger,* 20 Bradw. 227.

2825. REMEDY TO ENFORCE LIABILITY—*garnishee must be sued with corporation.* To render a stockholder liable under the statute, to the extent of his unpaid stock, for the debts of the corporation, proceedings must be instituted against him at the same time that the action is brought against the corporation. The remedy given by the statute is exclusive. *Peck* v. *Coalfield Coal Co.,* 3 Bradw. 619; *Robertson* v. *Noeninger,* 20 Bradw. 227.

2826. It is not essential that the stockholder shall be proceeded against at the same time the suit is brought against the corporation, as in garnishee proceedings under the attachment act. The intention is to give the remedy as ample and complete as in cases of garnishment, including process after judgment. *Coalfield Co.* v. *Peck,* 98 Ill. 139.

2827. DECLARATION—*in suit by receiver.* In a suit for the use of the receiver of an insolvent corporation against a stockholder to collect his subscription, or a note given therefor, no recovery can be had without an averment of the debts of the corporation. The declaration should also show that the capital stock paid in has been exhausted. *Lamar Ins. Co.* v. *Moore,* 84 Ill. 575.

2828. CREDITORS' BILL—*exhausting legal remedy—equitable attachment.* Under proper circumstances, creditors are not compelled to wait for the winding up of insolvent corporations, but may subject their unpaid subscriptions to the payment of their claims. They must first obtain judgment at law and have execution returned unsatisfied in whole or in part. In such case, the creditor is subrogated to the rights of the debtor corporation, and the proceeding is in the nature of an equitable attachment under which the debts due the corporation may be applied in the payment of its own liabilities. *Patterson v. Lynde*, 112 Ill. 196.

2829. COLLECTION OF SUBSCRIPTION—*compelled by mandamus or bill in equity.* A foreign insolvent corporation owing debts, if still in existence, may be compelled, by *mandamus* or by bill in equity, to collect its unpaid subscriptions wherever the stockholders may reside; and if it has ceased to exist, a receiver should be appointed, and the courts of other states, as a matter of comity, would recognize the right of the receiver, the same as they would the corporation itself, if still in existence, to prosecute actions at law for the recocovery of unpaid subscriptions. *Ib.*

2830. CREDITOR'S BILL—*joinder of plaintiff.* Two or more creditors of an insolvent corporation may unite in filing a creditor's bill against the corporation and its stockholders to reach unpaid subscriptions to the capital stock, and such bill is not multifarious. *Hickling v. Wilson*, 104 Ill. 54.

2831. SAME—*parties to.* To enforce the liability of a stockholder for his unpaid stock, it is indispensable that the corporation, (or, if it ceased to exist, that all of its stockholders and creditors) shall be before the court, so as to be bound by its orders and decrees, and so that complete justice may be meted out to all, and all conflicting rights and equities finally adjusted. *Patterson v. Lynde*, 112 Ill. 196.

2832. REMEDY WHERE CORPORATION IS DEFUNCT—*apportionment of burdens.* Where a corporation ceases to exist, its assets in excess of what is necessary to pay its debts belong to its stockholders, and the duty of the stockholder in such case, is only to pay his *pro rata* share of the amount needed to pay the debts. This duty is upon all owing for stock, and it is the duty of the court to adjust the equities between the different stockholders. *Ib.*

2833. DEFENSES — *judgment against corporation, fraudulent.* Under the act of 1872 making stockholders liable to creditors for unpaid stock, a stockholder when sued on his subscription, cannot attack the judgment against the corporation on the ground that it is collusive and unjust. If he can attack the judgment on that ground, he must do so in a court of equity. *Coalfield Co. v. Peck*, 98 Ill. 139.

2834. BANKRUPTCY OF CORPORATION—*who to collect unpaid stock.* Unpaid subscriptions to the capital stock of a corporation are a part of the assets of the company, and as such passes to its assignee in bankruptcy, who alone can sue for the same. *Lane v. Nickerson*, 99 Ill. 284.

(b.) TO AMOUNT OF STOCK, UNTIL WHOLE CAPITAL STOCK PAID IN, &C.

2835. CONSTRUCTION—*insurance law of 1861.* The words "trustees and corporators" in § 16 of the insurance act of 1861, making them severally liable for all debts and responsibilities of their companies, to the amount by him or them subscribed, includes stockholders. *Shufeldt v. Carver*, 8 Bradw. 545; *Gulliver v. Roelle*, 100 Ill. 141.

2836. The shareholders in all insurance companies subject to the insurance law of 1869, are liable for the debts of their companies to the full amount of their respective shares of stock, where the full amount subscribed has not been paid in. *Butler v. Walker*, 80 Ill. 345; *Tibballs v. Libby*, 87 Ill. 142.

2837. LIABILITY NOT RELEASED—*by payment of his stock.* A stockholder will not be relieved from this liability by the payment of his stock in full. Until the full capital stock is paid in, and a certificate of that fact made and recorded, he will be liable to the extent of his stock for the debts of the company. *Butler v. Walker*, 80 Ill. 345; *Tibballs v. Libby*, 87 Ill. 142; *Gulliver v. Roelle*, 100 Ill. 141.

2838. Until the entire capital stock shall be paid in and a certificate thereof filed with, and recorded by the county clerk, the stockholders are severally individually liable to the creditors to an amount equal to the stock held by them respectively for all debts of the company. *Baker* v. *Backus*, 32 Ill. 79.

2839. LIABILITY—*for debts of corporation—not for its torts.* The mere fact that a person is a stockholder or director, does not render him liable for the torts of the corporation, or its agents or servants. *Peck* v. *Cooper*, 8 Bradw. 403.

2840. SAME—*when under general law.* If an insurance company created by special charter increases its capital stock under the general law, this in effect is an incorporation under such law, and by subscribing to such stock, a party will incur the liability incurred by the general law. *Tibballs* v. *Libby*, 87 Ill. 142.

2841. INSURANCE LAW OF 1869—*its application to prior companies.* The provision of the insurance law of 1869 making stockholders of insurance companies liable for the debts of their companies, applies to companies organized before its passage, under general laws. *Arenz* v. *Weir*, 89 Ill. 25.

2842. § 16 of the insurance law of 1869 makes stockholders and directors of insurance companies organized under that law, severally liable for all debts of their respective companies, to the amount subscribed by them until the whole amount of the capital stock shall be paid in and a certified copy thereof recorded; and § 19 imposes the same liability on shareholders in companies organized under special charters and brought under the provisions of the general law. *Gulliver* v. *Roelle*, 100 Ill. 141.

2843. Under § 16 of the act of 1869 as well as § 2, art. 10, of the constitution of 1848, the word "corporators" is used in the sense of shareholders, and not in that of commissioners or promoters of the organization of the companies. *Ib.*

2844. The individual liability of a trustee or corporator of an insurance company to its creditors where it has not complied with the law, does not depend upon the fact that the creditor has sustained any actual loss or injury. The creditor is only bound to show that the company owes him, and that the whole amount of the capital stock of the company has not been paid in and a certificate thereof recorded. *Diversy* v. *Smith*, 103 Ill. 378.

2845. WHEN LIABILITY ATTACHES—*not until all the capital stock is taken.* A subscriber to the capital stock of a proposed corporation, until the full amount of stock fixed by law, or by the action of those connected therewith is subscribed, cannot be held individually liable for a debt of such corporation, unless for some cause he has estopped himself from alleging that the whole of the fixed capital stock was never subscribed. *Temple* v. *Lemon*, 112 Ill. 51.

2846. OF THE NATURE OF LIABILITY—*partnership.* As to claims against the corporation the stockholders stand in the relation of co-partners, and one cannot sue the other at law. *Meisser* v. *Thompson*, 9 Bradw. 368.

2847. A stockholder occupies the *status* of a partner to the extent of his individual liability, and as a partner he must answer to the amount of his stock for the debts of the corporation. *Gauch* v. *Harrison*, 12 Bradw. 457; *Fleischer* v. *Rentchler*, 17 Bradw. 402.

2848. The effect of a provision in the charter of a bank, making its stockholders liable to creditors of the bank on its default, to an amount equal to their stock, is to withdraw from the stockholders to the extent of their stock, the protection of the corporation, and leave them to that extent liable as partners. *Buchanan* v. *Meisser*, 105 Ill. 638; *Thompson* v. *Meisser*, 108 Ill. 359.

2849. Under a bank charter providing that, "whenever default shall be made in the payment of any debt or liability contracted by said corporation, the stockholders shall be held individually responsible for an amount equal to the amount of stock held by them respectively," &c., the liability of a stockholder for the debts of the bank is coeval with that of the bank, they both becoming bound at the same time and by the same contract. *Fleischer* v. *Rentchler*, 17 Brad. 402.

2850. WHETHER PRIMARY OR SECONDARY. Under the act of 1857 relating to private corporations, the liability of stockholders to creditors of the company, is not dependent upon a suit against the company and inability to collect, but such stockholders are primarily liable. *Culver* v. *Third Nat. Bank*, 64 Ill. 528.

2851. A bank charter provided that the stockholders should "be responsible, in their individual property, in an amount equal to the amount of stock held by them respectively, to make good losses to depositors or others": *Held*, that the individual liability was not in the nature of a penalty, and therefore enforceable only in a court of law, but was primary, and subject to the demands of depositors and other creditors equally with the assets of the bank. *Queenan* v. *Palmer*, 117 Ill. 619.

2852. HOW MADE LIABLE. The stockholders of a corporation can be held responsible only in the mode prescribed by the act under which they became a corporation. They are not individually liable, except under the circumstances and for the time specified in the act of incorporation. *Baker* v. *Backus*, 32 Ill. 79, 97, 99.

2853. WHEN SECONDARILY LIABLE—*law of 1849*. Under the act of 1849 relating to railway corporations, if a claim is owing by a railway company for services performed for it, a stockholder is not liable in an action therefor, until an execution shall be returned unsatisfied, in whole or in part, against the corporation, and then the amount due on such execution is the amount recoverable with costs, against the stockholder. *Cutright* v. *Stanford*, 81 Ill. 240.

2854. Stockholders in a corporation organized under a law making them liable individually "to the creditors" of the corporation, will not be required to pay any portion of the debts until the assets of the corporation are first exhausted. *Harper* v. *Union Manf. Co.*, 100 Ill. 225.

2855. LIABILITY—*restricted to debts of a certain class*. A bank charter provided, "the stockholders of this corporation shall, as to all funds deposited *as savings, and in trust with said covporation*, while they are stockholders, be individually liable to the extent of their stock," &c.: *Held*, as restricting the stockholders' liability to the particular class of deposits designated—those "as savings and in trust with said corporation," and not as embracing every deposit of money. The liability of the stockholder is the creature of the statute, and cannot be increased or enlarged beyond the express terms of the statute. *Bromley* v. *Goodwin*, 95 Ill. 118.

2856. LIABILITY TO MAKE LOANS GOOD—*what is a loss*. A charter or statute making the stockholders of a corporation individually responsible in an amount equal to their stock, to "make good losses to depositors or others," will be construed to make the stockholders' liable to all creditors who may suffer from the default or failure of the corporation to pay its indebtedness. The total or partial insolvency of the corporation and its neglect to pay, is a loss to the creditors in the sense of that word as used in the statute. *Queenan* v. *Palmer*, 117 Ill. 619.

2857. The charter of a bank contained this proviso: *"Provided, also,* that the stockholders in this corporation shall be individually liable, to the amount of their stock for all the debts of the corporation; and such liability shall continue for three months after the transfer of any stock on the books of the corporation": *Held,* that the stockholders were each individually liable to pay to the creditors of the bank, not merely the balance unpaid upon subscriptions for stock, but to the extent of the nominal or face value of the stock held by them, for debts of the bank. *Root* v. *Sinnock,* 120 Ill. 350.

2858. STOCKHOLDERS LIABILITY—*depending on time he acquired or parted with his stock.* The act of 1851 amendatory of the act of 1849, providing for the construction of plank roads, makes no distinction between original subscribers to the stock and subsequent purchasers, in regard to their individual liability for the debts of their companies, to the extent of the amount of their stock. *Gay* v. *Keys,* 30 Ill. 413.

2859. SAME—*hypothecation—or pledge.* Primarily a creditor of a national bank may proceed against the party in whom the legal title to the stock is vested. Where shares of stock in a banking corporation have been hypothecated and placed in the hands of the transferee, he will be subjected to all the liabilities of ordinary owners, for the reason the property is in his name and the legal ownership appears to be in him. *Wheelock* v. *Kost,* 77 Ill. 296.

2860. Thus where a party made a loan to a national bank, and made his promissory note, partly as an act of accommodation to the bank, to be held among their other assets, and fifty shares of its stock, equal in value to $5,000 were issued to him as security for his loans and as indemnity against liability on his note, it was *held,* that he was liable to the creditors of the bank as a stockholder, whatever might be his relation to the corporators of the bank. *Ib.*

2861. ASSIGNEE OF STOCK—*liability on informal transfer.* If a national bank issues certificates of shares to a purchaser in lieu of the certificates of the vendor, without observing its by-laws, so far as creditors of the bank are concerned, a party taking and holding them, will be subject to the liabilities imposed by § 5151 of the national banking law. *Laing* v. *Burley,* 101 Ill. 591.

2862. LIABILITY AS BETWEEN ASSIGNOR AND ASSIGNEE. There can be but one amount for which there is liability on account of the same share of stock, where that liability equals or exceeds the amount of such share, and for that amount both the assignor and assignee may be liable,—the former in case the debt was incurred by the corporation within three months after the assignment, where that is the limit of time during which the liability of the assignor shall continue, and the assignee, in case it was incurred after he became the holder of the stock; but there can be but one satisfaction. If the assignor is compelled to pay on account of debts of the corporation made within three months after his transfer of stock, he may have his action against the person owning such stock when the debt was created, and recover the sum so paid by him. *Thebus* v. *Smiley,* 110 Ill. 316.

2863. LIABILITY DEPENDING ON TIME OF BECOMING A STOCK HOLDER. Under a statutory provision making the stockholders liable for the debts of the corporation to the extent of their stock, for three months after the transfer of their stock on the books of the corporation, it is not essential to the stockholder's liability that he be such at the time the creditor's cause of action shall have accrued. It is sufficient if he is a stockholder when the suit is brought against him. *Root* v. *Sinnock,* 120 Ill. 350.

2864. The expression, "all stockholders," in the absence of any legis-

lative indication to the contrary, must be regarded as including not
only those who were such at the time the indebtedness was incurred,
but also all those who successively stand in their shoes in respect to
the same stock. *Ib.*

2865. LIABILITY, HOW DISCHARGED—*payment.* Stockholders
cannot evade the liability imposed on them by law by confessing judgments in favor of each other and paying the same. The liability is
created for the protection of the creditors of the bank, and not for the
stockholders. *Meisser* v. *Thompson*, 9 Bradw. 368.

2866. A stockholder may extinguish his individual liability by the
payment of debts of the corporation, but he will be allowed only the
sum actually paid for such claims, and not their face value. *Gauch*
v. *Harrison*, 12 Bradw. 457; *Kunkelman* v. *Rentchler*, 15 Bradw. 271.

2867. WHAT WILL DISCHARGE. Payment in full of the stock subscribed by a stockholder in a private corporation organized under the
act of 1857, will not discharge him from liability to creditors of the
corporation. To make it have that effect it must be shown that all
other shareholders have done the same thing, and a certificate of the
fact has been filed in the clerk's office as required in the 10th section of
the act. *Kipp* v. *Bell*, 86 Ill. 577.

2868. Where a stockholder has paid the corporation in full for his
stock, and has also paid a like sum to the creditors of the company,
he will be discharged from all liability for debts of the corporation
contracted thereafter. *Ib.*

2869. EXTINGUISHMENT OF THE LIABILITY. The recovery of a
judgment by a creditor of a corporation against a stockholder for a
sum equal to the amount of his stock, that being the limit of his liability for the corporation, will extinguish his liability. So, it is not
doubted, will a voluntary payment by him to such a creditor of the
corporation who has the right to sue him and recover a judgment.
Buchanan v. *Meisser*, 105 Ill. 638.

2870. But a payment of a sum equal to his stock to the firm of
which he is a member, in satisfaction of a debt due from the corporation to his firm, will not release him from his liability as a stockholder
of such corporation, or bar a suit by another creditor, as the firm could
not maintain an action at law against him. *Ib.*

2871. EXTINGUISHMENT. A stockholder individually liable for
the debts of his corporation, may discharge such liability by the payment, in good faith, of the amount of the same to any creditor who is
not also a stockholder. But he cannot discharge himself by buying
up debts of the corporation equal in amount to his liability, at a dis-
count. In such case if he retains such indebtedness so purchased by
him, he can only claim a discharge for the actual sum paid by him for
the same. *Thompson* v. *Meisser*, 108 Ill. 359.

2872. STOCK WHEN DISCHARGED FROM LIABILITY. Where a
judgment is recovered against a stockholder by a creditor of the corporation, under a statute making the former liable personally to
creditors for an amount equal to the stock held by him, which he
pays, his stock thereafter will be free from liability, and he may sell
and transfer the same, and his assignee will take such stock without
any liability on his part in consequence of his ownership of the
same. *Thebus* v. *Smiley*, 110 Ill. 316.

2873. SET OFF. In an action by a creditor of a corporation against
a stockholder to enforce his individual liability to creditors for an
amount equal to his stock, the stockholder will not be allowed to set
off against his liability an indebtedness of the corporation to him.
Thebus v. *Smiley*, 110 Ill. 316.

2874. In an action by a creditor of a corporation who is not also a stockholder to enforce the individual liability of a stockholder, the latter cannot set off a debt due from the corporation to himself. *Thompson* v. *Meisser*, 108 Ill. 359.

2875. In an action by a creditor of a corporation against a stockholder to enforce his individual liability, the latter cannot plead as a set off an indebtedness of the corporation to himself, as such debt is not that of the party suing. *Buchanan* v. *Meisser*, 105 Ill. 638.

2876. STOCKHOLDERS INDIVIDUAL LIABILITY—*effect of dissolution of corporation.* The dissolution of a corporation by decree of court does not affect the liability of the stockholder, or change it from that imposed by the statute. *Tarbell* v. *Page*, 24 Ill. 46.

2877. SAME—*effect of bankruptcy.* The right of a creditor of a corporation to recover against a stockholder is not taken away by the bankruptcy of the corporation. That fact fixes his liability. *Tibballs* v. *Libby*, 87 Ill. 142.

2878. The placing the assets of an insurance company into the hands of a receiver does not lessen the stockholder's individual liability to its creditors, but fixes the same. He is not under the control of the receiver, but holds a fund for the benefit of creditors. *Arenz* v. *Weir*, 89 Ill. 25.

2879. REMEDY—*whether at law or in equity.* The charter of a bank made the stockholders individually liable to depositors for the default of the corporation in making payment of any debt: *Held,* that the liability was purely legal, and the remedy against the stockholders was at law. The word "individually" as used in the charter, means separately, and an action will lie against a single stockholder. *Meisser* v. *Thompson*, 9 Bradw. 368.

2880. As the statute creates a legal liability upon stockholders to a certain extent for the debts incurred by their company, such liability is cognizable in a court of law, an implied promise being inferred from a legal liability. *Culver* v. *Third Nat. Bank*, 64 Ill. 528.

2881. Since the act of 1872 concerning corporations for pecuniary profit took effect, a court of law has no jurisdiction of a suit by a creditor of such a corporation against a stockholder, unless his debt accrued before the act of 1872 took effect. The remedy is in equity. *Richardson* v. *Akin*, 87 Ill. 138.

2882. The ruling of this court that an action at law by a single creditor will lie against any stockholder of an insolvent corporation to enforce his individual liability, is not to be taken as a denial of the right to seek relief in equity, where there are equitable grounds. Where the corporation is insolvent a court of equity may take jurisdiction for the purpose of marshaling the fund and making a ratable distribution. *Eames* v. *Doris*, 102 Ill. 350.

2883. The liability of stockholders under § 9 of the act of 1857, relating to corporations, is to the creditors of the corporation as a class, and not to each individual creditor. Therefore, the remedy of a creditor seeking to enforce the personal liability created by that section, is in a court of equity, and not at law. *Rounds* v. *McCormick*, 114 Ill. 252; *Harper* v. *Union Manf. Co.*, 100 Ill. 225; *Low* v. *Buchanan*, 94 Ill. 76; *Queenan* v. *Palmer*, 117 Ill. 619.

2884. This liability constitutes a common fund for the security of creditors, and a court of equity, aside from the ground of discovery, will have jurisdiction of a bill by a creditor, for himself and others, to enforce such penalty, and control the fund thus raised for their benefit, and distribute the same ratably among them, the remedy at law in such case being inadequate without bringing a multiplicity of suits. *Queenan* v. *Palmer*, 117 Ill. 619.

2885. REMEDY IN EQUITY—*parties plaintiff*. Under sec. 9 of the act of 1857, relating to manufacturing corporations, the stockholders are made severally and individually liable to the "creditors" of the company to the amount of stock held by them, for all debts, &c., made by such company prior to the time when the whole capital stock shall have been paid in. This liability cannot be enforced by a single creditor, suing in his own behalf alone. It can be enforced only upon a bill brought by, or at least, in behalf of all the creditors of the corporation. *Harper* v. *Union Manf. Co.*, 100 Ill. 225.

2886. SAME—*party defendant—assignee for creditors*. Where the stockholders are individually liable secondarily, and the assets of the corporation are in the hands of an assignee for the benefit of creditors, he will be a necessary party to a bill to enforce the stockholder's liability. *Ib.*

2887. SAME—*sufficiency of bill*. A bill by creditors of an insolvent banking company for and in behalf of complainants and all other creditors against the several stockholders of the company, alleged the insolvency of the bank, a deficiency of assets to pay its creditors, the personal liability of the stockholders under the charter to the depositors and creditors, the existence of some nine hundred unpaid depositors, some of whom were seeking by separate suits at law, to get an advantage over the others, and that such separate litigation would waste and exhaust the proceeds of this liability of stockholders, the only fund to which depositors could look for payment, and asking to have an account taken of all the liabilities of the bank and establish the amount for which the various stockholders were liable personally, and to have the amounts of the debts proven apportioned among the stockholders: *Held*, that the bill clearly showed a case for equitable relief and gave the court jurisdiction of the subject matter. *Tunesma* v. *Schuttler*, 114 Ill. 156.

2888. INSURANCE COMPANY—*party who may sue for penalty*. Although an action by a creditor against a stockholder to enforce a statutory liability is penal in character, yet the action may be brought in the name of the creditor. The provisions of § 24 of the insurance law do not apply in such case. *Gulliver* v. *Baird*, 9 Bradw. 421; *Felix* v. *Denton, Id.* 478.

2889. DECLARATION—*must show amount of stock held*. In an action under a law making the stockholders individually liable to creditors of the corporation, to a sum equal to the amount of stock held by them, the declaration should aver the amount of the defendant's stock. *Sherman* v. *Smith*, 20 Ill. 350.

2890. SAME—*sufficiency to admit proof of defendant being a stockholder*. In an action against a stockholder of a corporation to recover for debts of the company contracted in the summer of 1867, the declaration averred that he became a stockholder at some time anterior to December 1, 1868: *Held*, that proof was admissible to show that the defendant was a stockholder when the debt was contracted. *Culver* v. *Third Nat. Bank*, 64 Ill. 528.

2891. SAME—*one held sufficient*. In an action by a creditor of an insurance company against a stockholder to enforce his individual liability, the declaration averred that the defendant had subscribed for fifty shares of the capital stock of the company, and that the whole amount of the capital stock had not been paid in, and that no certificate of such payment had been given or recorded as required by the statute, but on the contrary, not more than one-half of said capital stock had ever been paid in to said company: *Held*, that the declaration showed a right of recovery. *Gulliver* v. *Roelle*, 100 Ill. 141.

2892. EVIDENCE—*proof of defendant being a stockholder.* To make one liable for the debts of a corporation, it must be clearly shown that he was a stockholder and within the purview of the law. The meaning of the statute cannot be so enlarged as to include cases not expressly within its provisions. Being a director is not sufficient. to make him liable. *Steele* v. *Dunne*, 65 Ill. 298; But see facts held sufficient to show a party to be a stockholder. *Corwith* v. *Culver*, 69 Ill. 502.

2893. EVIDENCE—*to show ownership of stock.* The plaintiff is not required to prove the ownership of stock by record evidence, but such fact may be shown by the defendant's admissions, and the testimony of the officers of the corporation. *Dows* v. *Naper*, 91 Ill. 44.

2894. EVIDENCE—*proof of such a debt as stockholder is liable for* In a suit against a stockholder of an insurance company based upon a decree against the company, no recovery can be had without proof of the execution of such a policy as is described in the declaration, and of a loss by fire. The recital in the decree of these facts is not evidence against the stockholder, if he was no party to that suit. *Chesnut* v. *Pennell*, 92 Ill. 55.

2895. In such a suit the admission of the loss by fire of the property insured, renders proof that notice of that fact was given to the company, wholly unnecessary, especially where judgment has been rendered against the company for the same loss. *Black* v. *Womer*, 100 Ill. 328.

2896. In such a suit it was admitted that the plaintiff had recovered judgment against the company for a loss on the policy issued by the company, and that the property insured was afterwards destroyed by fire, the plaintiff still owning the same: *Held*, that the admission was sufficient proof of the execution of the policy and of the loss. *Ib.*

2897. ESTOPPEL—*to deny liability.* Where a party acted as president of a private corporation, and held it out to the world as legally organized and acting, when in fact the whole of its capital was never subscribed: *Held*, in a suit by a creditor to enforce his individual liability as a stockholder, that he was estopped from showing such fact in avoidance of his obligation. *Corwith* v. *Culver*, 69 Ill. 502, 508.

2898. After acting under a charter or deriving a benefit therefrom, a stockholder will be estopped from setting up the unconstitutionality of the charter, or an amendment thereto, in avoidance of his individual liability for the debts of the corporation. *Dows* v. *Naper*, 91 Ill. 44.

2899. In a proceeding by a receiver to collect a note given by a stockholder for stock in an insurance company, the defendant cannot insist that an organization of the corporation must be shown in strict compliance with the statute. Organization *de facto* and *user* are sufficient. *Washburn* v. *Roesch*, 13 Bradw. 268.

2900. A suit and judgment against an imperfectly organized corporation, as between the plaintiff and defendant corporation, will operate as an estoppel to bar the same plaintiff from recovering from the members on their individual liability as partners in the same cause of action. *Cresswell* v. *Oberly*, 17 Bradw. 281.

2901. ABATEMENT OF ACTION—*death of stockholder.* An action under the statute to enforce a personal liability against a stockholder of an insurance company, is in the nature of a penal action, and dies with him. *Diversy* v. *Smith*, 9 Bradw. 437; same case, 103 Ill. 378

2902. EXTENT OF LIABILITY—*decree.* Where a decree on creditor's bill is taken against a stockholder of a national bank, on the basis his shares of stock bear to the whole stock of the bank, there will be no error. *Wheelock* v. *Kost*, 77 Ill. 296.

2903. LIEN OF CREDITOR—*equitable attachment.* The creditor first suing to enforce the individual liability of a stockholder, thereby acquires a preference over other creditors of the corporation, which neither they nor the stockholder can defeat, unless, possibly, by bringing a bill for a general closing up of the affairs of the corporation. Such action is in the nature of an equitable attachment of the stockholder's liability to the extent of the plaintiff's claim. After notice of such suit, the stockholder cannot defeat the action by paying other creditors to the extent of his liability. *Thebus* v. *Smiley,* 110 Ill. 316. But see *Chicago* v. *Hall,* 103 Ill. 342.

(c.) LIABILITY FOR DOUBLE THE STOCK.

2904. PRIMARY LIABILITY—*not lost by failure to sue in three months after transfer.* Under the charter of a bank which provided, "each stockholder shall be liable to double the amount of stock held or owned by him, and for three months after giving notice of transfer," &c., it was held that a stockholder assumed a primary liability to creditors of the bank to an amount double his stock, and not a secondary one; and having incurred such liability he was not released therefrom by his not being sued within three months after a transfer of his stock. *Fuller* v. *Ledden,* 87 Ill. 310.

2905. CONSTRUCTION OF CHARTER. The fair and reasonable construction of such clause in the charter is, that a stockholder is liable for debts incurred while a member, and also for such debts as the bank should contract for and during the three months after giving notice of a transfer of his stock. The clause does not relate to the time in which suit must be brought to enforce his liability. *Ib.*

2906. LIABILITY UNDER UNCONSTITUTIONAL CHARTER. Where persons become stockholders of a corporation, even under a charter repugnant to the organic law, which makes them liable for double the amount of their stock, it will operate as an agreement by each to become liable to creditors of the corporation according to the terms of the charter, and they cannot escape individual liability because of the unconstitutionality of the charter. *McCarthy* v. *Lavasche,* 89 Ill. 270.

2907. REMEDY—*at law—several liability.* Under a charter that "each stockholder shall be liable to double the amount of stock held or owned by him," a creditor of the corporation, will have an action in his own name and at law against any stockholder, for the sum due him, and each stockholder will be severally and individually liable. *McCarthy* v. *Lavasche,* 89 Ill. 270.

2908. Under the charter of a bank providing that "each stockholder shall be liable to double the amount of stock held or owned by him and for three months after giving notice of transfers," &c., a creditor of the corporation, to enforce the individual liability of a stockholder, is not compelled to sue in the name of the corporation for his use, or by bill in chancery, but may bring his action against the stockholder in his own name at law. *Hull* v. *Burtis,* 90 Ill. 213.

2909. The intention and effect of a clause in a charter making each stockholder thereof liable to double the amount of stock held or owned by him and for three months after notice of its transfer, is to charge the stockholders with every debt made by the corporation while they hold stock, and also such indebtedness as may be contracted during three months after notice that they have transferred their stock. The creditor whose debt was contracted within that time may maintain suit against a stockholder after the expiration of the three months after notice of a transfer. *Ib.*

2910. INTEREST. Interest is not recoverable in an action against a stockholder to enforce his liability to creditors of the corporation for double the amount of his stock. *Munger* v. *Jacobson*, 99 Ill. 349.

2911. DECREE—*before order of distribution.* Where the debts of the corporation exceed the total of its assets and all stock liabilities, so that the whole of the defendant's liabilities will be needed, there is no reason for deferring a decree against them until the final decree of distribution in the case. *Munger* v. *Jacobson*, 99 Ill. 349.

OF CONTRIBUTION BETWEEN STOCKHOLDERS.

2912. If one stockholder has been sued by a creditor of the corporation and paid the recovery he may have contribution from the other stockholders by proceeding in equity. *Meisser* v. *Thompson*, 9 Brad. 368.

2913. Where a stockholder in a corporation, the charter of which imposes an individual liability upon him for the debts of the corporation, has been sued and paid the recovery to a creditor, he will be entitled to contribution from all the other stockholders, and in enforcing that right it may be that a court of equity is the proper forum, as in it he can compel each shareholder to contribute *pro rata* according to the number of shares he may hold. *Wincock* v. *Turpin*, 96 Ill. 135.

2914. A stockholder of a bank who pays the amount of his individual liability to a firm in which he is a partner for a debt due such firm from the bank, thereby acquires an equitable right against his co-stockholders, cognizable and enforceable only in equity. *Buchanan* v. *Meisser*, 105 Ill. 638.

LIMITATION OF ACTION.

2915. In debt by a creditor of a corporation against a stockholder to enforce the individual liability of the latter created by § 16 of the insurance law of 1869, the liability sought to be enforced is in the nature of a penalty, and an action thereon is barred in two years. *Junker* v. *Kuhnen*, 18 Bradw. 478.

2916. A stockholder in a corporation formed under the act of 1849 is not liable as such to creditors of the corporation, unless suit is brought against the corporation within one year from the time the debt became due. *Tarbell* v. *Page*, 24 Ill. 46.

2917. It is apprehended that a plea by a stockholder, who has ceased to be such, that the cause of action did not accrue within two years after he had ceased to be a stockholder, that being the time prescribed in the act for the continuance of his liability, would be a good plea. *Baker* v. *Backus*, 32 Ill. 79, 100.

2918. The liability of the trustees and corporators of insurance companies arising under § 16 of the insurance act of 1869, is imposed, not as upon a contract, but by way of a satutory penalty only. So, a cause of action arising under that section will be barred within two years from the time it accrued. *Gridley* v. *Barnes*, 103 Ill. 211.

LIABILITY OF MANAGING OFFICERS.

FOR EXCESS OF DEBT ABOVE CAPITAL STOCK.

2919. The officers and directors assenting to debts above capital stock are made personally liable to the creditors of the corporation as a whole, and not to any individual creditor, and this liability is enforceable only in equity. *Buchanan* v. *Bartow Iron Co.*, 3 Bradw. 191; *Buchanan* v. *Low*, 3 Bradw. 202.

2920. DECLARATION. In an action to enforce this statutory liability the declaration must show that the indebtedness of the corporation exceeds the amount of the capital stock, and that the trustees assented thereto. *Sherman* v. *Smith,* 20 Ill. 350, 353.

2921. LIABILITY TO CREDITORS GENERALLY. Under the provisions of § 16, of chap. 32, R. S. 1874, the directors and officers of a stock corporation who assent to an indebtedness in excess of its capital stock, are made personally and individually liable for such excess to the creditors generally of such corporation, and not to any particular creditor. *Low* v. *Buchanan,* 94 Ill. 76.

2922. The object and purpose of this section is, that *all* claims arising under its provisions shall be regarded in the nature of a trust fund to be collected and divided *pro rata* among all the creditors, and this distribution can only be made in a court of equity. *Ib.*

2923. REMEDY—*in equity.* Where a stock corporation has incurred indebtedness in excess of its capital stock to various parties, the individual liability of its directors and officers assenting thereto cannot be enforced by action at law at the suit of a single creditor, but the remedy is in a court of equity, where the rights and liabilities of all may be determined and properly adjusted. *Ib.*

2924. If such an action can be maintained at law by a single creditor on the ground there are no other creditors, he must set forth by proper averments in his declaration, and prove on the trial, the special circumstances warranting such an action. *Ib.*

2925. In order to enforce penalties imposed upon stockholders of a corporation by its charter, which are not part of the assets of the company, the suit must be at law, in the name of the individual creditors, each for himself. *Lane* v. *Nickerson,* 99 Ill. 284; *Wincock* v. *Turpin,* 96 Ill. 135.

SALE OF STOCK ON EXECUTION.

CHAPTER 77.—TILL JUDGMENTS AND EXECUTIONS.

2926. SHARES OF STOCK IN CORPORATION—LIABLE TO SALE ON EXECUTION. § 52. The share or interest of a stockholder in any corporation may be taken on execution, and sold as hereinafter provided; but in all cases, where such share or interest has been sold or pledged in good faith for a valuable consideration, and the certificate thereof has been delivered upon such sale or pledge, such shares or interest shall not be liable to be taken on execution against the vendor, or pledgor, except for the excess of the value thereof over and above the sum for which the same may have been pledged and the certificate thereof delivered. [Laws of 1871–2, p. 505, § 52, as amended by the L. 1883, p. 110. See Laws 1861, p. 132, on subject. R. S. 1887, p. 809, § 52; S. & C., p. 1410, § 52; Cothran, p. 872, § 52.]

2927. STATUTORY REMEDY—*must be strictly pursued.* There being no authority at common law for the levy of an execution upon the defendant's interest in the capital stock of a corporation, and the proceeding being wholly statutory, the course pointed out in the statute must be strictly pursued. *Goss, &c.* v. *People,* 4 Bradw. 510.

2928. STEPS TO PERFECT LEVY—SALE SAME AS OF CHAT-

TELS. § 53. If the property has not been attached in the same suit, the officer shall leave an attested copy of the execution with the clerk, treasurer or cashier of the company, if there is any such officer, otherwise with any officer or person having the custody [of] the books and papers of the corporation; and the property shall be considered as seized on execution when the copy is so left, and shall be sold in like manner as goods and chattels. [R. S. 1887, p. 809, § 53; S. & C., p. 1410, § 53; Cothran, p. 872, § 53.]

2929. ATTESTED COPY OF EXECUTION. The attested copy of execution mentioned in the statute need not be verified by the clerk and attested by the seal of the court. The sheriff holding the writ may certify to the correctness of the copy. *People* v. *Goss, &c.*, 99 Ill. 355.

2930. An attested copy of the execution, regular on its face, must be left with officers of the company, or the person having custody of the books: and the sheriff's returns must show that this was done. *Goss, &c.* v. *People*, 4 Bradw. 510.

2931. WHO MAY ATTEST COPY—*sheriff*. The statute does not require the clerk of the court to verify such copy, and attest it by the seal of the court. The sheriff holding the execution may properly certify to the correctness of the copy. *People* v. *Goss, &c.*, 99 Ill. 355.

2932. VERIFICATION OF COPY. A copy of an execution directed to the sheriff, delivered by him to the clerk of a corporation having indorsed upon it the words: "the within is a true copy of the execution and fee bill in *my* hands, under which *I* have seized the shares of stock of the within named defendant," &c., but not signed by the sheriff, is officially verified or attested within the requirement of the statute. The language identifies the maker of the indorsement. *Ib.*

2933. LEVY ON STOCK—*when actual and complete*. An actual levy upon shares of stock held by a debtor in a corporation is accomplished by the sheriff, where he has exhibited to the keeper of the stock books of the corporation his execution, and on demand for the purpose of levy, has procured and received from the corporation "a certificate of the number of shares or amount of interest held by the judgment debtor," and has indorsed upon his execution a statement that the shares named in the certificate are taken in execution, or levied upon. When the sheriff delivers to the proper officer of the corporation an attested copy of the execution, the stock of the debtor shall be considered as seized on execution. This is only a constructive levy. *Ib.* As to the usual mode of levy, see *Powell* v. *Parker*, 38 Ga. 644; *Baily* v. *Strohecker*, *Id.* 259; *Mechanics & T. Bank* v. *Dakin*, 33 How. Pr. 316: S. C. 50 Barb. 587; *Kuhlman* v. *Orsen*, 5 Duer. 242; *Clarke* v. *Goodridge*, 41 N. Y. 210; *Drake* v. *Goodridge*, 54 Barb. 78.

2934. LEVY AND SALE—IN CASE OF ATTACHMENT. § 54. If the share is already attached in the same suit, the officer shall proceed in seizing and selling it on the execution, in the same manner as in selling goods and chattels. [R. S. 1887, p. 810, § 54; S. & C., p. 1411, § 54; Cothran, p. 872, § 54.]

2935. OFFICER OF CORPORATION—TO GIVE CERTIFICATE OF DEBTOR'S SHARES, &C.—LIABILITY FOR REFUSAL, &C. § 55. The officer of the company who keeps a record or account of the shares or interest of the stockholders therein, shall, upon the exhibiting to him of the execution, be bound to give a certificate of the number of shares or amount of the interest held by the judgment debtor. If he refuses to do so, or if he willfully gives a false certificate thereof, he shall be liable for double the amount of all damages occasioned by such refusal or false certificate, to be recovered in any proper action, unless the judgment is satisfied by the original defendant. [R. S. 1887, p. 810, § 55; S. & C., p. 1411, § 55; Cothran, p. 872, § 55.]

2936. LIABILITY OF OFFICER—*conditions to his liability, waiver of his rights.* The officer of the corporation before giving such certificate has the right to have not only an exhibition of the execution, but also an attested copy thereof as a voucher for his giving a certificate of the defendant's stock. But the right to such voucher may be waived by the corporation, and this is done by giving the certificate of the debtor's shares to the sheriff. The giving such certificate is a waiver of any defect in the attestation of the copy of the execution delivered. *People* v. *Goss Mfy. Co.,* 99 Ill. 355.

2937. CERTIFICATE OF SALE—*issue of certificate of stock.* § 56. An attested copy of the execution and of the return thereon shall, within fifteen days after the sale, be left with the officer of the company whose duty it is to record transfers of shares; and the purchaser shall thereupon be entitled to a certificate or certificates of the shares bought by him upon paying the fees therefor and for recording the transfer. [R. S. 1887, p. 810, § 56; S. & C., p. 1411, § 56; Cothran, p. 872, § 56.]

2938. DUTY TO TRANSFER ON BOOKS—*of shares sold on execution.* The purchaser of stock in a corporation at a sheriff's sale, has a right, under the statute, on leaving with the officer of the corporation whose duty it is to record transfers of shares, within fifteen days after the sale, an attested copy of the execution and of the return thereon, to have the corporation consent to hold possession of the stock for him, and to have his title made manifest by the necessary transfer upon the books, and by the issue of new stock certificates directly to him, for the shares sold to him. *People* v. *Goss, &c. Manuf. Co.,* 99 Ill. 355.

2939. RIGHTS OF PURCHASER—DIVIDENDS. § 57. If the shares or interest of the judgment debtor had been attached in the suit in which the execution issued, the purchaser shall be entitled to all of the dividends which have accrued after the attachment. [R. S. 1887, p. 810, § 57; S. & C., p. 1411, § 57; Cothran, p. 872, § 57.

CONSOLIDATION.

An act for an act to increase the powers of railroad corporations. Approved June 30, 1885. In force July 1, 1885. [L. 1885, p. 229; R. S. 1887, p. 1041; 3 S. & C., p. 447.]

2940. CONSOLIDATION OF RAILROAD CORPORATIONS. § 1. *Be it enacted by the people of the state of Illinois, represented in the general assembly,* That all railroad companies now organized, or hereafter to be organized, under the laws of this state, which now are, or hereafter may be in possession of, and operating in connection with, or extension of their own railway lines, any other railroad or railroads, in this state or in any other state or states, or owning and operating a railroad which connects at the boundary line of this state with a railroad in another state, are hereby authorized and empowered to purchase and hold in fee simple or otherwise, and to use and enjoy the railway property, corporate rights and franchises of the company or companies owning such other road or roads, upon such terms and conditions as may be agreed upon between the directors, and approved by the stockholders, owning not less than two-thirds in amount of the capital stock of the respective corporations becoming parties to such purchase and sale; such approval may be given at any annual or special meeting, upon sixty days' notice being given to all shareholders, of the question to be acted on, by publication in some newspaper published in the county where the principal business office of the corporation is situated: *Provided*, that notice of any special meeting called to act upon such question, shall be given to each shareholder whose postoffice address is known, by depositing in the postoffice, at least thirty days before the time appointed for such meeting, a notice properly addressed and stamped, signed by the secretary of the company, stating the time, place and object of such meeting: *And, provided further,* that no railroad corporation shall be permitted to purchase any railroad which is a parallel or competing line with any line owned or operated by such corporation.

2941. CONSOLIDATED COMPANY—BODY CORPORATE—POWER OF—ILLINOIS CENTRAL. § 2. Any railroad company now organized or hereafter to be organized under the laws of this state, shall have power from time to time to borrow such sums of money as may be necessary for the funding of its indebtedness paying for constructing, completing, improving or maintaining its lines of railroad, and to issue bonds therefor, and to mortgage its corporate property, rights, powers, privileges and franchises, including the right to be a corporation, to secure the payment of any debt contracted for such purposes; and to increase its capital stock to any amount required for the purposes aforesaid, not exceeding the cost of the roads and works owned or constructed and equipped by it; such increase of capital stock to be made in such manner and in accordance with and subject to such regulations, preferences, privileges and conditions as the company at any general or special meeting of its shareholders, held at the time such creation of new shares may be authorized, shall think fit: *Provided*, that no stock or bonds shall be issued, except for money, labor or property actually received and applied to the purposes for which such corporation was created;

nor shall the capital stock be increased for any purpose except upon giving sixty days' public notice in the manner provided in the first section of this act: *And, provided further*, that nothing contained in this act shall be held or construed to alter, modify, release or impair the rights of this state as now reserved to it in any railroad charter hertofore granted, or to affect in any way the rights or obligations of any railroad company derived from, or imposed by such charter: *And, provided further*, that nothing herein contained shall be so construed as to authorize or permit the Illinois Central Railroad company to sell the railway constructed under its charter, approved February 10, 1851, or to mortgage the same, except subject to the rights of the state under its contract with said company, contained in its said charter, or to dissolve its corporate existence, or to relieve itself or its corporate property from its obligations to this state, under the provisions of said charter; nor shall anything herein contained be so construed, as to in any manner, relieve or discharge any railroad company, organized under the laws of this state, from the duties or obligations imposed by virtue of any statute now in force or hereafter enacted: *And, provided further*, that nothing in this act shall be so construed as to authorize any corporation, other than those organized in and under the laws of this state, to purchase or otherwise become the owner, owners, lessee or lessees of any railroad within this state.

ELEVATED WAYS AND CONVEYORS.

An act in regard to elevated ways and conveyors. Approved April 7, 1875. In force July 1, 1875. [Laws 1875, p. 77- R. S. 1887, p. 342; S. & C., p. 1977; Cothran, p. 347.

2942. ORGANIZATION —ARTICLES OF. INCORPORATION. § 1. *Be it enacted by the people of the State of Illinois, represented in the general assembly:* Any company which has been or shall be incorporated under the general laws of this state, for the purpose of constructing, maintaining and operating any elevated way or conveyor, shall state in its articles of incorporation the places from and to which it is intended to construct the proposed elevated way or conveyor. And any such company may organize and become incorporated under the provisions of chapter (32) thirty-two of the revised statutes of 1874, concerning corporations for pecuniary profit, and shall be subject to the provisions of the laws of this state applicable to such corporations.

2943. RIGHT OF WAY—HOW OBTAINED. § 2. If any such corporation shall be unable to agree with the owner for the purchase of any real estate required for the purposes of its incorporation or the transaction of its business, or for its depots, station buildings, engine houses, or for right of way, or any other lawful purpose connected with or necessary to the construction, maintenance and operation of said elevated way or conveyor, such corporation may acquire such title in the manner that may be now or hereafter provided for by any law of eminent domain.

2944. MAY TAKE MATERIAL — COMPENSATION. § 3. Any such corporation may, by their agents and employes, enter upon and take from any land adjacent to its way, or road, or conveyor, earth, gravel, stone or other material, except fuel and wood, necessary for the construction of such elevated way, paying, if the owner of such land and the said corporation can agree thereto, the value of such material taken, and the amount of damage occasioned to any such land or its appurtenances; and if such owner and corporation cannot agree, then the value of such material and the damage occasioned to such real estate shall be ascertained, determined and paid in the manner that may now or hereafter be provided by any law of eminent domain; but the value of such material, and the damages to such real estate, shall be ascertained, determined and paid for before such corporation can enter upon and take the same.

2945. CAPITAL STOCK—INCREASE OF. § 4. In case the capital stock of any. such corporation shall be found insufficient for constructing and operating its elevated way or conveyor, such corporation may, with the concurrence of two-thirds, in value, of all its stock, increase its capital stock, from time to time, to any amount required for the purpose aforesaid.

2946. POWERS OF—RESTRICTION. § 5. Every corporation formed under this act shall, in addition to the powers hereinbefore conferred, have power—

First—To cause such examination and survey for its proposed elevated way to be made as may be necessary to the selection of the most advantageous route; and for this purpose, by its officers, agents or servants, may enter upon the lands or waters of any person or corporation, but subject to responsibility for all damages which shall be occasioned thereby.

Second—To lay out a strip of land, not exceeding sixty-six feet in width, on which to construct, maintain and operate said elevated way or conveyor; and for the purpose of cuttings and embankments, to take as much more land as may be necessary for the proper construction and security of the elevated way; to cut down any standing trees that may be in danger of falling upon and injuring such way, making compensation therefor in manner provided by law.

Third—To construct its way across, along or upon any stream of water, water-course, street, highway, plank-road,
—25

turnpike, canal or railroad, which the route of such elevated way shall intersect or touch; but such corporation shall restore the stream, water-course, street, highway, plank-road, turnpike and railroad thus intersected or touched, to its former state, or to such state as not unnecessarily to have impaired its usefulness, and keep such crossing in repair: *Provided*, that in no case shall any company construct its way without first constructing the necessary culverts and sluices, as the natural lay of the land requires for the necessary drainage thereof.

Nothing in this act contained shall be construed to authorize the erection of any bridge, or any other obstruction, across or over any stream navigated by steamboats, at the place where any bridge or other obstruction may be proposed to be placed, so as to prevent the navigation of such stream; nor to authorize the construction of any elevated way or conveyor upon or across any street in any city, or incorporated town, or village, without the assent of the corporation of such city, town or village: *Provided*, that in case of the construction of said elevated way or conveyor along highways, plankroads, turnpikes, canals or railroads, such company shall either first obtain the consent of the lawful authorities having control or jurisdiction of the same, or condemn the same under the provisions of any eminent domain law, now or hereafter in force in the state.

USE OF STREETS, ETC., BY ELEVATED RAILROADS.

An act in regard to the use of streets and alleys in incorporated cities and villages by elevated railroads and elevated ways and conveyors. Approved June 18, 1883. In force July 1, 1883. [L 1883, p. 126; R. S. 1887, p. 343; S. &. C., p. 1979; Cothran, p. 287j.]

2947. PETITION OF LAND-OWNERS. § 1. *Be it enacted by the people of the state of Illinois, represented in the general assembly,* That no person or persons, corporation or corporations, shall construct or maintain any elevated railroad or any elevated way or conveyor to be operated by steam power, or animal power or any other motive power, along any street or alley in any incorporated city or village, except by the permission of the city council or board of trustees of such city or village, granted upon a petition of the owners of the lands representing more than one-half of the frontage of the street or alley, or of so much thereof as is sought to be used for such elevated railroad or elevated way or conveyor; and the city council, or board of trustees, shall have no power to grant permission to use any street or alley, or part thereof, for any of the purposes aforesaid, except upon such petition of land-owners as is herein provided for.

2948. WHEN STREET MORE THAN ONE MILE. § 2. When the street or alley, or part thereof, sought to be used for any of the purposes aforesaid, shall be more than one mile in extent, no petition of land-owners shall be valid for the purposes of this act, unless the same shall be signed by the owners of the land representing more than one-half of the frontage of each mile and fractional part of a mile, of such street or alley or of the part thereof sought to be used for any of the purposes aforesaid.

2949. REPEAL. § 3. All acts and parts of acts inconsistent herewith are hereby repealed.

CHAPTER 31—CORONERS.

2950. LIABILITY OF RAILWAY, ETC.—FOR EXPENSES OF INQUEST AND BURIAL. § 22. When any railroad company, stage or any steamboat, propeller or other vessel engaged in whole or in part in carrying passengers for hire, brings the dead body of any person into this state, or any person dies upon any railroad car or any such stage, steamboat, propeller or other vessel in this state, or any person is killed by cars or machinery of any railroad company, or by accident thereto, or by accident to or upon any such stage, steamboat, propeller, or other vessel, or by accident to, in or about any mine, mill or manufactory, the company or person owning or operating such cars, machinery, stage, steamboat, propeller or other vessel, mine, mill or manufactory shall be liable to pay the expenses of the coroner's inquest upon and burial of the deceased, and the same may be recovered in the name of the county in any court of competent jurisdiction. [Laws of 1855, p. 170, §§ 1, 2, 3; R. S. 1887, p. 329, § 22; S. & C. p. 606, § 25; Cothran, p. 323, § 22. *Held* unconstitutional; see O. & M. Ry. v. Luckey, 78 Ill. 55.]

WAREHOUSES AND WAREHOUSE RECEIPTS.

2951. When the amount of grain of the different owners in a warehouse falls short, each owner is entitled to his proportion of what is left. *Sexton* v. *Graham.* 53 Iowa, 200.

2952. INTERMIXTURE. If there is a confusion of goods by reason of intermixture so that each party cannot distinguish his own, each will have a proportionate property in the whole. *Low* v. *Martin,* 18 Ill. 286.

2953. A party who consents that grain left with a warehouseman may be put in bulk with other grain, with the understanding that he shall receive a like quantity and quality, cannot maintain replevin for the grain. If the intermixture was without consent, or was the wrongful act of the warehouseman, it would be otherwise. *Ib.*

2954. LIEN. Warehousemen have a lien on grain stored with them for proper charges and may retain possession to secure their payment. *Low* v. *Martin,* 18 Ill. 286.

2955. The fraudulent issue of warehouse receipts for grain not in store, does not deprive the warehouseman of his lien for that which he has actually stored. *Ib.*

2956. LIABILITY IN CASE—*on fraudulent receipts.* An action on the case may be maintained upon fraudulent warehouse receipts purporting to have been given for produce in store, by a party who has advanced money upon the faith of them, and this whether the party has been deprived of the produce or his money. *Low* v. *Martin*, 18 Ill. 290.

2957. SALE OF GOODS—*right to surplus above charges.* Where goods erroneously shipped to a fictitious person are sold by the warehouseman, the surplus after paying charges belong to the shipper. *Boilvin* v. *Moore*, 22 Ill. 318.

2958. PURCHASER OF WAREHOUSE RECEIPT—*when takes subject to charges.* Where a party purchases a warehouse receipt for grain, with notice that it is subject to charges for storage, he will be liable for such charges, and the warehouseman will have a lien therefor. *Cole* y. *Tyng*, 24 Ill. 99.

2959. CHARGES NOT LOST BY DELIVERY OF GRAIN. If a warehouseman permits grain to be removed before his charges are paid, he will not thereby lose his recourse against the holder of the receipt. *Ib.*

2960. LIEN—*lost by delivery.* If a warehouseman or consignee delivers goods upon the receipt of the promissory note of the owner for charges, he will lose his lien, which will not revive should the goods accidentally be returned to his possession. *Hale* v. *Barrett*, 26 Ill. 195.

2961. LIEN—*for charges not on another's goods.* If goods belonging to different owners are shipped by one bill of lading, the consignee cannot hold the goods of one for the charges upon the goods of the other. Each owner is entitled to his goods on the payment of the appropriate charges thereon. *Ib.*

2962. STORAGE OF GRAIN—*degree of care required.* A warehouseman who receives the grain of another for storage, is only bound to ordinary care for its preservation. But where he purchases grain for another and has it in store, he takes the risk of any loss that may occur, until such delivery as will pass the title to the party for whom the grain was bought. *Myers* v. *Walker*, 31 Ill. 353.

2963. SAME—*compensation for.* A warehouse receipt was as follows: "Received in store for W. & K., and subject to their order, and free of all charges on board their boats, or any boats they may send for the same, thirty thousand bushels corn:" *Held*, that the warehouseman was bound to store the corn free of charge, only for a reasonable time; and if boats were not sent for the corn within such time, he would be entitled to compensation for storage and for any extra labor in delivery occasioned by the delay. But the right to charge for storage it seems would accrue only after notice to the owner to remove the grain. *Myers* v. *Walker*, 31 Ill. 353. Same case, 24 Ill. 133, 137.

2964. WAREHOUSE RECEIPTS—*rights of holder.* The holder of a warehouse receipt for grain has only the personal obligation of the warehouseman for the proper storage and delivery of his grain according to the terms of the receipt, or on default, to recover the damages growing out of a breach of the contract. *Dole* v. *Olmstead*, 36 Ill. 150.

2965. WAREHOUSE RECEIPT—*gives no lien in favor of holder.* The giving of a warehouse receipt creates no specific or general lien on the property of the warehouseman, although that should consist of grain put in the common bulk with that of the holder of the receipt. *Dole* v. *Olmstead*, 36 Ill. 150.

2966. CONFUSION OF PROPERTY—*rights of the several owners.* Where the grain of different owners has been intermingled in one common mass according to the usage of warehousemen, and without objection by the owners, it will become common property, owned by the several parties in the proportion in which each had contributed to the common stock. The several owners must sustain any loss *pro rata* which may occur by diminution, decay or otherwise. *Dole* v. *Olmstead*, 36 Ill. 150.

2967. SAME—*remedy in chancery.* Where the warehouseman assigns all the grain in store, including grain of his own, to a creditor, to secure a debt, to be held subject to the rights of others, the creditor will become a trustee for the benefit of all parties in interest, and where there is a deficiency of grain to satisfy all and the grain is intermixed, a court of equity will have jurisdiction. *Dole* v. *Olmstead*, 36 Ill. 150.

2968. CARE REQUIRED OF. A warehouseman must exercise reasonable care, but he is not an insurer against all losses except those arising from the act of God and the public enemy. He is only liable for losses which might have been guarded against by the exercise on his part of ordinary care and diligence. *St. L., A. & T. H. R. R.* v. *Montgomery*, 39 Ill. 335.

2969. WAREHOUSE RECEIPTS—*stand in place of property—negotiability.* Receipts given by a warehouseman for chattels stored with him, are not in a technical sense, negotiable instruments, but they merely stand in the place of the property itself, and a delivery of the receipts has the same effect in transferring the title as the delivery of the property, neither more nor less. *Burton* v. *Curyea*, 40 Ill. 320.

2970. SAME—*transfer by one having no title.* A purchaser of pork in warehouse, who takes warehouse receipts therefor, and then, to enable his vendor to withdraw the pork from the warehouse for the purpose of overhauling and re-packing it, delivers the receipts back to the vendor, who transfers them to a *bona fide* purchaser, still remains the owner of the pork and may maintain replevin for it against the warehouseman in whose possession it still remains. *Burton* v. *Curyea*, 40 Ill. 320.

2971. SAME—*negligence of rightful owner.* If the purchaser of warehouse receipts indorsed in blank should place them in the hands of his vendor for improper purposes, or be fairly chargeable with any negligence whereby the person having the receipts, was enabled to impose on an innocent purchaser, it may be a different rule might prevail. *Ib.*

2972. WAREHOUSE RECEIPTS—*negligence in respect to—notice of purchase.* The failure of a purchaser of pork in warehouse by the transfer in blank of the receipts therefor, to take new receipts in his own name, and putting them in the hands of his vendor instead of the original receipts, and his neglect to notify the warehouseman of his purchase, is not negligence on the part of the purchaser. *Burton* v. *Curyea*, 40 Ill. 320.

2973. WAREHOUSE RECEIPTS—*a good tender of grain.* An actual tender of warehouse receipts for grain stored by the vendor of grain in Chicago, is a good tender of the grain, unless the purchaser should insist on seeing it. *McPherson* v. *Gale*, 40 Ill. 368.

2974. LIABILITY—*conversion of grain.* Where the assignees of a warehouseman convert grain in store with them which they received from their assignor, and appropriate the money to their own use, they will at least be liable to account to the owners for the amount received, with interest from the date of sale. *Dole* v. *Olmstead*, 41 Ill. 344.

2975. ASSIGNEE OF WAREHOUSEMAN—*take no interest in grain of others in store.* Where a commission merchant having large amounts of grain on storage for others, makes a general assignment for the benefit of creditors, his assignees will take only the interest of the assignor, and cannot claim the grain of others so stored. *Ib.*

2976. INTERMIXTURE—*average of loss.* Where the grain of various parties in a warehouse is stored in a common mass by the consent of the owners, and the warehouseman makes an assignment for creditors, and there proves to be a loss in the quantity of the grain, the court should average the loss among all the owners; and if the grain has been sold by the assignees, each owner should be compensated in money in proportion to the grain he placed in store. *Ib.*

2977. DEGREE OF CARE. Where the carrier assumes the duties of warehouseman, he will be bound to ordinary care and diligence in the preservation of the property. The building in which the goods are stored must be a safe one, though it need not be fire proof. It should be under the charge of careful and competent servants, and in case of threatened danger from fire, ordinary diligence must be used to remove the property. *C. & A. R. R. v. Scott,* 42 Ill. 132.

2978. WAREHOUSE RECEIPT — *tender of grain sold by.* In an action by the vendor of grain to recover the price agreed to be paid, proof of the attendance of the plaintiff at the time and place agreed upon for its delivery, but in the absence of the purchaser, for the purpose of tendering warehouse receipts, is not a sufficient tender, without the further proof that such receipts were genuine and that the grain was not subject to charges. *McPherson v. Hall,* 44 Ill. 264.

2979. But a tender of the receipts to the defendant in person would have been good, if without objection, as the failure to object would impliedly admit that the receipts honestly represented the property. *Ib.*

2980. CONTRACT FOR STORAGE—*construction.* The plaintiff stored corn in the defendant's warehouse, taking from them the following agreement: "Feb. 9, 1860. We hereby agree to store ear corn for H. H. until the first of June next, for three cents per bushel; two cents for shelling, and receiving 75 pounds and deliver 58 pounds. If sold before the first of June, we are not to charge for shelling; if not sold by the first of June we are to charge one-half per cent per month till it is sold. The corn to be good and merchantable. C. & V.": *Held,* that the contract contemplated a storage beyond June, 1860. *Cushman v. Hayes,* 46 Ill. 145.

2981. Such a contract would not continue for an indefinite time wholly on the will of the owner of the corn. Although the contract provides the corn may remain in store by paying one-half per cent per month, until the corn is sold, there is nothing in the terms to prevent a termination of the contract by the defendants, on notice, where a necessity for so doing arises. *Ib.*

2982. WAREHOUSE RECEIPT—*negotiable.* Under the act incorporating the Chicago Dock company, a warehouse receipt issued by that company is made negotiable, and as such, absolutely vests in the holder the title to the property specified in it. *Ch. Dock Co. v. Foster,* 48 Ill. 507.

2983. WAREHOUSE RECEIPTS—*parol evidence to vary.* A warehouse receipt given for grain received in store, is the contract of the parties, and parol evidence is not admissible to vary its terms. *Leonard v. Dunton,* 51 Ill. 482.

2984. REMEDY—*against warehouseman refusing to deliver.* Where a warehouseman receives grain in store, and gives his receipt therefor, providing for a delivery of the grain on the order of the owner, while an action of trover might lie against the warehouseman on his refusal to deliver the grain on demand, yet assumpsit will also lie for the breach of the contract. *Leonard v. Dunton,* 51 Ill. 482.

2985. MEASURE OF DAMAGES. · In assumpsit against a warehouse-man for a refusal to deliver grain placed in store, on demand, according to his contract, the measure of damages is the value of the grain at the time it should have been delivered. *Leonard* v. *Dunton*, 51 Ill. 482.

2986. ACTION FOR NON-DELIVERY—*non-payment of storage no defense.* Where grain is stored in a warehouse to be kept a short time without charge, and to be delivered to the owner when demanded, the neglect of the owner to pay storage after such time, or to offer to do so, will not defeat his action against the warehouseman for the breach of the contract to deliver the grain on demand. The most the warehouseman could claim would be a reasonable deduction for storage after having given notice that storage would be charged. *Leonard* v. *Dunton*, 51 Ill. 482.

2987. PRIVATE WAREHOUSEMAN—*intermixture.* In case of a storage of grain by a private warehouseman, in the absence of any agreement on the subject, the inference would be that he was to keep it in the condition in which he received it, and if mixed with his own grain by consent of the owner, that it shall remain with the ware-houseman, until demanded. *Ives* v. *Hartley*, 51 Ill. 520.

2988. REMEDY. Where a person puts grain in a warehouse for the purpose of storage, and the warehouseman converts the same to his own use, the owner may waive the tort and recover from the ware-houseman in assumpsit for money had and received, for the value of the grain. *Ives* v. *Hartley*, 51 Ill. 520.

2989. RECEIPT—*whether a deposit or a sale.* The owner of wheat delivered the same to a miller, taking a receipt therefor as follows: "Received of A. B. to be stored 150 bushels wheat, to take market price when he sees fit to sell:" *Held*, that the form of the receipt implied a sale of the wheat and not merely a deposit for storage. *Ives* v. *Hartley*, 51 Ill. 520.

2990. WHETHER A SALE OR BAILMENT. Where grain was deposited in a warehouse on the understanding between the parties, not that the identical grain, or grain of like quality was to be returned, but the money value thereof to be ascertained by the market price on the day the depositor should choose to fix, the transaction was *held* to be a sale and not a bailment. *Lonergan* v. *Stewart*, 55 Ill. 44.

2991. LIEN—*how lost.* After the great fire in Chicago in 1871, the board of trade, acting in behalf of unknown owners and parties interested, and with the assent of the several warehousemen, took possession of the grain unconsumed and sold the same for the benefit of the owners. Previous to the sale, the warehousemen agreed in writing with the board of trade, that the latter might sell, the former to receive two cents per bushel as accrued storage thereon. After the sale they claimed a lien on the fund for charges over and above the sum stipulated: *Held*, that they had lost their lien for storage, except for two cents a bushel; and that the expense incurred in preserving the grain was a proper charge to be deducted from the fund. *Board of Trade* v. *Buckingham*, 65 Ill. 72.

2992. WAREHOUSE RECEIPT—*transfer of title by.* The transfer of a warehouse receipt, or bill of lading accompanied by a sale or pledge of the property specified in the receipt or bill, will have the same effect as the delivery of the property itself to the transferree. *W. U. R. R.* v. *Wagner*, 65 Ill. 197.

2993. SAME—*evidence of ownership.* Where one having an elevator and in the habit of purchasing grain for others, gave a warehouse receipt stating that he had received a lot of corn on storage for the holders of the receipt, in well covered cribs, and agreeing to hold the same for such holders, subject to their order, at the end of which were these words: "subject to their order, for all advances of money on the same:" *Held*, that the latter words did not convert the receipt into a mere pledge and render the corn liable to an execution against the party giving it, issued subsequently to the date of the receipt. *Cool* v. *Phillips*, 66 Ill. 216.

2994. ACT REGULATING—*constitution*. The act of April 25, 1871, to regulate public warehouses and the warehousing and inspection of grain, is not in contravention of § 22, art. 4, of the constitution of 1870. *Munn* v. *People*, 69 Ill. 80.

2995. The act of 1871 regulating public warehouses and the inspection of grain and to give effect to art. 13 of the constitution, and which provides a maximum rate of charges, is not unconstitutional. *Munn* v. *People*, 69 Ill. 80.

2996. CONTRACT TO INSURE. A warehouseman agreed to insure the property stored with him, which he did to their full value, and on a loss prosecuted the company in good faith on the policy, but was defeated on the ground he had given a receipt to the owner at his request: *Held*, that the warehouseman having complied with his contract, was not liable to the owner on the ground he failed to recover. *Cole* v. *Favorite*, 69 Ill. 457.

2997. WAREHOUSE RECEIPTS—*possession of, is possession of grain.* Usage has made the possession of warehouse receipts equivalent to the possession of the property they represent. *Broadwell* v. *Howard*, 77 Ill. 305.

2998. SAME—*given by the seller.* The law makes no distinction in respect to grain purchased or acquired by the holder of such receipts from others, and those acquired from the warehouseman himself. The law does not prohibit him from selling his property, and if he does so in good faith, he may become its future custodian; and the fact that he keeps a public warehouse. is sufficient to put parties on inquiry as to the ownership of grain stored. *Broadwell* v. *Howard*, 77 Ill. 305.

2999. CREDITORS OF WAREHOUSEMAN. Where a warehouseman purchased grain stored by him, for another person and with such other person's money, and took up his outstanding receipt, held by the vendor and issued a new receipt to the person for whom he bought, it was held that the grain was not liable thereafter to be taken in execution against the warehouseman. *Broadwell* v. *Howard*, 77 Ill. 305.

3000. INTERMIXTURE—*title in holder of receipt.* Where a consignee of grain stores the same in a warehouse, and the same is intermixed with other grain of like grade, and a receipt is taken for the amount, the grain being no longer capable of identification, the owner parts with his property in the same, and the consignee to whom the receipt is given, instead of being a bailee, becomes a debtor to the owner. *Bailey* v. *Bensley*, 87 Ill. 556.

3001. LIABILITY OF WAREHOUSEMAN—*trover—grain intermixed.* Where the grain is mingled with other grain of like character and grade belonging to different persons, so that its identity is lost, upon the refusal of the warehouseman to deliver upon presentation of the proper warehouse receipts, the quantity of the grain of the grade called for, the holder of the receipts may, in trover, recover damages according to the extent of his interest. *German Nat. Bank* v. *Meadowcroft*, 95 Ill. 124.

3002. LIABILITY—*transfer of warehouse.* If the warehouseman transfers the ownership and possession of the warehouse or elevator, the person succeeding to the possession of the warehouse and the grain stored therein, will be held to the same liability to the holders of warehouse receipts and subject to the same remedies as the original proprietor. *German Nat. Bank* v. *Meadowcroft*, 95 Ill. 124.

3003. FORWARDING. The business of warehousemen, when they forward goods, &c., ordinarily consists of storing produce for the owners thereof, and of shipping or forwarding the same for the owner. The legitimate income from such business is a compensation for storage and also the same for shipping or forwarding the produce. *Northrup* v. *Phillips*, 99 Ill. 449.

3004. What will bar charges for storage and insurance. *Bailey* v. *Bensley*, 87 Ill. 556.

3005. WAREHOUSE RECEIPTS—*negotiability.* The statute relating to negotiable instruments does not embrace warehouse receipts or bills of lading. They are not placed on the same footing as respects the title vested in the assignee of bills of exchange and notes. *Burton* v. *Curyea*, adhered to; *Canadian Bank* v. *McCrea*, 106 Ill. 281.

3006. A warehouse receipt is strictly speaking but the written evidence of a contract between the depositor and the warehouseman. The law implies certain duties from such receipt as devolving upon the warehouseman which becomes a part of the contract. *Ib.*

3007. SAME—*transfer of as passing title to grain.* The statute (§ 24, act 1871) makes the endorsement of a warehouse receipt evidence of the transfer of the grain it represents, the same as the actual delivery of the grain itself. But neither of these acts will pass the title to the grain which the seller or assignor does not possess. *Canadian Bank* v. *McCrea*, 106 Ill. 281.

3008. SAME. The receipt stands in the place of the grain it represents, and the possession of the receipt is regarded as the possession in law of the grain itself; and as the warehouseman is not required to surrender the grain until the return of the receipt and the payment of charges, one who obtains it under such circumstances as to charge him with notice of a want of title in his assignor, the real owner may recover of him in trover the value of the grain on his refusal to surrender the receipt to him. *Canadian Bank* v. *McCrea*, 106 Ill. 282.

3009. DELIVERY TO WRONG PERSON. A warehouseman will be liable to the party storing grain, if he delivers the same to any other person without authority from the owner, unless the latter has done some act or acts to estop him from denying permission to make a delivery. *P. & P. U. Ry.* v. *Buckley*, 114 Ill. 337.

3010. A sampler's ticket is not a warehouse receipt in the sense that term is used in the statute. *Ib.*

3011. NEGOTIABILITY. A warehouse receipt for a certain number of bushels of corn, to be delivered to the order of the person to whom the receipt is given, at a certain place, in sacks, in good order, free of charges, risk of fire excepted, is not a negotiable instrument under the law of Iowa. *M. & M. Bank* v. *Hewitt*, 3 Iowa 93.

3012. § 949 of Iowa Code, authorizes the assignee of receipt to sue in his own name, subject however to any defense or set off, legal or equitable which the maker had against the assignor, before notice of the assignment. *M. & M. Bank* v. *Hewitt*, 3 Iowa 93.

3013. SAME—*assignee takes subject to attachment.* A warehouseman who has given a receipt which entitles the holder to the goods stored upon presentation thereof, is liable to an attaching creditor of the bailor, if he surrenders the goods to a holder of such receipt, who purchased the same after the date of the attachment. *Smith* v. *Picket*, 7 Ga. 104.

3014. Under a statute making such receipts negotiable, a warehouse order for "corn to be loaded into sacks and when loaded to be

—26

sent down" was held not a receipt for storage, but merely an agreement for transportation. *Union Sav. Assoc.* v. *St. L. G. E. Co.*, 81 Mo. 341.

3015. Under such a statute a receipt by the overseer of a warehouseman is not a warehouse receipt so as to be negotiable. *Peoples' Bank* v. *Gagley*, 12 Phila. 183: *Troutman* v. *Peoples' Bank*, 12 *Id.* 276.

3016. In the absence of statutory provisions such receipts are not negotiable, but are assignable by transfer and indorsement, and such assignment will pass such title as the assignor had at the time thereof. *Solomon* v. *Bushnell*, 11 Or. 277; *Gibson* v. *Stevens*, 8 How. 384.

3017. Such a receipt cannot make the warehouseman a guarantor of the title of the property stored. *Mechanics & L. T. Co.* v. *Kiger*, 103 U. S. 352.

3018. The holder or assignee of such receipt takes no better title than if the goods were held by himself; their negotiability such as it is serving only to cut off any defenses the warehouseman may have. *Louisville Bank* v. *Boyce*, 78 Ky. 42.

3019. The transfer by indorsement and delivery of a warehouse receipt transfers the legal title and constructive possession of the property, and the warehouseman from the time of the transfer becomes the bailee of the transferree. *Gibson* v. *Stevens*, 8 How. 384; *Harris* v. *Bradley*, 2 Dill., 285; *McNeal* v. *Hill*, 1 Woolw. 96; *First Nat. Bank.* v *Bates*, 5 Cin. Law Bull. —.

3020. It is only after notice to the warehouseman who agrees to hold the property for the assignee that the title will vest absolutely in the latter. *Spangler* v. *Butterfield*, 6 Col. 356.

3021. Unless the warehouseman by his receipt agrees to deliver to the order of the bailor, the receipt will not pass title as against an attaching creditor before notice of the transfer to the warehouseman. *Hallgarten* v. *Oldham*, 135 Mass. 1.

3022. The transfer of the receipt clothes the transferree with constructive possession although the warehousman has no notice, and does not agree to hold for the transferree. *Durr* v. *Hervey*, 44 Ark. 301.

3023. In *Davis* v. *Russell*, 52 Cal. 611, it is said there is no reason why the same rule that is applied to bills of lading, making them transferable without notice, should not be followed as to warehouse receipts. In support of this see *Puckett* v. *Reed*, 31 Ark. 131; *Gibson* v. *Stevens*, 8 How. 384; *Burton* v. *Curyea*, 40 Ill. 320; *Second Nat. Bank* v. *Walbridge*, 11 Ohio St. 311; *Cool* v. *Phillips*, 66 Ill. 217; *Broadwell* v. *Howard*, 77 Ill. 305; *Cothran* v. *Ripy*, 13 Bush. 495; *Robson* v. *Swart*, 14 Minn. 370; *Hale* v. *Milwaukee Dock Co.*, 29 Wis. 482.

3024. PLEDGE OF RECEIPT. Warehouse receipts may be pledged, and an innocent pledgee will acquire title superior to the lien of the vendor of the goods represented by the receipt, where the latter permits his vendor to have possession of the receipt in such a manner as to enable him to pledge it. *Fourth Bank* v. *St. L. C. C. Co.*, 11 Mo. App. 333.

3025. A warehouseman having in store his own property may effectually pledge it to secure his own debt by transfer of his warehouse receipt. *Merchant's & M. Bank* v. *Hibbard*, 48 Mich. 118, By statute such a pledge in Iowa is made invalid.

3026. A pledge by delivery of a warehouse receipt will not give to the pledgee any general lien for debts not arising from the relation of pledgee. *J. M. Atherton Co.* v. *Ives*, 20 Fed. Rep. 894.

3027. A warehouse receipt may be transferred without indorsement so as to pass title to the property, if the owner makes the transfer with that intent, in cases where the receipt recites that the property therein mentioned is "deliverable to bearer." *Rice* v. *Cutler*, 17 Wis. 351.

3028. The Wisconsin statute providing that warehouse receipts may be transferred by indorsement, and what effect they shall have when so transferred, does not operate to prevent in all cases, a passing of title without indorsement, the language being permissive and not imperative, and the right existing independently of statute. The object of the statute is not to prevent the owner of property from passing the title in any manner previously effectual for that purpose, but to protect those dealing with persons who are intrusted with such evidence of title only as factors or agents. *Rice* v. *Cutler*, 17 Wis. 351.

3029. PURCHASER PROTECTED—*against fraud of vendor.* The fact that warehouse receipts are taken in discharge of prior indebtedness will not deprive the transferree of the protection to which he would otherwise be entitled as an innocent purchaser without notice that his vendor acquired title by fraud. *Rice* v. *Cutler*, 17 Wis. 351.

3030. TRANSFER OF TITLE BY. Where the evidence showed that grain had been delivered from the warehouse, and the warehouse receipts surrendered, an instruction to the jury to the effect, that if they believed from the evidence that the receipts in evidence were not held by the plaintiffs at the time of the levy of the execution offered in evidence, but had been surrendered to the warehouseman prior to that time, then the plaintiffs were not entitled to any of the property replevied by reason of their once having held such receipts: *Held* that the instruction was erroneous. If the reason of the surrender was the delivery to the plaintiffs of the grain mentioned in them, then they were most certainly entitled to the delivered grain, because they had once held the receipts and had surrendered them for grain delivered in exchange therefor. *Nelson* v. *McIntyre*, 1 Bradw. 603.

3031. RIGHTS OF HOLDER. The grain represented by the receipt need not be the identical grain stored, but as the mass of grain on hand is changed by successive storage and shipments, the title of the holder of the receipt passes by operation of law to that which remains in store, and he is entitled at any moment to assert his title by requiring a delivery to himself of the grain. *German Nat. Bank* v. *Meadowcroft*, 4 Bradw. 630.

3032. EFFECT OF THE TRANSFER. Upon the sale of property stored in a warehouse, the indorsement and delivery of the warehouse receipt has the effect, not only to transfer the title to the property to the indorsee, but also to give him a right of action for any breach of duty of which the warehouse company was guilty in respect thereto *at any time* during the bailment. *Sargent* v. *Central Warehouse Co.*, 15 Bradw. 553.

3033. RIGHT OF THE INDORSEE. An indorsement in blank of a warehouse receipt by the seller, authorizes the purchaser to write over such blank indorsement only a contract of mere assignment of the legal title, unlike the case of a negotiable promissory note. *Mida* v. *Geissman*, 17 Bradw. 207.

INDEX.

—27

ASSIGNMENT—*Continued.*
 liability of corporation, refusing to enter, 1202.
 new certificate not necessary, 1202d.
 individual liability of assignee, 2861.
 liability to creditors as between assignor and assignee, 2862.
 assignee for creditors, necessary party to bill, 2886.
 See also, STOCK.
of warehouse receipts.
 rights of assignee, 2971, 3012, 3013, 3029.
 See WAREHOUSE RECEIPTS.
for benefit of creditors by warehouseman.
 chancery jurisdiction, 2967.
 assignee, when liable for conversion, 2974.
 takes no title to grain of others, 2975.

ASSUMPSIT.
 for damages awarded, 955, 956.
 for value of land taken, 995.
 recovery of fine or penalty, 2781.
 against warehouseman, for non-delivery, 2984, 2988.
  measure of damages, 2985.

ATTACHMENT.
 of rolling stock—conditional sale, 1498.
 of witness for contempt, 2639.
 of grain in store before notice of transfer, 3021.
 of stock 2928, 2934.
 creditors, right to condemnation money, 1010.

ATTEMPTS, to injure railroad property, 173.

ATTORNEY.
 may sign petition to condemn, 402.
 power of president to employ, 1191.

ATTORNEY'S FEES.
 taxed as costs in lien cases, 1092.
 in action for killing stock, 1794-1799, 1518.
  notice of claim for, 1798.
 action for neglect as to scales, 2599, 2600.
 action for extortion, 2714.

ATTORNEY GENERAL.
 counsel to assist, when, 1474, 2719.
 to enforce repairs by railway, 2636.
 to prosecute under direction of railroad commissioners, 2642.
  consent to dismissal, 2719.

AUDITOR, annual reports to, 61, 1427.

AUCTION, sale of damaged grain, at, 2766.

AUTOMATIC COUPLINGS.
 for cars, 2444.
 signals at railroad crossings, 2726, 2727.
 railroad commissioner may order disuse of, 2726.

AVERAGE.
 of loss of mixed grain in warehouse, 2976.
 of the evidence by jury, 761, 762.

AWARD OF COMPENSATION.
 award construed, 1026.
 enforcement of, 1027.
 acquiescence in, 1032.

AWNING, of station house, too near track, 2137.

AX, to be kept in passenger cars, 2443.

AYES AND NOES, call on passage of ordinance 140.

BADGE.
 officers of railway to wear, 2338.
  not to exercise powers without, 2338.

BAGGAGE.
 malicious mischief to, 176.
 smashing, fine, &c., for, 2274.
 checks to be given for, 2236.
  penalty for refusing to give, 2236.
 checks as evidence, 2237, 2238, 2246, 2248, 2266.
 what included in, 2242, 2243, 2245, 2251, 2252, 2255, 2256, 2259, 2260, 2262.
 liability for lost, 2237-2273.
 liability for over other lines, 2249.
 contract must be shown, 2244.
 when may store, and be liable only as warehouseman, 2257, 2258, 2263-2265, 2269-2273.
 owner of sleeping-car not liable, 2261.
 See COMMON CARRIER.

BONDS OF RAILWAY COMPANIES—*Continued.*
  retiring, by taking lots, 1367.
  purchaser takes subject to consolidation, 1407.

BOOKS OF CORPORATION.
  to be kept at office in state, 61, 1174.
  where to be kept and what to show, 61, 1174.
  right of inspection, 61, 1186.
  record of capital stock, 61, 1174.
  to show corporate acts, 1166.
  prima facie evidence of incorporation, 1166.
    to show organization, 1167.
  of subscription for union depots, 1515.

BOOKS OF STOCK.
  open to public inspection, 1174.
  to show amount of capital stock, 1174.
  to show names of owners and amount paid, 1171, 1174.
    transfers of stock, 1174.
    names and residence of officers, 1174.
  right of stockholders to examine, 1186, 1471.
  for registry and transfer of stock, 1471.
  liability for refusing to enter transfer, 1202.
    not if stock is void, 1381.
  transfer on, not necessary to pass equitable title, 1200f.
    not on, subject to execution, 1201d, 1201f, 1201b.
    not on, good *inter partes*, 1201e.
  of warehouse, open to inspection, 105.
  See STOCK—STOCKHOLDERS.

BORROWING MONEY.
  power to issue mortgage bonds for, 1338.
  power to borrow, 1467-1470, 2941.
  power of union depots, 1514.

BOULEVARD—condemnation for, 348, 950.

BRANCH ROAD—power to condemn for, 330, 353, 363.

BRAKES.
  operated by steam, 2229, 2234.
  penalty for neglect to apply, 2233.

BRAKEMEN.
  one for every two passenger cars, 2229.
  number for freight trains, 2234.
  damages and penalty for neglect, 2235.
  contributory negligence in respect to, 2230-2232.
  to wear badge, 2338.
  injury to, from defective coupling, 2445-2449.

BRIDGE.
  removing signal light from, 171.
  injury to, 172, 173, 174.
  condemnation for abutment of, 263.
  over another railroad, 554.
  must allow the water to pass, 867.
  overflowing land by catching drift, 1243.
  liability for obstructing water by, 1398.
  in street—action by adjacent owners, 826.
    liability of city for, 834, 839, 840.
  approaches to in street, liability of city, 835, 900, 901.
  on navigable streams, 1235.
  over other streams, 1236, 1236c.
  when treated as built by city, 1244.
  for cars—duty as to connections, 1487.
  over highway, when, 2097a.
  stopping trains at, 2101-2110.
  powers of railroad commissioners over, 2636.

BRIDGE COMPANY—in what county sued, 1097.

BUCKETS—leather for passenger cars, 2443.

BUILDING.
  conspiring to injure railway, 172.
  destroyed—measure of damages, 611.
    value of *debris* when deducted, 611.
  value of land from, 658.
  injury to railway in street, 902.
    from bridge in street, 826, 835, 839, 840, 900, 901.
  removal from land after condemnation, 1024.
  of jail, no action by adjacent owner, 805.
  of railway—condemnation for, 1213.
  stipulation as to depot on question of damages, 732.

CHARITABLE.
  corporations, 46.
  institutions, 325, 1072.
CHARTER.
  amending by special laws, 3, 14.
  amendment of, 3, 14, 46, 1192a.
  extension of, 46, 1172.
  effect of constitution on, 45, 58.
  of private corporations, 52-55.
  repeal of, 56.
  time limited for organization, 58, 1150.
  construed as to taking land in public use, 262.
  powers in, construed, 860, 1267.
    reservations in, construed, 1210b-1211a.
  of East St. Louis, construed, 1265.
  construed as to stockholder's liability, 2904, 2905, 2909.
  contract by, 55, 78, 81, 82, 1210a-1211a.
  subject to implied conditions, 1434.
    limitations as to charges, 2648, 2652.
  subject to law as to unjust discrimination, 2654.
  does not prevent state from fixing rates, 2723.
  stockholder's liability under, unconstitutional, 2906.
  limitation as.to duration, 1172.
  renewal of, 1172.

CHECKS—See BAGGAGE. EVIDENCE.
CHIEF INSPECTOR OF GRAIN.
  appointment and qualification, 2756.
  duties of, 2756.
  appointment of assistants, 2756.
  oath and bond of, 2756.
  removal—vacancy, 2756.
  right to retain fees, 2759.
  liability on bond, 2761-2763.
  See INSPECTION OF GRAIN. INSPECTOR OF GRAIN.
CHILD.
  injuries to, 1970, 1952, 2039, 2062, 2088.
  negligence, 1952, 1970, 1606, 2124.
CIRCUIT COURT.
  condemnation in, 325.
  petition to for incorporation, 1508.
  issue of warehouse licenses, 2737.
  filing justice's transcript in, 1091.
CITIES AND VILLAGES.
  subject to legislative control, 57.
  act for incorporation of, 18.
  charters not abrogated by new constitution, 45.
  city election law, valid, 31.
  apportioning taxes between it and county, 188.
  of the powers of.
    to condemn for street, 325, 327, 340, 358, 364, 365, 525, 505, 526, 392.
    limitation on power, 342.
    not for a city prison, 343.
    to condemn for a sewer, 361.
    to condemn a street-crossing over railroad, 150.
    to condemn for a boulevard, 348.
    cannot confer power of eminent domain, 213, 346.
    as to streets, 117, 136.
    may grant right to use of street, 40.
    connecting tracks and switches in, 72, 73.
    railroad tracks in street, 118.
    location, grade and crossing of, 117, 1258, 1280.
    control over tracks in, 118-136, 1270, 1280.
    mode of assenting to track in, 119.
    delegation of power over, void, 121, 123, 132, 1258, 1258a, 1262, 1262a.
    ordinance granting right, 135, 144.
    grant of privilege in street, 137, 1273.
      construed, 139, 140.
    alone may question right, 138.
    may bind public by grant, 141, 1253, 1254, 1260.
    grant of use, a contract, 143.
    estoppel to dispute right to street, 291.
    consent to track in street, necessary, 60, 1148, 1276.
      to depot and track in, 1513.
      not necessary except as to streets, 359.
    cannot grant exclusive use of, 1287, 1274.
    vacation of street, 140.
  powers in respect to railroads, 117-143, 145, 147, 150, 151, 1250.

—29

    —29

CONTRACT—*Continued.*

evidence of acceptance by new corporation, 1463.
    law applies only to railways, adopting, 1464.
to carry coal at reduced rates, 1464.
for leasing roads of other states, 1476.
for operating other roads, 1476.
for lands for depot, 1476.
for purchase of lands needed, 1476.
of horse railways, 1477.
sale of tickets over other lines, 1479, 1480.
joint for operating roads, 1484, 1484a.
power to make, 1484b.
to stop trains at places, 1484c.
conditional for rolling stock, 1493.
for connections, 1485.
like those of natural persons, 1484b.
of owner to fence track, 1721, 1724, 1725 1727.
limiting carrier's liability. See COMMON CARRIERS.
of carrier, no release of duty. 2582.
for rebate on fare, void, 2680.
for delivery of grain, contrary to orders, 2775.

CONTRACTORS.

corporation liable for their acts, 1216–1219a, 1481, 2458–2460, 2165, 2466, 2468.
road in hands of—no excuse for not fencing, 1545.
included in the word corporation, 2457.
liability of private owner, for acts of, 2463.
taking posts for road, 2465.
railway not liable to servant of, 2467.
taking materials, 182, 1215.
reserving money to protect against lien, 1086.

CONTRIBUTION.

between stockholders, 2912–2914.
in equity only, 2914.

CONTRIBUTORY NEGLIGENCE. See NEGLIGENCE.

CONVENIENCE—not enough for condemnation, 297, 376, 384.

CONVERSION.

of grain by carrier, 2774.
of grain by warehouseman, 2988.

COPY.

of contract to be served in lien case, 1087, 1089, 1090.
service of process by, 1099, 1116, 1118, 1119.
of articles of incorporation, evidence, 1155.
of by-laws, recording, 1173.

CORONER—railway to pay expenses of inquest, 2950.

CORPORATE EXISTENCE.

when brought into existence, 1155.
when complete, 1156, 1156f.
who may question, 1156f.
when a *de facto*, 1147.
uses under general law, 1169.
evidence and proof of, 512–516, 528, 529.
evidence of, as to consolidated company, 1394, 1424.
proof of by admission, 1170.
not until all the stock is taken, 1198.
estoppel to deny, 2897–2900.
de facto and *user*, when sufficient, 2899.
when it ceases, 1462.
See CORPORATION. INCORPORATION.

CORPORATE PROPERTY—how far private, 255, 256.

CORPORATIONS.

special legislation, 3, 46.
general laws for, 42, 46.
curing defects in organization, 47.
subject to police power, 50–53.
reservation in charters, 54.
charter, a contract, 55, 78, 81, 82, 1210a–1211a.
    repeal of, 54, 56.
limitation as to time for organizing, 58.
election of directors, 59, 1187.
regulation of charges, 78.
property, &c., of, subject to eminent domain, 91, 215, 242–256e.
discovery by, 114.
answer by, 115.
delegation of power to, 300, 379, 380.
powers granted—a judicial question, 300.

COMULATIVE REMEDY.
  as to fraudulent and false warehouse receipts, 111.
  eminent domain law, 1071.
  for killing stock for want of fence, 1612.
  in respect to warehouses, 2724.

CUMULATIVE VOTING—election of directors by, 59, 1459.

CURATIVE LEGISLATION; 47.

CURING DEFECTS.
  in incorporating, 1463-1466.
  errors, 736.
  errors by verdict, 1569, 1701.

CUSTOM.
  limiting carrier's liability, 2426, 2597.
  as to place receiving grain, 2601.
  to deliver grain at elevator, 2606.
  to permit connections—proof of, 2612.

DAMAGES.
  to contiguous property, none of which is taken, 799-808.
      under old constitution, 846.
      new burdens, 847, 848.
      for what injuries, 849.
      construction of new constitution, 235, 850, 804-808.
      laying a highway, 799, 800.
      must be real—not speculative, 802, 850, 903, 626.
      depreciation in value, 802.
      must be in excess of that to public, 804, 843-845.
      for change of street grade, 809, 810, 813-815, 818-820, 822, 824, 886.
      defective sewer, 811, 817.
      injury from street drainage, 812, 817.
      depriving of sidewalk, 816, 828.
      gutter out of repair, 821.
      making levee of street, 823.
      excavation in street, 825, 829, 830, 911, 917, 918.
      bridge in street, 826.
      bridge approach in, 835, 901.
      viaduct or bridge, 839, 840.
      structure in street, 833, 834.
      tunnel in street, 836, 837.
      water tank in, 838.
      railroad in street, 827, 832, 842, 845, 846-866, 883-910.
      grantee of city, takes subject to action, 851, 857.
      obstructing access to lots, 831, 866, 237, 646, 677, 825, 830, 834, 835, 840-842, 865, 904.
      obstructing access to place of business, 911, 915.
      additional tracks in street, 865.
      overflowing land, 852, 875, 905, 906.
      turning water and mud on lot, 852.
      bridge on river, 898.
      right to have condemnation, 853, 861.
      injunction, as a remedy, 846, 854, 856, 858, 862, 863.
      obstructing street, as an element, 234, 891, 892.
      elements of, 902.
      direct and physical, 851, 859, 903, 803.
      when too remote, 883, 884.
      depreciation in value, 626, 899, 903, 907.
      special—not in common, 890, 894, 895, 898.
      right to have assessed, 853, 861.
      trespass not necessary, 904.
      matters not actionable, 909.
      under ordinance requiring payment, 911-920.
          moving place of business, 912, 916.
          loss of business and profits, 912, 913, 916.
      vacation of street, 237.
      embraces past, present and future, 407, 921-923.

  to railway from crossing its track.
      property adapted to a special use, 561.
      when it has no market value, 561.
      loss and damage from proper construction, 554.
      cost of bridge and keeping in repair, 555.
      evidence—expectations of contractors, 556.
      opinions of witness on matter of law, 557.
      opinions of experts as to damages, 558.
      when new road is through an embankment, 560.
      use of road as an entirety, 562.
      to part not taken, 563.
      to business and operation of road, 564.
      elements of, 554-580.
      direct and remote, 565, 567, 568.

—30

DIRECTORS—*Continued.*
 to require payment of subscription, 1192.
  not until whole stock is taken, 1198.
 removal of, 1185.
 powers in general, 1207b.
  to make bonds convertible into stock, 1338, 1467.
 estoppel to deny consolidation. 1408.
 must agree to terms of consolidation, 1422.
 minority representation, 1459.
 cumulative voting, 1459.
 to call stockholder's meeting, 1468.
 mode of voting for, 1459.
 election for union depot, 1507, 1516.
  term of office and notice, 1516.
 examination by railroad commissioners, 2638.
 liability for debts in excess of capital stock, 2919-2925.

DISCHARGE—of stockholder's individual liability, 2876.

DISCOVERY.
 bill for, against corporation, 114-116.
  party to bill, 114, 115.

DISCRETION—as to same jury assessing several tracts, 463.

DISCRIMINATION.
 unjust, prohibited, 95, 96, 98.
 in rules and regulations, 1157s.
 on account of color, 1331, 1332, 2308, 2309.
 in fare—want of ticket, 2296.
 as to carrying grain, 2562.
 as to shippers, 2563, 2564, 2577.
 taking goods out of order, 2581, 2582.
 must be shown to be unjust, 2698.
 by warehousemen, 2739.
 See EXTORTION.   UNJUST DISCRIMINATION.

DISMISSAL.
 of petition, on administrators' motion, 425,
 after filing cross-bill, 459.
 on appeal, 1038, 1049.
 of cross petition—appeal lies, 1061.
 of suit for fines, 1474.
 consent of railroad commissioners necessary, 2719.

DISORDERLY CONDUCT—expulsion of passenger for, 2275, 2281, 2317, 2321-2323, 2550.

DISSOLUTION.
 of insolvent corporation, 24.
 of corporation by legislature, void, 55, 56.
 failure to elect directors, not a, 1187.
 of railway by consolidation, 1409, 1420, 1421.
 not building road in time, 1462.
 no release of stockholders, 2876.

DISTANCE OF HAUL—on question of unjust discrimination, 2670.

DITCHES,
 water and ice in, as negligence, 1689, 148.
 made necessary—element of damages, 601.
 power of city to compel opening of, 145.
 new use of street—compensation, 269.
 liability for negligence in construction, 879-882.

DIVERSION.
 of stream on farm, 767.
 of trade and travel, 826.
 of a stream, 868, 869.

DIVESTITURE OF TITLE.
 of title by condemnation, 192, 205, 1008.
  strict compliance required, 321.

DIVIDENDS.
 on fictitious stock, 87.
 must be general on stock, 1385a.
 to purchaser of stock on execution, 2939.

DIVIDING FARM.
 element of damage, 606, 609. 621-623, 628, 636, 640, 653, 667, 699, 716.
 of warehouse receipts, 2743.

DOCKS—damages to, 899.

DOCKET ENTRY—evidence, 451.

DOMESTIC CORPORATIONS.
 consolidation with foreign, 1413, 1414, 1415, 1422.
 power to contract with foreign, 1476.

EVIDENCE—*Continued.*

GRADE.
of grain, mixing, 104, 2740.
    for inspection to be fixed, 2796.
    notice before change of, 2796.
of streets,—profile, evidence, 760, 887.
    liability for changing, 809, 810, 813–815, 818–820, 822, 886, 1250.
    by condemnation, 820.
of railways in cities, 117, 145.
See Cities and Villages. Damages.

GRANTEE.
action by, for injury to land, 422, 423, 1025.
damages to, 758, 792.
ejectment by, 1001.

GRASS AND WEEDS—on right of way, negligence, 1687, 1800–1806, 2496, 2497, 2499, 2502, 2508, 2509, 2520, 2522, 2531–2536, 2542.

GUARD.
as to fences and gates, 1630, 1631, 1635, 1636, 1638, 1639.
See Patrol. Fencing. Railways.

GUARDIAN.
party to condemn, 325.
liability for stock, 1208.

HACKMEN—injury to baggage, 176.

HEARING.
of condemnation, 497–529.
setting of time, 432.
in vacation, 453.
in term, 457, 458.
assessment—laying road, 451–453.

HEATING OF GRAIN—liability of warehouseman, 2766.

HEIR.
appeal by, 1063.
necessary party to condemn, 424, 425,

HIGHWAY.
how far a railway is, 68–74.
condemnation for, 284.
expediency of, 374.
notice to land-owner, 450.
fixing time of hearing, 451.
continuance by justice, 451.
petition, description, 452.
claim of damages, 497, 498.
damages must be adjusted, 501–504.
compensation not in benefits, 631, 632.
measure of damages, 645, 698, 706.
right to abandon proceeding, 957.
order for, necessary before taking, 1040.
appeals, 1038–1047.
act construed with eminent domain act, 706.
powers of commissioners over, 1247.
telegraph in, 232, 241, 351, 1003.
obstructing access to, 677.
right of way, is property, 268.
grant of use, joint--not exclusive, 1245.
    by state construed, 1246.
power of railway to build road over, 1235, 1278.
railway in—damage to contiguous property, 231.
private switch, not a highway, 1312.

HIGHWAY CROSSINGS.
change of by railway, 1294, 1294a.
equitable interference with change, 1294, 1294a.
change of place of intersection, 1295a.
duty to make and maintain, 1296, 1297, 1518.
    as to new street, 1296.
restoring to former usefulness, 1297, 1298.
what is, a crossing, 1583.
proof that road is a public one, 1903.
cattle guards at, 1625.
no fencing at, 1518.
duty as to approaches, 2089–2097a.
notice to make or repair, 2098.
neglect to make or repair, 2099.
    penalty for, 2100.
obstructing with cars, 2111.
    liability under ordinance, 2113–2117a.
    penalty for, 2119–2121.

JURY—*Continued.*

swearing—waiver of objection, 487, 488.
questions before, 371, 385, 517–519, 524, 720.
evidence not proper for, 528.
view of premises, 489–496.
not bound by opinions of witnesses, 745–749.
may act on their own view, 744.
averaging the evidence, 751, 762.
adding and dividing, 762.
parol evidence to show basis adopted, 1029.
amending verdict, 467.
necessary in condemning for highway, 284, 285,
trial by, in prosecutions, 1472, 2710.

JUSTICE OF THE PEACE.

jurisdiction in actions for injury to animals, 144.
   in suits for penalties, 2106, 2126, 2599.
    uniting causes to defeat, 2105.
filing transcript of, 1091.
district of, 17.
condemnation for highway, 450, 452, 471.
   selection of jury by, 482.
   error in swearing jury, 488.
summons and service, 1115.

JUSTIFICATION—when condemnation is, 924, 932, 935.

LABOR AND LABORER.

lien for labor, 1073, 1076.
when under contract, 1085.

LADIES.

car, 1157t.
waiting-room, 1158b, 1504, 1336.

LANDING.

for water craft—no condemnation, 1489.
connecting with railway, 1275.

LANDLORD AND TENANT.

apportionment of compensation between, 1022, 1023.
condemnation does not release tenant, 1021.

LARCENY—of railway—punishable, 169, 170.

LATERAL ROAD.

as to length of, 329.
condemnation for, 330.
   not for private use, 353, 354.
charter must authorize, 363.

LATERAL SUPPORT—of the soil, 829, 830.

LAW.

questions of, 1158g, 1158h, 1568, 1857, 1860–1864.
when remedy is at, 1401.
what governs, 309, 312, 315–318, 320.

LAY-OVER TICKET—holder bound by conditions, 2318.

LEAKAGE—service to make good, 2562.

LEAPING—from moving train, 2211, 2215–2223.

LEASE.

power of railway to, 1172, 1403.
condemnation does not extinguish, 1021.
unauthorized—no release of subscription, 1482.
remedy of stockholder for an unlawful, 1483.
of roads of other states, 1476.
power to contract for, 1484.
of rolling stock, till paid for, 1493, 1494.
of tracks in street, 1264.
contract held not a, 1313, 1404.
legislation necessary for lease to foreign company, 1414, 1415.

LEASEHOLD.

subject to condemnation, 275.
protected, 272.
damage to entirety, 659.
apportionment of compensation, 1022.

LEGAL TITLE—when passes as to stock, 1200f, 1201–1201b.

LEGISLATION.

validating incorporation, 1463–1466.
as to railway, applies to prior companies, 1434.

LEGISLATIVE.

control over union depots, 1515.
discretion, 1523.

LEGISLATIVE—*Continued.*
    grant, in restraint of eminent domain, 229.
    function, 229.
    powers, 304.
        embraces eminent domain, 304, 219.
        when conferred, 219.
    questions, 373, 376–378, 381, 391, 394.
    recognition of corporate rights, 345.
        of consolidation, 1402.
        of power to build road in city, 1269, 1269.

LEGISLATURE.
    power to fix individual liability of stockholders, 1210.
    reservation of power over corporation, 1210a, 1210c.
    control over railways, 1429–1458.
    power to fix rates of charges, 1428.
    approval of route, etc., of railway, 335.
    constitution a limitation on, 180, 219, 220, 40, 46.
    may delegate powers, 200.

LESSEE.
    condemnation, damage to future profits, 647, 660, 661.
        payment to before taking, 1017, 1018.
        compensation for improvements, 1017–1018.
        when a necessary party, 272, 431.

LESSEE OF RAILWAY.
    must conform to charter of lessor, 1478.
    lessor liable for acts of, 1481, 1217–1219a, 2459, 2464.
    liable for injury to stock, for want of fence, 1602, 1604, 1628, 1629.
    notice to as to fencing, 1816.
    duty as to flagmen, 1888.
    treated same as the corporation, 2457.
    liable for acts of those using road, 2474.
    liable for defects in road, 2477, 2478.
        for escape of fire, 2484.
        for grass on right of way, 2534.
    examination of by railroad commissioners, 2638.
    penalty for neglect to report, 2641.
    may purchase foreign roads, 1426.
    may condemn in lessor's name, 267.
    penalty as to, of warehouse, 2775, 2776.

LEVEE—action for converting street into, 823, 824.

LEX LOCI—governs as to carrier's liability, 2379, 2406, 3437.

LICENSE.
    of county board to make telegraph poles in highway, 351.
        no protection against action, 351.
    structure on land by, 665.
    to lay track in street, no defense to action, 849.
        where city has but an easement, 849.
    presumption of, 983.
    revocation of, for right of way, 997.
    as to speed of train, ordinance is not, 2190.
    of warehouseman—revoking, 2737.
        limitation as to re-licensing, 2737.
        of class A, how obtained, 2737.
        revocation, 2737, 2766.
        renewal, 2739.

LIEN.
    of carrier, 168.
    on railroad, 1073, 1096.
    for what given, 1073.
    as against mortgage and other liens, 1073.
    of what date, 1073.
    where it is given, 1074–1080.
    in whose favor, 1081.
    does not extend beyond sub-contractor, 1082.
    petition, what to show, 1083.
    release of, by contractor, 1084.
    relation of parties, 1085.
    not to exceed contract price, 1081.
    notice of, 1087–1089.
        service of, 1087.
    list of persons claiming, 1090.
    action to enforce, 1091.
    petition to enforce, 1093.
    attorney's fee taxed, 1092.
    limitation of action, 1094, 1095.
    on sale of rolling stock, 1498.

LIEN—*Continued.*
    of warehouse for storage, 2748-2751, 2754, 2755.
        purchaser of receipt, takes subject to, 2748a.
        when lost, 2749-2751, 2991.
        by agreement, 2752.
        on grain for charges, 2774, 2954, 2958.
    of creditor on stockholder's liability, 2903.
        not lost by issue of fraudulent receipt, 2955.
        on stock, 1203a, 1204.
    of warehouse, not on goods of another, 2961.
    warehouse receipt, gives none to holder, 2965.
    priority of, 1073, 1079, 1081.

LIGHTING DEPOT. See DEPOTS AND GROUND.

LIMITATION.
    of action, to enforce municipal subscription, 25.
    ☐for penalty—extortion, 97, 2705, 2715.
        to enforce lien, 1073, 1077, 1079.
        for penalty, 2147a.
        Texas cattle, 2148.
        against stockholders, 2915-2918.
    of time for organizing under old charters, 58.
        for commencing and completing road, 1462.
    on issue of bonds and stock, 87-90.
        increase of capital stock, 87-90, 1206.
    of rates and charges, 68.
    of right to lay track in streets, 118.
    of carrier's liability, 162, 2339-2442.
    of legislature as to eminent domain, 180, 183.
    of power of eminent domain, 199, 219, 220, 390.
    as to place of side track, 352.
    as to conditional sales of rolling stock, 1493.
    on right to select route in city, 1257, 1258, 1262a.
    on right granted by city, 1264-1266.
        to enter city, 1268, 1288.
    on police power, 1437, 1439, 1445.
    of warehouseman's liability, 2744.
    on transfers of stock, 1200.

LIMITED USE —is property, 562.

LOCAL LAWS.
    how far prohibited, 3.
    effect of prohibition, 9.
    depending on local option, 10.
    limited in object to which it applies, 11.
    laws held local or not, 11-38.
    not prohibited by old constitution, 29.

LOCAL OPTION—laws dependent on, 10.

LOCATION.
    of land as affecting value, 751.
    of highway and streets—compensation, 497-500, 645.
    of depot, opinion as to benefits, 733.

LOCATION OF RAILWAY.
    power given, 1220, 1223, 1288.
    no power to change, 1225a.
    benefits from not considered, 582, 713.
    benefits from, of a park, 591.
    change after municipal aid, 1422.
    of points of intersection, 1225.
    in cities.
        power of city to regulate, 117-120, 125, 128, 136, 137, 1258-1270, 1288.
        is subject to assent of city, 118-120, 1257.
        city's assent, a limitation of power, 118, 1257, 1270, 1288.
            not till exercised, 133, 1288.
            delegation of city, power, 120-123, 132, 1258, 1258a.
            injunction, till city assents, 124.
            how obtained, 118, 119, 125-127.
            petition for leave, 151-161.
            sufficiency of ordinance for, 121, 131.
        railway alone can locate, 128.
            may select route without assent, 118, 128-130, 1257, 1261, 1261a, 1288.
            ordinance not necessary for, 359.
    limitation as to time for, 332.
    See RAILWAYS. SELECTION.

MACHINE SHOPS.
    injury to, 172, 174.
    condemnation for, 1213.

NAVIGABLE STREAM.
>bridges over, 1235, 1244-1244b.
>water-craft over, by railway, 1488.
>>no condemnation of landing, 1489.
>railway to receive and deliver at intersections, 2624.

NAVIGATION—obstructing, 1235, 1244-1244b.

NECESSITY.
>eminent domain founded on, 228.
>for exercise of, 228.
>must be pressing, 297.
>reasons for, not material, 367.
>fixes amount of land taken, 366-372.
>who may determine, 873-894.
>essential to exercise of right, 384.
>delegation of power to decide, 379, 380.

NEGATIVE EVIDENCE. See Affirmative evidence.

NEGATIVING EXCEPTIONS. See Declaration.

NEGLIGENCE.
>as to culverts, 1236a-1236c.
>as to drains, 879-882.
>in construction of road, 878.
>criminal, liability, 166, 167.
>defective floor in platform, 1500.
>obstructing passage to eating house, 1500a.
>in passing over depot ground, 1502.
>allowing cattle guards to fill up, 1591a.
>cattle guards out of repair, 1596, 1611, 1625.
>neglect to repair, 1608, 1609.
>allowing fence, &c., to get out of repair, 1610, 1618, 1623, 1626.
>neglect to keep gates closed, 1630-1634.
>>to discover breaches, 1635-1643.
>leaving gate or bars open, 1728, 1729.
>ice and water in ditches, 1689.
>defective repair of fence, 1731, 1732.
>allowing grass, &c., to obstruct view, 1687.
>failing to stop train, 1688.
>in management and running train, evidence, 1766.
>speed of train, 2153, 2154, 2158-2163, 2169, 2170-2183, 2186, 2187, 2189-2191.
>>as evidence of negligence, 1685, 2198, 2199, 2000, 2194.
>>in excess of ordinance, 2152-2200.
>>in absence of ordinance, 2169-2203, 2197.
>>as negligence, 1878-1880, 1888.
>>regulation as to, 1453.
>>at road crossing, 2188.
>whether speed is, a question of fact, 2175, 2177, 2183, 2192.
>>must have caused the injury, 2163.
>>instruction ignoring whether the cause, 2201.
>starting train suddenly, 2226, 2227.
>>without signal, 2087-2088a.
>want of brakeman, 2233, 2235.
>failure to apply brakes, 2233, 2235.
>explosion of boiler, 2078.
>*signals on approaching crossing, failure to give*, 1889, 1890, 2185, 1827.
>>whether negligence, question of fact, 1855, 1856, 1862-1864, 1867, 1870-1873, 1882, 1930.
>>need not apprise, 1874, 1878, 1879, 1880.
>>must have caused the injury, 1836, 1837, 1840-1849, 1866, 1869, 1870.
>>when not required, 1856, 1883, 1887.
>>ordinance requiring, 1888.
>>evidence relating to, 1904, 1905.
>>burden of proof, 1838, 1839.
>>at other places than crossings, 1883, 1887.
>in escape of fire, 2484-2548.
>>*prima facie* negligence, 2484, 2510, 2516, 2525, 2535, 2540, 2546.
>>>but not conclusive, 2498, 2525.
>>when inferred or presumed, 2486, 2485, 2495, 2498.
>>burden of proof, 2486, 2487.
>>use of too much steam, 2488, 2494, 2504.
>>wood in coal burner, 2518.
>>amount of sparks emitted, 2511.
>>rebutting presumption, 2516, 2521, 2533, 2540, 2546.
>>failure to employ best appliances, 2492, 2503, 2504.
>>grass on right of way, as evidence, 2496, 2497, 2499-2502, 2508, 2509, 2520, 2522, 2531-2536, 2542.
>>See, also, Fire.
>railway may not contract against gross, 2340.
>proof of delay does not show, 2417.
>of owner of warehouse receipt, 2971.

RAILWAYS—*Continued.*

to take and negotiate notes, 1162.
to lease and take lease, 1163–1165.
to make stock transferable, 1155.
to make by-laws, rules and regulations, 1155, 1157†–1161, 1320–1337.
to make by-laws, 1157–1157g.
to transport persons and property, 1316.
to make rules and regulations for, 1320.
to borrow money and issue bonds, 1467, 1468, 2941.
to contract for lease and use of road, 1484.
to form partnerships, 1484a.
to make contracts, 1484b.
to own and use watercraft, 1488.
to purchase roads of other states, 2940.
to extend lines beyond state, 2940.
limitation on power to issue bonds, 87–90.
limitation on increase of debts, 87.
to fix route and termini of road, 352, 1463.
to mortgage, not without statute, 1342.
to enter land, survey and locate, 1220.
to lay out and construct road, 1231.
altering route—further damages, 1234.
how much land it may take, 1231.
to build over streams, 1235–1236c.
right to connect with other roads, 1485, 1304.
    with rail on bridge, 1487.
right of way over school land, 1492.
    to use union depot 1517.
to buy its own stock, 1200, 1203.
lessees may purchase, 1426.
limitation as to time of beginning work, 1462.
    as to time of completing, 1462
contract for reduced rates, 1460, 1461.
to intersect and unite with other roads, 1304.
property of—what real and what personal, 62–66.
power of legislature over, 1428–1433.
subject to general laws, 1429–1445.
vested in board of directors, 1175.
consolidation. See Consolidation.
to increase capital. See Capital—increase of.

powers as to right of way and construction.

power to take by condemnation, 1213, 325, 330, 334.
to take materials necessary, 1214, 1215, 182.
width it may take, 1231–1233, 366–372.
cutting trees near right of way, 1231.
as to crossings, connections, &c., 1304.
under law of 1849, 181.
    as to the fee, 214.
for lateral road, 329–331, 353, 354, 363.
    length of, 329.
no limitation as to switches, 352.
not exhausted by exercise, 336, 349.
for work and paint shops, 337, 338.
lumber sheds and depot grounds, 338.
switches, turn-outs and side-tracks, 352, 356.
additional tracks in city, 356.
de facto railway may, 362.
extent of land taken, 366–372.
    width of right of way, 366, 367, 369–371.
    for depots and side-tracks, 370.
for union depots, 1512.
for elevated ways, 2943, 2944.
of ground used by consent, 350.
taking public property, 341.
taking for a public use, 289, 292, 295, 296, 302.
    how far private, 292.
railway property subject to, 245–251, 255–256a.
may condemn property of another, 231, 244–251.
part of another road in length, 256a.
right to take railway already in public use, 256a–256e.
    limited to crossing and connections, 256d, 256e, 1304, 1305.
of rival road, 259–261.
presumption as to right to take property in public use, 262.
power to condemn in city, 339, 346, 355.
    by implication, 345.

power to build road in city and in streets
right to bring road into city, 1267–1269.
legislative recognition of right, 1259.

RAILWAYS—*Continued.*

    power of city as to location, grade and crossing of streets, 117, 118, 120.

    location subject to assent of city, 118.

        a limitation of power of railway, 118, 1262a.

        but not till used by city, 133.

    consent of city necessary, 60, 118.

        how obtained, 118, 119, 125-127.

        how made, 1255.

        sufficiency of ordinance for, 121-123, 127, 131, 1261b.

        delegation of power, 120, 123, 132, 1262.

    enjoined until city assents, 124.

    grant of city binding, 1254, 1255, 1260.

        passes to successor, 1256.

        petition for, 151, 156.

        is subject to rights of lot-owner to sue, 154.

        conditioned for use of track by other roads, 134, 135, 1263a.

        must clearly appear, 1281, 1282.

    right to lay track in streets, 1259.

        by charter, 860, 1251.

        who may question, 138, 1252, 1279.

    power of city to give leave, 137, 125, 1253.

    may select route without city's, assent, 128-130.

    may cross street without leave, 1261, 1278.

    connecting tracks in street, 72, 73.

duties in respect to the public.

    to keep public office in state, 61, 1471.

        fine for neglect of duty, 1472, 1473.

    books open to inspection, 61.

        what to show, 61.

        subject to examination, 2638.

    annual reports by, 1427.

        penalty for not making, 2641.

        prosecution for neglect, 2642, 2643.

duties and liabilities.

    injury by entry to survey, etc., 1221.

    duty to unite and form intersections, 1304.

        compensation for, 1304.

    on contracts after taking benefits of, 1463.

    to repair on recommendation by railroad commissioners, 2636,

        mandamus to compel, 2636.

duty to stop passenger trains at county seat, 2204.

    statute a proper police regulation, 53, 1456, 2224.

        not a regulation of inter-state commerce, 2224.

        what are passenger trains, 2225.

        all its passenger trains to stop at, 2228.

        not at new depot, out of town, 2228.

    awning too close to track, 2137.

    badges, what employes to wear, 2338.

        not to exercise powers without, 2338.

    baggage, checks for, 2236.

    baggage smashing, 176, 2274.

    bell to be rung at crossings, 1827-1830.

    brakemen, required, damages, 2229-2235.

    buckets in passenger coaches, 2443.

    cars, provision for supply, 2130, 2130a, 2130b, 2140, 2141.

    combustibles, on right of way, 1800-1806.

    conductors, police powers of, 2519a, 2551.

    connections by union depots, 1513.

        facilities for to be allowed, 1304.

        of tracks in streets, 72, 73, 1275, 1314.

    corporation, defined, 49, 2457, 2725.

        for construction, 2457.

    couplings, for cars, 2444-2449.

    crossings, street, powers of city to require, 144.

        duty as to new streets, 149.

        duty as to approaches, 1296.

        duty as to highway, 2089-2097a.

        binds its successor, 2092.

        neglect to make, 2098.

        notice to make, 2099.

        penalty for neglect, 2100.

        stopping a railway at, 2101.

        penalty—limitation, 2103.

        effect of change in law, 2104.

        actions for penalty, 2105-2110.

    animals, cruelty to, 168.

        willful injury to, 175.

    accommodation at stations, 2130, 2130a, 2130b, 2140, 2141.

STREETS—*Continued.*
require fencing of track, 1280.
steam power in by railway, 1250.
new use, operation of freight trains, 1289.
dummy railway in, 1286.
dedication—acceptance necessary, 1249.
title to vested in city, 1248, 1249.
injunction of improper use, 1290-1293.
restoring to former usefulness, 1297.
condemnation of property in, 268-272.
damages for new burden, 269.
easement in protected, 268-270.
right to locate depot in, 1513.
right to lay tracks in, 1513.
cattle guards in, 1582, 1596.
use of by elevated ways, 2947, 2948.

STREET CROSSINGS.
city may compel railway to construct, 144, 1280.
duty of railway as to, 2089-2097a.
change of, 2090.
leaving safe, 2091.
duty on successor, 2092.
notice to make, 2093.
penalty for not making, 2100.
no duty as to new street, 2096, 149.
warning boards at, 1825, 1826.
signal by bell, etc., at, 1827.
requiring flagman at, 2450.
See CROSSINGS OF HIGHWAYS AND STREETS.

STREET RAILWAYS.
general railroad law not applicable to, 2483.
liability for injury from sudden starting, 2227.

STRICT COMPLIANCE—when required, 306, 319, 321, 1199, 1822, 2927.

STRICT CONSTRUCTION.
when applied, 323, 328, 2147, 2116.
See CONSTRUCTION.

STRIKES.
of railway employes, 2552.
of servants, no excuse for delay, 2587, 2594.

STRIKING OUT—pleas to petition to condemn, 419-422.

STRIP OF LAND CUT OFF.
in condemnation—value, 606.
relative value as to whole, 686, 687.

STRUCTURES ON LAND.
on right of way, 1238.
by consent—value on condemnation, 630, 665-667.
in street, 826, 833-866.

SUB-CONTRACTOR—lien for labor and materials to, 1081-1085, 1089.

SUBJECT MATTER—jurisdiction of, 395-401.

SUBPOENA—of witnesses by railroad commissioners, 2639, 2640.

SUBSCRIPTION.
municipal, limited, 25.
municipal, not under eminent domain, 187.
of capital stock, necessary, 1156g.
directors may require payment, 1192.
forfeiture of stock, 1192.
fraud, as a defense, 1192k.
when released, 1192a-1192j.
alteration of charter, 1192a.
amendment of charter, 1192b-1192h.
when collection enjoined, 1192i.
estoppel, 1192j.
who liable to call, 1193.
identity of corporation, 1194.
release, void as to creditors, 1195-1197.
whole capital must be taken, 1198.
strict compliance as to incorporation, 1199.
to union depot, 1515.
for reduced rates, 1460, 1461.
unauthorized lease in defense, 1482.
how enforced by creditors, 2829.
assignee in bankruptcy to collect, 2834.
—36

VERDICT—*Continued.*
    sufficiency of description of land, 538.
    gross sum as to compensation and damages—presumptive, 539, 540.
    construed as to allowance of benefits, 709. ·
    mode of finding—addition and division, 762.
    recording, 1070.
    curing defects, 1569, 1701, 1710.

VIADUCT—in street, action for, 839, 840.

VIEW.
    of premises by jury, 490-496.
        as evidence, 740, 768.
    of animal on or near track, 1802, 1806.
    obstructing, of approaching train by brush, etc., 1803, 1805.

VILLAGE.
    what is, under fencing law, 1575-1577.
    laying street—ordinance, 364, 365.1
        penalty to, for neglect as to depot, 1506.

VIOLENT ENTRY—not evidence on condemnation, 736.

VOID.
    allowance of benefits against compensation, 632.
    when judgment is, 485.

VOLUNTARY GRANT—to railway authorized, 1226.

VOTE.
    required to vacate street, 140.
    to increase capital stock, 1206.
    to mortgage railway, 1338.
    to consolidation of roads, 1422.
    regulating by by-laws, 1157a.
    by proxy allowed, 1206, 1459, 1468.

WAITING-ROOM—at depot, 1158b, 1504.

WAIVER.
    of trial by jury, 288.
    of objections, 487.
    of damages, 497-500.
    of proof, 514.
    as to separate finding, 537.
    of rights, 992.
    of lien, mechanic's, 1080.
    of right to increased capital, 1207c.
    of tort, money had and received, 2988.
    of prepayment of freight, 2568.

WALKING ON TRACK. See 1808-1814, 1832, 1897, 1896, 1932, 1933, 1959, 1961, 2024.

WANTON INJURY—to stock, 1670.

WAREHOUSE AND WAREHOUSEMEN.
    liability for defects in approach, 2097a.
    storing baggage in, 2270-2273.
    action by, for not delivering grain to, 2600.
    delivery to, if on railway's line, 2603.
    track of, when part of railway, 2604.
    statements, by to commissioners, 2633.
    licensees—cancellation, 2637.
    delivery of grain, after revocation, 2637.
    re-licensing—limitation, 2637.
    licenses contrary to law, void, 2637.
    act, not special legislation, 20.
        not unconstitutional, 85.
    what a public, 99.
    posting statements, 104.
    mixing grain, 104.
    inspection of books and property of, 105.
    receipts to be delivered, 106.
    fraudulent receipts, 111.
    connections to reach, 107-110, 137, 1275, 1308-1310.
    inspection of grain regulated, 112.
    delivery of grain at, 107-110.
    examination of books of, 2638.
    act regulating, is constitutional, 2732, 2733.
    classification of, 2734, 2735.
    license to class A—revocation, 2737.
    bond of licensees, 2738.
    penalty for doing business without, 2739.
    renewing license, 2739.
    not to discriminate or mix grain, 2740.
    grain in, to be inspected, 2740.
    receipt for grain in separate bin, 2740.

    —37